AF388936

Tuning the Physicochemical Properties of Nanostructured Materials through Advanced Preparation Methods

Tuning the Physicochemical Properties of Nanostructured Materials through Advanced Preparation Methods

Editors

John Vakros
George Avgouropoulos

MDPI • Basel • Beijing • Wuhan • Barcelona • Belgrade • Manchester • Tokyo • Cluj • Tianjin

Editors
John Vakros
University of Patras
Greece

George Avgouropoulos
University of Patras
University of Patras

Editorial Office
MDPI
St. Alban-Anlage 66
4052 Basel, Switzerland

This is a reprint of articles from the Special Issue published online in the open access journal *Nanomaterials* (ISSN 2079-4991) (available at: https://www.mdpi.com/journal/nanomaterials/special_issues/physicochemical).

For citation purposes, cite each article independently as indicated on the article page online and as indicated below:

LastName, A.A.; LastName, B.B.; LastName, C.C. Article Title. *Journal Name* **Year**, *Volume Number*, Page Range.

ISBN 978-3-0365-3659-0 (Hbk)
ISBN 978-3-0365-3660-6 (PDF)

Contents

About the Editors

John Vakros now serves as a researcher in the Department of Chemical Engineering at the University of Patras, Greece, where he also obtained a diploma in chemistry in 1992, as well as a Ph.D. in 1997. His teaching activities (Hellenic Open University, University of Patras and University of Peloponnese) cover general chemistry, physical chemistry, chemical instrumental analysis, and catalysis at undergraduate and postgraduate levels. His research activities focus on heterogeneous catalytic processes, catalyst preparation, acid–base behavior of solids, the preparation and evaluation of biochars and, over the last five years, the circular economy, with emphasis on catalysts preparation, catalytic measurements for environmental applications, the determination of interactions of reactants with catalytic surface groups and tuning the physicochemical properties of supported catalysts.

George Avgouropoulos is an associate professor at the Materials Science Department, owning an M.Sc. diploma in energy and the environment and a Ph.D. diploma in chemical engineering from the University of Patras, Greece. He has extensive and high expertise in heterogeneous catalysis (the design, synthesis and characterization of mixed oxides, spinels, precious metals and nanostructured materials), H2 production through low-temperature methanol reforming, H2 purification via PROX and WGS processes, reactor and fuel cell engineering and high-temperature PEM fuel cells. He has also developed research activities concerning Li- and Na-ion batteries. He has coauthored 64 research papers and received more than 4300 citations (h-index = 29). He has presented more than 60 research papers in scientific conferences, while co-holding two international patents. In addition, three of his papers have been recognized as the "Top-50 most cited articles" by Elsevier. He is the coeditor of an RSC book entitled *"Environmental catalysis over gold-based materials"*. He has been involved as a researcher and coordinator in various national and international research projects. He is a member of the Greek Committee for the national hydrogen strategy.

Preface to "Tuning the Physicochemical Properties of Nanostructured Materials through Advanced Preparation Methods"

Over the last few decades, nanotechnology has gained huge interest due to its extensive application in various fields including catalysis, electronics, optics, energy, and environment. The design and controlled synthesis of advanced nanomaterials with unique properties make them highly attractive in these fields.

Nanomaterials can be classified into one-, two-, and three-dimensional materials. The main characteristic of nanostructured materials is their surface reactivity due to their active surface functional groups. The control of the size, shape, and nature of nanoparticles is strongly influenced by the synthetic route applied during the preparation step (i.e. hydrothermal, solvothermal, combustion, sol-gel).

The Special Issue of Nanomaterials, entitled, "Tuning the physicochemical properties of nanostructured materials through advanced preparation methods," contains the contribution of research groups from different fields and discuss the recent developments in nanomaterials with regard to the preparation method used.

John Vakros, George Avgouropoulos
Editors

nanomaterials

MDPI

Editorial

Tuning the Physicochemical Properties of Nanostructured Materials through Advanced Preparation Methods

John Vakros [1],* and George Avgouropoulos [2]

[1] Department of Chemical Engineering, University of Patras, Caratheodory 1, University Campus, GR-26504 Patras, Greece

[2] Department of Materials Science, University of Patras, University Campus, GR-26504 Patras, Greece; geoavg@upatras.gr

* Correspondence: vakros@chemistry.upatras.gr

Citation: Vakros, J.; Avgouropoulos, G. Tuning the Physicochemical Properties of Nanostructured Materials through Advanced Preparation Methods. *Nanomaterials* **2022**, *12*, 956. https://doi.org/10.3390/nano12060956

Received: 18 February 2022

Accepted: 8 March 2022

Published: 14 March 2022

Publisher's Note: MDPI stays neutral with regard to jurisdictional claims in published maps and institutional affiliations.

Over the last few decades, nanotechnology has received a huge level of interest due to its extensive applications in various fields, including catalysis, electronics, optics, energy, and the environment. The main characteristic of nanostructured materials is their surface reactivity due to their active surface functional groups. The control of the size, shape, and nature of nanoparticles is strongly influenced by the synthetic route applied during the preparation step (i.e., hydrothermal, solvothermal, combustion, sol-gel). The design and controlled synthesis of advanced nanomaterials with unique properties make them highly attractive in these fields.

Today, nanotechnology, providing the capability to precisely manufacture small structures, is being promoted as a national policy in many countries. The progress of nanotechnology allows information and functions to be integrated in smaller spaces. There are two main categories for the synthesis and preparation of nanomaterials. The first deals with top-down methods, where the desired nanostructure is produced by decreasing the size of a larger substrate, while the second follows bottom-up approaches that assemble nanostructures from atoms and molecules. Nanomaterials can be classified into one-, two-, and three-dimensional materials [1]. The review by Ebina et al. [1] summarizes the research on different techniques developed to form one-, two-, and three-dimensional connected structures (CSs) of metal nanoclusters (NCs) through self-assembly. Metal nanoclusters (NCs) consisting from a few to about 100 metal atoms have gained significant interest due to their unique properties. Generally, these properties, in terms of surface reactivity, are different than the corresponding bulk metal. This is important, since the NCs are of an easy-to-handle size. Connected structures of 1D, 2D, and 3D can be synthesized through newly designed synthetic routes, which determine their geometrical structures and physical/chemical properties. The 1D, 2D, and 3D CSs are discussed in detail in the first sections of this review article and the authors give a brief outlook on their future.

An important class of materials for optical applications is semiconductor quantum dots (QDs) [2]. They have strong quantum confinement effects, as the size of QDs is close to the Bohr radius of the exciton. Among other QDs, PbS QDs is a representative zero-dimensional material. Compared to 1D and 2D nanomaterials, PbS QDs possess strong absorption and an adjustable bandgap in the near IR region. In the contribution of L. Yun and W. Zhao [2], fiber-based PbS QDs were used as saturable absorbers and tested for dual-wavelength ultrafast pulse generation. Specifically, PbS QDs, fabricated via a modified hot-injection method, have the advantages of fast relaxation time, wide bandwidth, large modulation depth, and thermal damage. With the introduction of PbS QDs into an erbium-doped fiber laser, the laser can simultaneously generate dual-wavelength conventional solitons with central wavelengths of 1532 and 1559 nm, with 3 dB bandwidths of 2.8 and 2.5 nm, respectively, and it can be adopted as a broadband SA for application in pulsed lasers.

The applications of nanoparticles are almost unlimited; for example, ceramic nanoparticles can be used as a reinforcement of lightweight alloys. In a recently published work [3], yttria is widely used in the ceramic technology. Specifically, solution combustion synthesis (SCS) was applied for the preparation of nanoyttria. This technique is based on the high-energy reaction between metal nitrates (yttrium nitrate, in this case) and a reducing agent such as glycine, leading to well-formed nanoparticles. The properties of yttria nanopowder were studied by several physicochemical techniques, including differential thermal analysis coupled with FT-IR spectrometry, for the online SCS monitoring, Scanning Electron Microscopy, measurements of their specific surface area with BET method, and particle size distribution. Interestingly, the obtained powders showed nanoscale structures after calcination at 1100 °C. The obtained nanopowder showed very high sintering activity as the shrinkage onset was detected at a temperature of about 1150 °C.

A new method was applied for the preparation of yttrium-doped barium cerate (BCY15) [4]. This material was used as an anode ceramic matrix for synthesis of the Ni-based cermet anode in a proton-conducting solid oxide fuel cell (pSOFC). Although SOFCs are very promising for clean energy production and have many advantages, they are difficult to find commercial applications for, due to the fact that their operation temperature is significantly high—between 800–1000 °C with a commercial goal of 600–700 °C. SOFCs require a reduction step for the supported metal, usually Ni. A modification on the preparation route with the application of hydrazine wet-chemical synthesis can be a low-cost alternative. This method promotes 'in situ' introduction of metallic Ni particles in the BCY15 matrix. Two different Ni/BCY15 cermets were synthesized using either water or ethylene glycol (EG). The samples were characterized using a variety of physicochemical techniques and resulted in a more active and stable Ni cermet with well-dispersed nanoscale metal Ni particles, due to stronger interactions between the Ni and BCY15 matrix. These factors contributed to better performance, higher stability and lower degradation rate during operation of a pSOFC.

The physicochemical properties of nanomaterials are greatly influenced by their dimensions. Thus, the dispersion media are very important. In the absence of a dispersion medium, the nanoparticles tend to agglomerate and lose their interesting properties. The aggregation of nanoparticles depends on both their characteristics and those of the different media. The dispersion medium can be a solid support, in the case of solid catalyst, a liquid medium, or even cell culture media; thus, the effect of the various biological dispersion media on the state of aggregation of the nanoparticles has been extensively investigated in the literature and reviewed in [5]. Generally, nanomaterials interact with the surrounding environment and an interface is formed, whose properties depend on the physicochemical interactions and on colloidal forces. Parameters such as size, shape, surface chemistry, surface charge of the nanomaterials, and properties of the dispersion medium affect the behavior in a test medium. The relationship between the nanomaterials' properties and their practical use is defined as Functionality. It is important to understand this relationship for the safe use of these nanomaterials, since it can play an important role on the safe design of manufactured nanomaterials (MNMs), thus, reducing the possible health and environmental risks early in the innovation process, where the functionality of a nanomaterial and its toxicity/safety will be taken into account in an integrated way. In this mini review, the authors attempt to identify the key parameters of the nanomaterials and establish a relationship between those and the main functionalities of the nanomaterials. Finally, the review aims to contribute to the decision tree strategy for the optimum design of safe nanomaterials.

In the field of catalysis, the preparation of the supported catalyst is, in many cases, crucial for the performance of the catalyst. The main physicochemical characteristics of the supported catalyst can be defined by the preparation route and the composition of the active phase [6]. For example, monoclinic zirconia-supported platinum (Pt/m-ZrO$_2$) catalysts are active for the ethanol steam-reforming reaction. The reaction proceeds via ethanol dissociation to ethoxy species. The next step is oxidative dehydrogenation to acetate followed by acetate decomposition. This last step strongly depends on the catalysts' composition. The desired step is the decarboxylation pathway, which tends to produce higher overall hydrogen selectivity, and can be promoted with the addition of alkali metal ions in the catalyst. Indeed, it was found that high loadings of K^+ or Rb^+ ions can promote hydrogen production, while for catalysts which are undoped or doped with low alkali loadings, decarbonylation is the preferable route. Thus, the overall hydrogen selectivity is significantly different between the two cases. Detailed studies with in situ diffuse reflectance infrared Fourier transform spectroscopy (DRIFTS) and the temperature-programmed reaction of ethanol steam reforming show that alkali doping promotes forward acetate decomposition while exposed metallic sites tend to facilitate decarbonylation [6].

Ceria (CeO_2) is among the most extensively reducible supports that have been studied. Nanoceria exhibits interesting properties, among them high capacity to store and release oxygen, different surface ratio of Ce(III)/Ce(IV), bulk oxygen vacancies, and redox properties. These unique properties find significant applications in the field of catalysis. Especially when ceria is used as support for deposition of atomically dispersed Au, the corresponding catalyst exhibits superb performance in Water Gas Shift reaction (WGS). In this contribution from Prof. Tabakova's group [7], layered double hydroxides NiAl (NiAl LDH) prepared via co-precipitation method were used as support for Au clusters, resulting in a catalyst with significant activity. By direct deposition of ceria on NiAl LDH and precipitation with NaOH, the authors were able to obtain the CeO_2 phase and to preserve the NiAl layered structure by avoiding the calcination treatment. Using the deposition–precipitation method, Au was deposited in the modified support. The WGS performance of Au/NiAl catalysts was significantly affected by the addition of CeO_2, as the CO conversion increased from 83.4% to 98.8% for the modified catalyst.

Hybrid biochar-ceria nanomaterials were synthesized using biochar (BC) from spent malt rootlets used as the template [8]. These hybrid materials were tested for the activation of persulfates (SPS) and subsequent degradation of sulfamethoxazole (SMX). Using wet impregnation, a simple preparation method, cerium nitrate was deposited on BC, and by regulating the calcination temperature, a tuning in the content of BC was achieved for the hybrid materials. The CeO_2-BC hybrid materials were characterized, and it was found that calcination temperature affects the biochar content and the physicochemical properties of the hybrid materials. The most active materials were obtained with calcination temperatures of 300–350 °C, exhibiting high specific surface area, intense interactions between CeO_2 and BC, and could degrade 500 μg/L SMX at about 60% within 2 h. Their activity was higher than the starting BC, which is a good candidate for the activation of SPS. Concerning the degradation process, it can be concluded that it takes place through different pathways, including the oxidation of SMX by sulfate and hydroxyl radicals and singlet oxygen. It should be noted that commercial CeO_2 or CeO_2-BC calcined at higher temperature was rather inactive. This method is simple and low cost and can be used to obtain hybrid materials with interesting properties.

The application of nanomaterials is not limited only to inorganic compounds; it can be expanded also to polymers. Aromatic polyimides present a variety of excellent physico-chemical properties, and they have found significant applications. Polyimides can be easily tuned and different units, such as azobenzene units, can be introduced to their structure. These azo derivatives are optically active and, under the action of linearly polarized light, undergo multiple reversible trans to cis photo-isomerization processes. Moreover, the cyclic photo-isomerization can lead to a large-scale mass transport of the polymer chains, which appear as a surface relief grating (SRG). In the contribution of Iuliana Stoica and Ion Sava group [9], Atomic Force Microscopy (AFM) was used for the evolution of local mechanism and chemical properties under Pulsed UV Laser-Nanoinduced Patterns on Azo-Naphthalene-Based Polyimide Films. Specifically, AFM was applied to determine morphological, statistical, local mechanical, and chemical properties combined with the molecular modeling. Interestingly, the properties were different in various regions due to reorganization of the matter by azo-naphthalene dipoles orientation and trans-cis isomerization of the azo-segments. It was found that polymers with 50% azo groups in cis had either a maximum or a minimum peak of the calculated parameters. Confocal Raman measurements confirm that the cis isomer evolution is mainly responsible for the observed differences.

Medium density fiberboard (MDF) is a natural timber panel, produced from lingo cellulosic fibers and binders under pressure and temperature [10]. MDF can find many applications in furniture industries and interior constructions. MDF has poor physical properties and it cannot be used in moist and hot environments. An option to improve the MDF properties is the addition of multiwall carbon nanotubes (MWCNTs). The embedment of MWCNTs urea formaldehyde resin at concentrations from 0–5% was investigated in [10]. The MWCNTs can deeply penetrate into the wood, effectively altering its surface chemistry and resulting in a high degree of improvement in physical and mechanical strength. It was found that this addition enhanced thermal conductivity by 24.2%, reduced curing time by 20%, and controlled formaldehyde emission by 59.4%. Moreover, properties such as internal bonding, modulus elasticity, modulus of rupture, thickness swelling, and water absorption were significantly improved.

Author Contributions: J.V. and G.A. writing-reviewing. All authors have read and agreed to the published version of the manuscript.

Funding: This research received no external funding.

Acknowledgments: Erika Zhao is gratefully acknowledged for her valuable help in all the stages of the special issue.

Conflicts of Interest: The authors declare no conflict of interest.

References

1. Ebina, A.; Hossain, S.; Horihata, H.; Ozaki, S.; Kato, S.; Kawawaki, T.; Negishi, Y. One-, Two-, and Three-Dimensional Self-Assembly of Atomically Precise Metal Nanoclusters. *Nanomaterials* **2020**, *10*, 1105. [CrossRef] [PubMed]
2. Yun, L.; Zhao, W. PbS Quantum Dots Saturable Absorber for Dual-Wavelength Solitons Generation. *Nanomaterials* **2021**, *11*, 2561. [CrossRef] [PubMed]
3. Gizowska, M.; Piątek, M.; Perkowski, K.; Konopka, G.; Witosławska, I. Fabrication of Nanoyttria by Method of Solution Combustion Synthesis. *Nanomaterials* **2020**, *10*, 831. [CrossRef] [PubMed]
4. Nikolova, D.; Gabrovska, M.; Raikova, G.; Mladenova, E.; Vladikova, D.; Kostov, K.L.; Karakirova, Y. New Insights on the Nickel State Deposited by Hydrazine Wet-Chemical Synthesis Route in the Ni/BCY15 Proton-Conducting SOFC Anode. *Nanomaterials* **2021**, *11*, 3224. [CrossRef] [PubMed]
5. Anastasiadis, S.H.; Chrissopoulou, K.; Stratakis, E.; Kavatzikidou, P.; Kaklamani, G.; Ranella, A. How the Physicochemical Properties of Manufactured Nanomaterials Affect Their Performance in Dispersion and Their Applications in Biomedicine: A Review. *Nanomaterials* **2022**, *12*, 552. [CrossRef] [PubMed]
6. Martinelli, M.; Garcia, R.; Watson, C.D.; Cronauer, D.C.; Kropf, A.J.; Jacobs, G. Promoting the Selectivity of Pt/m-ZrO$_2$ Ethanol Steam Reforming Catalysts with K and Rb Dopants. *Nanomaterials* **2021**, *11*, 2233. [CrossRef] [PubMed]
7. Gabrovska, M.; Ivanov, I.; Nikolova, D.; Krstić, J.; Venezia, A.M.; Crişan, D.; Crişan, M.; Tenchev, K.; Idakiev, V.; Tabakova, T. Improved Water–Gas Shift Performance of Au/NiAl LDHs Nanostructured Catalysts via CeO$_2$ Addition. *Nanomaterials* **2021**, *11*, 366. [CrossRef] [PubMed]
8. Papatheodorou, G.; Ntzoufra, P.; Hapeshi, E.; Vakros, J.; Mantzavinos, D. Hybrid Biochar/Ceria Nanomaterials: Synthesis, Characterization and Activity Assessment for the Persulfate-Induced Degradation of Antibiotic Sulfamethoxazole. *Nanomaterials* **2022**, *12*, 194. [CrossRef] [PubMed]
9. Stoica, I.; Epure, E.-L.; Constantin, C.-P.; Damaceanu, M.-D.; Ursu, E.-L.; Mihaila, I.; Sava, I. Evaluation of Local Mechanical and Chemical Properties via AFM as a Tool for Understanding the Formation Mechanism of Pulsed UV Laser-Nanoinduced Patterns on Azo-Naphthalene-Based Polyimide Films. *Nanomaterials* **2021**, *11*, 812. [CrossRef] [PubMed]
10. Gul, W.; Alrobei, H.; Shah, S.R.A.; Khan, A.; Hussain, A.; Asiri, A.M.; Kim, J. Effect of Embedment of MWCNTs for Enhancement of Physical and Mechanical Performance of Medium Density Fiberboard. *Nanomaterials* **2021**, *11*, 29. [CrossRef] [PubMed]

Review

One-, Two-, and Three-Dimensional Self-Assembly of Atomically Precise Metal Nanoclusters

Ayano Ebina [1], Sakiat Hossain [1], Hikaru Horihata [1], Shuhei Ozaki [1], Shun Kato [1], Tokuhisa Kawawaki [1,2,3] and Yuichi Negishi [1,2,3,*]

[1]   Department of Applied Chemistry, Faculty of Science, Tokyo University of Science, Kagurazaka, Shinjuku-ku, Tokyo 162-8601, Japan; ayano0may9@gmail.com (A.E.); hisakiat@gmail.com (S.H.); konpei1082saikou@gmail.com (H.H.); shuheiozaki@icloud.com (S.O.); syunpeeee9@gmail.com (S.K.); kawawaki@rs.tus.ac.jp (T.K.)

[2]   Research Institute for Science & Technology, Tokyo University of Science, Shinjuku-ku, Tokyo 162-8601, Japan

[3]   Photocatalysis International Research Center, Tokyo University of Science, 2641 Yamazaki, Noda, Chiba 278-8510, Japan

*   Correspondence: negishi@rs.tus.ac.jp

Received: 14 May 2020; Accepted: 27 May 2020; Published: 3 June 2020

Abstract: Metal nanoclusters (NCs), which consist of several, to about one hundred, metal atoms, have attracted much attention as functional nanomaterials for use in nanotechnology. Because of their fine particle size, metal NCs exhibit physical/chemical properties and functions different from those of the corresponding bulk metal. In recent years, many techniques to precisely synthesize metal NCs have been developed. However, to apply these metal NCs in devices and as next-generation materials, it is necessary to assemble metal NCs to a size that is easy to handle. Recently, multiple techniques have been developed to form one-, two-, and three-dimensional connected structures (CSs) of metal NCs through self-assembly. Further progress of these techniques will promote the development of nanomaterials that take advantage of the characteristics of metal NCs. This review summarizes previous research on the CSs of metal NCs. We hope that this review will allow readers to obtain a general understanding of the formation and functions of CSs and that the obtained knowledge will help to establish clear design guidelines for fabricating new CSs with desired functions in the future.

Keywords: metal cluster; one-dimensional connected structure; two-dimensional connected structure; three-dimensional connected structure; metal–organic framework; photoluminescence; electrical conductivity

---

1. Introduction

1.1. Metal Nanoclusters for Nanotechnology

Nanotechnology is technology to precisely manufacture small structures. In many countries, nanotechnology is being promoted as a national policy. Progress of nanotechnology allows information and functions to be integrated in smaller spaces, making it possible to manufacture devices with more functions at the same scale. In addition, since nanotechnology makes it possible to integrate the same function in a smaller volume than is the case for current devices, it is expected that devices will be downsized, which will increase their portability. This eliminates the need for the user to be dependent on the device location, which could solve the problems such as crowding and traffic jams and allow users to manage their time more effectively. In addition, the progress of nanotechnology has many advantages, such as saving resources and energy and decreasing waste and environmental damage [1].

Techniques to fabricate small materials can be roughly divided into two categories. One category is top-down methods, in which a desired structure is produced by decreasing the size of a larger substrate (Figure 1). Nanotechnology has been supported by the development of top-down methods. For example, increased functionality and miniaturization of electronic devices have been realized by the progress of top-down techniques. However, when a fine structure is manufactured by using tools (light, electron beam, scanning probe microscope, etc.), it is difficult to manufacture nanostructures with finer accuracy than that of the tools. Therefore, in recent years, bottom-up methods that assemble nanostructures from atoms and molecules have attracted attention (Figure 1) [1].

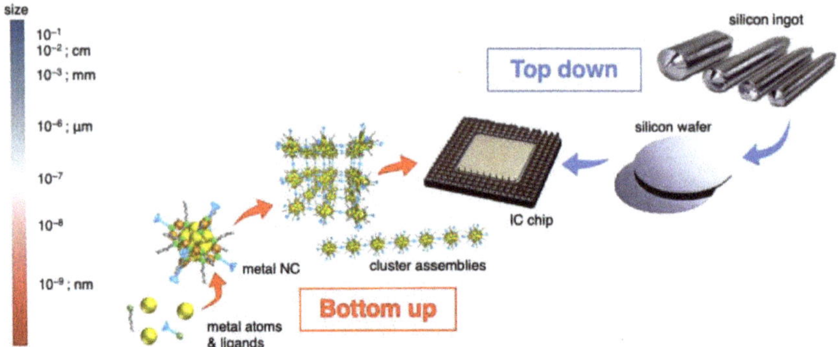

Figure 1. Fine processing techniques, including top-down (cyan) and bottom-up (red) methods. IC = integrated circuit.

Metal nanoclusters (NCs), which consist of several, up to about one hundred, metal atoms [1–10], are nanomaterials that can be synthesized by bottom-up methods. Metal NCs are not only small (<2 nm in size), but they also show physical/chemical properties and functions which are different from those of the corresponding bulk metals [11–35]. Furthermore, the physical/chemical properties and functions of metal NCs change considerably depending on the number of constituent atoms [36–65]. Therefore, if the number of constituent atoms of metal NCs is controlled, it is possible to produce various physical/chemical properties and functions by using only one type of metal element. If several types of elements can be used, it becomes possible to obtain more functionalities [66–82]. For these reasons, metal NCs have been attracting considerable attention as a central material in nanotechnology. In recent years, it has become possible to synthesize such metal NCs precisely at the atomic and molecular level by using thiolate (SR) [2,66], alkyne [59,83], phosphine [9,84–92], carbon monoxide [93–100], and dendrimers [7] as protective organic molecules. Investigation of the obtained precise metal NCs has revealed their geometrical structure (aggregation pattern of metal atoms) and the influences of miniaturization [1–65] and alloying [66–82] on the electronic structures and physical/chemical properties of metal NCs. In parallel, research on the applications of the optical properties and catalytic activity of metal NCs is being actively conducted [81,101–108].

1.2. Controlled Assembly of Metal Nanoclusters

As mentioned above, metal NCs show promise as constituent units of functional nanomaterials. Gold (Au) NCs have already been put to practical use in the fields of sensors, catalysts, and paints. On the other hand, at present, electronic devices are manufactured by top-down methods. Therefore, to replace the components of current devices with metal NCs, it is necessary to grow metal NCs to a size that allows their combination with structures manufactured by top-down methods. Moreover, in other applications, the small size of metal NCs often makes them difficult to handle. To realize nanodevices and next-generation materials with the advantageous characteristics of metal NCs, it is essential to establish techniques to assemble metal NCs to a size that makes them easy to handle.

To form a one-dimensional (1D) arrangement of metal NCs, templates [109,110] and host–guest interactions [111] are extremely effective. Two-dimensional (2D) and three-dimensional (3D) arrays of metal NCs can be fabricated by Langmuir–Blodgett [112] and alternate adsorption methods. There are many reports in which metal NCs are arranged in one, two, and three dimensions by using these methods. However, in the structures produced by these methods, the metal NCs are not regularly arranged in a strict sense. Since the conductivity (σ) of metal NCs changes exponentially with the length of the insulating organic ligands [113], the existence of a distribution in the distances between NCs (i.e., the length of the insulating part) is undesirable when applying the assembled metal NCs in electronic devices. Furthermore, although the Langmuir–Blodgett and alternate adsorption methods are suitable for arranging metal NCs in a wide size area, they are not suitable for arranging those in a tiny area.

Metal NCs are regularly arranged in single crystals. Therefore, it is possible to produce precise structures in which metal NCs are regularly connected in one, two, and three dimensions by crystallizing them while including an ingenious means to connect metal NCs. In fact, many crystals in which metal NCs are regularly linked by such a method have been reported in recent years. These structures contain strong bonds, such as Au–Au, Au–silver (Ag), Ag–oxygen (O), Ag–sulfur (S), Ag–chloride (Cl), Ag–nitrogen (N), cesium (Cs)–S, or hydrogen (H) bonds [114], and weak interactions, such as π–π, anion–π, cation$\cdots\pi$, aryl CH$\cdots$Cl, and van der Waals interactions [115] (Figure 2). Similar to supramolecules, molecular assemblies [116], and metal–organic frameworks (MOFs) [117] that are self-assembled from metal ions and organic molecules, connected structures (CSs) of metal NCs can be formed by self-assembly during crystallization, using these bonds and interactions. Such structures, which could also be called "suprametal NC crystals", exhibit physical/chemical properties different from those of an individual metal NC. Thus, the formation of CSs not only increases the size of the structure, but also enables the application of NCs in new fields.

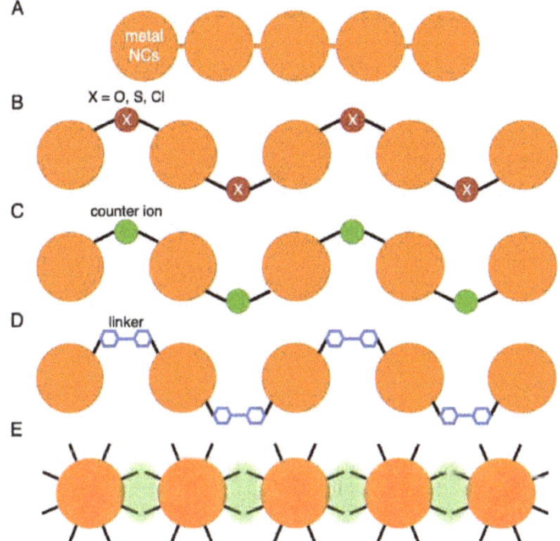

Figure 2. Representative methods for connecting metal NCs: (**A**) formation of metal–metal bond; (**B**) formation of Ag–O, Ag–S, and Ag–Cl bonds; (**C**) control of counterions; (**D**) introduction of linker molecules; and (**E**) use of inter-ligand interactions. In this review, 1D, 2D, and 3D CSs formed by inter-ligand interactions (**E**) are not introduced.

1.3. Contents of This Review

In recent years, it has become possible to control not only the geometrical structure of metal NCs, but also 1D, 2D, and 3D CSs of metal NCs. Further development of these techniques will lead to novel nanomaterials possessing the characteristics of metal NCs. Such development may enable a future in which metal NCs are applied in devices. However, since these studies have been initiated in recent years, there have been few review articles focusing on 1D, 2D, and 3D CSs of metal NCs [118,119].

In this review, we summarize the existing research, with the purposes of understanding the current situation regarding these structures and giving perspective regarding clear design for producing new 1D, 2D, and 3D CSs with desired functions.

This review is structured as follows. Section 2 outlines the fabrication of 1D CSs consisting of metal NCs, their geometrical structures, and physical/chemical properties. Then, Sections 3 and 4 present research on 2D and 3D CSs, respectively. After summarizing this review article in Section 5, a brief future outlook is described in Section 6.

It should be noted that 1D, 2D, and 3D CSs of metal NCs can be formed by methods other than crystallization [120,121]; for example, hydrophilic Au NCs have been arranged in 1D and 3D form by the Xie's group, although these NCs have not been crystallized. However, in this review, only the CSs of metal NCs in crystals are summarized, because our focus is on the regularly CSs in a strict sense. In addition, we described the synthesis methods only for the several examples. Thus, we recommend the readers who want to know the detail of synthesis methods for each example to refer to each original paper.

2. One-Dimensional Structures

The formation of 1D CSs composed of precise metal NCs is important from the viewpoint of the fabrication of controlled nanodevices by bottom-up methods. In this section, we introduce some typical examples of the construction of 1D CSs by the formation of metal−metal bonds (Figure 2A), formation of Ag−O bonds (Figure 2B), control of counterions (Figure 2C), and introduction of linker molecules (Figure 2D). The connection methods, NCs, linkers, year reported, and reference numbers of 1D CSs are summarized in Table 1. Chemical structures of some of the ligands used in these studies are shown in Scheme 1. The chemical structures of organic molecules used as linkers are illustrated in Scheme 2.

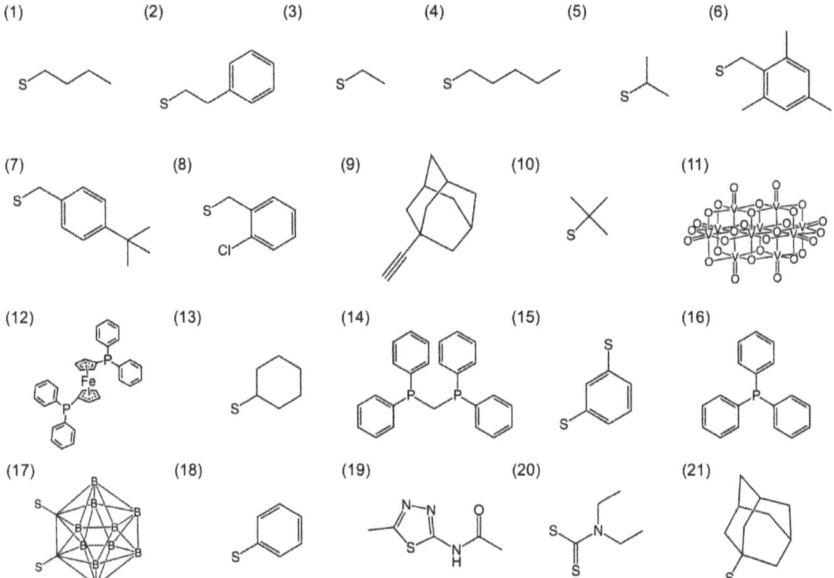

Scheme 1. Molecules used in the synthesis of metal NCs: **(1)** S-Bu, **(2)** PET, **(3)** S-Et, **(4)** S-Pen, **(5)** S-*i*Pr, **(6)** SCH$_2$Ph(CH$_3$)$_3$, **(7)** SCH$_2$Ph^{t}Bu, **(8)** SCH$_2$PhCl, **(9)** A-Adm, **(10)** S-^{t}Bu, **(11)** V$_{10}$O$_{28}$$^{6-}$, **(12)** dppf, **(13)** S-*c*-C$_6$H$_{11}$, **(14)** DPPM, **(15)** BDT, **(16)** PPh$_3$, **(17)** DT-*o*-C, **(18)** S-Ph, **(19)** N-L, **(20)** C$_5$NS$_2$H$_{10}$, and **(21)** S-Adm.

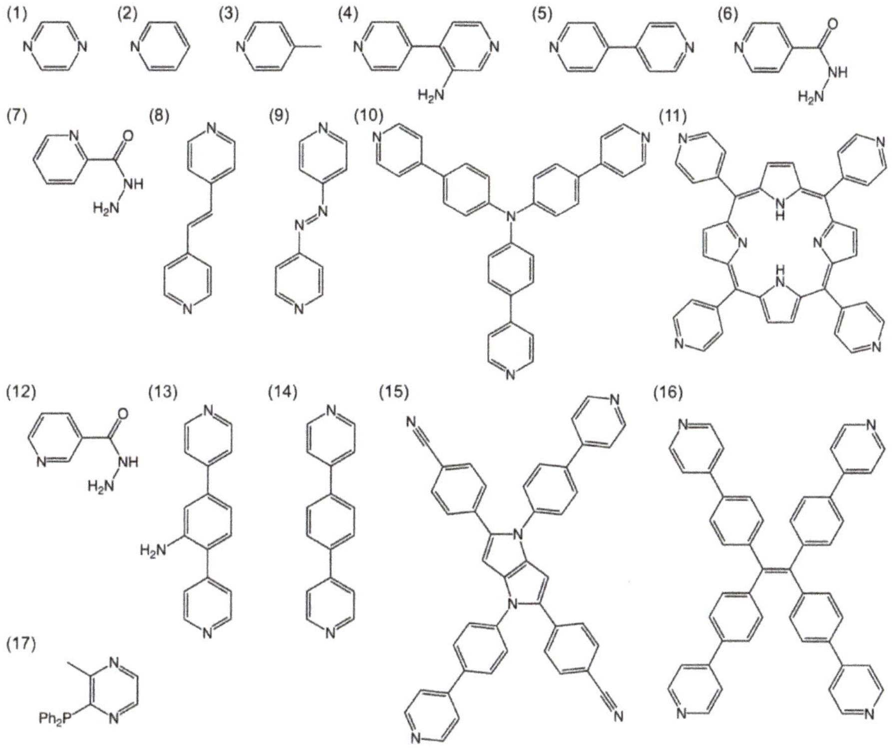

Scheme 2. Linker molecules used to connect metal NCs: (**1**) pyrazine, (**2**) pyridine, (**3**) *p*-methylpyridine, (**4**) bpy-NH$_2$, (**5**) bpy, (**6**) *p*-iah, (**7**) *o*-iah, (**8**) bpe, (**9**) dipyridin-4-yl-diazene, (**10**) TPPA, (**11**) TPyP, (**12**) *m*-iah, (**13**) bpz-NH$_2$, (**14**) 1,4-bis(4-pyridyl)benzene, (**15**) CPPP, (**16**) tppe, and (**17**) mdppz.

2.1. Direct Connection via Metal−Metal Bonds

In 2014, Maran et al. [122] fabricated a 1D CS composed of an SR-protected Au 25-atom NC ([Au$_{25}$(SR)$_{18}$]^0). Since Au$_{25}$(SR)$_{18}$ NCs exhibit high stability among Au$_n$(SR)$_m$ NCs, their geometrical/electronic structures and physical/chemical properties have been studied extensively [18,27,123–139]. However, because most of the studies were conducted on Au$_{25}$(SR)$_{18}$ in solution and there had been few studies on Au$_{25}$(SR)$_{18}$ in the solid phase, Maran's group studied the behavior of Au$_{25}$(SR)$_{18}$ in the solid state.

In their study, butanethiolate (S-Bu, Scheme 1(**1**)) was used as the SR ligand. First, [Au$_{25}$(S-Bu)$_{18}$]$^-$ anion was synthesized by reducing the Au(I)- S-Bu complex by using sodium borohydride (NaBH$_4$). Then, [Au$_{25}$(S-Bu)$_{18}$]$^-$ anion was oxidized into neutral [Au$_{25}$(S-Bu)$_{18}$]^0 in open column packed with silica gel, and single crystals were grown by slow evaporation. Figure 3A(a) shows the geometrical structure of [Au$_{25}$(S-Bu)$_{18}$]^0 obtained by single-crystal X-ray diffraction (SC-XRD). Each [Au$_{25}$(S-Bu)$_{18}$]^0 has almost the same framework structure as that of [Au$_{25}$(SR)$_{18}$]^0 protected by other SR ligands; e.g., phenylethanethiolate (PET, Scheme 1(**2**)) and ethanethiolate (S-Et, Scheme 1(**3**)) (Figure 3A(b)) [27,53]. However, Au−Au bonds formed between adjacent NCs in the [Au$_{25}$(S-Bu)$_{18}$]^0 crystal, unlike the case for other [Au$_{25}$(SR)$_{18}$]^0 crystals (Figure 3A(a),B(a)). This indicates that [Au$_{25}$(S-Bu)$_{18}$]^0 is a suitable structural unit to form 1D CSs. The Au−Au distance between adjacent NCs of 3.15 Å was within the range of aurophilic interactions (2.9–3.5 Å) and shorter than the non-bonding Au−Au distance (3.80 Å) estimated from the van der Waals radius of Au. This result indicates that a 1D CS was formed in the crystal structure of [Au$_{25}$(S-Bu)$_{18}$]^0 via Au−Au bonds. To form such a 1D CS, it was considered that the repulsion between the ligands was suppressed and an attractive force between the ligands was induced because the adjacent NCs twisted and approached each other (twist-and-lock mechanism). It was suggested that 1D CSs did not form when S-Et and PET were used (Figure 3A(b),B(b)) because S-Et has a short alkyl group that leads to a weak attractive force between ligands and PET with a bulky functional group has large steric repulsion between ligands. In 2017, these researchers also succeeded in forming a 1D CS of $[Au_{25}(S\text{-}Pen)_{18}]^0$ (S-Pen = pentanethiolate, Scheme 1(4)) [140]. In the same paper, they reported that the distance between NCs was shorter in this 1D CS than in the 1D CS of $[Au_{25}(S\text{-}Bu)_{18}]^0$ (Figure 3C). In addition, in 2019, they formed 1D CSs of $[Au_{24}Hg(S\text{-}Bu)_{18}]^0$ (Hg = mercury) and $[Au_{24}Cd(S\text{-}Bu)_{18}]^0$ (Cd = cadmium), in which one Au of $[Au_{25}(S\text{-}Bu)_{18}]^0$ was replaced with Hg or Cd [141].

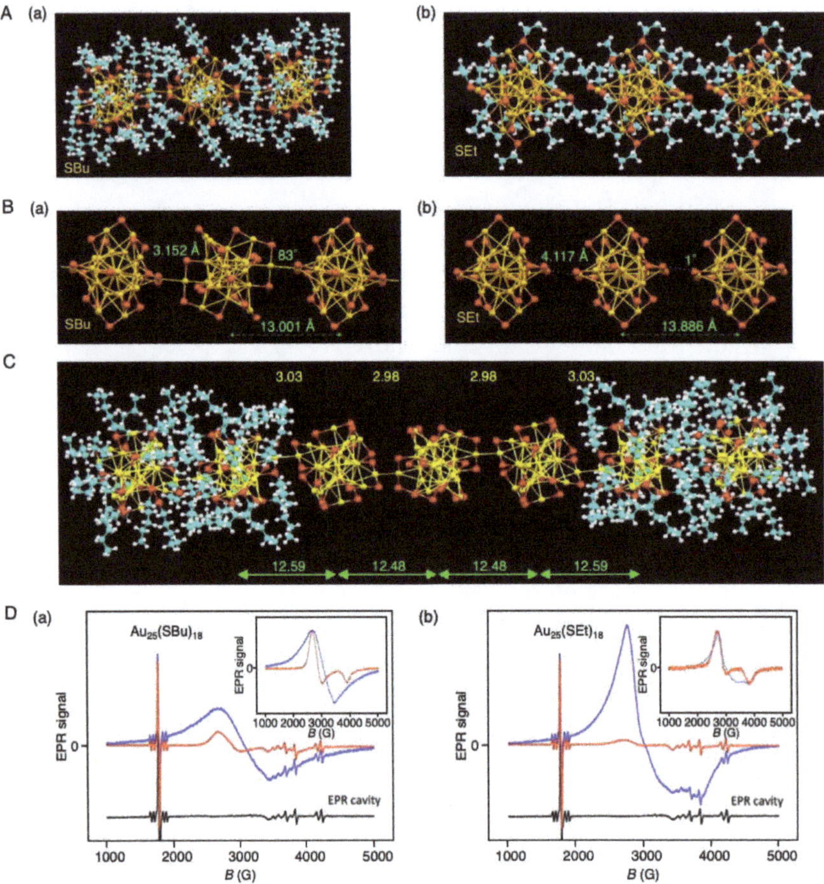

Figure 3. (**A,B**) Crystal structures of (a) $[Au_{25}(S\text{-}Bu)_{18}]^0$ and (b) $[Au_{25}(S\text{-}Et)_{18}]^0$. In (**B**), R groups are omitted for clarity. (**C**) Crystal structure of $[Au_{25}(S\text{-}Pen)_{18}]^0$. In (**A–C**), Au = yellow, S = red, C = light blue, and H = white. (**D**) Comparison of the continuous wave-electron paramagnetic resonance (EPR) spectra of solid (blue traces) and frozen toluene solution (red traces) for (a) $[Au_{25}(S\text{-}Bu)_{18}]^0$ and (b) $[Au_{25}(S\text{-}Et)_{18}]^0$ at −253 °C. The inset shows the same spectra with normalized peak intensity. The black curve corresponds to the EPR cavity signal, which is subtracted in the inset for clarity. All spectra were obtained by using the following parameters: microwave frequency = 9.733 GHz; microwave power = 150 µW; amplitude modulation = 1 G. Reproduced with permission from References [122,140]. Copyright 2014 American Chemical Society and 2017 American Chemical Society.

The same group also revealed that the 1D CSs had electronic structures and physical properties different from those of individual NCs. Since $[Au_{25}(S\text{-}Bu)_{18}]^0$ has unpaired electrons, it exhibits paramagnetism in solution. Conversely, the 1D CS of $[Au_{25}(S\text{-}Bu)_{18}]^0$ was non-magnetic (Figure 3D) [122]. This change was mainly ascribed to the formation of the 1D CS, which led to the close proximity of NCs, allowing the unpaired electrons of adjacent NCs to form electron pairs. Because of the formation of such electron pairs, the conduction band of the 1D CS was full and its valence band was empty, so the obtained 1D CS was predicted to have the properties of a semiconductor [122].

In 2020, we [142] conducted a detailed study on the factors responsible for the formation of 1D CSs via Au−Au bonds by using $[Au_4Pt_2(SR)_8]^0$ (Pt = platinum) as the NC. A similar NC, $[Au_4Pd_2(PET)_8]^0$ (Pd = palladium), was reported by Wu and colleagues in 2017 (Figure 4) [143]. Although it was not mentioned in their paper, $[Au_4Pd_2(PET)_8]^0$ formed a 1D CS in its crystal (Figure 4). Because this type of metal NC has a smaller metal core than that of $[Au_{25}(SR)_{18}]^0$ described above (Figure 4), the distribution of the ligands in this type of NC should change depending on the ligand structure. Moreover, Au and Pt form a stronger bond than that between Au and Pd [144]. Therefore, it was expected that changing Pd to Pt would increase the stability of the NC [145], thereby expanding the variety of ligand functional group structures that can be used in 1D CSs. For these reasons, we chose $[Au_4Pt_2(SR)_8]^0$ as the building block of their 1D CS. The SR ligands shown in Scheme 1(2),(5)–(8) were used. Because the functional group structures of these SR ligands differ greatly, it was expected that there would be different ligand–ligand interactions between the resulting NCs.

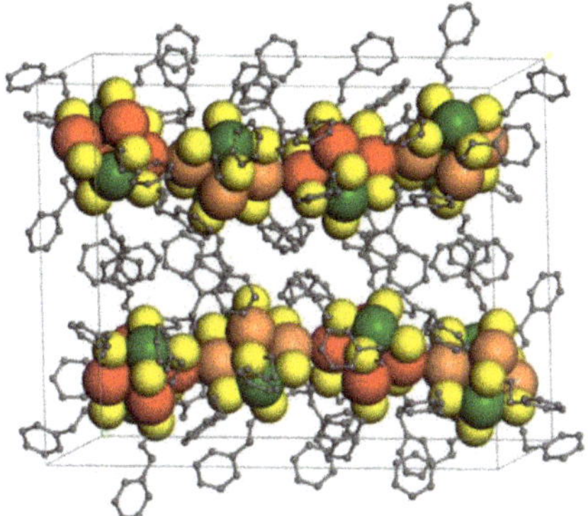

Figure 4. Crystal unit cell of $[Au_4Pd_2(PET)_8]^0$. S = yellow, Au = red and orange, Pd = olive, C = gray. Reproduced with permission from Reference [143]. Copyright 2017 Wiley-VCH.

In the experiment, $[Au_4Pt_2(SR)_8]^0$ NCs with different SRs were precisely synthesized by reducing the metal–SR complex with $NaBH_4$. Each $[Au_4Pt_2(SR)_8]^0$ NC was separated from by-products, using open column chromatography, and then single crystals were grown by vapor diffusion. The SC-XRD of the series of $[Au_4Pt_2(SR)_8]^0$ crystals revealed the following three points for $[Au_4Pt_2(SR)_8]^0$: (1) $[Au_4Pt_2(SR)_8]^0$ is a metal NC that can become a structural unit of 1D CSs via Au−Au bond formation (Figure 5A); (2) although all $[Au_4Pt_2(SR)_8]^0$ NCs have similar structures, the intra-cluster ligand interactions vary depending on the ligand structure. As a result, the distribution of the ligands in $[Au_4Pt_2(SR)_8]^0$ changes depending on the ligand structure; (3) the differences in the ligand distributions influence the inter-cluster ligand interactions, which in turn affect the formation of 1D CSs and change their structure (Figure 5B). These results demonstrate that we need to design intra-cluster ligand interactions, to produce 1D CSs with desired configurations. This study also explored the effects of 1D CS formation on the electronic structure of NCs. The results revealed that the formation of the 1D CS caused the band gap of the NCs to decrease (Figure 5C,D) [142].

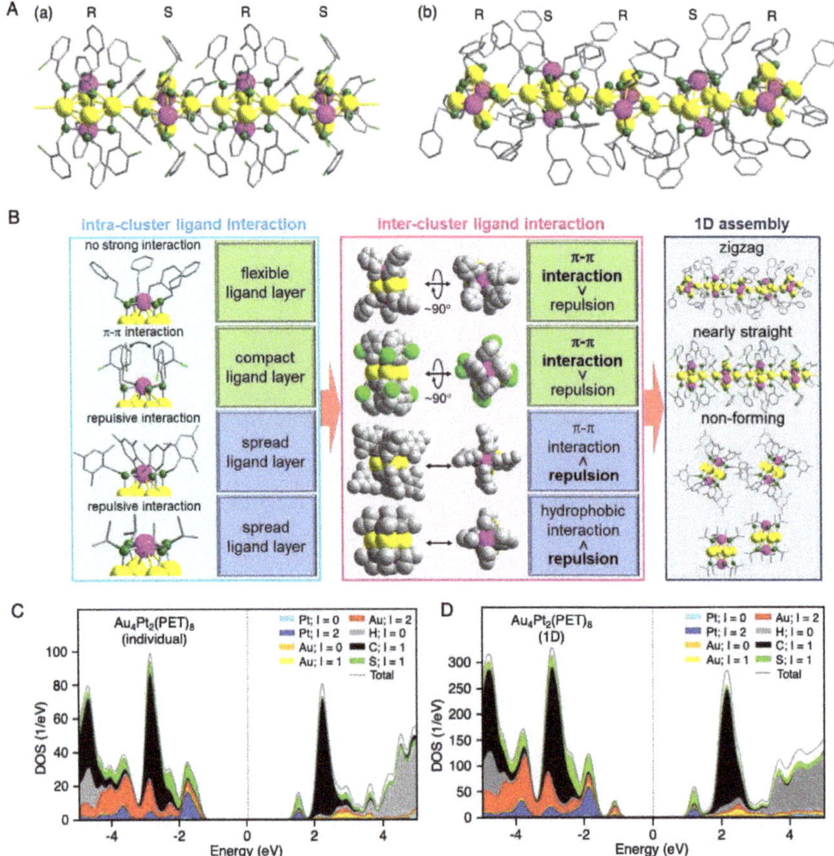

Figure 5. (**A**) Crystal unit cells of (a) [Au$_4$Pt$_2$(SCH$_2$PhCl)$_8$]^0 and (b) [Au$_4$Pt$_2$(PET)$_8$]^0. Au = yellow, Pt = magenta, S = green, Cl = light green, C = gray. R and S indicate two enantiomers in each NC. (**B**) Relationships between intra-cluster ligand interactions, which are related to the distribution of the ligands within each cluster, inter-cluster ligand interactions, and 1D assembly. Projected density of states of (**C**) an individual [Au$_4$Pt$_2$(PET)$_8$]^0 NC and (**D**) the 1D CS of [Au$_4$Pt$_2$(PET)$_8$]^0. Reproduced with permission from Reference [142]. Copyright 2020 Royal Society of Chemistry.

Zhang et al. also very recently reported the formation of 1D CS consisting of the (AuAg)$_{34}$(A-Adm)$_{20}$ alloy NCs (A-Adm = 1-ethynyladamantane; Scheme 1(9)) [146]. For (AuAg)$_{34}$(A-Adm)$_{20}$ alloy NCs, either monomeric NC or 1D CSs were formed depending on the solvent. Monomeric NC could be converted to 1D CSs by dissolving in an appropriate solvent. In the 1D CS, NCs were connected each other via Ag–Au–Ag bond (Figure 6A,B). They studied the electronic structure of the obtained 1D CS by density functional theory (DFT) calculations, which predicted that the single crystals of 1D CS have a band gap of about 1.3 eV (Figure 6C). Field-effect transistors (FETs) fabricated with single crystals of 1D CS (Figure 6D) showed highly anisotropic *p*-type semiconductor properties with ~1800-fold conductivity in the direction of the polymer as compared to cross directions (Figure 6E), hole mobility of ≈0.02 cm^2/Vs, and an ON/OFF ratio up to ~4000. They noted that the conductivity (1.49 × 10^{-5} S/m) of these crystals in the *c*-crystallographic axis is one-to-three orders of magnitude higher than the values reported for 1D CS consisting of Au$_{21}$ clusters, where 1D CS was formed by modulating the weak interactions in the ligand layers (see Section 2.3). It was interpreted that the conductivity and charge carrier mobility was increased by several orders of magnitude in their 1D CS via direct linking of the metal NCs by the –Ag–Au–Ag– chains in the crystal. They described in this paper that this result holds promise for further design of functional cluster-based materials with highly anisotropic semiconducting properties.

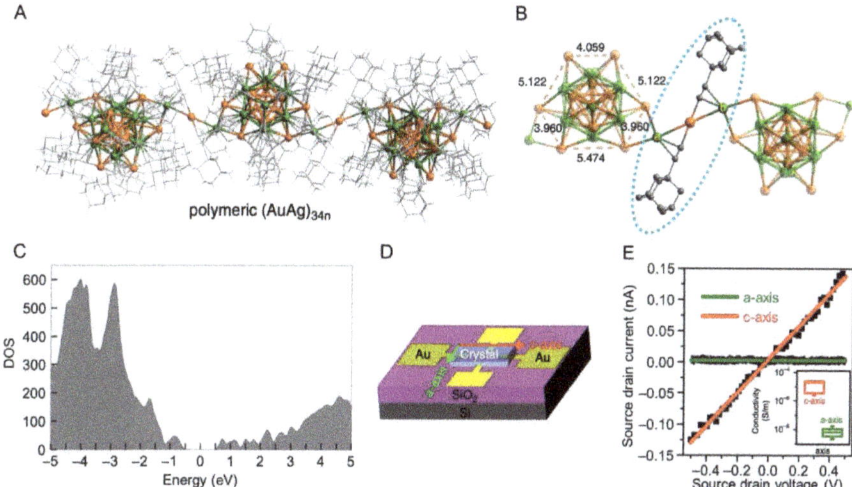

Figure 6. (**A**) Structures of the cluster polymer (approximately orthogonal to the *c*-axis). (**B**) Au–Au distances in the distorted Au$_6$ hexagon and Ag–Ag distance in the "Ag–Au–Ag" unit of between alloy NCs. Au/Ag = golden and green, C = gray. All hydrogen atoms are omitted for clarity. (**C**) DFT-computed electronic density of states (DOS) of the cluster polymer crystal. Cluster model was used to build the periodic crystal, and the integration over the Brilloin zone was done in a 4 × 4 × 4 Monkhorst–Pack *k*-point mesh. The band gap is centered around zero. (**D,E**) Electrical transport properties of the cluster polymer crystals; (**D**) structure of the polymer crystal FET; (**E**) *I–V* plot of the polymer crystal along *a*-axis and *c*-axis, respectively, with the range of corresponding conductivity values shown in the inset. Reproduced with permission from Reference [146]. Copyright 2020 Springer-Nature.

2.2. Connection via Ag–O Bonds

When Ag NCs contain acetic acid ions (CH$_3$COO$^-$), trifluoroacetic acid ions (CF$_3$COO$^-$), or nitrate ions (NO$_3^-$) in the ligand layer, it is possible to connect Ag NCs by forming Ag–O bonds. Su et al. [147] first reported the formation of such a 1D CS in 2014. In this study, the 1D CS was obtained by crystallization of the product which was obtained by the reaction between AgS-^{t}Bu, (NH$_4$)$_3$[CrMo$_6$O$_{24}$H$_6$] (Cr = chromium, Mo = molybdenum), Ni(CH$_3$COO)$_2$ (Ni = nickel), AgCF$_3$COO, and AgBF$_4$. The each Ag NC had a chemical composition of Ag$_{20}$(CO$_3$)(S-^{t}Bu)$_{10}$(CH$_3$COO)$_8$(DMF)$_2$ (CO$_3^{2-}$ = carbonate anion; S-^{t}Bu = *tert*-butylthiolate, Scheme 1 (10), DMF = *N,N*-dimethylformamide). The Ag$_{20}$(CO$_3$) core of the NC was formed by the aggregation of Ag around CO$_3^{2-}$ (Figure 7A) as an anion template. In the crystal, the Ag NCs were connected in one dimension via two Ag–O–Ag bonds (Figure 7B). The obtained 1D CS was stable in both solid and solution states, had a bandgap of 3.22 eV, and exhibited reversible thermochromic emission.

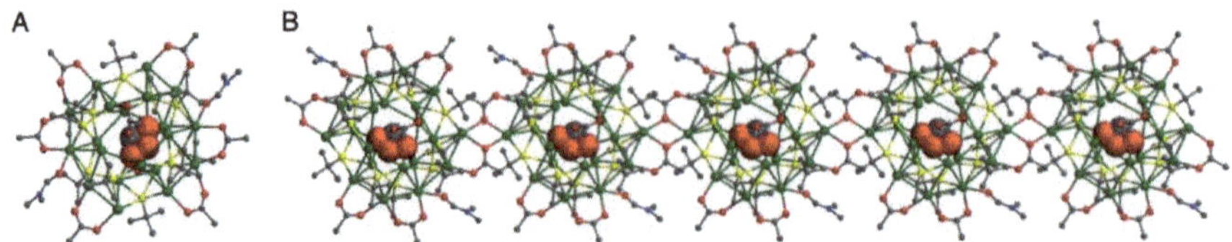

Figure 7. (**A**) Structure of $Ag_{20}(CO_3)(S\text{-}^tBu)_{10}(CH_3COO)_8(DMF)_2$. (**B**) Ball-and-stick view of the 1D chain of $Ag_{20}(CO_3)(S\text{-}^tBu)_{10}(CH_3COO)_8(DMF)_2$. Ag = green, S = yellow, N = blue, O = red, C = gray. Reproduced with permission from Reference [147]. Copyright 2014 Royal Society of Chemistry.

Formation of 1D CSs based on a similar principle was also reported by Mak and co-workers in 2017 [148]. In their report, $Ag_{18}(CO_3)(S\text{-}^tBu)_{10}(NO_3)_6(DMF)_4$ was linked by the formation of Ag–O bonds (Figure 8). The $Ag_{18}(CO_3)$ core contained CO_3^{2-} as an anion template at the center, like $Ag_{20}(CO_3)$. As described in Section 3.1, this group also succeeded in forming a 2D CS of $Ag_{20}(CO_3)$ by changing the SR structure.

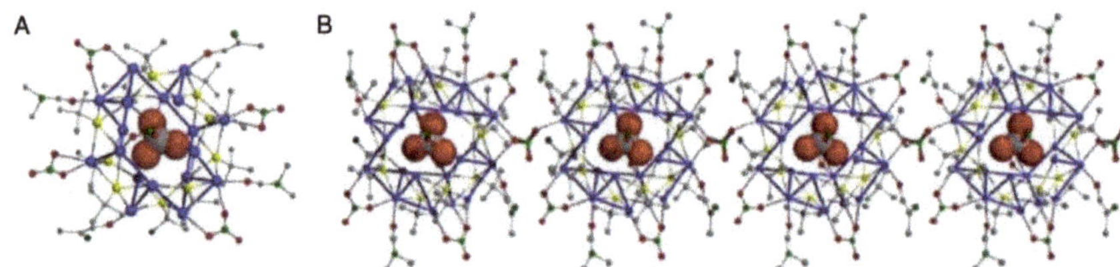

Figure 8. (**A**) Structure of $Ag_{18}(CO_3)(S\text{-}^tBu)_{10}(NO_3)_6(DMF)_4$. (**B**) Ball-and-stick view of the 1D chain of $Ag_{18}(CO_3)(S\text{-}^tBu)_{10}(NO_3)_6(DMF)_4$. Ag = blue, S = yellow, O = red, C = gray, N = green. Reproduced with permission from Reference [148]. Copyright 2017 Wiley-VCH.

Recently, Sun et al. [149] succeeded in the connection of $Ag_{44}(V_{10}O_{28})(S\text{-}Et)_{20}(PhSO_3)_{18}(H_2O)_2$ ($V_{10}O_{28}^{6-}$, Scheme 1(11), $PhSO_3^- $ = benzenesulfonic acid ion). In this 1D CS, the $Ag_{44}(V_{10}O_{28})$ core was formed by using the polyoxometalate (POM) $V_{10}O_{28}^{6-}$ as an anion template (Figure 9A). This was the first report of the formation of a structure in which $V_{10}O_{28}^{6-}$ was covered with an Ag NC with SR as a ligand. This 1D CS assembled because two Ag–O bonds were formed between two $PhSO_3^-$ in the ligand layer and one $Ag_{44}(V_{10}O_{28})(S\text{-}Et)_{20}$ NC (Figure 9B).

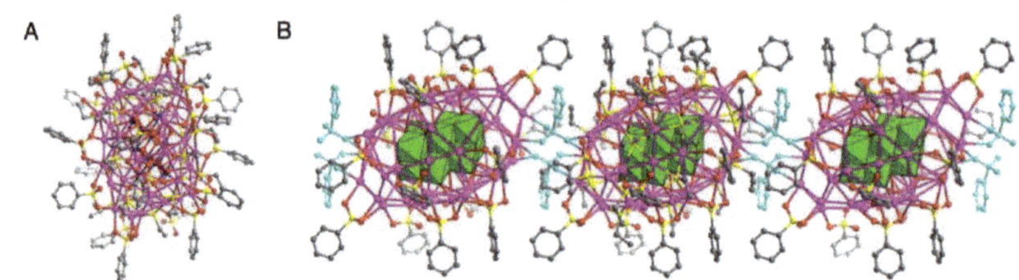

Figure 9. (**A**) Structure of $Ag_{44}(V_{10}O_{28})(S\text{-}Et)_{20}(PhSO_3)_{18}(H_2O)_2$ (**B**) 1D chain structure of $Ag_{44}(V_{10}O_{28})(S\text{-}Et)_{20}(PhSO_3)_{18}(H_2O)_2$ with all bridging $PhSO_3^-$ ligands highlighted in cyan and $V_{10}O_{28}^{6-}$ anions shown as green polyhedra. Ag = purple, V = dark blue, S = yellow, C = gray, O = red. All H atoms are omitted. Reproduced with permission from Reference [149]. Copyright 2019 Royal Society of Chemistry.

In the 1D CS of $[Au_7Ag_9(dppf)_3(CF_3COO)_7BF_4]_n$ (dppf = 1,1′-bis(diphenylphosphino)ferrocene, Scheme 1(12), BF_4 = tetrafluoroboric acid) reported by Wang et al. [150] in 2019, each NC was also connected via an Ag–O bond (Figure 10A), although this was not direct connection of metal NCs. In the above three NCs (i.e., $Ag_{20}(CO_3)(S\text{-}^tBu)_{10}(CH_3COO)_8(DMF)_2$, $Ag_{18}(CO_3)(S\text{-}^tBu)_{10}(NO_3)_6(DMF)_4$, and $Ag_{44}(V_{10}O_{28})(S\text{-}Et)_{20}(PhSO_3)_{18}(H_2O)_2)$, the anion template was contained in the center, whereas $Au_7Ag_8(dppf)_3(CF_3COO)_7$ had an icosahedral metal core composed of Au_7Ag_8. Such an icosahedral core structure is often seen in metal NCs [27,53]. The 1D CS of $[Au_7Ag_9(dppf)_3(CF_3COO)_7BF_4]_n$ was synthesized in one pot. Probably, excess Ag binds to CF_3COO^- in the ligand layer during synthesis, resulting in the formation of a 1D CS composed of $Au_7Ag_8(dppf)_3(CF_3COO)_7$ NCs. The researchers also revealed that this 1D CS possessed a band gap of 2.18 eV (Figure 10B).

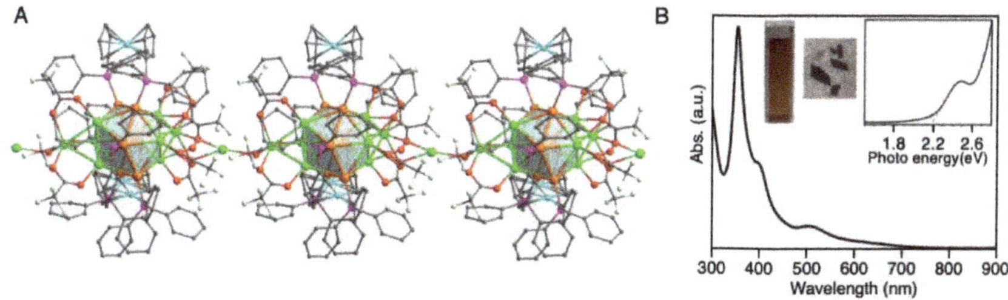

Figure 10. (**A**) View of the whole structure of $[Au_7Ag_9(dppf)_3(CF_3COO)_7BF_4]_n$ (anions and H atoms are omitted for clarity). Ag = green, Au = orange, O = red, P = purple, C = gray, Fe = blue. (**B**) Absorption spectrum of $[Au_7Ag_9(dppf)_3(CF_3COO)_7BF_4]_n$ in CH_2Cl_2 solution. Inset: absorption spectrum on the energy scale (eV) and photographs showing actual colors of $[Au_7Ag_9(dppf)_3(CF_3COO)_7BF_4]_n$ in CH_2Cl_2 and the crystalline state. Reproduced with permission from Reference [150]. Copyright 2019 Royal Society of Chemistry.

2.3. Control of Counterions

For certain metal NCs, the total number of valence electrons satisfies the number for the closed-shell electronic structure when it is a cation, so they are generated as a cation [151]. For example, $[Au_{21}(S\text{-}c\text{-}C_6H_{11})_{12}(DPPM)]^+$ ($S\text{-}c\text{-}C_6H_{11}$ = cyclohexanethiolate, Scheme 1(13), and DPPM = bis(diphenylphosphinomethane), Scheme 1(14)) is synthesized as a cation. In 2018, Jin et al. [152] revealed that $[Au_{21}(S\text{-}c\text{-}C_6H_{11})_{12}(DPPM)]^+$ formed a 1D CS in a crystal by assembling as a pair with the counter anion and that the structure of the 1D CS changed depending on the counterion (Figure 11A,B).

In this study, $[Au_{21}(S\text{-}c\text{-}C_6H_{11})_{12}(DPPM)_2]^+[AgCl_2]^-$ and $[Au_{21}(S\text{-}c\text{-}C_6H_{11})_{12}(DPPM)_2]^+[Cl]^-$ were precisely synthesized and single crystals were grown. As shown in Figure 11B, a 1D CS was formed by the alternating connection of $[Au_{21}(S\text{-}c\text{-}C_6H_{11})_{12}(DPPM)_2]^+$ and $[AgCl_2]^-$ in the $[Au_{21}(S\text{-}c\text{-}C_6H_{11})_{12}(DPPM)_2]^+[AgCl_2]^-$ crystal. This 1D CS was considered to assemble via π-π, anion-π, and aryl C-H···Cl interactions. The connection pattern of $[Au_{21}(S\text{-}c\text{-}C_6H_{11})_{12}(DPPM)_2]^+$ in the 1D CS changed slightly when the counterion was Cl^- rather than $[AgCl_2]^-$. It was considered that the connection pattern of $[Au_{21}(S\text{-}c\text{-}C_6H_{11})_{12}(DPPM)_2]^+$ changed because the arrangement of phenyl groups in the NC was affected by the counterion (Figure 11A).

The obtained 1D CSs of $[Au_{21}(S\text{-}c\text{-}C_6H_{11})_{12}(DPPM)_2]^+$ had different electron transport properties depending on the counter anion. The 1D CS of $[Au_{21}(S\text{-}c\text{-}C_6H_{11})_{12}(DPPM)_2]^+[AgCl_2]^-$ had a σ of only ~1.44×10^{-8} S/m, whereas that of $[Au_{21}(S\text{-}c\text{-}C_6H_{11})_{12}(DPPM)_2]^+[Cl]^-$ was σ ~2.38×10^{-6} S/m (Figure 11C). Changing the counter anion from $[AgCl_2]^-$ to $[Cl]^-$ shortened the distance between NCs from 16.80 to 16.39 Å and formed an intra-cluster π-stacking structure that allowed electricity to flow easily (Figure 11B). These two reasons explained why σ of $[Au_{21}(S\text{-}c\text{-}C_6H_{11})_{12}(DPPM)_2]^+[Cl]^-$ was two orders of magnitude higher than that of $[Au_{21}(S\text{-}c\text{-}C_6H_{11})_{12}(DPPM)_2]^+[AgCl_2]^-$ (Figure 11D).

There are not only metal NCs synthesized as cations but also metal NCs synthesized as anions. Because the total number of valence electrons of $[Ag_{29}(BDT)_{12}(PPh_3)_4]^{3-}$ (BDT = 1,3-benzenedithiolate, Scheme 1(15); PPh_3 = triphenylphosphine, Scheme 1(16)) satisfies the number for a closed-shell electronic structure as the anion, it is generated as the anion [153]. In 2019, Zhu et al. [154] reported that mixing this NC with Cs acetate in DMF induced Cs^+ attachment to the NC and PPh_3 desorption from the NC, resulting in the formation of $[Cs_3Ag_{29}(BDT)_{12}(DMF)_x]^0$ (x = 5 or 6) and that the obtained $[Cs_3Ag_{29}(BDT)_{12}(DMF)_x]^0$ formed a 1D CS in its crystal. Figure 12A shows the resulting 1D CS. In the crystal, $[Cs_3Ag_{29}(BDT)_{12}(DMF)_x]^0$ was connected by a series of bonds consisting of $-Cs^+-DMF-Cs^+-S-Ag-Ag-S-$. This 1D CS was considered to be formed because of the electrostatic attraction between $[Ag_{29}(BDT)_{12}(DMF)_x]^{3-}$ and Cs^+, Cs–S bond formation, and Cs···π interactions (Figure 12B).

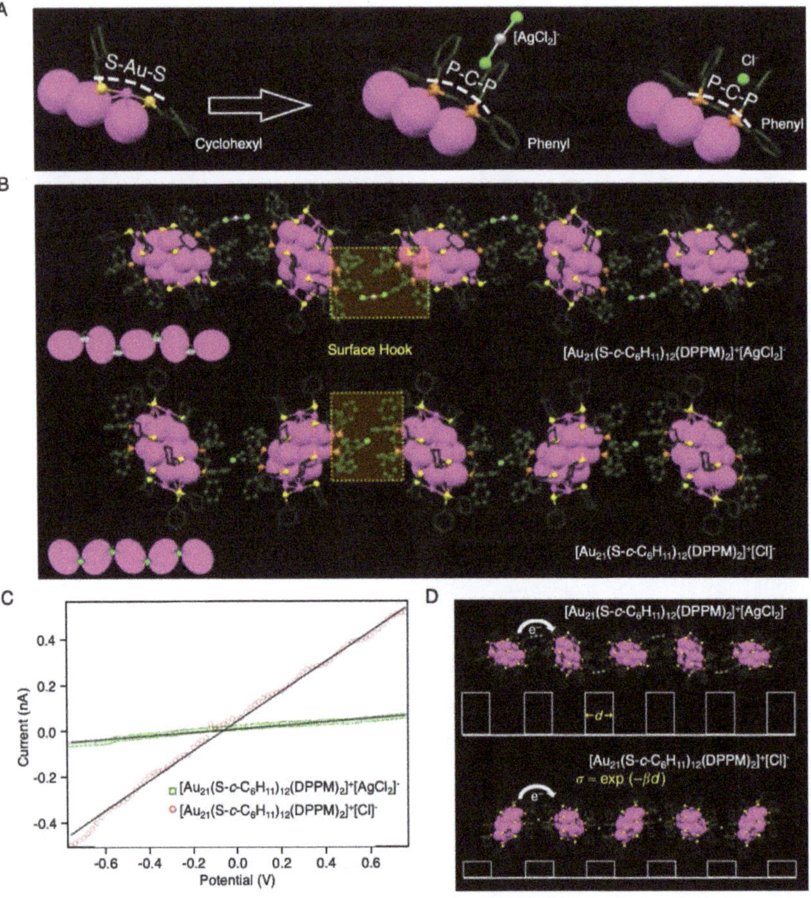

Figure 11. (**A**) Site-specific tailoring of the surface motifs and associated counterions of Au NCs. The two RS-Au-SR (R is cyclohexyl) surface motifs in Au_{23} (precursor NC) were replaced by two DPPM motifs in Au_{21}. Each P atom was connected to two phenyl rings. Dashed lines indicate the motifs. (**B**) Packing of $[Au_{21}(S\text{-}c\text{-}C_6H_{11})_{12}(DPPM)_2]^+[AgCl_2]^-$ and $[Au_{21}(S\text{-}c\text{-}C_6H_{11})_{12}(DPPM)_2]^+[Cl]^-$ in their 1D assemblies. The orientation of Au NCs is modulated by the counterion. Au = magenta, Ag = gray, Cl = light green, S = yellow, P = orange, C = green. All H atoms are omitted for clarity. Yellow areas are the surface hooks connecting neighboring NCs. (**C**) Room-temperature conductivity of single crystals of $[Au_{21}(S\text{-}c\text{-}C_6H_{11})_{12}(DPPM)_2]^+[AgCl_2]^-$ (green) and $[Au_{21}(S\text{-}c\text{-}C_6H_{11})_{12}(DPPM)_2]^+[Cl]^-$ (red). (**D**) Schematic diagram of electron hopping in Au_{21} NC assemblies. Different configurations of the interacting π–π pairs result in tunneling barriers of different heights (white solid squares), thus changing the electron conductivity (e^- represents an electron, σ is the conductivity, d is the interparticle distance, and β is the tunneling decay constant). Reproduced with permission from Reference [152]. Copyright 2018 Springer-Nature.

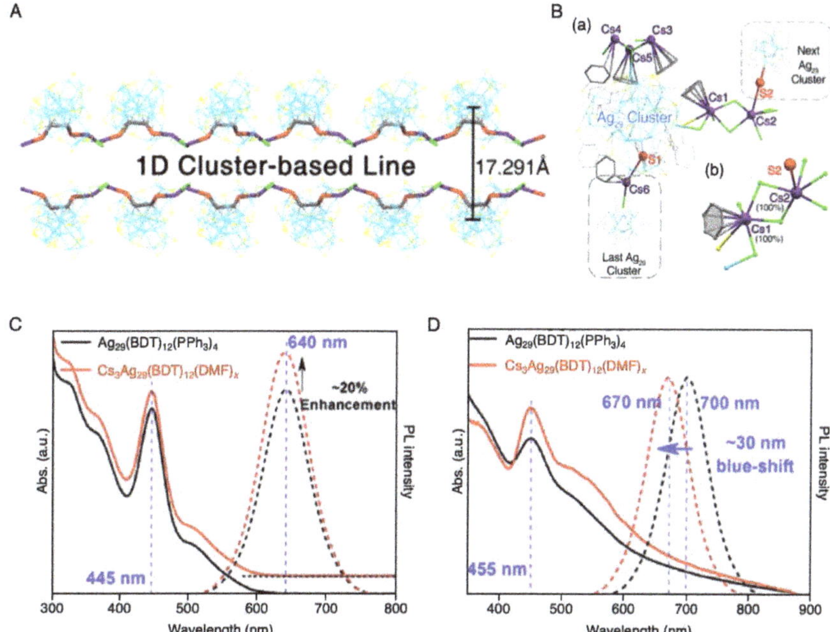

Figure 12. (**A**) 1D linear assembly of $[Cs_3Ag_{29}(BDT)_{12}(DMF)_x]^0$ in the crystal lattice. Ag = light blue/gray, Cs = dark purple, S = yellow and red, O = green. For clarity, all H, C, and N atoms, some Cs^+, and DMF molecules are omitted. Each O atom represents a DMF molecule. (**B**) (a) Overall surface structure of $[Cs_3Ag_{29}(BDT)_{12}(DMF)_x]^0$ and (b) interactions between $Ag_{29}(BDT)_{12}$, Cs_1, Cs_2, and DMF. (**C**) Comparison of optical absorption and emission spectra of $Ag_{29}(BDT)_{12}(PPh_3)_4$ (black) and $Cs_3Ag_{29}(BDT)_{12}(DMF)_x$ (red) NCs dissolved in DMF. (**D**) Comparison of optical absorption and emission spectra of $Ag_{29}(BDT)_{12}(PPh_3)_4$ (black) and $Cs_3Ag_{29}(BDT)_{12}(DMF)_x$ (red) NCs in crystalline films. Reproduced with permission from Reference [154]. Copyright 2019 American Chemical Society.

Both $[Ag_{29}(BDT)_{12}(PPh_3)_4]^{3-}$ and $[Cs_3Ag_{29}(BDT)_{12}(DMF)_x]^0$ solutions showed similar absorption and photoluminescence (PL) spectra (Figure 12C). This indicates that Cs^+ attachment and PPh_3 desorption did not markedly change the electronic structure of the $Ag_{29}(BDT)_{12}$ NCs. In contrast, the absorption and PL spectra of $[Ag_{29}(BDT)_{12}(PPh_3)_4]^{3-}$ and $[Cs_3Ag_{29}(BDT)_{12}(DMF)_x]^0$ in the crystalline state were quite different (Figure 12D). The 1D CS of $[Cs_3Ag_{29}(BDT)_{12}(DMF)_x]^0$ was considered to show different optical behavior from that of the individual $[Ag_{29}(BDT)_{12}(PPh_3)_4]^{3-}$ because of the electronic interactions between adjacent NCs in the 1D CS of $[Cs_3Ag_{29}(BDT)_{12}(DMF)_x]^0$.

2.4. Introduction of Linker Molecules

In the examples described in Section 2.3, 1D CSs were formed by the counterion acting as a linker. When an organic molecule is used as the linker, the distance between NCs in a 1D CS can be freely controlled because the design of the structure of organic molecules is well understood. In fact, the geometry of an MOF is controlled by the design of the linker organic molecule [155]. In recent years, several 1D CSs of metal NCs with organic molecules as linkers have also been reported.

Table 1. Connection modes, NCs, linkers, publication years, and references for 1D CS.

Connection Mode	NC	Linker	Year	Ref.
Formation of metal–metal bond (Figure 2A)	$[Au_{25}(S\text{-}Bu)_{18}]^0$	-	2014	[122]
	$[Au_{25}(S\text{-}Pen)_{18}]^0$	-	2017	[140]
	$[Au_{24}Hg(S\text{-}Bu)_{18}]^0$	-	2019	[141]
	$[Au_{24}Cd(S\text{-}Bu)_{18}]^0$	-		
	$[Au_4Pd_2(PET)_8]^0$	-	2017	[143]
	$[Au_4Pt_2(SCH_2PhCl)_8]^0$	-	2020	[142]
	$[Au_4Pt_2(PET)_8]^0$	-		
	$(AuAg)_{34}(A\text{-}Adm)_{20}$	-	2020	[146]
Formation of Ag–O, Ag–S, Ag–Cl bond, etc. (Figure 2B)	$Ag_{20}(CO_3)(S\text{-}^tBu)_{10}(CH_3COO)_8(DMF)_2$	-	2014	[147]
	$Ag_{18}(CO_3)(S\text{-}^tBu)_{10}(NO_3)_6(DMF)_4$	-	2017	[148]
	$Ag_{44}(V_{10}O_{28})(S\text{-}Et)_{20}(PhSO_3)_{18}(H_2O)_2$	-	2019	[149]
	$Au_7Ag_8(dppf)_3(CF_3COO)_7$	-	2019	[150]
Control of counter ion (Figure 2C)	$[Au_{21}(S\text{-}c\text{-}C_6H_{11})_{12}(DPPM)_2]^+$	$[AgCl_2]^-$	2018	[152]
	$[Au_{21}(S\text{-}c\text{-}C_6H_{11})_{12}(DPPM)_2]^+$	$[Cl]^-$		
	$[Ag_{29}(BDT)_{12}(PPh_3)_4]^{3-}$	$[Cs]^+$	2019	[154]
Introduction of linker molecule (Figure 2D)	$Ag_{14}(DT\text{-}o\text{-}C)_6$	pyrazine [a]	2018	[156]
	$Ag_{18}(PhPO_3)(S\text{-}^tBu)_{10}(CF_3COO)_2(PhPO_3H)_4$	bpy-NH$_2$ [a]	2019	[157]
	$Ag_{15}Cl(S\text{-}^tBu)_8(CF_3COO)_{5.67}(NO_3)_{0.33}(DMF)_2$	bpy [a]	2019	[158]
	$Ag_{10}(CF_3COO)_4(S\text{-}^tBu)_6(CH_3CN)_2$	*p*-iah [a]	2019	[159]
	$Ag_{10}(CF_3COO)_4(S\text{-}^tBu)_6(CH_3CN)$	*o*-iah [a]		
	$Cd_6Ag_4(S\text{-}Ph)_{16}(DMF)_3(CH_3OH)$	bpe [a]	2020	[160]

[a] See Scheme 2.

For example, in 2018, Zang et al. [156] reported a 1D CS in which $Ag_{14}(DT\text{-}o\text{-}C)_6$ NCs (DT-*o*-C = 1,2-dithiolate-*o*-carborane, Scheme 1(17)) were linked by pyrazine (Scheme 2(1)). In this study, first, $[Ag_{14}(DT\text{-}o\text{-}C)_6(pyridine/p\text{-}methylpyridine)_8]$ (Scheme 2(2),(3)) were identified as Ag NCs with high thermal stability that maintained their framework structure even at 150 °C or higher in air (Figure 13A). Then, they attempted to synthesize Ag NCs in which the pyridine or *p*-methylpyridine ligands of these Ag NCs were replaced by pyrazine. As a result, they obtained a 1D CS in which $Ag_{14}(DT\text{-}o\text{-}C)_6$ NCs were linked by pyrazine (Figure 13B). In the obtained structure, pyrazine was coordinated to each $Ag_{14}(DT\text{-}o\text{-}C)_6$ NC at a diagonal position, which caused the 1D CS to rotate in the clockwise direction with respect to the (001) axis. In this study, the researchers also succeeded in forming 2D and 3D CSs composed of $Ag_{14}(DT\text{-}o\text{-}C)_6$ NCs by changing the structure of the bipyridine ligand, as described later in Sections 3.2 and 4.3, respectively.

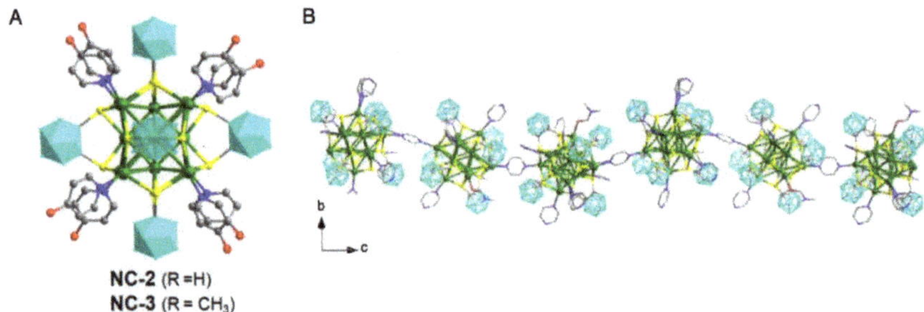

Figure 13. (**A**) Structure of $Ag_{14}(DT\text{-}o\text{-}C)_6(pyridine/p\text{-}methylpyridine)_8$. (**B**) 1D helix of $Ag_{14}(DT\text{-}o\text{-}C)_6$ NC. Ag = green and pink, S = yellow, C = gray, N = blue, carborane = turquoise. Reproduced with permission from Reference [156]. Copyright 2018 American Chemical Society.

Because N readily coordinates to Ag, bipyridine is often used to connect NCs containing Ag. In 2019, Zang and colleagues formed a 1D CS composed of $Ag_{18}(PhPO_3)(S\text{-}^tBu)_{10}(CF_3COO)_2(PhPO_3H)_4$ ($PhPO_3^{2-}$ = phenylphosphinic diion; $PhPO_3H^-$ = phenylphosphinic acid ion) NC nodes with bipyridine(3-amino-4,4′-bipyridine (bpy-NH$_2$, Scheme 2(4)) linkers [157]. In this experiment, NC synthesis, ligation, and crystallization were performed simultaneously in one pot (Figure 14A). The node $Ag_{18}(PhPO_3)(S-^tBu)_{10}$ NCs contained $PhPO_3^{2-}$ as an anion template in their center. In the crystal, adjacent $Ag_{18}(PhPO_3)(S-^tBu)_{10}$ NCs were linked by two bpy-NH$_2$ to form a 1D CS (Figure 14B).

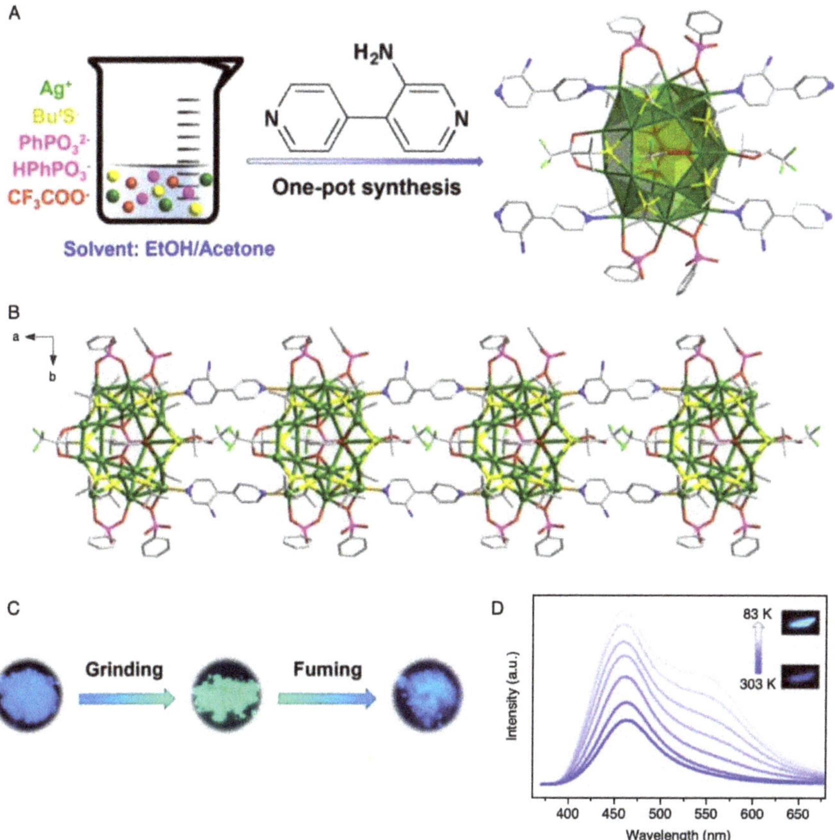

Figure 14. (**A**) Schematic representation of the one-pot synthesis of $[Ag_{18}(PhPO_3)(S-^tBu)_{10}(CF_3COO)_2(PhPO_3H)_4(bpy-NH_2)_2]\cdot(PhPO_3H_2)$. (**B**) 1D structure of $[Ag_{18}(PhPO_3)(S-^tBu)_{10}(CF_3COO)_2(PhPO_3H)_4(bpy-NH_2)_2]\cdot(PhPO_3H_2)$. Ag = green, S = yellow, C = gray, N = blue, O = red, F = light green, P = purple. H atoms are omitted for clarity. (**C**) Luminescent images of the as-synthesized, ground, and fumed $[Ag_{18}(PhPO_3)(S-^tBu)_{10}(CF_3COO)_2(PhPO_3H)_4(bpy-NH_2)_2]$ under ultraviolet light irradiation. (**D**) Temperature-dependent luminescence spectra of $[Ag_{18}(PhPO_3)(S-^tBu)_{10}(CF_3COO)_2(PhPO_3H)_4(bpy-NH_2)_2]$ from 30 to −190 °C in the solid state. The inset photographs show the emission of $[Ag_{18}(PhPO_3)(S-^tBu)_{10}(CF_3COO)_2(PhPO_3H)_4(bpy-NH_2)_2]$ in the solid state under ultraviolet light irradiation at room temperature and liquid nitrogen temperature. Reproduced with permission from Reference [157]. Copyright 2019 Wiley-VCH.

The obtained 1D CS was stable up to 110 °C in a nitrogen (N$_2$) atmosphere. When mechanical stimulation was applied to the 1D CS, its PL wavelength changed. When the 1D CS sample subjected to mechanical stimulation was recrystallized, its PL wavelength returned to the original value (Figure 14C). Thus, the PL of the 1D CS composed of $Ag_{18}(PhPO_3)(S-^tBu)_{10}$ NCs exhibited reversible mechanochromism. Because this 1D CS emitted light at two wavelengths and its PL intensity ratio changed with temperature (thermochromism; Figure 14D), the authors suggested that this 1D CS could be applied as a thermometer.

In 2019, Bakr et al. [158] also reported the connection of Ag NCs by bipyridine. In this study, a 1D CS was synthesized in one pot (Figure 15A). $Ag_{15}Cl(S-^tBu)_8(CF_3COO)_{5.67}(NO_3)_{0.33}(DMF)_2$ was used as the node, and 4,4'-bipyridine (bpy, Scheme 2(5)) was used as the linker. In $Ag_{15}Cl(S-^tBu)_8(CF_3COO)_{5.67}(NO_3)_{0.33}(DMF)_2$, Cl^- acted as an anion template. The core of the $Ag_{15}Cl$ NC had a geometry in which one Ag was lost from the $Ag_{16}Cl$ core of the $Ag_{16}Cl(S-^tBu)_8(CF_3COO)_7(DMF)_4(H_2O)$ NC, which did not form a 1D CS, namely individual $Ag_{16}Cl(S-^tBu)_8(CF_3COO)_7(DMF)_4(H_2O)$ NCs (Figure 15B). The 1D CS with a ladder structure was formed by combining these $Ag_{15}Cl$ NCs with three adjacent $Ag_{15}Cl$ NCs via four bpy molecules (Figure 15C). It was found that the 1D CS displayed slightly higher thermal stability than that of the $Ag_{16}Cl$ NCs (Figure 15D).

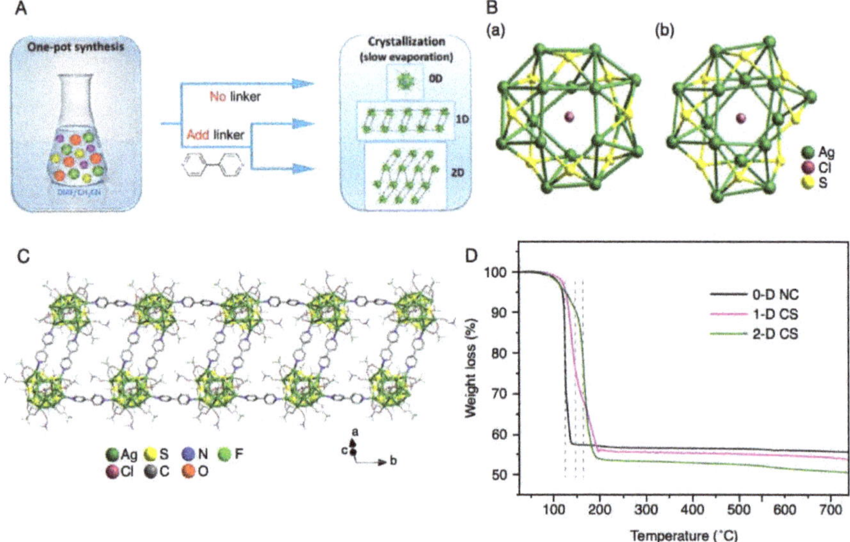

Figure 15. (**A**) Synthesis of Ag NCs and NC-based frameworks. (**B**) Top views of the core structures of (a) $Ag_{16}Cl(S-^tBu)_8(CF_3COO)_7(DMF)_4(H_2O)$ and (b) $Ag_{15}Cl(S-^tBu)_8(CF_3COO)_{5.67}(NO_3)_{0.33}(DMF)_2$. (**C**) Crystal structure of the corresponding 1D CS. Free (co-crystallized) DMF molecules are not shown. The green semitransparent spheres in the Ag clusters are shown as a visual guide. H atoms were omitted for clarity. (**D**) Thermogravimetric analysis curves of NCs, 1D CS, and 2D CS (see Section 3.2). Reproduced with permission from Reference [158]. Copyright 2019 American Chemical Society.

In the above three studies, the bipyridines had a rigid framework structure. In 2019, Cao et al. formed 1D CSs of Ag NCs by using pyridine derivatives (*p*-iah = 4-pyridine carboxylic hydrazide, Scheme 2(6); *o*-iah = 2-carboxylic hydrazide, Scheme 2(7)) that contained N in both the rigid pyridine framework and flexible substituents [159]. In this study, the 1D CSs were obtained by reacting $Ag_{12}(S-^tBu)_6(CF_3COO)_6(CH_3CN)_6$ (CH_3CN = acetonitrile) with the above-mentioned pyridine derivatives. The SC-XRD analysis of the products revealed that $Ag_{10}(CF_3COO)_4(S-^tBu)_6(CH_3CN)_2$ and $Ag_{10}(CF_3COO)_4(S-^tBu)_6(CH_3CN)$ were the nodes in the 1D CSs with *p*-iah and *o*-iah, respectively (Figure 16A(a),B(a)). These 1D CSs containing *p*-iah and *o*-iah had cross-helical and parallel chain structures, respectively (Figure 16A(b),B(b)). The latter structure also contained hydrogen bonds (N–H···O) between the parallel linker molecules. It is interesting that different 1D CSs formed depending on the position of N in the linker molecule (Scheme 2(6),(7)).

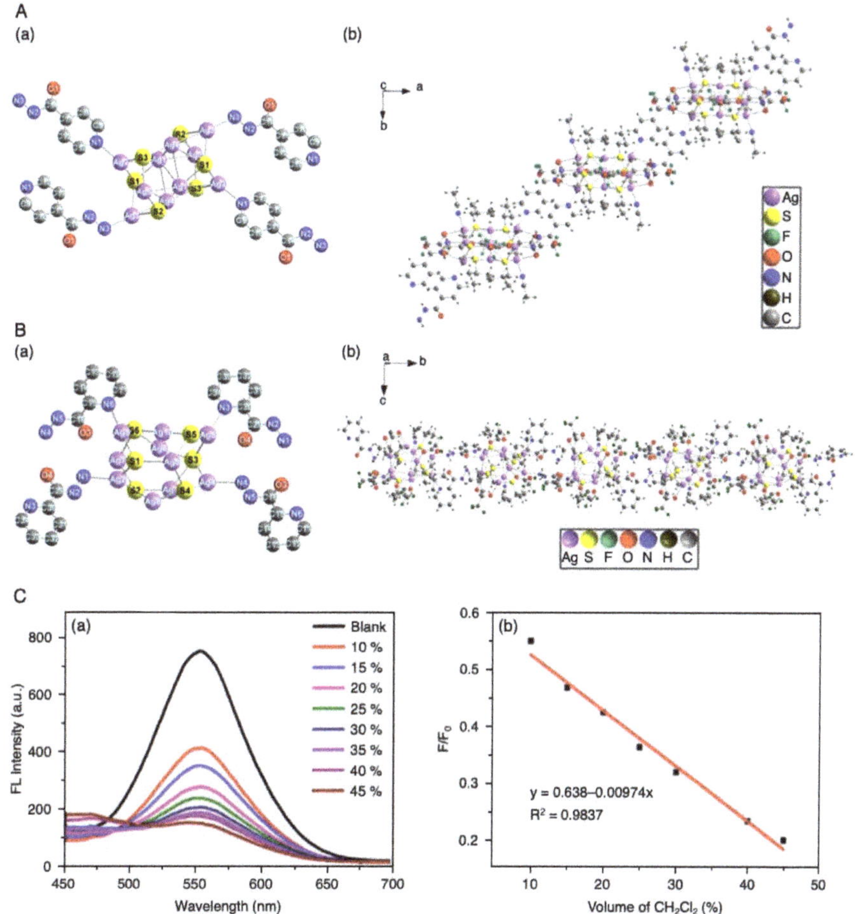

Figure 16. (a) Structural units and (b) spatial stacking diagrams of (**A**) $Ag_{10}(CF_3COO)_4(S$-$^tBu)_6(CH_3CN)_2(p$-iah$)_2$ and (**B**) $Ag_{10}(CF_3COO)_4(S$-$^tBu)_6(CH_3CN)(o$-iah$)_2$. (**C**) (a) PL spectra of $Ag_{10}(CF_3COO)_4(S$-$^tBu)_6(CH_3CN)_2(o$-iah$)_4$ in CCl_4 with various volume fractions of CH_2Cl_2. (b) Linear plot of fluorescence intensity against the volume fraction of CH_2Cl_2 in CCl_4. Reproduced with permission from Reference [159]. Copyright 2019 American Chemical Society.

Both 1D CSs showed PL and that with *o*-iah as the linker exhibited weak green PL in highly polar solvents and strong yellow PL in solvents with low polarity. Based on these characteristics, the authors suggested that the 1D CS with *o*-iah could be used to measure the concentration of dichloromethane (CH_2Cl_2, Figure 16C) or trichloromethane ($CHCl_3$) in tetrachloromethane (CCl_4).

Bipyridines can also be used as linkers to form 1D CSs of Ag chalcogenide NCs. Very recently, Xu and co-workers reported the formation of a zigzag-type of 1D CS with $Cd_6Ag_4(S$-$Ph)_{16}(DMF)(H_2O)$ (S-Ph = benzenethiolate, Scheme 1(18) and Figure 17A) as a node and *trans*-1,2-bis(4-pyridyl)ethylene (bpe, Scheme 2(8)) as a linker (Figure 17B) [160]. This 1D CS was obtained by the reaction of $Cd_6Ag_4(S$-$Ph)_{16}(DMF)_3(CH_3OH)$ (CH_3OH = methanol) with bpe. In the obtained 1D CS, N of bpe coordinated to Cd not Ag (Figure 17A). Such a coordination pattern has also been observed in 2D and 3D CSs composed of $Cd_6Ag_4(S$-$Ph)_{16}$ and bpe previously reported by Zhang et al. [161,162].

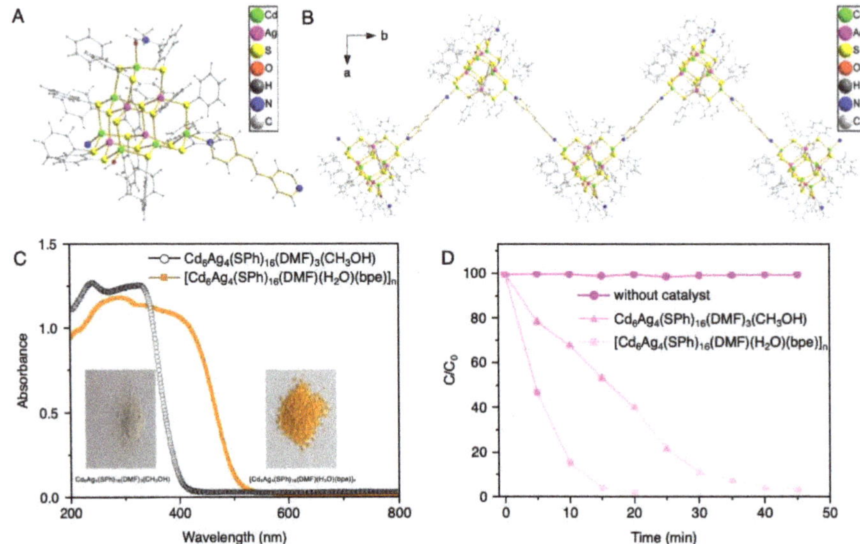

Figure 17. **(A)** Structures of $Cd_6Ag_4(S\text{-}Ph)_{16}(DMF)(H_2O)$ and **(B)** 1D CS of $Cd_6Ag_4(S\text{-}Ph)_{16}(DMF)(H_2O)(bpe)$. **(C)** Solid-state ultraviolet-visible diffuse reflectance spectra of the discrete $Cd_6Ag_4(SPh)_{16}(DMF)_3(CH_3OH)$ (open circles) and 1D CS of $Cd_6Ag_4(S\text{-}Ph)_{16}(DMF)(H_2O)(bpe)$ (filled squares). **(D)** Comparison of the photocatalytic-degradation efficiencies of the discrete $Cd_6Ag_4(S\text{-}Ph)_{16}(DMF)_3(CH_3OH)$ (triangles), 1D CS of $Cd_6Ag_4(S\text{-}Ph)_{16}(DMF)(H_2O)(bpe)$ (squares), and without a catalyst (circles). Reproduced with permission from Reference [160]. Copyright 2020 American Chemical Society.

They compared the electronic structures of the resulting 1D CS and individual $Cd_6Ag_4(S\text{-}Ph)_{16}(DMF)_3(CH_3OH)$ NCs. The results revealed that the band gap of the NCs was narrowed by the formation of the 1D CS (Figure 17C). In the 1D CS, the optical absorption onset was redshifted to the visible region. They used the 1D CS as a visible light (>420 nm)-responsive photocatalyst to decompose the organic dye Rhodamine B in water. The 1D CS exhibited higher photocatalytic activity toward Rhodamine B degradation than that of the $Cd_6Ag_4(S\text{-}Ph)_{16}$ NCs (Figure 17D) and high stability during the photocatalytic reaction.

3. Two-Dimensional Structures

To date, Ag NCs have been used as the building blocks in almost all 2D CSs. Like Au NCs, Ag NCs have unique electronic and optical properties [68,163–166] and are expected to be applied in various fields. However, Ag NCs are less stable than Au NCs against external stimuli, such as light and solvents. Therefore, studies have been actively conducted to improve the stability of Ag NCs by assembly of CSs and thereby improve their physical properties. In Sections 3.1 and 3.2 we focus on the connection of NCs by Ag–O bond formation (Figure 2B) and introducing linker molecules (Figure 2D), respectively. Table 2 summarizes the connection methods, NCs, linkers, reported years, and reference numbers of the relevant literature. Several of the ligands used in these studies are shown in Scheme 1. The organic molecules used as linkers are depicted in Scheme 2.

3.1. Connection via Ag–O Bonds

In 2017, Mak et al. [148] reported the formation of 2D CSs with $Ag_{20}(CO_3)(S\text{-}iPr)_{10}(CF_3COO)_9(CF_3COOH)(CH_3OH)_2$ (S-iPr = isopropylthiolate, Scheme 1(5)) or $Ag_{20}(CO_3)(S\text{-}c\text{-}C_6H_{11})_{10}(CF_3COO)_{10}(CF_3COOH)_2(H_2O)_2$ as building blocks in their paper on the formation of 1D

CSs. These NCs contained $CO_3{}^{2-}$ as an anion template at the center of their cores (Figure 18A,B). The structures of the SR in these two types of $Ag_{20}(CO_3)(SR)_{10}$ NCs were different (S-*i*Pr vs. S-*c*-C_6H_{11}), which influenced the formation angle and bond distance between adjacent NCs in the 2D CSs. The O^- of CF_3COO^- and Ag of an adjacent NC were directly linked via an Ag–O bond with a length of 14 Å. In addition, isolated Ag was trapped between CF_3COO^- of adjacent NCs, so the adjacent NCs were connected via an O–Ag–O bond with a length of 17 Å. The 2D CS consisting of $Ag_{20}(CO_3)(S\text{-}i\mathrm{Pr})_{10}(CF_3COO)_9(CF_3COOH)(CH_3OH)_2$ showed dual emission at room temperature. Because both zero-dimensional $[Ag_{20}(S\text{-}^t\mathrm{Bu})_{10}(CF_3COO)_2]Cl\cdot(CF_3COO)_7\cdot5CH_3OH$ NCs and the 1D CS of $Ag_{20}(CO_3)(S\text{-}^t\mathrm{Bu})_{10}(CH_3COO)_8(DMF)_2$ emit only single emission peaks, it was speculated that the formation of the 2D CS was related to the observed dual emission.

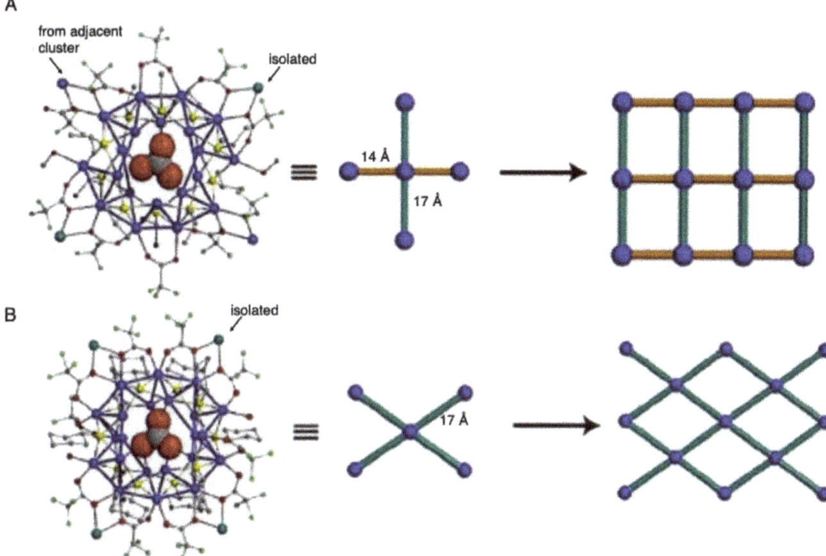

Figure 18. (**A**) The molecular building block in $Ag_{20}(CO_3)(S\text{-}i\mathrm{Pr})_{10}(CF_3COO)_9(CF_3COOH)(CH_3OH)_2$ with its four linking sites, and condensation of blocks into a 4,4-net. (**B**) Similar condensation of molecular building blocks of $Ag_{20}(CO_3)(S\text{-}c\text{-}C_6H_{11})_{10}(CF_3COO)_{10}(CF_3COOH)_2(H_2O)_2$. Note the difference between the 'arms' with lengths of 14 Å (orange) and 17 Å (cyan). Ag = blue and cyan, S = yellow, O = red, C = gray, F = green. Reproduced with permission from Reference [148]. Copyright 2017 Wiley-VCH.

In 2019, Sun and co-workers also reported the formation of a 2D CS, using Ag NCs [149]. In this study, a 2D CS consisting of $Ag_{46}(V_{10}O_{28})(S\text{-}Et)_{23}(PhSO_3)_{15}(CO_3)$ was formed (Figure 19), using a different solvent from the case of $Ag_{44}(V_{10}O_{28})(S\text{-}Et)_{20}(PhSO_3)_{18}(H_2O)_2$, which formed a 1D CS (Figure 9). These Ag NCs have the same total number of ligands (S-Et and $PhSO_3^-$) of 38 but different ratios of the ligand types. They considered that the ligand ratio affected the number of Ag atoms in the core and also the connection mode between adjacent NCs.

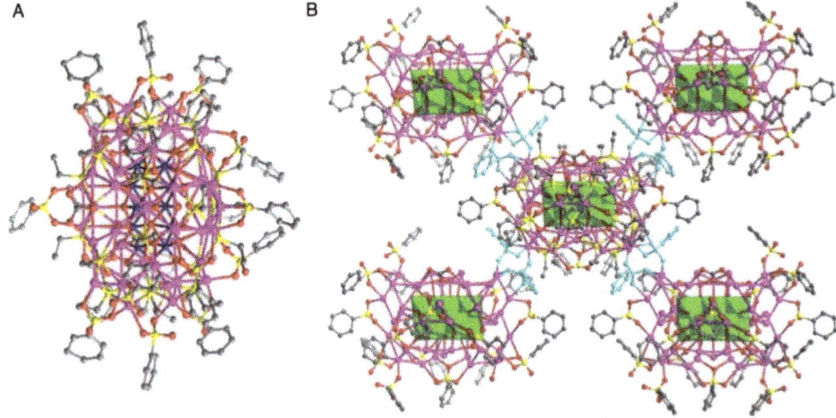

Figure 19. (**A**) Structure and (**B**) 2D extended layer structure of $Ag_{46}(V_{10}O_{28})(S\text{-}Et)_{23}(PhSO_3)_{15}(CO_3)$ with all bridging $PhSO_3^-$ ligands highlighted in cyan and $V_{10}O_{28}^{6-}$ shown as green polyhedra. Ag = purple, V = dark blue, S = yellow, C = gray, O = red. All H atoms are omitted. Reproduced with permission from Reference [149]. Copyright 2019 Royal Society of Chemistry.

Xu et al. [166], also in 2019, reported 2D CSs of $Ag_{11}Cl(N\text{-}L)_8(CF_3COO)_2\cdot2CHCl_3$ (Figure 20), $Ag_{11}Cl(N\text{-}L)_8(NO_3)_2\cdot2CHCl_3$, and $Ag_{11}Cl(N\text{-}L)_8(CF_3SO_3)_2\cdot2CHCl_3$ (N-L = 2-acetamido-5-methyl-1,3,4-thiadiazole, Scheme 1(19)), in which adjacent NCs are linked by Ag–O bonds. SR, alkyne, or phosphine ligands are generally used in metal NCs. In this study, their aim was to synthesize Ag NCs by using an N-donor ligand, which is not appropriate based on the hard/soft acid/base theory [167], and form corresponding CSs. The three kinds of 2D CSs obtained had similar frameworks regardless of the coordination ions (CF_3COO^-, NO_3^-, or $CF_3SO_3^-$), which means that the framework structure shown in Figure 20B is very rigid. It was found that $Ag_{11}Cl(N\text{-}L)_8(CF_3COO)_2\cdot2CHCl_3$ showed dual-emission behavior and that its PL peaks had different optimal excitation wavelengths.

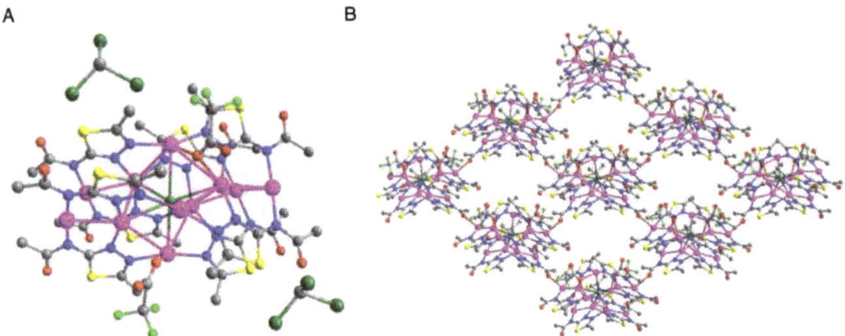

Figure 20. (**A**) Structure of $Ag_{11}Cl(N\text{-}L)_8(CF_3COO)_2\cdot2CHCl_3$ and (**B**) simplified 2D network featuring a four-connected topology. Ag = purple, S = yellow, O = red, C = gray, Cl = dark green, N = blue, F = neon green. Reproduced with permission from Reference [166]. Copyright 2019 Royal Society of Chemistry.

3.2. Introduction of Linker Molecules

As described below, in Section 4.3, in 2017, Zang et al. [168] reported the formation of a 3D CS in which $Ag_{12}(S\text{-}^tBu)_8(CF_3COO)_4$ NCs were linked by bpy. This was a pioneering study on the formation of an MOF, using Ag NCs as nodes, and has greatly influenced subsequent studies. In 2018, they synthesized a 2D CS consisting of Ag_{12} NCs and bpy [169]. The core structure of the node was changed (isomerized) by dissolving an $Ag_{12}(S^{-t}Bu)_6(CF_3COO)_6$ NC MOF crystal in a mixed solvent consisting of *N,N*-dimethylethanamide (DMAC) and toluene, which changed the geometrical structure of the entire CS from 3D to 2D (Figure 21A,B). This result indicates that the solvent selection is important in the design of the structure of metal NCs and their CSs. In the 2D CS consisting of newly formed $Ag_{12}(S^{-t}Bu)_6(CF_3COO)_6$ NCs, each $Ag_{12}(S^{-t}Bu)_6(CF_3COO)_6$ NC was linked to six adjacent $Ag_{12}(S^{-t}Bu)_6(CF_3COO)_6$ NCs via linkers to produce a highly symmetric 2D CS (Figure 21C). The layers were separated by 7.23 Å, with weak interactions between them. It was also revealed that the reversible structural transformation between 3D and 2D CSs could be induced by appropriate solvent selection (Figure 21A,B).

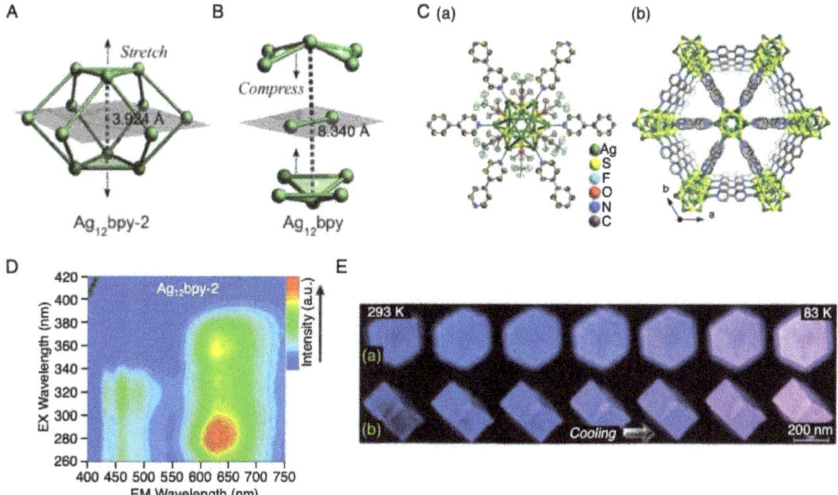

Figure 21. Comparison of the Ag_{12} core structures in (**A**) $Ag_{12}(S^{-t}Bu)_6(CF_3COO)_6(bpy)_3$ (Ag_{12}bpy-2) and (**B**) $Ag_{12}(S^{-t}Bu)_8(CF_3COO)_4(bpy)_4$ (Ag_{12}bpy). (**C**) (a) Perspective view of an $Ag_{12}S_6$ node with six pendant bpy linkers (ORTEP drawing at the 50% probability level). (b) Stacking of the 2D network structure of $Ag_{12}(S^{-t}Bu)_6(CF_3COO)_6(bpy)_3$ viewed along the crystallographic *c*-axis. Ag = green, S = yellow, C = gray, N = blue; CF_3COO^-, ^tBu, and H atoms are omitted for clarity. (**D**) 3D-excitation emission matrix of $Ag_{12}(S^{-t}Bu)_6(CF_3COO)_6(bpy)_3$ measured at −190 °C. (**E**) Thermochromic images of the (a) exterior {001} surfaces and (b) exposed interior {010}/{100} planes of $Ag_{12}(S^{-t}Bu)_6(CF_3COO)_6(bpy)_3/NH_2 \cdot (bpy:bpy-NH_2 = 20:1)$ solvated single crystals under ultraviolet light irradiation. Reproduced with permission from Reference [169]. Copyright 2018 Wiley-VCH.

Structural deformation of the CSs also induced changes in their electronic structure and PL properties. For example, the 3D CS only showed PL at a single wavelength, regardless of temperature and excitation wavelength, whereas the 2D CS exhibited PL of two colors (blue and red), depending on the excitation wavelength (Figure 21D). To enhance the blue emission of the 2D CS, the researchers introduced bpy-NH_2, which itself emits blue light, as a linker, to fabricate a 2D CS containing two types of linkers, bpy and bpy-NH_2. The intensity ratio of the red and blue PL signals depended on the mixing ratio of linker molecules. At the optimum linker mixing ratio, the PL intensity ratio of the red and blue peaks depended on temperature. Therefore, this 2D CS containing two types of linkers could be used as a temperature sensor (Figure 21E).

In 2018, the same group also reported the formation of a 2D CS consisting of $Ag_{14}(DT\text{-}o\text{-}C)_6$ NCs [156]. The 2D CS with $Ag_{14}(DT\text{-}o\text{-}C)_6$, as a structural unit (Figure 22A), was fabricated by changing the linker structure from that used to form the 1D CS (Section 2.4 and Figure 13) to dipyridin-4-yl-diazene (Scheme 2(9)). The obtained 2D CS possessed a rhombic network structure with $Ag_{14}(DT\text{-}o\text{-}C)_6(CH_3CN)_4$ as nodes (Figure 22B).

Table 2. Connection modes, NCs, linkers, publication years, and references for 2D CS.

Connection Mode	NC	Linker	Year	Ref.
Formation of Ag–O, Ag–S, Ag–Cl bond, etc. (Figure 2B)	$Ag_{20}(CO_3)(S\text{-}iPr)_{10}(CF_3COO)_9(CF_3COOH)(CH_3OH)_2$	-	2017	[148]
	$Ag_{20}(CO_3)(S\text{-}Cy)_{10}(CF_3COO)_{10}(CF_3COOH)_2(H_2O)_2$	-		
	$Ag_{46}(V_{10}O_{28})(S\text{-}Et)_{23}(PhSO_3)_{15}(CO_3)$	-	2019	[149]
	$Ag_{11}Cl(N\text{-}L)_8(CF_3COO)_2 \cdot 2CHCl_3$	-		
	$Ag_{11}Cl(N\text{-}L)_8(NO_3)_2 \cdot 2CHCl_3$	-	2019	[166]
	$Ag_{11}Cl(N\text{-}L)_8(CF_3SO_3)_2 \cdot 2CHCl_3$	-		
Introduction of linker molecule (Figure 2D)	$Ag_{12}(S\text{-}^tBu)_6(CF_3COO)_6$	bpy [a]	2018	[169]
	$Ag_{14}(DT\text{-}o\text{-}C)_6$	dipyridin-4-yl-diazene [a]	2018	[156]
	$Ag_{12}(S\text{-}^tBu)_6(CF_3COO)_6$	TPPA [a]	2018	[170]
	$Ag_{12}(S\text{-}^tBu)_6(CF_3COO)_3$	TPyP [a]	2019	[171]
	$Ag_{14}Cl(S\text{-}^tBu)_8(CF_3COO)_5(DMF)$	bpy [a]	2019	[158]
	$Ag_{10}(CF_3COO)_4(S\text{-}^tBu)_6(CH_3CN)_4$	*m*-iah [a]	2019	[159]
	$Ag_{12}(S\text{-}^tBu)_6(CF_3COO)_6(CH_3CN)_6$	bpz-NH$_2$ [a]	2019	[172]

a See Scheme 2.

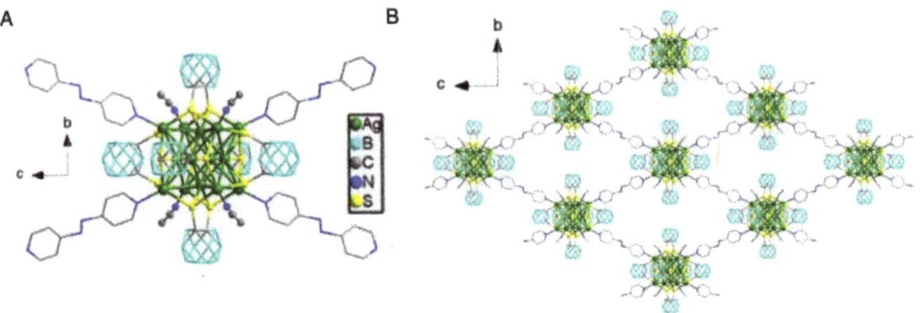

Figure 22. (**A**) Structural unit and (**B**) 2D CS of $Ag_{14}(DT\text{-}o\text{-}C)_6(CH_3CN)_4(\text{dipyridin-4-yl-diazene})_2$. Reproduced with permission from Reference [156]. Copyright 2018 American Chemical Society.

In 2018, Zang et al. [170] reported the formation of a 2D CS consisting of $Ag_{12}(S\text{-}^tBu)_6(CF_3COO)_6$ and tris(4-pyridylphenyl)-amine (TPPA; Scheme 2(10) and Figure 23A). This structure is interesting because the distance between the 2D layers can be changed. In this 2D CS, DMAC used as a solvent existed between layers immediately after the synthesis and the 2D layers overlapped, as shown in Figure 23B(a),(d). When the DMAC was partially removed from this structure, the overlap of the 2D layers changed, as illustrated in Figure 23B(b),(e). Furthermore, when this structure was immersed again in the mother liquor, the structure changed to that depicted in Figure 23B(c),(f). It was also found that the size of the crystal and its emission characteristics changed in accordance with the overlap manner in the 2D CS.

Recently, this group also formed a 2D CS with $Ag_{12}(S^tBu)_6(CF_3COO)_3$ NCs as nodes by using 5,10,15,20-tetra(4-pyridyl)porphyrin (TPyP, Scheme 2(11)) as a linker (Figure 24A) [171]. TPyP has a photosensitizing effect. Therefore, the ability of the 2D CS to degrade the toxic substance 2-chloroethyl ethyl sulfide (CEES), also called mustard gas, was studied (Figure 24B). The obtained 2D CS showed higher photocatalytic activity than that of a reported MOF. This high photocatalytic activity was ascribed to the synergistic effect of Ag NCs and TPyP, promoting the production of singlet oxygen, which induced the degradation of CEES (Figure 24B). The 2D CS maintained its crystallinity after the photocatalytic reaction and was able to be used repeatedly. The authors pointed out that photocatalytic activity could be further increased by selecting appropriate Ag NCs and organic molecular linkers.

Two other types of 2D CSs were also reported in 2019. In their paper on 1D CS formation (Figure 15 and Section 2.4), Bakr et al. [158] also reported that a 2D CS with $Ag_{14}Cl(S\text{-}^tBu)_8(CF_3COO)_5(DMF)$ as nodes was formed by changing the concentration of bpy during synthesis (Figure 25A). This $Ag_{14}Cl$ core contained Cl^- at its center as an anion template. Compared with the $Ag_{15}Cl(S\text{-}^tBu)_8(CF_3COO)_{5.67}(NO_3)_{0.33}(DMF)_2$

node of the 1D CS, the node of the 2D CS (Ag$_{14}$Cl(S-^{t}Bu)$_8$(CF$_3$COO)$_5$(DMF)) had one less Ag atom. However, the frameworks of these NCs were similar to each other (Figure 25B). The 2D CS showed higher thermal stability than those of individual Ag NCs and the 1D CS (Figure 15D). Unlike individual Ag NCs and the 1D CS, the 2D CS emitted green light with a strong intensity, even at room temperature (Figure 25C). Based on the results of a DFT calculation, it was interpreted that the enhancement of PL intensity was caused by a linker-to-cluster charge transfer excitation. In addition, Gao et al. [159] recently formed a 2D CS (Figure 26) with Ag$_{10}$(CF$_3$COO)$_4$(S-^{t}Bu)$_6$(CH$_3$CN)$_4$ nodes, using 3-pyridine carboxylic hydrazide (*m*-iah, Scheme 2(12)) as a linker, which is the *meta* equivalent of *p*-iah and *o*-iah used to form 1D CSs (Figure 16).

Thus, 2D CS formation by using a linker leads to the assembly of a structure with high stability and quantum yield (QY). However, 2D CS formation decreases the solubility of NCs, which limits their processability and device applicability. Therefore, Zang et al. [172] recently established a method to polymerize 2D CS to overcome this problem and provide materials suitable for practical use (Figure 27A). In this study, Ag$_{12}$(S-^{t}Bu)$_6$(CF$_3$COO)$_6$ NCs reported in their previous work (Figure 21) [168] were used as nodes. Moreover, 1,4-bis (pyrid-4-yl)benzenamine (bpz-NH$_2$, Scheme 2(13)) was used as the linker. The amino group of the bpz-NH$_2$ linker played an important role in polymerization. First, 2D CS crystals consisting of Ag$_{12}$(S-^{t}Bu)$_6$(CF$_3$COO)$_6$ and bpz-NH$_2$ were fabricated (Figure 27B). The crystal size was limited to about 200–300 nm by quenching the reaction within 1 min. A 2D CS film was obtained by reacting the crystals with methacrylic anhydride (MA). MA bound to the amino group of bpz-NH$_2$ (Figure 27C) and was then polymerized with acrylate monomers butyl methacrylate (BMA) and triethylene glycol dimethacrylate (TEGDMA), as shown in Figure 27D.

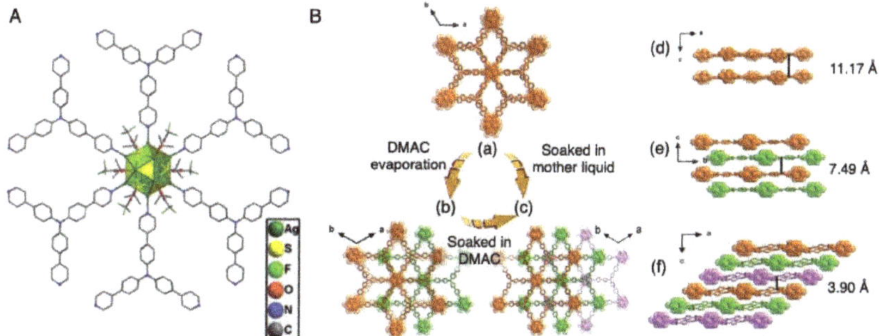

Figure 23. (**A**) Perspective view of an Ag$_{12}$S$_6$ subunit with six pendant TPPA linkers in each layer. (**B**) Stacking of the 2D network structure of Ag$_{12}$(S-^{t}Bu)$_6$(CF$_3$COO)$_6$ by TPPA. AA stacking viewed along the (a) *c*-axis and (d) *b*-axis, AB stacking viewed along the (b) *c*-axis and (e) *a*-axis, and ABC stacking viewed along the (c) *c*-axis and (f) *a*-axis. Reproduced with permission from Reference [170]. Copyright 2018 Royal Society of Chemistry.

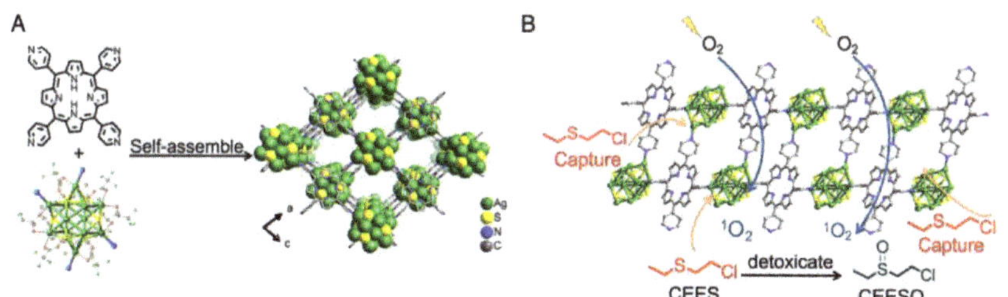

Figure 24. (**A**) Synthesis of Ag$_{12}$(S-^{t}Bu)$_6$(CF$_3$COO)$_3$(TPyP). (**B**) Schematic illustration of the capture and photodetoxification of CEES by Ag$_{12}$(S-^{t}Bu)$_6$(CF$_3$COO)$_3$(TPyP). Reproduced with permission from Reference [171]. Copyright 2019 American Chemical Society.

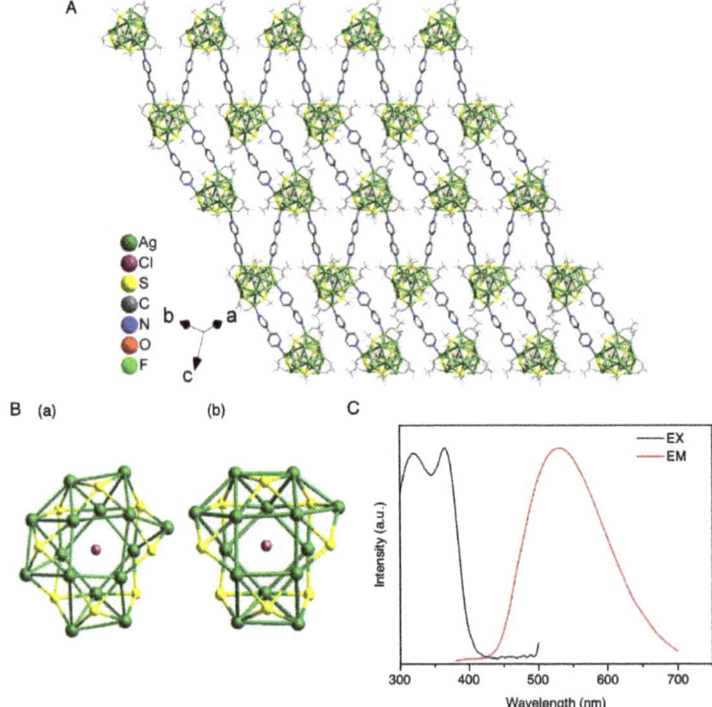

Figure 25. (**A**) Structure of $Ag_{14}Cl(S\text{-}^tBu)_8(CF_3COO)_5(DMF)(bpy)_2$. Free (co-crystallized) DMF molecules are not shown. (**B**) Top views of the core structures of (a) $Ag_{15}Cl(S\text{-}^tBu)_8(CF_3COO)_{5.67}(NO_3)_{0.33}(DMF)_2$ and (b) $Ag_{14}Cl(S\text{-}^tBu)_8(CF_3COO)_5(DMF)$. The green semitransparent spheres in the Ag NCs are shown as a visual guide. H atoms have been omitted for clarity. (**C**) Steady-state PL and excitation spectra of 2D CS crystals measured at room temperature (~25 °C). Emission spectra were measured under 365 nm excitation. Reproduced with permission from Reference [158]. Copyright 2019 American Chemical Society.

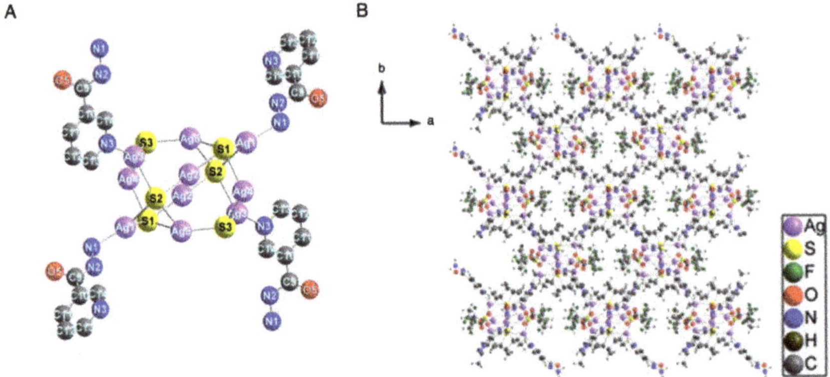

Figure 26. (**A**) Structural unit and (**B**) spatial stacking diagram of $Ag_{10}(CF_3COO)_4(S\text{-}^tBu)_6(CH_3CN)_4(m\text{-iah})_4$. Reproduced with permission from Reference [159]. Copyright 2019 American Chemical Society.

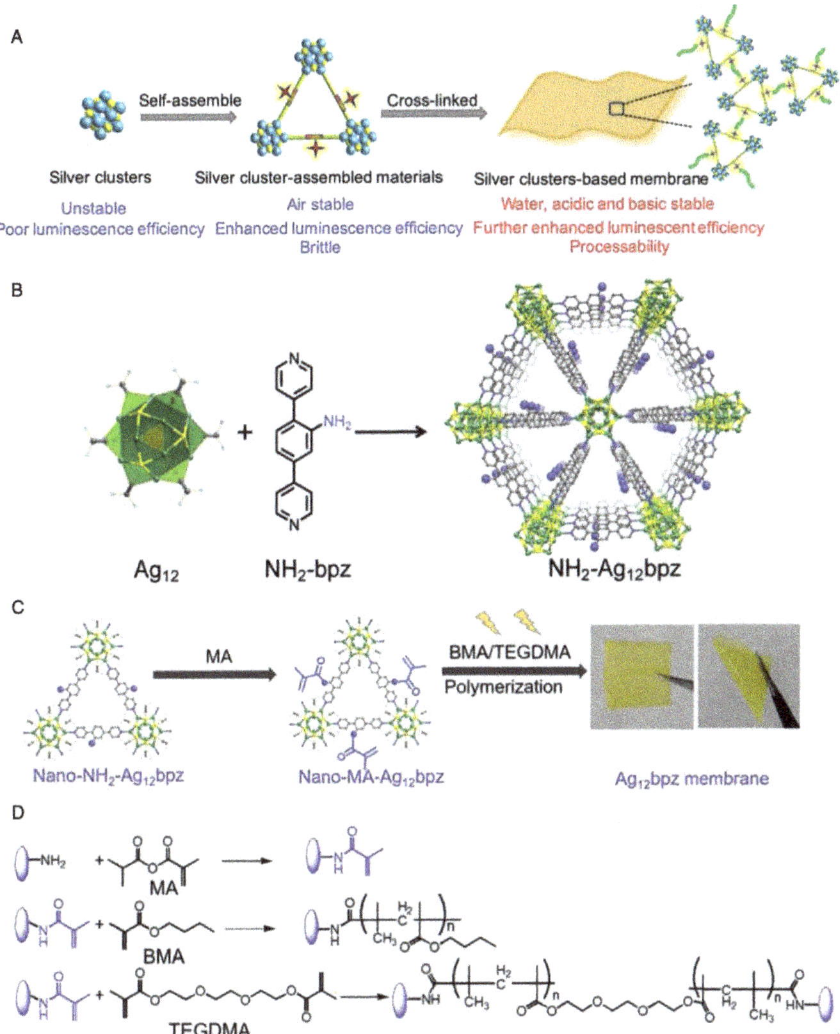

Figure 27. (**A**) Schematic illustration of the fabrication process of an Ag NC-based membrane. (**B**) Structure views of $Ag_{12}(S\text{-}^{t}Bu)_6(CF_3COO)_6(bpz\text{-}NH_2)_3$. (**C**) Fabrication process of the membrane. (**D**) Chemical reactions in the post-modification and cross-linking steps. Reproduced with permission from Reference [172]. Copyright 2019 Royal Society of Chemistry.

The resulting membrane exhibited PL with a QY of 14.8%, which was higher than that of the unpolymerized 2D CS crystals (Figure 28A(a)). The increased PL intensity was ascribed to the polymerization strengthening the structure of the 2D CS, which suppressed molecular vibrations and thus nonradiative decay. The membrane was stable in water regardless of pH (Figure 28A(b)). The researchers also attempted to use the membrane to sense the harmful substance nitrobenzene in solution. The results revealed that the membrane was able to detect nitrobenzene with a sensitivity of 3.19 ppb (Figure 28B). This membrane also displayed high reusability (Figure 28C). These results indicate that polymerizing 2D CSs is an effective approach to obtain Ag NCs suitable for applications.

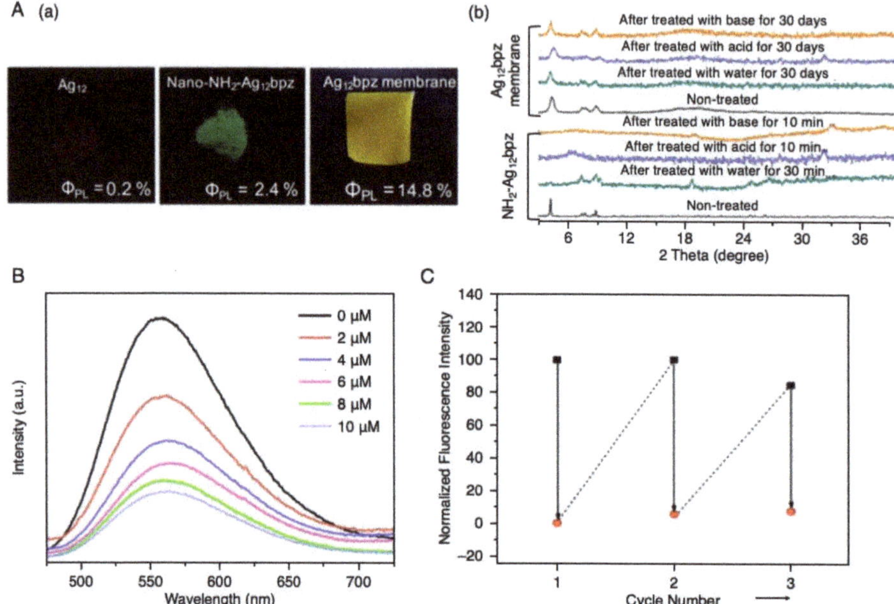

Figure 28. (**A**) (a) Photographs of the Ag_{12} clusters, nano-NH_2-Ag_{12}bpz, and an Ag_{12}bpz membrane under 365 nm ultraviolet light irradiation. (b) PXRD patterns of NH_2-Ag_{12}bpz and the Ag_{12}bpz membrane upon treatment with water, base, and acid for different periods. (**B**) Fluorescence spectra showing the response of the Ag_{12}bpz membrane to the incremental addition of a nitrobenzene solution. (**C**) Cycling test of the Ag_{12}bpz membrane upon exposure to nitrobenzene vapor. Reproduced with permission from Reference [172]. Copyright 2019 Royal Society of Chemistry.

4. Three-Dimensional Structures

Ag NCs are often used as nodes in 3D CSs. Because 3D CSs generally possess stronger frameworks than those of 2D CSs, the formation of 3D CSs is effective to enhance the stability of Ag NCs and thereby improve their physical properties. In 3D CS formation, the principles of NC assembly are similar to those in 1D and 2D CS formation, although the ligands used are often different. In Sections 4.1–4.3, we focus on the assembly of 3D CSs via the formation of Ag–O, Ag–S, or Ag–Cl bonds (Figure 2B), control of counterions (Figure 2C), and the introduction of linker molecules (Figure 2D), respectively. The metal NCs, connection modes, linker molecules, year reported, and reference numbers for these studies are summarized in Table 3. Several of the ligands used in 3D CSs are shown in Scheme 1. The organic molecules used as linkers are illustrated in Scheme 2.

4.1. Connection via Ag–O, Ag–S, or Ag–Cl Bonds

In 2017, Mak et al. [148] formed a 3D CS consisting of $Ag_{14}(S\text{-}iPr)_6(CF_3COO)_{11}(H_2O)_3(CH_3OH)$ NCs (Figure 29A). The NCs were connected via O–Ag–O bonds formed between CF_3COO^- and Ag ions (Figure 29B,C). Each Ag NC was connected to six other NCs, thereby forming a distorted octahedral-like coordination structure. It was speculated that this 3D CS formed by the assembly of NCs after NC generation.

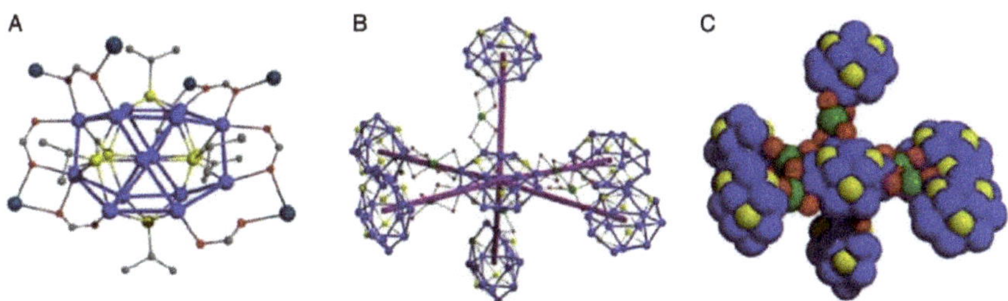

Figure 29. (**A**) Perspective view of $Ag_{14}(S\text{-}iPr)_6(CF_3COO)_{11}(H_2O)_3(CH_3OH)Ag_3$. (**B**) Ball-and-stick and (**C**) space-filling diagrams showing the spatial arrangement of a central cluster surrounded by six adjacent clusters. Ag = blue, cyan, and green, S = yellow, O = red, C = gray. Reproduced with permission from Reference [148]. Copyright 2017 Wiley-VCH.

In 2019, Sun and colleagues also produced a 3D CS, and theirs consisted of $Ag_{44}(Mo_6O_{19})(S\text{-}Et)_{24}(SCl_4)_3$ NCs (Figure 30A) containing a POM as an anion template (Figure 30B); these NCs were reported in their paper on 1D CS (Figure 9) and 2D CS (Figure 19) formation [149]. In this 3D CS, $Mo_6O_{19}{}^{2-}$ was used as an anion template, which was different from the case of the 1D and 2D CSs, in which the POM $V_{10}O_{28}{}^{6-}$ was located in the center of the cluster. This was the first report in which $Mo_6O_{19}{}^{2-}$ was used as an anion template of Ag NCs. $Mo_6O_{19}{}^{2-}$ has octahedral symmetry, and thereby the outer $Ag_{44}(S\text{-}Et)_{24}$ layer also displayed high symmetry (Figure 30C). The Ag_{44} shell had six quadrangles and 24 pentagonal faces. Ag at the vertices of these six quadrangles was connected with one S atom and four Cl atoms, leading to the formation of a 3D CS consisting of $Ag_{44}(Mo_6O_{19})(S\text{-}Et)_{24}(SCl_4)_3$ (Figure 30D). The S and Cl atoms used in the connections were generated by the decomposition of S-Et and CH_2Cl_2 during the CS synthesis.

The 3D CS in which Ag NCs are linked by dithiocarb, reported in 2019 by Gao et al. [173], should also be included in this category. The researchers first synthesized an $Ag_{11}S(C_5NS_2H_{10})_9$ precursor ($C_5NS_2H_{10}$ = diethyldithiocarbamate, Scheme 1(20)) [174]. The obtained precursor was reacted under high pressure, in an autoclave, to form a 3D CS with $Ag_{17}(C_5NS_2H_{10})_{14}$ as a repeating unit. This structure consists of Ag_9 NCs bound to twelve $C_5NS_2H_{10}$ (Figure 31A) and Ag_5 NCs containing six $C_5NS_2H_{10}$ (Figure 31B). The 3D CS (Figure 31C) was formed by the sharing of S between the two types of NCs. In these structures, the Ag_9 NCs were the nodes for three-point bridges and the Ag_5 NCs were the nodes for four-point bridges.

Table 3. Connection modes, NCs, linkers, publication years, and references for 3D CS.

Connection Mode	NC or Metal Ion	Linker	Year	Ref.
Formation of Ag–O, Ag–S, Ag–Cl bond, etc. (Figure 2B)	$Ag_{14}(S\text{-}iPr)_6(CF_3COO)_{11}(H_2O)_3(CH_3OH)$	-	2017	[148]
	$Ag_{44}(Mo_6O_{19})(S\text{-}Et)_{24}(SCl_4)_3$	-	2019	[149]
	$Ag_{17}(C_5NS_2H_{10})_{14}$	-	2019	[173]
Control of counter ion (Figure 2C)	$[Au_1Ag_{22}(S\text{-}Adm)_{12}Cl]^{2+}$	$SbF_6{}^-$	2020	[175]
	$[Au_1Ag_{22}(S\text{-}Adm)_{12}]^{3+}$	$SbF_6{}^-$		
Introduction of linker molecule (Figure 2D)	$Ag_{12}(S\text{-}{}^tBu)_8(CF_3COO)_4$	bpy [a]	2017	[168]
	$Ag_{14}(DT\text{-}o\text{-}C)_6$	1,4-bis(4- pyridyl)benzene [a]	2018	[156]
	$Ag_{10}(S\text{-}{}^tBu)_6(CF_3COO)_2(PhPO_3H)_2$	bpy [a]	2018	[176]
	$Ag_{12}(S\text{-}{}^tBu)_6(CF_3COO)_6$	CPPP [a]	2019	[177]
	$Ag_{12}(S\text{-}{}^tBu)_6(CF_3COO)_6$ $Ag_8(S\text{-}{}^tBu)_4(CF_3COO)_4$	tppe [a]	2019	[178]
	Ag^+	$[C(Au\text{-}mdppz)_6](BF_4)_2$ [a]	2014	[179]

[a] See Scheme 2.

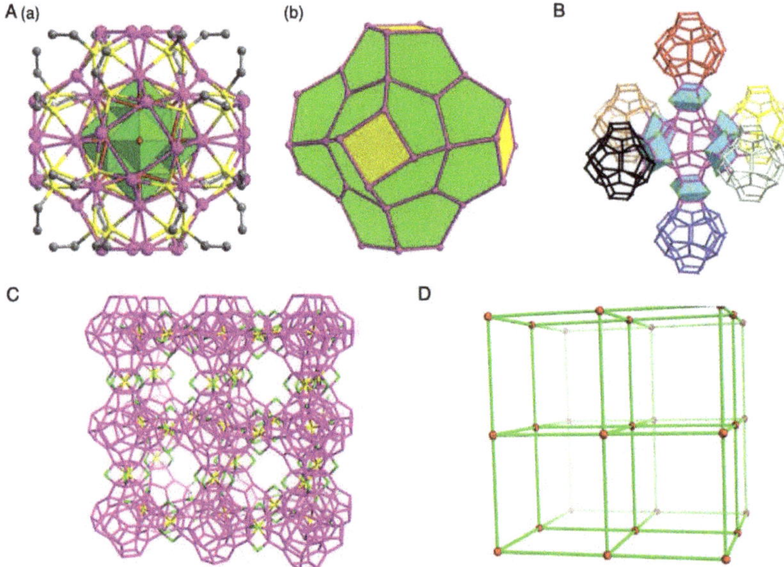

Figure 30. (**A**) Structure of (a) $Ag_{44}(Mo_6O_{19})(S\text{-Et})_{24}(SCl_4)_3$ and (b) the $Mo_6O_{19}^{2-}$ anion template. Tetragons (yellow) and pentagons (green) in an Ag_{44} cage are shown. (**B**) Connections (cyan polyhedra) between Ag_{44} subunits (highlighted in different colors) in the 3D framework. (**C**) Framework and (**D**) simplified primitive cubic topology with the Ag_{44} subunit as a node (represented as red balls) of the $Ag_{44}(Mo_6O_{19})(S\text{-Et})_{24}(SCl_4)_3$ 3D CS. Reproduced with permission from Reference [149]. Copyright 2019 Royal Society of Chemistry.

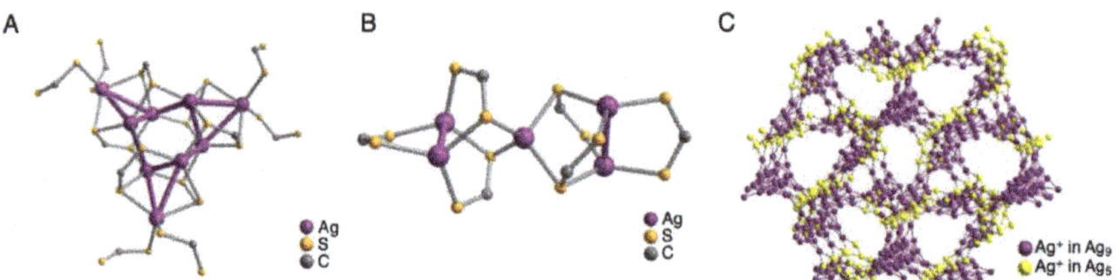

Figure 31. (**A**) Structures of the Ag_9 cluster, (**B**) Ag_5 subunits, and (**C**) 3D framework of $Ag_{17}(C_5NS_2H_{10})_{14}$. Reproduced with permission from Reference [173]. Copyright 2019 Royal Society of Chemistry.

4.2. Control of Counterions

Very recently, Zhu's group synthesized two types of 3D CSs in which $[Au_1Ag_{22}(S\text{-Adm})_{12}]^{3+}$ NCs (S-Adm = 1-adamantanethiolate, Scheme 1(21)) were connected in three dimensions via hexafluoroantimonate ions (SbF_6^-) [175]. The $[Au_1Ag_{22}(S\text{-Adm})_{12}]^{3+}$ node had a geometric structure in which an icosahedral Au_1Ag_{12} alloy core (Figure 32A(a)) was surrounded by an oligomer with a chemical composition of $Ag_{10}(S\text{-Adm})_{12}$ (Figure 32A(b),(c)). The $[Au_1Ag_{22}(S\text{-Adm})_{12}]^{3+}$ NCs formed as a pair of optical isomers, depending on the winding method of the oligomer (Figure 32A(b),(c)). In the first 3D CS, $[Au_1Ag_{22}(S\text{-Adm})_{12}](SbF_6)_2Cl$ was a structural unit, and two SbF_6^- were connected to $[Au_1Ag_{22}(S\text{-Adm})_{12}]^{3+}$ via an Ag–F–Ag bond, to form the 3D CS (Figure 32A(d),(e)). This 3D CS possessed a diamondlike structure (Figure 32A(f)) consisting of interpenetrating clockwise and counterclockwise optical isomers (Figure 32A(g)), which led to a small pore diameter of 6.2 Å (Figure 32A(h)). In the second 3D CS, $[Au_1Ag_{22}(S\text{-Adm})_{12}](SbF_6)_3$ was a structural unit, and the

3D CS was formed by connecting $[Au_1Ag_{22}(S\text{-}Adm)_{12}]^{3+}$ to three SbF_6^- via Ag–F–Ag bonds (Figure 32B(a)–(c)). This structure only contained clockwise or counterclockwise optical isomers (Figure 32B(d)). As a result, this 3D CS had a larger pore diameter (15 Å, Figure 32B(e)) than that of the first 3D CS (6.2 Å, Figure 32A).

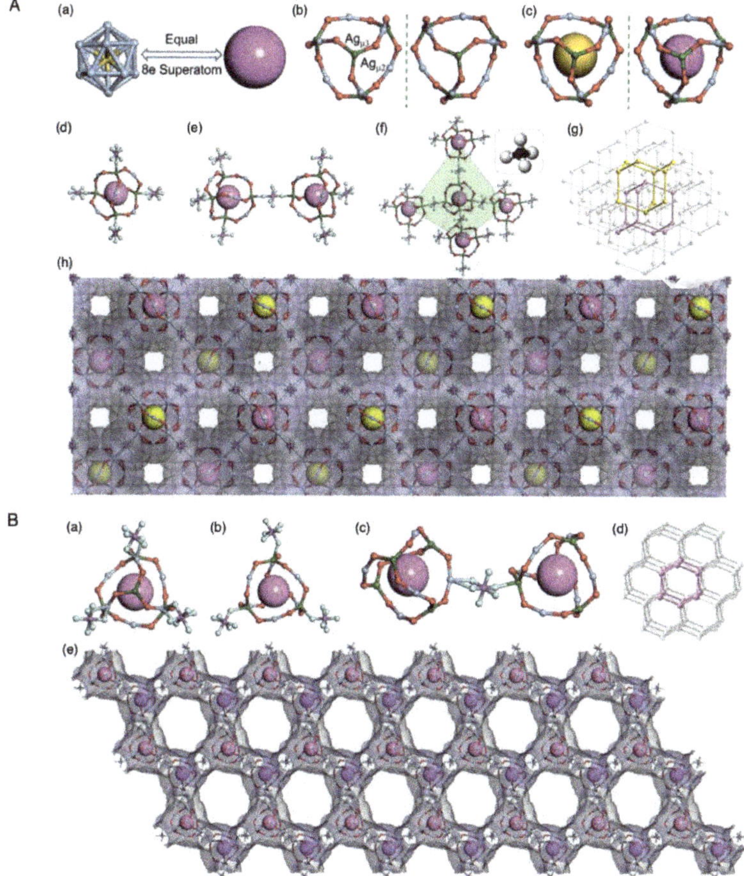

Figure 32. (**A**) Structure of the Au_1Ag_{22} superatom complex and interpenetrating 3D channel framework assembled from alloy NC monomers (for clarity, C and H atoms are omitted). (a) Icosahedral Au_1Ag_{12} core, (b) cage-like $Ag_{10}(SR)_{12}$ complex shell, (c) a pair of Au_1Ag_{22} isomers, (d) the connection of SbF_6^- and alloy NCs, (e) two alloy NCs connected by SbF_6^-, (f) tetrahedral structure of NC monomers (the inset shows methane), (g) topology of the diamond-like structure, and (h) interconnected channels of Au_1Ag_{22} along the z-axis. The left- and right-handed enantiomers in (c), (g), and (h) are highlighted in pink and yellow, respectively. Atoms are denoted in conventional colors: Au = gold, Ag in core and the Agμ2 motif = pale blue, Ag in the Agμ3 motif = green, S = red, F = light turquoise, Sb = purple. (**B**) Crystal and channel structure of left-handed chiral 3D channel framework (C and H atoms are omitted for clarity). (a) The connection of Ag and SbF_6^-, (b) the connection of Ag and SbF_6^-, (c) two alloy NCs linked by SbF_6^-, (d) illustration of the hexagonal network structure, and (e) schematic of the large hexagonal channel structure. Note that the packing pattern of the right-handed chiral 3D channel framework was the same as that of the left-handed chiral 3D channel framework. Reproduced with permission from Reference [175]. Copyright 2020 Wiley-VCH.

Study of the physical and chemical properties of the 3D CSs revealed that both exhibited red PL in the presence of polar solvents such as CH_3OH, ethanol, and water, which disappeared when the solvent was evaporated (Figure 33A). This behavior indicates that the obtained 3D CSs can function as sensors for polar solvents. The 3D CS composed of only the right- or left-handed enantiomer exhibited circularly polarized luminescence (CPL) (Figure 33B).

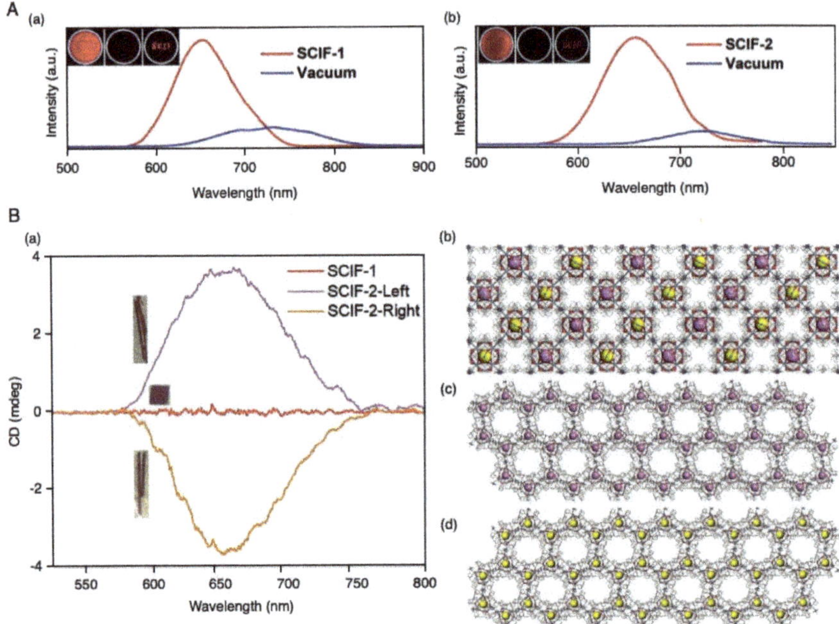

Figure 33. (**A**) Luminescence switching response to protic solvents. Fluorescence of films coated with (a) the interpenetrating 3D CS (SCIF-1) and (b) the left-handed 3D CS (SCIF-2) before and after solvent evaporation. Inset images show solvatochromic photographs of the corresponding films excited with 365 nm ultraviolet light before (left) and after desolvation (middle), and fluorescence recovery after treatment with *n*-hexane containing 5% ethanol (right). (**B**) CPL spectra of SCIF-1, SCIF-2 (left-handed), and SCIF-2 (right-handed) single crystals and the superstructures of these three crystal samples. (a) CPL spectra of SCIF-1, SCIF-2 (left-handed), and SCIF-2 (right-handed). Insets show photographs of the corresponding crystals. (b) Crystal structure of the SCIF-1 framework. (c) Crystal structure of the SCIF-2 (left-handed framework). (d) Crystal structure of the SCIF-2 (right-handed) framework. Reproduced with permission from Reference [175]. Copyright 2020 Wiley-VCH.

4.3. Introduction of Linker Molecules

The formation of 3D CSs by using linker molecules is a technique often used to fabricate molecular assemblies and MOFs. When a 3D CS composed of metal NCs is formed by such a method, in addition to increasing the stability of the NCs, it is also expected to adsorb gas molecules within its pores and behave as a catalyst with high selectivity because of the narrow pores. Furthermore, because the metal NCs, which are used as nodes, have more diversity in terms of coordination direction than that of metal ions, metal NC-based MOFs may have different connection modes from those of normal MOFs formed by using metal ions as nodes, and thereby they construct novel framework structures. Thus, metal NC-based MOFs possess not only the characteristics of individual metal NCs and MOFs, but also the possibility to produce new functions through synergistic effects.

Zang's group have been energetically researching 3D CSs with linkers, as well as the cases of 1D and 2D CSs. First, in 2017, Zang et al. [168] formed an Ag_{12} NC-based MOF ($Ag_{12}(S\text{-}^tBu)_8(CF_3COO)_4(bpy)_4$)

in which $Ag_{12}(S\text{-}^tBu)_8(CF_3COO)_4$ was bridged by bpy (Figure 34A). The obtained Ag_{12} NC-based MOF possessed a bilayer structure (Figure 34B). The formation of such a 3D CS markedly improved the stability of the Ag_{12} NCs. For example, a crystal of the individual $Ag_{12}(S\text{-}^tBu)_6(CF_3COO)_6(CH_3CN)_6$ NCs discolored in just 30 min when left in the atmosphere. In contrast, the Ag_{12} NC-based MOF showed almost no change in crystallinity, even when left in the air for one year (Figure 34C). The 3D CS also showed high stability during long-term gas adsorption and irradiation with visible light for several hours.

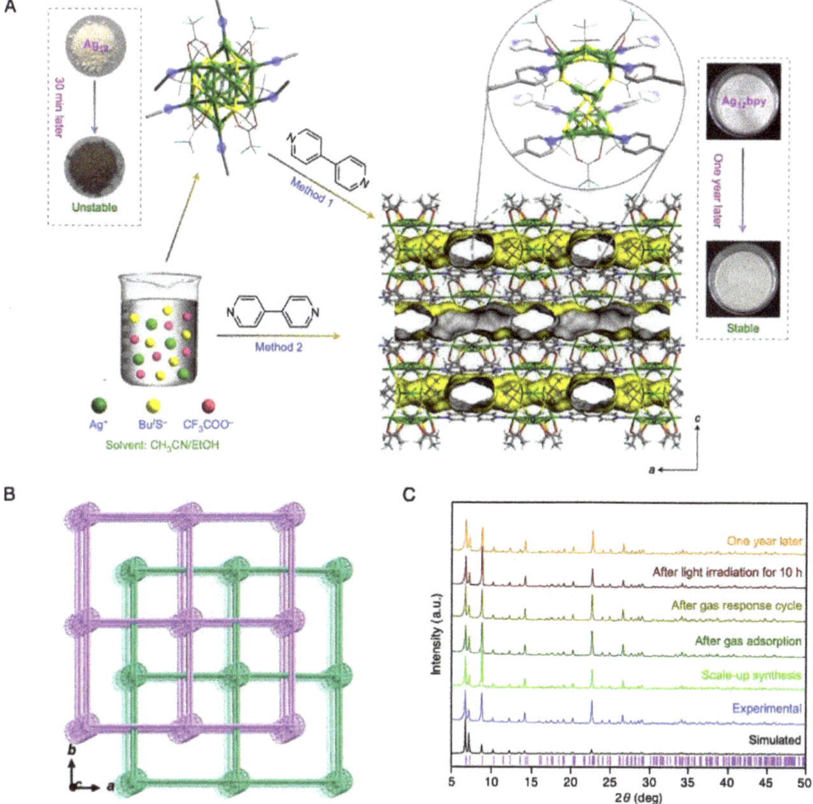

Figure 34. (**A**) Schematic representation of the ligand-exchange strategy used to obtain $Ag_{12}(S\text{-}^tBu)_8(CF_3COO)_4(bpy)_4$ crystals (Method 1, giving low yield) and one-pot synthesis (Method 2, for gram-quantity production) under identical conditions. Interconnected channels of $Ag_{12}(S\text{-}^tBu)_8(CF_3COO)_4(bpy)_4$ viewed along the *a* and *b* axes, where the yellow surface represents the pore surface. Ag = green, C = gray, O = red, S = yellow, F = turquoise. H atoms are omitted for clarity. Inset are photographs showing the changes of $Ag_{12}(S\text{-}^tBu)_6(CF_3COO)_6(CH_3CN)_6$ and $Ag_{12}(S\text{-}^tBu)_8(CF_3COO)_4(bpy)_4$ crystals under ambient conditions. (**B**) Schematic representation of the topology of $Ag_{12}(S\text{-}^tBu)_8(CF_3COO)_4(bpy)_4$ along the *c*-axis. The square windows in one double layer are blocked by Ag–S cluster nodes of adjacent layers. (**C**) PXRD patterns of $Ag_{12}(S\text{-}^tBu)_8(CF_3COO)_4(bpy)_4$ (in ascending order): simulated, experimental, gram-scale synthesis, after gas adsorption experiments (O_2, N_2, ethanol), after sensing experiments (O_2/N_2, O_2/vacuum, ethanol/air), after 10 h of visible-light irradiation under a Xe lamp equipped with a 420 nm cutoff filter, and the sample vial after one year under ambient conditions. Reproduced with permission from Reference [168]. Copyright 2017 Springer-Nature.

The formation of the 3D CS greatly changed the PL properties of the NCs. The individual Ag_{12} NCs exhibited red PL with low QY. Conversely, the Ag_{12} NC-based MOF exhibited green PL under vacuum, which was quenched by O_2 in the atmosphere (Figure 35A). The PL emission wavelength of the 3D CS under vacuum was independent of temperature and excitation wavelength, and its QY was 60 times higher than that of individual Ag_{12} NCs. The authors ascribed this high QY to the efficient suppression of nonradiative decay in the 3D CS. Moreover, the fact that the PL of the 3D CS is quenched by O_2 means that it is highly sensitive to O_2. The 3D CS showed a fast response to O_2 in experiments in which the atmosphere was repeatedly switched between air and N_2. No such O_2 response was observed for the individual Ag_{12} NCs. Based on these results, they suggested that the Ag_{12} NC-based MOF can be applied as an O_2 sensor. In addition, the Ag_{12} NC-based MOF was able to adsorb volatile organic compounds (VOCs) in its pores. The VOC-containing Ag_{12} NC-based MOF exhibited different PL colors, depending on the kind of VOC (Figure 35B). This indicates that the Ag_{12} NC-based MOF displays solvatochromism and therefore can be used for VOC detection.

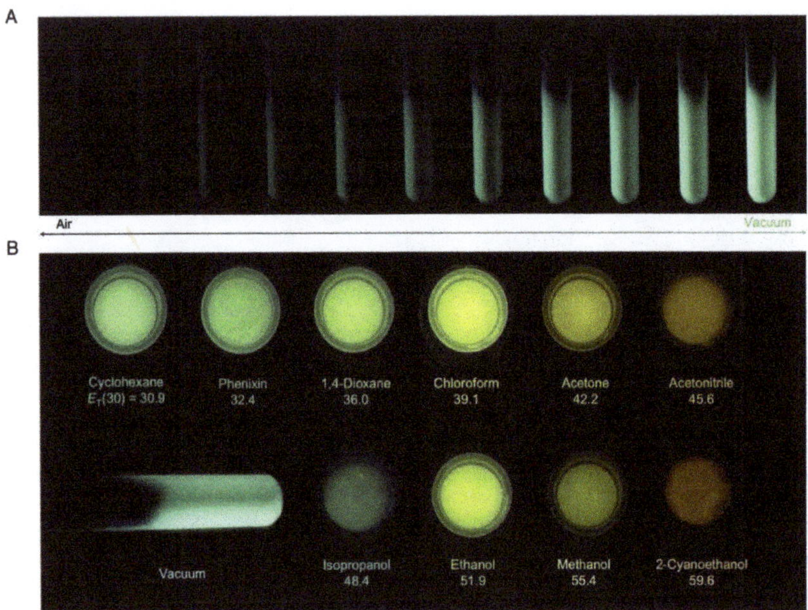

Figure 35. (**A**) Photographs of $Ag_{12}(S-^tBu)_8(CF_3COO)_4(bpy)_4$ excited by 365 nm light in a glass tube, beginning under vacuum and then filling with air (from left to right). (**B**) Photographs of the luminescence responses of $Ag_{12}(S-^tBu)_8(CF_3COO)_4(bpy)_4$ to different VOCs under 365 nm ultraviolet light irradiation. Reproduced with permission from Reference [168]. Copyright 2017 Springer-Nature.

In a paper on $Ag_{14}(DT-o-C)_6$ NCs published in 2018 (Figures 13 and 22), the same group reported that a 3D CS containing $Ag_{14}(DT-o-C)_6$ NCs formed when bpy was used as a linker (Figure 36A) [156]. The Ag_{14} NC nodes in the 3D CS had a face-centered cubic structure like that of other Ag_{14} NCs. This 3D CS was formed by connecting the eight vertices of each Ag_{14} NC with bpy linker molecules. However, this 3D CS was not stable after solvent evaporation, similar to the case of the corresponding 1D CS (Figure 13) and 2D CS (Figure 22) [156]. Therefore, they synthesized a 3D CS with an interpenetrating framework by using 1,4-bis(4-pyridyl)benzene (Scheme 2(14)) as a linker, in order to decrease the pore size and form a 3D CS with a strong framework (Figure 36B). The obtained 3D CS showed high thermal stability, remaining intact up to 220 °C (Figure 36C), and possessed pores with a diameter of about 1.12 nm (Figure 36D). This 3D CS showed optical absorption over a wide wavelength range and thermochromism (Figure 36E).

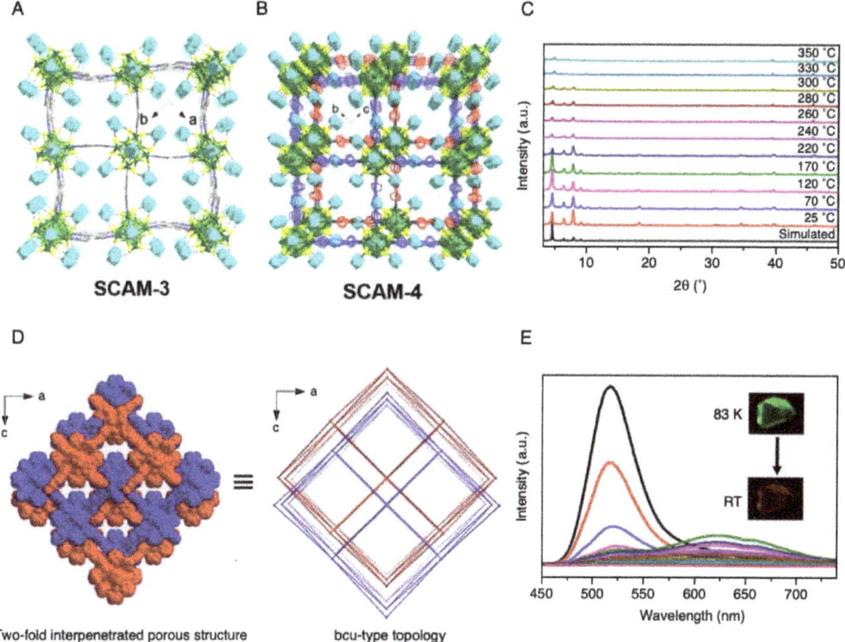

Figure 36. Three-dimensional CSs composed of (**A**) $Ag_{14}(DT\text{-}o\text{-}C)_6$ NCs and bpy (SCAM-3) and (**B**) $Ag_{14}(DT\text{-}o\text{-}C)_6$ NCs and 1,4-bis(4-pyridyl)benzene (SCAM-4). (**C**) Variable-temperature PXRD of SCAM-4 from 25 to 350 °C. (**D**) Different view of SCAM-4. (**E**) Evacuated SCAM-4 (excited at 380 nm) from −190 to 25 °C in air. Inset are corresponding photographs of SCAM-4 under 365 nm ultraviolet light irradiation. Reproduced with permission from Reference [156]. Copyright 2018 American Chemical Society.

Zang et al. [176] have also produced several other functional Ag NC-based MOFs. For example, in 2018, they reported the synthesis of a flexible Ag NC-based MOF. This structure consisted of 2D layers of $Ag_{10}(S\text{-}^tBu)_6(CF_3COO)_2(PhPO_3H)_2$ NCs linked via bpy, which were stacked through hydrogen bond (O-H···O) and C-H···O interactions, to form the 3D CS (Figure 37A,B). The 2D CS layers were thus linked by weak interactions, which facilitated the sliding of the layers, allowing the 3D CS to undergo structural deformation in response to guest organic molecules (Figure 37C). This Ag_{10} NC-based MOF exhibited green PL in air. Upon inclusion of guest organic molecules, it exhibited PL with an emission color depending on the guest organic molecule (Figure 37D). As such, this Ag_{10} NC-based MOF has potential as a sensor for distinguishing VOCs by its PL color. In 2019, they also formed a 3D CS consisting of $Ag_{12}(S\text{-}^tBu)_6(CF_3COO)_6$ NC nodes and 2,5-bis(4-cyanophenyl)-1,4-bis(4-(pyridine-4-yl)-phenyl)-1,4-dihydropyrrolo[3,2-b]pyrrole (CPPP, Scheme 2(15)) as a linker (Figure 38A) [177]. This was the first report in which a nitrile group (-CN) was used to link Ag NCs. The obtained 3D CS exhibited PL with a higher QY (61%) than that of CPPP in solution and solid states because the aggregation-induced quenching of CPPP was suppressed in the 3D CS (Figure 38B).

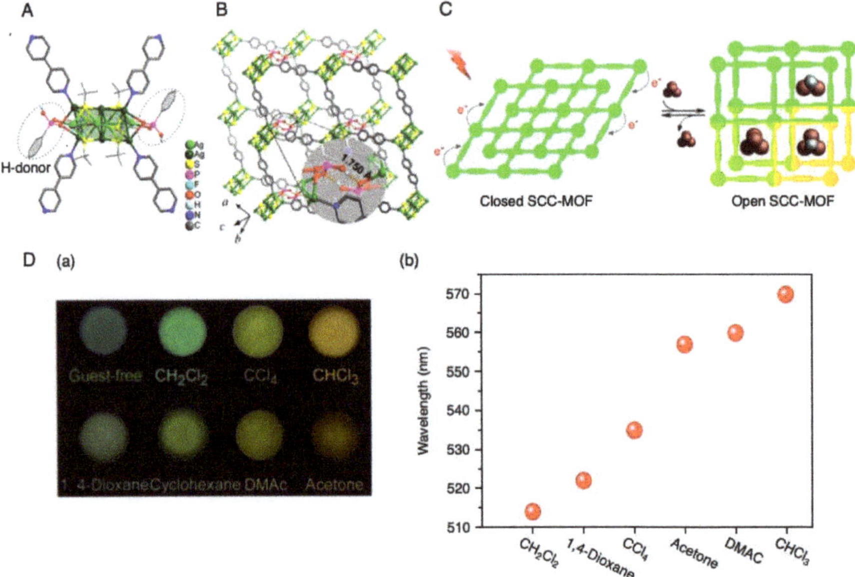

Figure 37. (**A**) Perspective view of the coordination environment of the $Ag_{10}(S-^tBu)_6$ core in $Ag_{10}(S-^tBu)_6(CF_3COO)_2(PhPO_3H)_2(bpy)_2$ at $-173\ °C$. (**B**) Two-layer stack of the host framework of $Ag_{10}(S-^tBu)_6(CF_3COO)_2(PhPO_3H)_2(bpy)_2$ with complementary hydrogen bonding (O–H···O; the H···O distance is 1.750 Å) between interlayer $-PO_2OH$ moieties. (**C**) Illustration of reversible pore open/closed structural transformation induced by CH_2Cl_2, $CHCl_3$, and CCl_4 (represented as space-filling models) and switchable solvatochromism. (**D**) (a) Luminescence images of $Ag_{10}(S-^tBu)_6(CF_3COO)_2(PhPO_3H)_2(bpy)_2$/solvent (guest free, CH_2Cl_2, $CHCl_3$, CCl_4, 1,4-dioxane, cyclohexane, DMAC, and acetone) combinations under 365 nm ultraviolet light irradiation. (b) Emission maxima of various $Ag_{10}(S-^tBu)_6(CF_3COO)_2(PhPO_3H)_2(bpy)_2$/solvent combinations at room temperature. Reproduced with permission from Reference [176]. Copyright 2018 American Chemical Society.

In the above 3D CSs, the nodes consisted of only one kind of Ag NCs. Tang et al. [178] synthesized an Ag NC-based MOF that had two types of Ag NCs as nodes. This Ag NC-based MOF was composed of 1,1,2,2-tetrakis(4-(pyridin-4-yl)phenyl)-ethene (tppe, Scheme 2(16)), $Ag_{12}(S-^tBu)_6(CF_3COO)_6$, and $Ag_8(S-^tBu)_4(CF_3COO)_4$ (Figure 39). In the 3D CS, one $Ag_{12}(S-^tBu)_6(CF_3COO)_6$ NC and three $Ag_8(S-^tBu)_4(CF_3COO)_4$ NCs were bound to the four N atoms of tppe. The estimated pore volume of this Ag NC-based MOF was 40.9%. DMAC, which was used as a solvent in the synthesis, was present in the pores of the 3D CS immediately after synthesis (Figure 40A). When the obtained 3D CS was exposed to the atmosphere, DMAC was removed, while maintaining the framework of the 3D CS. The Ag NC-based MOF thus obtained exhibited PL in the visible region because tppe is a light-emitting molecule. The PL wavelength of the 3D CS depended on the presence or absence of DMAC in its pores (Figure 40B). When DMAC was present in the pores of the framework, intramolecular rotation of tppe was suppressed, which changed the excited-state dynamics of the 3D CS. The change in the emission behavior of the Ag NC-based MOF induced by DMAC was ascribed to this change of its excited-state characteristics (Figure 40C).

Wang and colleagues reported the formation of a 3D CS, using metal NCs as linkers, before the use of metal NCs as nodes was developed [179]. In 2014, they succeeded in forming an NbO-type MOF by using Ag ions as nodes and $[C(Au-mdppz)_6](BF_4)_2$ (mdppz = 2-(3-methylpyrazinyl)diphenylphosphine, Scheme 2(17)) NCs as linkers (Figure 41). $[C(Au-mdppz)_6](BF_4)_2$ has a framework with C in the center

(Figure 41A) and is luminescent. The 3D CS was formed by the outer N atom of mdppz (Scheme 2(17)) binding to an Ag ion (Figure 41B,C). The obtained 3D CS consisted of two interpenetrating frameworks (Figure 41D,E) with a 1D channel in the *c*-axis direction (Figure 41C). Because a luminescent NC was used as the linker, the obtained MOF also showed green PL. The 3D CS displayed a PL QY of 25.6%, which was much higher than that of the luminescent NCs (1.5%). This increase of QY was caused by the strengthening of the framework of the linker NCs upon MOF formation and the excited-state perturbation induced by the coordination of Ag ions.

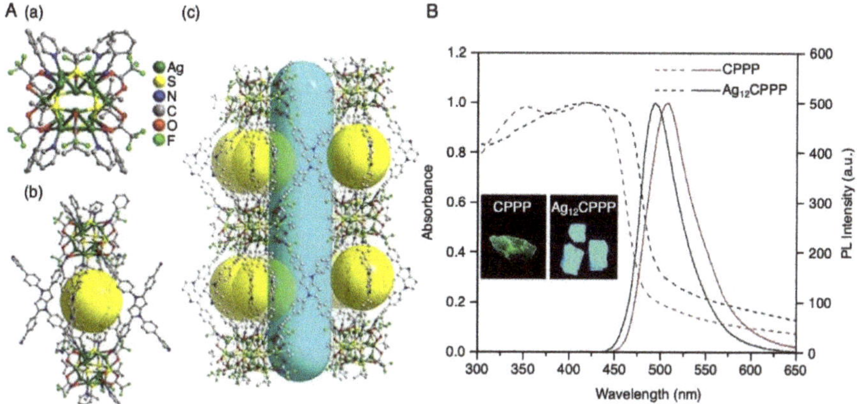

Figure 38. (**A**) Structures of (a) Ag chalcogenolate cluster nodes, (b) cage in $Ag_{12}(S\text{-}^tBu)_6(CF_3COO)_6(CPPP)_2$ ($Ag_{12}CPPP$), and (c) distribution of the cages in $Ag_{12}CPPP$. All H atoms and guest solvent molecules are omitted for clarity. (**B**) Solid-state absorption (dashed lines) and emission (solid lines) spectra of $Ag_{12}CPPP$ and CPPP at room temperature. Inset are photographs of the crystals of CPPP and $Ag_{12}CPPP$ under 365 nm ultraviolet light irradiation. Reproduced with permission from Reference [177]. Copyright 2019 Wiley-VCH.

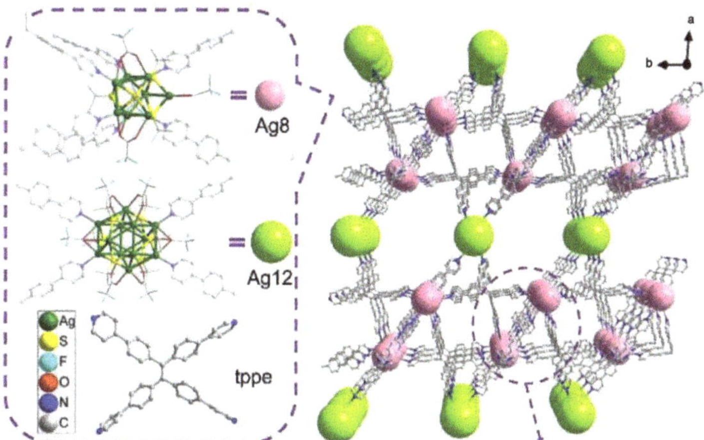

Figure 39. Structure of the Ag_8 cluster, Ag_{12} cluster, and tppe ligand, and single net of the $[Ag_{12}(S\text{-}^tBu)_6(CF_3COO)_6]_{0.5}[Ag_8(S\text{-}^tBu)_4(CF_3COO)_4](tppe)_2(DMAC)_{10}$ framework viewed along the *c*-axis. DMAC molecules are omitted for clarity. Reproduced with permission from Reference [178]. Copyright 2019 Wiley-VCH.

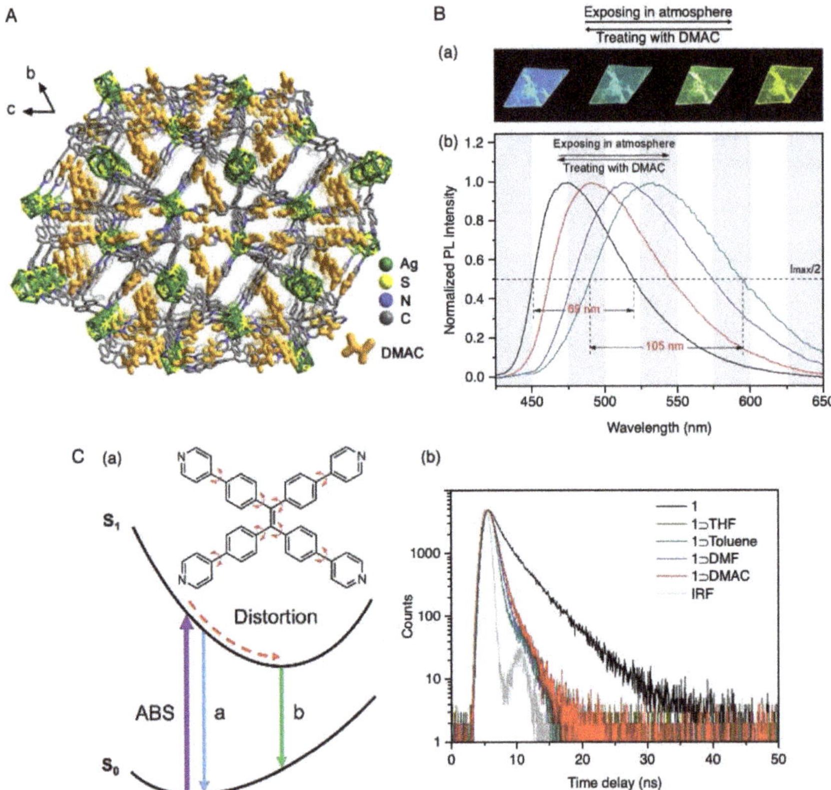

Figure 40. (**A**) Distribution of DMAC molecules in the $[Ag_{12}(S\text{-}^tBu)_6(CF_3COO)_6]_{0.5}[Ag_8(S\text{-}^tBu)_4(CF_3COO)_4](tppe)_2(DMAC)_{10}$ framework. (**B**) (a) Gradual fluorescence changes of the same $[Ag_{12}(S\text{-}^tBu)_6(CF_3COO)_6]_{0.5}[Ag_8(S\text{-}^tBu)_4(CF_3COO)_4](tppe)_2(DMAC)_{10}$ crystal under atmospheric exposure and (b) normalized fluorescence spectra of $[Ag_{12}(S\text{-}^tBu)_6(CF_3COO)_6]_{0.5}[Ag_8(S\text{-}^tBu)_4(CF_3COO)_4](tppe)_2(DMAC)_{10}$. (**C**) (a) Proposed fluorescence decay paths in $[Ag_{12}(S\text{-}^tBu)_6(CF_3COO)_6]_{0.5}[Ag_8(S\text{-}^tBu)_4(CF_3COO)_4](tppe)_2(DMAC)_{10}$ (path a) and $[Ag_{12}(S\text{-}^tBu)_6(CF_3COO)_6]_{0.5}[Ag_8(S\text{-}^tBu)_4(CF_3COO)_4](tppe)_2$ (**1**) (path b) and (b) fluorescence decay profiles of $[Ag_{12}(S\text{-}^tBu)_6(CF_3COO)_6]_{0.5}[Ag_8(S\text{-}^tBu)_4(CF_3COO)_4](tppe)_2$ in DMAC, THF, toluene, and DMF (denoted as **1**⊃ ∩DMAC, **1**⊃ ∩THF, **1**⊃ ∩Toluene, and **1**⊃ ∩DMF, respectively). Reproduced with permission from Reference [178]. Copyright 2019 Wiley-VCH.

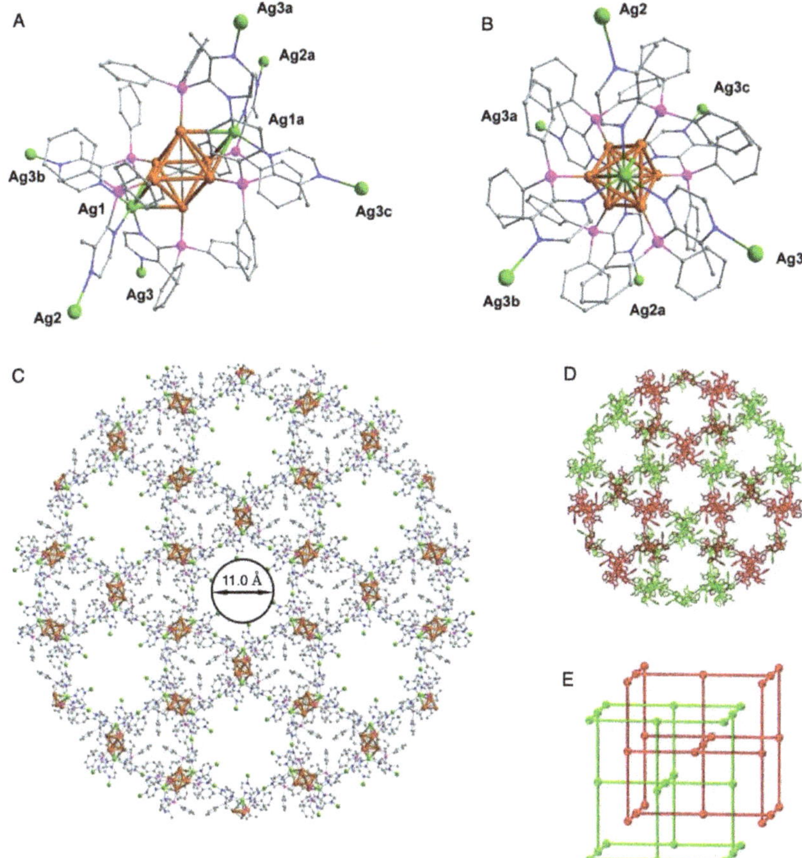

Figure 41. (**A**) Structure of [C(Au-mdppz)$_6$](BF$_4$)$_2$. (**B**) Four extensions: Ag3, Ag3a, Ag3b, and Ag3c. Note that Ag2 and Ag2a are not involved in the structural extension. (**C**) Perspective view of the 3D CS along the *c* direction. Au = orange, Ag = green, P = purple, N = blue, C = gray. (**D**) Two interpenetrating nets shown in different colors; anions and solvent molecules are omitted for clarity. (**E**) Schematic representation of NbO topology in the 3D CS. Reproduced with permission from Reference [179]. Copyright 2014 Wiley-VCH.

5. Summary

In this review, representative studies on the formation of 1D, 2D, and 3D CSs in which metal NCs were self-organized and regularly linked were summarized. From this summary, the following points became clear.

(1)   **Methods.** The methods to connect metal NCs that have been reported to date can be roughly divided into the following five categories: (i) direct connection by formation of metal−metal bonds (Figure 2A); (ii) connection by Ag−O, Ag−S, or Ag−Cl bond formation (Figure 2B); (iii) connection by counterions (Figure 2C); (iv) connection by linker molecules (Figure 2D); and (v) connection by inter-ligand interactions (Figure 2E; not introduced in this review).

(2)   **Diversity.** Among CSs produced by the above methods, there are many examples of the formation of 1D, 2D, and 3D CSs through the use of methods (ii) and (iv). An important point when constructing CSs by these methods is the design of the ligand of the NCs and linker,

respectively. It is presumed that the control of these species is relatively easy, which has led to the wider utilization of methods (ii) and (iv) than of the other methods. In particular, for method (iv), existing knowledge obtained in the study of normal MOFs can be considered.

(3) **Metal element.** To directly connect metal NCs, it is effective to use Au as a main element because it forms strong aurophilic interactions (intermetal interactions). In the connections involving metal−O or metal−Cl bonds, it is effective to use Ag as a main element because it readily bonds with O or Cl. Moreover, in the connections using bpy as a linker, Ag is attractive as the main element because of the high affinity of N and Ag.

(4) **Stability.** The formation of a CS generally improves the thermal stability of the component metal NCs regardless of the connection mode.

(5) **Electronic structure.** The formation of a CS often causes the band gap of the NC to narrow. This means that CS formation allows the use of a broader wavelength range of light, opening up the possibility of visible-light-driven photocatalysis by using CSs.

(6) **PL properties.** For 2D and 3D CS using linkers, CS formation often leads to an increase in PL emission intensity. When metal NCs are the PL source, there are many cases in which dual emission peaks appear upon connection with a linker. In addition, the PL color of a CS often changes depending on the kind of VOC trapped in its pores.

(7) **Electrical conduction.** The electron conductivity of CSs changes dramatically depending on the distance between each metal NC and the mode of connection; 1D CS formed by the direct connection via metal−metal bond shows the higher conductivity than 1D CS connected through counter ion.

(8) **Possible applications.** The reported CSs have potential applications in fields such as electronic devices, luminescent devices, gas and temperature sensing, and photocatalysis.

This review allowed us to obtain a common understanding of the CSs reported to date and their functions. We hope that the knowledge thus clarified will lead to clear design guidelines for developing new CSs with desired functions in the future.

6. Outlook

It is expected that the following studies will be conducted in the future, leading to new CSs.

(1) **Use of other metal elements.** At present, mostly Au and Ag are used as the metal element in CSs. This is largely related to the high stability of Au and Ag NCs. For Ag NCs, the good connectivity between Ag and linker molecules is also related to this fact. On the other hand, several syntheses of individual copper (Cu) NCs have been reported recently [10,180–183]. In addition, other metal ions are often used in normal MOFs with metal ions as nodes [153]. CS formation of NCs based on Cu or other metals may also lead to materials with high thermal stability. In the future, it is expected that many elements will be used in CSs, thereby realizing various functions and decreased cost of such materials.

(2) **Use of the alloying effect.** At present, there are few examples in which alloy NCs are connected to form CSs [142,143,146]. Mixing different elements leads to NCs with physical/chemical properties and functions that are different from those of monometal NCs. In fact, for individual metal NCs, many cases have been reported in which new physical properties/functions appeared because of mixing/synergistic effects [66,184–194]. The previous studies have established basic techniques for the formation of CSs consisting of Ag NCs. In the future, it is expected that more functional materials will be created by extending such CS formation techniques to Ag-based alloy NCs.

(3) **Connection of reported metal NCs.** Ag NC-based MOFs are interesting because they can be synthesized by a one-pot process. However, in CSs formed by such a method, metal NCs that are stable only in the CS are often found as nodes. For individual metal NCs, many NCs have already been synthesized with atomic precision [1–65]. In addition, much information has been

obtained on methods to generate novel functions in such NCs, including alloying [66–68]. In the future, it is expected that a method to more effectively utilize the reported metal NCs in CSs will be found. To achieve this, it may be necessary to establish new connection methods different from those described in this review (Figure 2).

(4) **Elucidation of electronic conductivity.** We believe that 1D CSs may be applied as nanodevices. However, at present, few experiments on the conductivity of 1D CSs have been reported [146,152]. In the future, it is expected that the conductivity of 1D CSs will be measured as a basic physical property. It is anticipated that the accumulation of such information will eventually lead to the production of nanodevices based on 1D CSs of metal NCs.

(5) **Exploration of other possible applications of connected structures.** Various applications, such as gas storage, gas separation, gas conversion, and reaction-selective catalysis, have been studied for normal MOFs with metal ions as nodes [117]. It has also been reported that, in the case of self-assembled complexes, a reaction different from that in the case of using an ordinary flask proceeds in the cage structure (i.e., the cage behaves as a nanoflask) [116]. In the future, it is expected that these possibilities will be investigated for metal NC-based MOFs and that their functions will be much higher than those of conventional MOFs and self-assembled complexes.

As mentioned in Section 1, the advance of bottom-up technology is essential for the further development of nanotechnology in the future. In previous research, multiple techniques to generate metal atoms on the molar scale in a solution and to self-organize them to form nanomaterials with the same number of constituent atoms and the same number of molecules have been studied; that is, precise synthesis techniques of metal NCs have been developed. However, to apply these metal NCs as devices and next-generation materials, developing techniques to assemble metal NCs to a size that is easy to handle is necessary (Figure 1). We hope that technologies that allow the self-organization of regularly arranged CSs composed of metal NCs will be further developed in the future. Ultimately, such nanotechnology is expected to enable resource conservation, energy conservation, decreased waste and environmental load, and the better use of time by human society.

Author Contributions: Y.N. constructed the structure of this review; T.K., S.O., and S.K. wrote Section 1, Section 5, and Section 6 and compiled figures and tables; A.E. wrote Sections 3 and 4; H.H. wrote Section 2; Y.N. and S.H. revised the entire draft before submission. All authors have read and agreed to the published version of the manuscript.

Funding: This work was supported by the Japan Society for the Promotion of Science (JSPS) KAKENHI (grant numbers JP16H04099, 16K21402, 20H02698, and 20H02552), Scientific Research on Innovative Areas "Coordination Asymmetry" (grant numbers 17H05385 and 19H04595), and Scientific Research on Innovative Areas "Innovations for Light-Energy Conversion" (grant numbers 18H05178 and 20H05115). Funding from the Asahi Glass Foundation, TEPCO Memorial Foundation Research Grant (Basic Research), and Kato Foundation for Promotion of Science (grant number KJ-2904) is also gratefully acknowledged.

Conflicts of Interest: There are no conflicts to declare.

References

1. Pradeep, T. *Nano: The Essentials*, 3rd ed.; Tata McGraw-Hill Education: New Delhi, India, 2008.
2. Jin, R.; Zeng, C.; Zhou, M.; Chen, Y. Atomically Precise Colloidal Metal Nanoclusters and Nanoparticles: Fundamentals and Opportunities. *Chem. Rev.* **2016**, *116*, 10346–10413. [CrossRef] [PubMed]
3. Nasaruddin, R.R.; Chen, T.; Yan, N.; Xie, J. Roles of Thiolate Ligands in the Synthesis, Properties and Catalytic Application of Gold Nanoclusters. *Coord. Chem. Rev.* **2018**, *368*, 60–79. [CrossRef]
4. Du, Y.; Sheng, H.; Astruc, D.; Zhu, M. Atomically Precise Noble Metal Nanoclusters as Efficient Catalysts: A Bridge between Structure and Properties. *Chem. Rev.* **2020**, *120*, 526–622. [CrossRef] [PubMed]
5. Sakthivel, N.A.; Dass, A. Aromatic Thiolate-Protected Series of Gold Nanomolecules and a Contrary Structural Trend in Size Evolution. *Acc. Chem. Res.* **2018**, *51*, 1774–1783. [CrossRef]
6. Kawasaki, H.; Kumar, S.; Li, G.; Zeng, C.; Kauffman, D.R.; Yoshimoto, J.; Iwasaki, Y.; Jin, R. Generation of Singlet Oxygen by Photoexcited $Au_{25}(SR)_{18}$ Clusters. *Chem. Mater.* **2014**, *26*, 2777–2788. [CrossRef]

7. Yamamoto, K.; Imaoka, T.; Tanabe, M.; Kambe, T. New Horizon of Nanoparticle and Cluster Catalysis with Dendrimers. *Chem. Rev.* **2020**, *120*, 1397–1437. [CrossRef]

8. Whetten, R.L.; Weissker, H.-C.; Pelayo, J.J.; Mullins, S.M.; López-Lozano, X.; Garzón, I.L. Chiral-Icosahedral (*I*) Symmetry in Ubiquitous Metallic Cluster Compounds (145A,60X): Structure and Bonding Principles. *Acc. Chem. Res.* **2019**, *52*, 34–43. [CrossRef] [PubMed]

9. Bakar, M.A.; Sugiuchi, M.; Iwasaki, M.; Shichibu, Y.; Konishi, K. Hydrogen Bonds to Au Atoms in Coordinated Gold Clusters. *Nat. Commun.* **2017**, *8*, 576. [CrossRef]

10. Sharma, S.; Chakrahari, K.K.; Saillard, J.-Y.; Liu, C.W. Structurally Precise Dichalcogenolate-Protected Copper and Silver Superatomic Nanoclusters and Their Alloys. *Acc. Chem. Res.* **2018**, *51*, 2475–2483. [CrossRef]

11. Brust, M.; Walker, M.; Bethell, D.; Schiffrin, D.J.; Whyman, R. Synthesis of Thiol-derivatised Gold Nanoparticles in a Two-Phase Liquid-Liquid System. *J. Chem. Soc. Chem. Commun.* **1994**, *801–802*. [CrossRef]

12. Whetten, R.L.; Khoury, J.T.; Alvarez, M.M.; Murthy, S.; Vezmar, I.; Wang, Z.L.; Stephens, P.W.; Cleveland, C.L.; Luedtke, W.D.; Landman, U. Nanocrystal Gold Molecules. *Adv. Mater.* **1996**, *8*, 428–433. [CrossRef]

13. Schaaff, T.G.; Shafigullin, M.N.; Khoury, J.T.; Vezmar, I.; Whetten, R.L.; Cullen, W.G.; First, P.N.; Gutiérrez-Wing, C.; Ascensio, J.; Jose-Yacamán, M.J. Isolation of Smaller Nanocrystal Au Molecules: Robust Quantum Effects in Optical Spectra. *J. Phys. Chem. B* **1997**, *101*, 7885–7891. [CrossRef]

14. Donkers, R.L.; Lee, D.; Murray, R.W. Synthesis and Isolation of the Molecule-Like Cluster Au$_{38}$(PhCH$_2$CH$_2$S)$_{24}$. *Langmuir* **2004**, *20*, 1945–1952. [CrossRef]

15. Ingram, R.S.; Hostetler, M.J.; Murray, R.W.; Schaaff, T.G.; Khoury, J.T.; Whetten, R.L.; Bigioni, T.P.; Guthrie, D.K.; First, P.N. 28 kDa Alkanethiolate-Protected Au Clusters Give Analogous Solution Electrochemistry and STM Coulomb Staircases. *J. Am. Chem. Soc.* **1997**, *119*, 9279–9280. [CrossRef]

16. Schaaff, T.G.; Shafigullin, M.N.; Khoury, J.T.; Vezmar, I.; Whetten, R.L. Properties of a Ubiquitous 29 kDa Au:SR Cluster Compound. *J. Phys. Chem. B* **2001**, *105*, 8785–8796. [CrossRef]

17. Murthy, S.; Bigioni, T.P.; Wang, Z.L.; Khoury, J.T.; Whetten, R.L. Liquid-Phase Synthesis of Thiol-Derivatized Silver Nanocrystals. *Mater. Lett.* **1997**, *30*, 321–325. [CrossRef]

18. Link, S.; Beeby, A.; FitzGerald, S.; El-Sayed, M.A.; Schaaff, T.G.; Whetten, R.L. Visible to Infrared Luminescence from a 28-Atom Gold Cluster. *J. Phys. Chem. B* **2002**, *106*, 3410–3415. [CrossRef]

19. Hostetler, M.J.; Wingate, J.E.; Zhong, C.-J.; Harris, J.E.; Vachet, R.W.; Clark, M.R.; Londono, J.D.; Green, S.J.; Stokes, J.J.; Wignall, G.D.; et al. Alkanethiolate Gold Cluster Molecules with Core Diameters from 1.5 to 5.2 nm: Core and Monolayer Properties as a Function of Core Size. *Langmuir* **1998**, *14*, 17–30. [CrossRef]

20. Hostetler, M.J.; Stokes, J.J.; Murray, R.W. Infrared Spectroscopy of Three-Dimensional Self-Assembled Monolayers: *N*-Alkanethiolate Monolayers on Gold Cluster Compounds. *Langmuir* **1996**, *12*, 3604–3612. [CrossRef]

21. Terrill, R.H.; Postlethwaite, T.A.; Chen, C.-H.; Poon, C.-D.; Terzis, A.; Chen, A.; Hutchison, J.E.; Clark, M.R.; Wignall, G.; Londono, J.D.; et al. Monolayers in Three Dimensions: NMR, SAXS, Thermal, and Electron Hopping Studies of Alkanethiol Stabilized Gold Clusters. *J. Am. Chem. Soc.* **1995**, *117*, 12537–12548. [CrossRef]

22. Schaaff, T.G.; Knight, G.; Shafigullin, M.N.; Borkman, R.F.; Whetten, R.L. Isolation and Selected Properties of a 10.4 kDa Gold: Glutathione Cluster Compound. *J. Phys. Chem. B* **1998**, *102*, 10643–10646. [CrossRef]

23. Schaaff, T.G.; Whetten, R.L. Giant Gold–Glutathione Cluster Compounds: Intense Optical Activity in Metal-Based Transitions. *J. Phys. Chem. B* **2000**, *104*, 2630–2641. [CrossRef]

24. Alvarez, M.M.; Chen, J.; Plascencia-Villa, G.; Black, D.M.; Griffith, W.P.; Garzón, I.L.; José-Yacamán, M.; Demeler, B.; Whetten, R.L. Hidden Components in Aqueous "Gold-144" Fractionated by PAGE: High-Resolution Orbitrap ESI-MS Identifies the Gold-102 and Higher All-Aromatic Au-*p*MBA Cluster Compounds. *J. Phys. Chem. B* **2016**, *120*, 6430–6438. [CrossRef]

25. Plascencia-Villa, G.; Demeler, B.; Whetten, R.L.; Griffith, W.P.; Alvarez, M.; Black, D.M.; José-Yacamán, M. Analytical Characterization of Size-Dependent Properties of Larger Aqueous Gold Nanoclusters. *J. Phys. Chem. C* **2016**, *120*, 8950–8958. [CrossRef]

26. Bootharaju, M.S.; Burlakov, V.M.; Besong, T.M.D.; Joshi, C.P.; AbdulHalim, L.G.; Black, D.M.; Whetten, R.L.; Goriely, A.; Bakr, O.M. Reversible Size Control of Silver Nanoclusters via Ligand-Exchange. *Chem. Mater.* **2015**, *27*, 4289–4297. [CrossRef]

27. Heaven, M.W.; Dass, A.; White, P.S.; Holt, K.M.; Murray, R.W. Crystal Structure of the Gold Nanoparticle [N(C$_8$H$_{17}$)$_4$] [Au$_{25}$(SCH$_2$CH$_2$Ph)$_{18}$]. *J. Am. Chem. Soc.* **2008**, *130*, 3754–3755. [CrossRef] [PubMed]

28. Negishi, Y.; Sakamoto, C.; Ohyama, T.; Tsukuda, T. Synthesis and the Origin of the Stability of Thiolate-Protected Au_{130} and Au_{187} Clusters. *J. Phys. Chem. Lett.* **2012**, *3*, 1624–1628. [CrossRef]

29. Negishi, Y.; Nakazaki, T.; Malola, S.; Takano, S.; Niihori, Y.; Kurashige, W.; Yamazoe, S.; Tsukuda, T.; Häkkinen, H. A Critical Size for Emergence of Nonbulk Electronic and Geometric Structures in Dodecanethiolate-Protected Au Clusters. *J. Am. Chem. Soc.* **2015**, *137*, 1206–1212. [CrossRef]

30. Negishi, Y.; Kurashige, W.; Niihori, Y.; Iwasa, T.; Nobusada, K. Isolation, Structure, and Stability of a Dodecanethiolate-Protected Pd_1Au_{24} Cluster. *Phys. Chem. Chem. Phys.* **2010**, *12*, 6219–6225. [CrossRef]

31. Negishi, Y.; Iwai, T.; Ide, M. Continuous Modulation of Electronic Structure of Stable Thiolate-Protected Au_{25} Cluster by Ag Doping. *Chem. Commun.* **2010**, *46*, 4713–4715. [CrossRef]

32. Negishi, Y.; Kurashige, W.; Kobayashi, Y.; Yamazoe, S.; Kojima, N.; Seto, M.; Tsukuda, T. Formation of a $Pd@Au_{12}$ Superatomic Core in $Au_{24}Pd_1(SC_{12}H_{25})_{18}$ Probed by ^{197}Au Mössbauer and Pd K-Edge EXAFS Spectroscopy. *J. Phys. Chem. Lett.* **2013**, *4*, 3579–3583. [CrossRef]

33. Negishi, Y.; Munakata, K.; Ohgake, W.; Nobusada, K. Effect of Copper Doping on Electronic Structure, Geometric Structure, and Stability of Thiolate-Protected Au_{25} Nanoclusters. *J. Phys. Chem. Lett.* **2012**, *3*, 2209–2214. [CrossRef] [PubMed]

34. Niihori, Y.; Matsuzaki, M.; Pradeep, T.; Negishi, Y. Separation of Precise Compositions of Noble Metal Clusters Protected with Mixed Ligands. *J. Am. Chem. Soc.* **2013**, *135*, 4946–4949. [CrossRef] [PubMed]

35. Niihori, Y.; Matsuzaki, M.; Uchida, C.; Negishi, Y. Advanced Use of High-Performance Liquid Chromatography for Synthesis of Controlled Metal Clusters. *Nanoscale* **2014**, *6*, 7889–7896. [CrossRef]

36. Niihori, Y.; Kikuchi, Y.; Kato, A.; Matsuzaki, M.; Negishi, Y. Understanding Ligand Exchange Reactions on Thiolate-Protected Gold Clusters by Probing Isomer Distributions Using Reversed-Phase High-Performance Liquid Chromatography. *ACS Nano* **2015**, *9*, 9347–9356. [CrossRef]

37. Niihori, Y.; Eguro, M.; Kato, A.; Sharma, S.; Kumar, B.; Kurashige, W.; Nobusada, K.; Negishi, Y. Improvements in the Ligand-Exchange Reactivity of Phenylethanethiolate-Protected Au_{25} Nanocluster by Ag or Cu Incorporation. *J. Phys. Chem. C* **2016**, *120*, 14301–14309. [CrossRef]

38. Niihori, Y.; Koyama, Y.; Watanabe, S.; Hashimoto, S.; Hossain, S.; Nair, L.V.; Kumar, B.; Kurashige, W.; Negishi, Y. Atomic and Isomeric Separation of Thiolate-Protected Alloy Clusters. *J. Phys. Chem. Lett.* **2018**, *9*, 4930–4934. [CrossRef] [PubMed]

39. Niihori, Y.; Hashimoto, S.; Koyama, Y.; Hossain, S.; Kurashige, W.; Negishi, Y. Dynamic Behavior of Thiolate-Protected Gold–Silver 38-Atom Alloy Clusters in Solution. *J. Phys. Chem. C* **2019**, *123*, 13324–13329. [CrossRef]

40. Niihori, Y.; Kikuchi, Y.; Shima, D.; Uchida, C.; Sharma, S.; Hossain, S.; Kurashige, W.; Negishi, Y. Separation of Glutathionate-Protected Gold Clusters by Reversed-Phase Ion-Pair High-Performance Liquid Chromatography. *Ind. Eng. Chem. Res.* **2017**, *56*, 1029–1035. [CrossRef]

41. Niihori, Y.; Shima, D.; Yoshida, K.; Hamada, K.; Nair, L.V.; Hossain, S.; Kurashige, W.; Negishi, Y. High-Performance Liquid Chromatography Mass Spectrometry of Gold and Alloy Clusters Protected by Hydrophilic Thiolates. *Nanoscale* **2018**, *10*, 1641–1649. [CrossRef]

42. Murayama, H.; Narushima, T.; Negishi, Y.; Tsukuda, T. Structures and Stabilities of Alkanethiolate Monolayers on Palladium Clusters As Studied by Gel Permeation Chromatography. *J. Phys. Chem. B* **2004**, *108*, 3496–3503. [CrossRef]

43. Tsunoyama, H.; Negishi, Y.; Tsukuda, T. Chromatographic Isolation of "Missing" Au_{55} Clusters Protected by Alkanethiolates. *J. Am. Chem. Soc.* **2006**, *128*, 6036–6037. [CrossRef] [PubMed]

44. Negishi, Y.; Arai, R.; Niihori, Y.; Tsukuda, T. Isolation and Structural Characterization of Magic Silver Clusters Protected by 4-(*tert*-butyl) benzyl mercaptan. *Chem. Commun.* **2011**, *47*, 5693–5695. [CrossRef] [PubMed]

45. Kurashige, W.; Yamazoe, S.; Yamaguchi, M.; Nishido, K.; Nobusada, K.; Tsukuda, T.; Negishi, Y. Au_{25} Clusters Containing Unoxidized Tellurolates in the Ligand Shell. *J. Phys. Chem. Lett.* **2014**, *5*, 2072–2076. [CrossRef]

46. Negishi, Y.; Takasugi, Y.; Sato, S.; Yao, H.; Kimura, K.; Tsukuda, T. Structures, Stabilities and Physicochemical Properties of Organometallic Hybrid Clusters. *J. Am. Chem. Soc.* **2004**, *126*, 6518–6519. [CrossRef]

47. Shichibu, Y.; Negishi, Y.; Tsunoyama, H.; Kanehara, M.; Teranishi, T.; Tsukuda, T. Extremely High Stability of Glutathionate-Protected Au_{25} Clusters Against Core Etching. *Small* **2007**, *3*, 835–839. [CrossRef]

48. Shichibu, Y.; Negishi, Y.; Tsukuda, T.; Teranishi, T. Large-Scale Synthesis of Thiolated Au_{25} Clusters via Ligand Exchange Reactions of Phosphine-Stabilized Au_{11} Clusters. *J. Am. Chem. Soc.* **2005**, *127*, 13464–13465. [CrossRef]

49. Ikeda, K.; Kobayashi, Y.; Negishi, Y.; Seto, M.; Iwasa, T.; Nobusada, K.; Tsukuda, T.; Kojima, N. Thiolate-Induced Structural Reconstruction of Gold Clusters Probed by [197]Au Mössbauer Spectroscopy. *J. Am. Chem. Soc.* **2007**, *129*, 7230–7231. [CrossRef]

50. Negishi, Y.; Takasugi, Y.; Sato, S.; Yao, H.; Kimura, K.; Tsukuda, T. Kinetic Stabilization of Growing Gold Clusters by Passivation with Thiolates. *J. Phys. Chem. B* **2006**, *110*, 12218–12221. [CrossRef]

51. Omoda, T.; Takano, S.; Yamazoe, S.; Koyasu, K.; Negishi, Y.; Tsukuda, T. An $Au_{25}(SR)_{18}$ Cluster with a Face-Centered Cubic Core. *J. Phys. Chem. C* **2018**, *122*, 13199–13204. [CrossRef]

52. Jadzinsky, P.D.; Calero, G.; Ackerson, C.J.; Bushnell, D.A.; Kornberg, R.D. Structure of a Thiol Monolayer-Protected Gold Nanoparticle at 1.1 Å Resolution. *Science* **2007**, *318*, 430–433. [CrossRef] [PubMed]

53. Zhu, M.; Aikens, C.M.; Hollander, F.J.; Schatz, G.C.; Jin, R. Correlating the Crystal Structure of a Thiol-Protected Au_{25} Cluster and Optical Properties. *J. Am. Chem. Soc.* **2008**, *130*, 5883–5885. [CrossRef]

54. Qian, H.; Eckenhoff, W.T.; Zhu, Y.; Pintauer, T.; Jin, R. Total Structure Determination of Thiolate-Protected Au_{38} Nanoparticles. *J. Am. Chem. Soc.* **2010**, *132*, 8280–8281. [CrossRef] [PubMed]

55. Chen, Y.; Zeng, C.; Liu, C.; Kirschbaum, K.; Gayathri, C.; Gil, R.R.; Rosi, N.L.; Jin, R. Crystal Structure of Barrel-Shaped Chiral $Au_{130}(p\text{-}MBT)_{50}$ Nanocluster. *J. Am. Chem. Soc.* **2015**, *137*, 10076–10079. [CrossRef]

56. Desireddy, A.; Conn, B.E.; Guo, J.; Yoon, B.; Barnett, R.N.; Monahan, B.M.; Kirschbaum, K.; Griffith, W.P.; Whetten, R.L.; Landman, U.; et al. Ultrastable Silver Nanoparticles. *Nature* **2013**, *501*, 399–402. [CrossRef] [PubMed]

57. Yang, H.; Wang, Y.; Huang, H.; Gell, L.; Lehtovaara, L.; Malola, S.; Häkkinen, H.; Zheng, N. All-Thiol-Stabilized Ag_{44} and $Au_{12}Ag_{32}$ Nanoparticles with Single-Crystal Structures. *Nat. Commun.* **2013**, *4*, 2422. [CrossRef]

58. Joshi, C.P.; Bootharaju, M.S.; Alhilaly, M.J.; Bakr, O.M. $[Ag_{25}(SR)_{18}]^-$: The "Golden" Silver Nanoparticle. *J. Am. Chem. Soc.* **2015**, *137*, 11578–11581. [CrossRef]

59. Lei, Z.; Wan, X.-K.; Yuan, S.-F.; Guan, Z.-J.; Wang, Q.-M. Alkynyl Approach toward the Protection of Metal Nanoclusters. *Acc. Chem. Res.* **2018**, *51*, 2465–2474. [CrossRef]

60. Kwak, K.; Lee, D. Electrochemistry of Atomically Precise Metal Nanoclusters. *Acc. Chem. Res.* **2019**, *52*, 12–22. [CrossRef]

61. Kurashige, W.; Hayashi, R.; Wakamatsu, K.; Kataoka, Y.; Hossain, S.; Iwase, A.; Kudo, A.; Yamazoe, S.; Negishi, Y. Atomic-Level Understanding of the Effect of Heteroatom Doping of the Cocatalyst on Water-Splitting Activity in AuPd or AuPt Alloy Cluster-Loaded $BaLa_4Ti_4O_{15}$. *ACS Appl. Energy Mater.* **2019**, *2*, 4175–4187. [CrossRef]

62. Wang, S.; Meng, X.; Das, A.; Li, T.; Song, Y.; Cao, T.; Zhu, X.; Zhu, M.; Jin, R. A 200-Fold Quantum Yield Boost in the Photoluminescence of Silver-Doped Ag_xAu_{25-x} Nanoclusters: The 13 th Silver Atom Matters. *Angew. Chem. Int. Ed.* **2014**, *53*, 2376–2380. [CrossRef]

63. Li, Z.; Yang, X.; Liu, C.; Wang, J.; Li, G. Effects of Doping in 25-Atom Bimetallic Nanocluster Catalysts for Carbon–Carbon Coupling Reaction of Iodoanisole and Phenylacetylene. *Prog. Nat. Sci. Mater. Int.* **2016**, *26*, 477–482. [CrossRef]

64. Liu, Y.; Chai, X.; Cai, X.; Chen, M.; Jin, R.; Ding, W.; Zhu, Y. Central Doping of a Foreign Atom into the Silver Cluster for Catalytic Conversion of CO_2 toward C–C Bond Formation. *Angew. Chem. Int. Ed.* **2018**, *57*, 9775–9779. [CrossRef]

65. Tsukuda, T. Toward an Atomic-Level Understanding of Size-Specific Properties of Protected and Stabilized Gold Clusters. *Bull. Chem. Soc. Jpn.* **2012**, *85*, 151–168. [CrossRef]

66. Hossain, S.; Niihori, Y.; Nair, L.V.; Kumar, B.; Kurashige, W.; Negishi, Y. Alloy Clusters: Precise Synthesis and Mixing Effects. *Acc. Chem. Res.* **2018**, *51*, 3114–3124. [CrossRef]

67. Takano, S.; Hasegawa, S.; Suyama, M.; Tsukuda, T. Hydride Doping of Chemically Modified Gold-Based Superatoms. *Acc. Chem. Res.* **2018**, *51*, 3074–3083. [CrossRef]

68. Yan, J.; Teo, B.K.; Zheng, N. Surface Chemistry of Atomically Precise Coinage—Metal Nanoclusters: From Structural Control to Surface Reactivity and Catalysis. *Acc. Chem. Res.* **2018**, *51*, 3084–3093. [CrossRef]

69. Gan, Z.; Xia, N.; Wu, Z. Discovery, Mechanism, and Application of Antigalvanic Reaction. *Acc. Chem. Res.* **2018**, *51*, 2774–2783. [CrossRef]

70. Niihori, Y.; Kurashige, W.; Matsuzaki, M.; Negishi, Y. Remarkable Enhancement in Ligand-Exchange Reactivity of Thiolate-Protected Au_{25} Nanoclusters by Single Pd Atom Doping. *Nanoscale* **2013**, *5*, 508–512. [CrossRef]

71. Xie, S.; Tsunoyama, H.; Kurashige, W.; Negishi, Y.; Tsukuda, T. Enhancement in Aerobic Alcohol Oxidation Catalysis of Au_{25} Clusters by Single Pd Atom Doping. *ACS Catal.* **2012**, *2*, 1519–1523. [CrossRef]

72. Negishi, Y.; Igarashi, K.; Munakata, K.; Ohgake, W.; Nobusada, K. Palladium Doping of Magic Gold Cluster $Au_{38}(SC_2H_4Ph)_{24}$: Formation of $Pd_2Au_{36}(SC_2H_4Ph)_{24}$ with Higher Stability than $Au_{38}(SC_2H_4Ph)_{24}$. *Chem. Commun.* **2012**, *48*, 660–662. [CrossRef]

73. Yamazoe, S.; Kurashige, W.; Nobusada, K.; Negishi, Y.; Tsukuda, T. Preferential Location of Coinage Metal Dopants (M = Ag or Cu) in $[Au_{25-x}M_x(SC_2H_4Ph)_{18}]^-$ (x~1) As Determined by Extended X-ray Absorption Fine Structure and Density Functional Theory Calculations. *J. Phys. Chem. C* **2014**, *118*, 25284–25290. [CrossRef]

74. Hossain, S.; Ono, T.; Yoshioka, M.; Hu, G.; Hosoi, M.; Chen, Z.; Nair, L.V.; Niihori, Y.; Kurashige, W.; Jiang, D.-E.; et al. Thiolate-Protected Trimetallic $Au_{\sim20}Ag_{\sim4}Pd$ and $Au_{\sim20}Ag_{\sim4}Pt$ Alloy Clusters with Controlled Chemical Composition and Metal Positions. *J. Phys. Chem. Lett.* **2018**, *9*, 2590–2594. [CrossRef]

75. Sharma, S.; Kurashige, W.; Nobusada, K.; Negishi, Y. Effect of Trimetallization in Thiolate-Protected $Au_{24-n}Cu_nPd$ Clusters. *Nanoscale* **2015**, *7*, 10606–10612. [CrossRef]

76. Sharma, S.; Yamazoe, S.; Ono, T.; Kurashige, W.; Niihori, Y.; Nobusada, K.; Tsukuda, T.; Negishi, Y. Tuning the Electronic Structure of Thiolate-Protected 25-Atom Clusters by Co-Substitution with Metals Having Different Preferential Sites. *Dalton Trans.* **2016**, *45*, 18064–18068. [CrossRef]

77. Nair, L.V.; Hossain, S.; Takagi, S.; Imai, Y.; Hu, G.; Wakayama, S.; Kumar, B.; Kurashige, W.; Jiang, D.-E.; Negishi, Y. Hetero-Biicosahedral $[Au_{24}Pd(PPh_3)_{10}(SC_2H_4Ph)_5Cl_2]^+$ Nanocluster: Selective Synthesis and Optical and Electrochemical Properties. *Nanoscale* **2018**, *10*, 18969–18979. [CrossRef] [PubMed]

78. Kurashige, W.; Yamaguchi, M.; Nobusada, K.; Negishi, Y. Ligand-Induced Stability of Gold Nanoclusters: Thiolate Versus Selenolate. *J. Phys. Chem. Lett.* **2012**, *3*, 2649–2652. [CrossRef]

79. Kurashige, W.; Munakata, K.; Nobusada, K.; Negishi, Y. Synthesis of Stable Cu_nAu_{25-n} Nanoclusters (n = 1–9) Using Selenolate Ligands. *Chem. Commun.* **2013**, *49*, 5447–5449. [CrossRef]

80. Hossain, S.; Imai, Y.; Negishi, Y. Precise Synthesis of Platinum and Alloy Clusters and Elucidation of Their Structures. *AIP Conf. Proc.* **2019**, *2186*, 030018.

81. Kawawaki, T.; Negishi, Y.; Kawasaki, H. Photo/Electrocatalysis and Photosensitization Using Metal Nanoclusters for Green Energy and Medical Applications. *Nanoscale Adv.* **2020**, *2*, 17–36. [CrossRef]

82. Hossain, S.; Imai, Y.; Suzuki, D.; Choi, W.; Chen, Z.; Suzuki, T.; Yoshioka, M.; Kawawaki, T.; Lee, D.; Negishi, Y. Elucidating Ligand Effects in Thiolate-Protected Metal Clusters Using $Au_{24}Pt(TBBT)_{18}$ as a Model Cluster. *Nanoscale* **2019**, *11*, 22089–22098. [CrossRef]

83. Ito, S.; Takano, S.; Tsukuda, T. Alkynyl-Protected $Au_{22}(C\equiv CR)_{18}$ Clusters Featuring New Interfacial Motifs and R-Dependent Photoluminescence. *J. Phys. Chem. Lett.* **2019**, *10*, 6892–6896. [CrossRef]

84. Briant, C.E.; Theobald, B.R.C.; White, J.W.; Bell, L.K.; Mingos, D.M.P.; Welch, A.J. Synthesis and X-ray Structural Characterization of the Centred Icosahedral Gold Cluster Compound $[Au_{13}(PMe_2Ph)_{10}Cl_2](PF_6)_3$; the Realization of a Theoretical Prediction. *J. Chem. Soc. Chem. Commun.* **1981**, 201–202. [CrossRef]

85. Schmid, G.; Pfeil, R.; Boese, R.; Bandermann, F.; Meyer, S.; Calis, G.H.M.; Van der Velden, J.W.A. $Au_{55}[P(C_6H_5)_3]_{12}CI_6$-Ein Goldcluster Ungewöhnlicher Größe. *Chem. Ber.* **1981**, *114*, 3634–3642. [CrossRef]

86. Kurasov, S.S.; Eremenko, N.K.; Slovokhotov, Y.L.; Struchkov, Y.T. High-Nuclearity Icosahedral Carbonylphosphineplatinum Clusters: Synthesis and Crystal Structure of $Pt_{17}(\mu_2\text{-}CO)_4(CO)_8(PEt_3)_8$. *J. Organomet. Chem.* **1989**, *361*, 405–408. [CrossRef]

87. McPartlin, M.; Mason, R.; Malatesta, L. Novel Cluster Complexes of Gold(0)-Gold(I). *J. Chem. Soc. D* **1969**, *7*, 334. [CrossRef]

88. Mednikov, E.G.; Dahl, L.F. Syntheses, Structures and Properties of Primarily Nanosized Homo/Heterometallic Palladium CO/PR3-Ligated Clusters. *Philos. Trans. R. Soc. A* **2010**, *368*, 1301–1332. [CrossRef]

89. Schmid, G. Large Clusters and Colloids. Metals in the Embryonic State. *Chem. Rev.* **1992**, *92*, 1709–1727. [CrossRef]

90. Schulz-Dobrick, M.; Jansen, M. Characterization of Gold Clusters by Crystallization with Polyoxometalates: The Intercluster Compounds $[Au_9(dpph)_4]$ $[Mo_8O_{26}]$, $[Au_9(dpph)_4]$ $[PW_{12}O_{40}]$ and $[Au_{11}(PPh_3)_8Cl_2]_2[W_6O_{19}]$. *Z. Anorg. Allg. Chem.* **2007**, *633*, 2326–2331. [CrossRef]

91. Teo, B.K.; Shi, X.; Zhang, H. Pure Gold Cluster of 1:9:9:1:9:9:1 Layered Structure: A Novel 39-Metal-Atom Cluster $[(Ph_3P)_{14}Au_{39}Cl_6]Cl_2$ with an Interstitial Gold Atom in a Hexagonal Antiprismatic Cage. *J. Am. Chem. Soc.* **1992**, *114*, 2743–2745. [CrossRef]

92. Vollenbroek, F.A.; Bour, J.J.; van der Veden, J.W.A. *Gold-Phosphine Cluster Compounds*: The Reactions of $[Au_9L_8]^{3+}$ (L = PPh$_3$) with L, SCN$^-$ and Cl$^-$ to $[Au_8L_8]^{2+}$ (Au$_{11}$L$_8$(SCN)$_2]^+$ and $[Au_{11}L_8Cl_2]^+$. *Recueil des Travaux Chimiques des Pays-Bas* **1980**, *99*, 137–141. [CrossRef]

93. Chini, P. Large Metal Carbonyl Clusters (LMCC). *J. Organomet. Chem.* **1980**, *200*, 37–61. [CrossRef]

94. Roth, J.D.; Lewis, G.J.; Safford, L.K.; Jiang, X.; Dahl, L.F.; Weaver, M.J. Exploration of the Ionizable Metal Cluster-Electrode Surface Analogy: Infrared Spectroelectrochemistry of $[Pt_{24}(CO)_{30}]^n$, $[Pt_{26}(CO)_{32}]^n$, and $[Pt_{38}(CO)_{44}]^n$ ($n = 0$ to -10) and Comparisons with Potential-Dependent Spectra of CO Adlayers on Platinum Surfaces. *J. Am. Chem. Soc.* **1992**, *114*, 6159–6169. [CrossRef]

95. Ceriotti, A.; Masciocchi, N.; Macchi, P.; Longoni, G. $[Pt_{19}(CO)_{21}(NO)]^{3-}$ and $[Pt_{38}(CO)_{44}]^{2-}$: Nitrosyl Bending through Intramolecular Electron Transfer as an Intermediate Step in the Nucleation Process from Polydecker to *ccp* Platinum Carbonyl Clusters. *Angew. Chem. Int. Ed.* **1999**, *38*, 3724–3727. [CrossRef]

96. Ciabatti, I.; Femoni, C.; Iapalucci, M.C.; Longoni, G.; Zacchini, S. Platinum Carbonyl Clusters Chemistry: Four Decades of Challenging Nanoscience. *J. Clust. Sci.* **2014**, *25*, 115–146. [CrossRef]

97. Negishi, Y.; Shimizu, N.; Funai, K.; Kanako, R.; Wakamatsu, K.; Harasawa, A.; Hossain, S.; Schuster, M.E.; Ozkaya, D.; Kurashige, W.; et al. γ-Alumina-Supported Pt$_{17}$ Cluster: Controlled Loading, Geometrical Structure, and Size-Specific Catalytic Activity for Carbon Monoxide and Propylene Oxidation. *Nanoscale Adv.* **2020**, *2*, 669–678. [CrossRef]

98. Hao, L.; Spivak, G.J.; Xiao, J.; Vittal, J.J.; Puddephatt, R.J. First Octahedral Platinum Cluster: Structure as a Function of Electron Count in Pt$_6$ Clusters. *J. Am. Chem. Soc.* **1995**, *117*, 7011–7012. [CrossRef]

99. Cattabriga, E.; Ciabatti, I.; Femoni, C.; Funaioli, T.; Iapalucci, M.C.; Zacchini, S. Syntheses, Structures, and Electrochemistry of the Defective *ccp* $[Pt_{33}(CO)_{38}]^{2-}$ and the *bcc* $[Pt_{40}(CO)_{40}]^{6-}$ Molecular Nanoclusters. *Inorg. Chem.* **2016**, *55*, 6068–6079. [CrossRef]

100. Cesari, C.; Ciabatti, I.; Femoni, C.; Iapalucci, M.C.; Mancini, F.; Zacchini, S. Heteroleptic Chini-Type Platinum Clusters: Synthesis and Characterization of Bis-Phospine Derivatives of $[Pt_{3n}(CO)_{6n}]^{2-}$ ($n = 2$–4). *Inorg. Chem.* **2017**, *56*, 1655–1668. [CrossRef]

101. Kawawaki, T.; Negishi, Y. Gold Nanoclusters as Electrocatalysts for Energy Conversion. *Nanomaterials* **2020**, *10*, 238. [CrossRef]

102. Xie, J.; Zheng, Y.; Ying, J.Y. Highly selective and ultrasensitive detection of Hg^{2+} based on fluorescence quenching of Au nanoclusters by Hg^{2+}–Au$^+$ interactions. *Chem. Commun.* **2010**, *46*, 961–963. [CrossRef]

103. Li, G.; Jin, R. Atomically Precise Gold Nanoclusters as New Model Catalysts. *Acc. Chem. Res.* **2013**, *46*, 1749–1758. [CrossRef]

104. Kurashige, W.; Kumazawa, R.; Ishii, D.; Hayashi, R.; Niihori, Y.; Hossain, S.; Nair, L.V.; Takayama, T.; Iwase, A.; Yamazoe, S.; et al. Au$_{25}$-Loaded BaLa$_4$Ti$_4$O$_{15}$ Water-Splitting Photocatalyst with Enhanced Activity and Durability Produced Using New Chromium Oxide Shell Formation Method. *J. Phys. Chem. C* **2018**, *122*, 13669–13681. [CrossRef]

105. Negishi, Y.; Matsuura, Y.; Tomizawa, R.; Kurashige, W.; Niihori, Y.; Takayama, T.; Iwase, A.; Kudo, A. Controlled Loading of Small Au$_n$ Clusters ($n = 10$–39) onto BaLa$_4$Ti$_4$O$_{15}$ Photocatalysts: Toward an Understanding of Size Effect of Cocatalyst on Water-Splitting Photocatalytic Activity. *J. Phys. Chem. C* **2015**, *119*, 11224–11232. [CrossRef]

106. Kurashige, W.; Mori, Y.; Ozaki, S.; Kawachi, M.; Hossain, S.; Kawawaki, T.; Shearer, C.J.; Iwase, A.; Metha, G.F.; Yamazoe, S.; et al. Activation of Water-Splitting Photocatalysts by Loading with Ultrafine Rh–Cr Mixed-Oxide Cocatalyst Nanoparticles. *Angew. Chem. Int. Ed.* **2020**, *59*, 7076–7082. [CrossRef]

107. Chen, Y.-S.; Choi, H.; Kamat, P.V. Metal-Cluster-Sensitized Solar Cells. A New Class of Thiolated Gold Sensitizers Delivering Efficiency Greater Than 2%. *J. Am. Chem. Soc.* **2013**, *135*, 8822–8825. [CrossRef]

108. Sakai, N.; Tatsuma, T. Photovoltaic Properties of Glutathione-Protected Gold Clusters Adsorbed on TiO$_2$ Electrodes. *Adv. Mater.* **2010**, *22*, 3185–3188. [CrossRef]

109. Teranishi, T.; Sugawara, A.; Shimizu, T.; Miyake, M. Planar Array of 1D Gold Nanoparticles on Ridge-and-Valley Structured Carbon. *J. Am. Chem. Soc.* **2002**, *124*, 4210–4211. [CrossRef]

110. Yonezawa, T.; Onoue, S.-Y.; Kimizuka, N. Metal Coating of DNA Molecules by Cationic, Metastable Gold Nanoparticles. *Chem. Lett.* **2002**, *31*, 1172–1173. [CrossRef]

111. Negishi, Y.; Tsunoyama, H.; Yanagimoto, Y.; Tsukuda, T. Subnanometer-sized Gold Clusters with Dual Molecular Receptors: Synthesis and Assembly in One-dimensional Arrangements. *Chem. Lett.* **2005**, *34*, 1638–1639. [CrossRef]

112. Yokoyama, T.; Hirata, N.; Tsunoyama, H.; Negishi, Y.; Nakajima, A. Characterization of Floating-Gate Memory Device with Thiolate-Protected Gold and Gold-Palladium Nanoclusters. *AIP Adv.* **2018**, *8*, 065002. [CrossRef]

113. Zhang, H.; Yasutake, Y.; Shichibu, Y.; Teranishi, T.; Majima, Y. Tunneling Resistance of Double-Barrier Tunneling Structures with an Alkanethiol-Protected Au Nanoparticle. *Phys. Rev. B* **2005**, *72*, 205441. [CrossRef]

114. Yoon, B.; Luedtke, W.D.; Barnett, R.N.; Gao, J.; Desireddy, A.; Conn, B.E.; Bigioni, T.; Landman, U. Hydrogen-Bonded Structure and Mechanical Chiral Response of a Silver Nanoparticle Superlattice. *Nat. Mater.* **2014**, *13*, 807–811. [CrossRef]

115. Zeng, C.; Chen, Y.; Kirschbaum, K.; Lambright, K.J.; Jin, R. Emergence of Hierarchical Structural Complexities in Nanoparticles and Their Assembly. *Science* **2016**, *354*, 1580–1584. [CrossRef]

116. Yoshizawa, M.; Tamura, M.; Fujita, M. Diels-Alder in Aqueous Molecular Hosts: Unusual Regioselectivity and Efficient Catalysis. *Science* **2006**, *312*, 251–254. [CrossRef]

117. Kitao, T.; Zhang, Y.; Kitagawa, S.; Wang, B.; Uemura, T. Hybridization of MOFs and Polymers. *Chem. Soc. Rev.* **2017**, *46*, 3108–3133. [CrossRef]

118. Kang, X.; Zhu, M. Intra-Cluster Growth Meets Inter-Cluster Assembly: The Molecular and Supramolecular Chemistry of Atomically Precise Nanoclusters. *Coord. Chem. Rev.* **2019**, *394*, 1–38. [CrossRef]

119. Wu, Z.; Yao, Q.; Zang, S.; Xie, J. Directed Self-Assembly of Ultrasmall Metal Nanoclusters. *ACS Mater. Lett.* **2019**, *1*, 237–248. [CrossRef]

120. Wu, Z.; Du, Y.; Liu, J.; Yao, Q.; Chen, T.; Cao, Y.; Zhang, H.; Xie, J. Aurophilic Interactions in the Self-Assembly of Gold Nanoclusters into Nanoribbons with Enhanced Luminescence. *Angew. Chem. Int. Ed.* **2019**, *58*, 8139–8144. [CrossRef]

121. Goswami, N.; Lin, F.; Liu, Y.; Leong, D.T.; Xie, J. Highly Luminescent Thiolated Gold Nanoclusters Impregnated in Nanogel. *Chem. Mater.* **2016**, *28*, 4009–4016. [CrossRef]

122. Nardi, M.D.; Antonello, S.; Jiang, D.-E.; Pan, F.; Rissanen, K.; Ruzzi, M.; Venzo, A.; Zoleo, A.; Maran, F. Gold Nanowired: A Linear $(Au_{25})_n$ Polymer from Au_{25} Molecular Clusters. *ACS Nano* **2014**, *8*, 8505–8512. [CrossRef] [PubMed]

123. Negishi, Y.; Nobusada, K.; Tsukuda, T. Glutathione-Protected Gold Clusters Revisited: Bridging the Gap between Gold (I)–Thiolate Complexes and Thiolate-Protected Gold Nanocrystals. *J. Am. Chem. Soc.* **2005**, *127*, 5261–5270. [CrossRef]

124. Kang, X.; Chong, H.; Zhu, M. $Au_{25}(SR)_{18}$: The Captain of the Great Nanocluster Ship. *Nanoscale* **2018**, *10*, 10758–10834. [CrossRef]

125. Luo, Z.; Nachammai, V.; Zhang, B.; Yan, N.; Leong, D.T.; Jiang, D.-E.; Xie, J. Toward Understanding the Growth Mechanism: Tracing All Stable Intermediate Species from Reduction of Au(I)–Thiolate Complexes to Evolution of Au_{25} Nanoclusters. *J. Am. Chem. Soc.* **2014**, *136*, 10577–10580. [CrossRef]

126. Dharmaratne, A.C.; Krick, T.; Dass, A. Nanocluster Size Evolution Studied by Mass Spectrometry in Room Temperature $Au_{25}(SR)_{18}$ Synthesis. *J. Am. Chem. Soc.* **2009**, *131*, 13604–13605. [CrossRef] [PubMed]

127. Parker, J.F.; Fields-Zinna, C.A.; Murray, R.W. The Story of a Monodisperse Gold Nanoparticle: $Au_{25}L_{18}$. *Acc. Chem. Res.* **2010**, *43*, 1289–1296. [CrossRef] [PubMed]

128. Shibu, E.S.; Muhammed, M.A.H.; Tsukuda, T.; Pradeep, T. Ligand Exchange of $Au_{25}SG_{18}$ Leading to Functionalized Gold Clusters: Spectroscopy, Kinetics, and Luminescence. *J. Phys. Chem. C* **2008**, *112*, 12168–12176. [CrossRef]

129. Ni, T.W.; Tofanelli, M.A.; Phillips, B.D.; Ackerson, C.J. Structural Basis for Ligand Exchange on $Au_{25}(SR)_{18}$. *Inorg. Chem.* **2014**, *53*, 6500–6502. [CrossRef]

130. Dainese, T.; Antonello, S.; Gascón, J.A.; Pan, F.; Perera, N.V.; Ruzzi, M.; Venzo, A.; Zoleo, A.; Rissanen, K.; Maran, F. $Au_{25}(SEt)_{18}$, a Nearly Naked Thiolate-Protected Au_{25} Cluster: Structural Analysis by Single Crystal X-ray Crystallography and Electron Nuclear Double Resonance. *ACS Nano* **2014**, *8*, 3904–3912. [CrossRef]

131. Jiang, D.-E.; Kühn, M.; Tang, Q.; Weigend, F. Superatomic Orbitals under Spin–Orbit Coupling. *J. Phys. Chem. Lett.* **2014**, *5*, 3286–3289. [CrossRef]

132. Liu, C.; Lin, S.; Pei, Y.; Zeng, X.C. Semiring Chemistry of $Au_{25}(SR)_{18}$: Fragmentation Pathway and Catalytic Active site. *J. Am. Chem. Soc.* **2013**, *135*, 18067–18079. [CrossRef] [PubMed]

133. Tlahuice-Flores, A.; Whetten, R.L.; Jose-Yacaman, M. Ligand Effects on the Structure and the Electronic Optical Properties of Anionic $Au_{25}(SR)_{18}$ Clusters. *J. Phys. Chem. C* **2013**, *117*, 20867–20875. [CrossRef]

134. Zhang, P. X-ray Spectroscopy of Gold–Thiolate Nanoclusters. *J. Phys. Chem. C* **2014**, *118*, 25291–25299. [CrossRef]

135. Chong, H.; Li, P.; Wang, S.; Fu, F.; Xiang, J.; Zhu, M.; Li, Y. Au_{25} Clusters as Electron-Transfer Catalysts Induced the Intramolecular Cascade Reaction of 2-nitrobenzonitrile. *Sci. Rep.* **2013**, *3*, 3214. [CrossRef]

136. Kwak, K.; Kumar, S.S.; Pyo, K.; Lee, D. Ionic Liquid of a Gold Nanocluster: A Versatile Matrix for Electrochemical Biosensors. *ACS Nano* **2014**, *8*, 671–679. [CrossRef]

137. Wu, Z.; Jiang, D.-E.; Mann, A.K.P.; Mullins, D.R.; Qiao, Z.-A.; Allard, L.F.; Zeng, C.; Jin, R.; Overbury, S.H. Thiolate Ligands as a Double-Edged Sword for CO Oxidation on CeO_2 Supported $Au_{25}(SCH_2CH_2Ph)_{18}$ Nanoclusters. *J. Am. Chem. Soc.* **2014**, *136*, 6111–6122. [CrossRef] [PubMed]

138. Stamplecoskie, K.G.; Kamat, P.V. Size-Dependent Excited State Behavior of Glutathione-Capped Gold Clusters and Their Light-Harvesting Capacity. *J. Am. Chem. Soc.* **2014**, *136*, 11093–11099. [CrossRef] [PubMed]

139. Negishi, Y. Toward the Creation of Functionalized Metal Nanoclusters and Highly Active Photocatalytic Materials Using Thiolate-Protected Magic Gold Clusters. *Bull. Chem. Soc. Jpn.* **2014**, *87*, 375–389. [CrossRef]

140. Antonello, S.; Dainese, T.; Pan, F.; Rissanen, K.; Maran, F. Electrocrystallization of Monolayer-Protected Gold Clusters: Opening the Door to Quality, Quantity, and New Structures. *J. Am. Chem. Soc.* **2017**, *139*, 4168–4174. [CrossRef]

141. Fei, W.; Antonello, S.; Dainese, T.; Dolmella, A.; Lahtinen, M.; Rissanen, K.; Venzo, A.; Maran, F. Metal Doping of $Au_{25}(SR)_{18}^-$ Clusters: Insights and Hindsights. *J. Am. Chem. Soc.* **2019**, *141*, 16033–16045. [CrossRef]

142. Hossain, S.; Imai, Y.; Motohashi, Y.; Chen, Z.; Suzuki, D.; Suzuki, T.; Kataoka, Y.; Hirata, M.; Ono, T.; Kurashige, W.; et al. Understanding and Designing One-Dimensional Assemblies of Ligand-Protected Metal Nanoclusters. *Mater. Horiz.* **2020**, *7*, 796–803. [CrossRef]

143. Chen, J.; Liu, L.; Liu, X.; Liao, L.; Zhuang, S.; Zhou, S.; Yang, J.; Wu, Z. Gold-Doping of Double-Crown Pd Nanoclusters. *Chem. Eur. J.* **2017**, *23*, 18187–18192. [CrossRef]

144. Jiang, D.-E.; Dai, S. From Superatomic $Au_{25}(SR)_{18}^-$ to Superatomic $M@Au_{24}(SR)_{18}^q$ Core–Shell Clusters. *Inorg. Chem.* **2009**, *48*, 2720–2722. [CrossRef] [PubMed]

145. Qian, H.; Jiang, D.-E.; Li, G.; Gayathri, C.; Das, A.; Gil, R.R.; Jin, R. Monoplatinum Doping of Gold Nanoclusters and Catalytic Application. *J. Am. Chem. Soc.* **2012**, *134*, 16159–16162. [CrossRef] [PubMed]

146. Yuan, P.; Zhang, R.; Selenius, E.; Ruan, P.; Yao, Y.; Zhou, Y.; Malola, S.; Häkkinen, H.; Teo, B.K.; Cao, Y.; et al. Solvent-Mediated Assembly of Atom-Precise Gold–Silver Nanoclusters to Semiconducting One-Dimensional Materials. *Nat. Commun.* **2020**, *11*, 2229. [CrossRef]

147. Zhou, K.; Qin, C.; Wang, X.-L.; Shao, K.-Z.; Yan, L.-K.; Su, Z.-M. Unexpected 1D Self-Assembly of Carbonate-Templated Sandwich-Like Macrocycle-Based $Ag_{20}S_{10}$ Luminescent Nanoclusters. *CrystEngComm* **2014**, *16*, 7860–7864. [CrossRef]

148. Chen, Z.-Y.; Tam, D.Y.S.; Zhang, L.L.-M.; Mak, T.C.W. Silver Thiolate Nano-Sized Molecular Clusters and Their Supramolecular Covalent Frameworks: An Approach Toward Pre-Templated Synthesis. *Chem. Asian J.* **2017**, *12*, 2763–2769. [CrossRef]

149. Wang, Z.; Sun, Y.-M.; Qu, Q.-P.; Liang, Y.-X.; Wang, X.-P.; Liu, Q.-Y.; Kurmoo, M.; Su, H.-F.; Tung, C.-H.; Sun, D. Enclosing Classical Polyoxometallates in Silver Nanoclusters. *Nanoscale* **2019**, *11*, 10927–10931. [CrossRef]

150. Wen, Z.-R.; Guan, Z.-J.; Zhang, Y.; Lin, Y.-M.; Wang, Q.-M. $[Au_7Ag_9(dppf)_3(CF_3CO_2)_7BF_4]_n$: A Linear Nanocluster Polymer from Molecular Au_7Ag_8 Clusters Covalently Linked by Silver Atoms. *Chem. Commun.* **2019**, *55*, 12992–12995. [CrossRef]

151. Li, Q.; Luo, T.-Y.; Taylor, M.G.; Wang, S.; Zhu, X.; Song, Y.; Mpourmpakis, G.; Rosi, N.L.; Jin, R. Molecular "Surgery" on a 23-Gold-Atom Nanoparticle. *Sci. Adv.* **2017**, *3*, e1603193. [CrossRef]

152. Li, Q.; Russell, J.C.; Luo, T.-Y.; Roy, X.; Rosi, N.L.; Zhu, Y.; Jin, R. Modulating the Hierarchical Fibrous Assembly of Au Nanoparticles with Atomic Precision. *Nat. Commun.* **2018**, *9*, 3871. [CrossRef] [PubMed]

153. AbdulHalim, L.G.; Bootharaju, M.S.; Tang, Q.; Gobbo, S.D.; AbdulHalim, R.G.; Eddaoudi, M.; Jiang, D.-E.; Bakr, O.M. $Ag_{29}(BDT)_{12}(TPP)_4$: A Tetravalent Nanocluster. *J. Am. Chem. Soc.* **2015**, *137*, 11970–11975. [CrossRef] [PubMed]

154. Wei, X.; Kang, X.; Yuan, Q.; Qin, C.; Jin, S.; Wang, S.; Zhu, M. Capture of Cesium Ions with Nanoclusters: Effects on Inter- and Intramolecular Assembly. *Chem. Mater.* **2019**, *31*, 4945–4952. [CrossRef]

155. Kitagawa, S.; Kitaura, R.; Noro, S.-I. Functional Porous Coordination Polymers. *Angew. Chem. Int. Ed.* **2004**, *43*, 2334–2375. [CrossRef]

156. Wang, Z.-Y.; Wang, M.-Q.; Li, Y.-L.; Luo, P.; Jia, T.-T.; Huang, R.-W.; Zang, S.-Q.; Mak, T.C.W. Atomically Precise Site-Specific Tailoring and Directional Assembly of Superatomic Silver Nanoclusters. *J. Am. Chem. Soc.* **2018**, *140*, 1069–1076. [CrossRef]

157. Ma, X.-H.; Wang, J.-Y.; Guo, J.-J.; Wang, Z.-Y.; Zang, S.-Q. Reversible Wide-Range Tuneable Luminescence of a Dual-Stimuli-Responsive Silver Cluster-Assembled Material. *Chin. J. Chem.* **2019**, *37*, 1120–1124. [CrossRef]

158. Alhilaly, M.J.; Huang, R.-W.; Naphade, R.; Alamer, B.; Hedhili, M.N.; Emwas, A.-H.; Maity, P.; Yin, J.; Shkurenko, A.; Mohammed, O.F.; et al. Assembly of Atomically Precise Silver Nanoclusters into Nanocluster-Based Frameworks. *J. Am. Chem. Soc.* **2019**, *141*, 9585–9592. [CrossRef]

159. Lu, S.-H.; Li, Y.; Yang, S.-X.; Zhao, R.-D.; Lu, Z.-X.; Liu, X.-L.; Qin, Y.; Zheng, L.-Y.; Cao, Q.-E. Three Silver Coordination Polymers with Diverse Architectures Constructed from Pyridine Carboxylic Hydrazide Ligands. *Inorg. Chem.* **2019**, *58*, 11793–11800. [CrossRef]

160. Wang, Z.-K.; Sheng, M.-M.; Qin, S.-S.; Shi, H.-T.; Strømme, M.; Zhang, Q.-F.; Xu, C. Assembly of Discrete Chalcogenolate Clusters into a One-Dimensional Coordination Polymer with Enhanced Photocatalytic Activity and Stability. *Inorg. Chem.* **2020**, *59*, 2121–2126. [CrossRef]

161. Xu, C.; Hedin, N.; Shi, H.-T.; Zhang, Q.-F. A Semiconducting Microporous Framework of $Cd_6Ag_4(SPh)_{16}$ Clusters Interlinked Using Rigid and Conjugated Bipyridines. *Chem. Commun.* **2014**, *50*, 3710–3712. [CrossRef]

162. Xu, C.; Sheng, M.-M.; Shi, H.-T.; Strømme, M.; Zhang, Q.-F. Interlinking Supertetrahedral Chalcogenolate Clusters with Bipyridines to Form Two-Dimensional Coordination Polymers for Photocatalytic Degradation of Organic Dye. *Dalton Trans.* **2019**, *48*, 5505–5510. [CrossRef]

163. Ghosh, A.; Mohammed, O.F.; Bakr, O.M. Atomic-Level Doping of Metal Clusters. *Acc. Chem. Res.* **2018**, *51*, 3094–3103. [CrossRef] [PubMed]

164. Bhattarai, B.; Zaker, Y.; Atnagulov, A.; Yoon, B.; Landman, U.; Bigioni, T.P. Chemistry and Structure of Silver Molecular Nanoparticles. *Acc. Chem. Res.* **2018**, *51*, 3104–3113. [CrossRef] [PubMed]

165. Chakraborty, I.; Pradeep, T. Atomically Precise Clusters of Noble Metals: Emerging Link between Atoms and Nanoparticles. *Chem. Rev.* **2017**, *117*, 8208–8271. [CrossRef]

166. Li, X.-Y.; Su, H.-F.; Xu, J. A 2D Layer Network Assembled from an Open Dendritic Silver Cluster $Cl@Ag_{11}N_{24}$ and an N-Donor Ligand. *Inorg. Chem. Front.* **2019**, *6*, 3539–3544. [CrossRef]

167. Pearson, R.G. Hard and Soft Acids and Bases. *J. Am. Chem. Soc.* **1963**, *85*, 3533–3539. [CrossRef]

168. Huang, R.-W.; Wei, Y.-S.; Dong, X.-Y.; Wu, X.-H.; Du, C.-X.; Zang, S.-Q.; Mak, T.C.W. Hypersensitive Dual-Function Luminescence Switching of a Silver-Chalcogenolate Cluster-Based Metal–Organic Framework. *Nat. Chem.* **2017**, *9*, 689–697. [CrossRef]

169. Huang, R.-W.; Dong, X.-Y.; Yan, B.-J.; Du, X.-S.; Wei, D.-H.; Zang, S.-Q.; Mak, T.C.W. Tandem Silver Cluster Isomerism and Mixed Linkers to Modulate the Photoluminescence of Cluster-Assembled Materials. *Angew. Chem. Int. Ed.* **2018**, *57*, 8560–8566. [CrossRef]

170. Du, X.-S.; Yan, B.-J.; Wang, J.-Y.; Xi, X.-J.; Wang, Z.-Y.; Zang, S.-Q. Layer-Sliding-Driven Crystal Size and Photoluminescence Change in a Novel SCC-MOF. *Chem. Commun.* **2018**, *54*, 5361–5364. [CrossRef]

171. Cao, M.; Pang, R.; Wang, Q.-Y.; Han, Z.; Wang, Z.-Y.; Dong, X.-Y.; Li, S.-F.; Zang, S.-Q.; Mak, T.C.W. Porphyrinic Silver Cluster Assembled Material for Simultaneous Capture and Photocatalysis of Mustard-Gas Simulant. *J. Am. Chem. Soc.* **2019**, *141*, 14505–14509. [CrossRef]

172. Wang, Y.-M.; Zhang, J.-W.; Wang, Q.-Y.; Li, H.-Y.; Dong, X.-Y.; Wang, S.; Zang, S.-Q. Fabrication of Silver Chalcogenolate Cluster Hybrid Membranes with Enhanced Structural Stability and Luminescence Efficiency. *Chem. Commun.* **2019**, *55*, 14677–14680. [CrossRef] [PubMed]

173. Wu, T.; Yin, D.; Hu, X.; Yang, B.; Liu, H.; Xie, Y.-P.; Liu, S.-X.; Ma, L.; Gao, G.-G. A Disulfur Ligand Stabilization Approach to Construct a Silver(I)-Cluster-Based Porous Framework as a Sensitive SERS Substrate. *Nanoscale* **2019**, *11*, 16293–16298. [CrossRef]

174. Su, W.; Hong, M.; Jiang, F.; Liu, H.; Zhou, Z.; Wu, D.; Mak, T.C.W. A Cleavage of the S–C Bond in 2-aminothiophenal: Synthesis and Crystal Structure of $[Ag_{11}(\mu_5\text{-}S)(\mu_4\text{-}S_2CNEt_2)_6(\mu_3\text{-}S_2CNEt_2)_3]$. *Polyhedron* **1996**, *15*, 4047–4051. [CrossRef]

175. Chen, S.; Du, W.; Qin, C.; Liu, D.; Tang, L.; Liu, Y.; Wang, S.; Zhu, M. Assembly of the Thiolated $[Au_1Ag_{22}(S\text{-}Adm)_{12}]^{3+}$ Superatom Complex into a Framework Material through Direct Linkage by SbF_6^- Anions. *Angew. Chem. Int. Ed.* **2020**, *59*, 7542–7547. [CrossRef]

176. Dong, X.-Y.; Huang, H.-L.; Wang, J.-Y.; Li, H.-Y.; Zang, S.-Q. A Flexible Fluorescent SCC-MOF for Switchable Molecule Identification and Temperature. *Chem. Mater.* **2018**, *30*, 2160–2167. [CrossRef]

177. Wei, Z.; Wu, X.-H.; Luo, P.; Wang, J.-Y.; Li, K.; Zang, S.-Q. Matrix Coordination Induced Emission in a Three-Dimensional Silver Cluster-Assembled Material. *Chem. Eur. J.* **2019**, *25*, 2750–2756. [CrossRef]

178. Wu, X.-H.; Luo, P.; Wei, Z.; Li, Y.-Y.; Huang, R.-W.; Dong, X.-Y.; Li, K.; Zang, S.-Q.; Tang, B.Z. Guest-Triggered Aggregation-Induced Emission in Silver Chalcogenolate Cluster Metal–Organic Frameworks. *Adv. Sci.* **2019**, *6*, 1801304. [CrossRef]

179. Lei, Z.; Pei, X.-L.; Jiang, Z.-G.; Wang, Q.-M. Cluster Linker Approach: Preparation of a Luminescent Porous Framework with NbO Topology by Linking Silver Ions with Gold(I) Clusters. *Angew. Chem. Int. Ed.* **2014**, *53*, 12771–12775. [CrossRef]

180. Nguyen, T.-A.D.; Jones, Z.R.; Goldsmith, B.R.; Buratto, W.R.; Wu, G.; Scott, S.L.; Hayton, T.W. A Cu_{25} Nanocluster with Partial Cu(0) Character. *J. Am. Chem. Soc.* **2015**, *137*, 13319–13324. [CrossRef]

181. Cook, A.W.; Jones, Z.R.; Wu, G.; Scott, S.L.; Hayton, T.W. An Organometallic Cu_{20} Nanocluster: Synthesis, Characterization, Immobilization on Silica, and "Click" Chemistry. *J. Am. Chem. Soc.* **2018**, *140*, 394–400. [CrossRef]

182. Nguyen, T.-A.D.; Jones, Z.R.; Leto, D.F.; Wu, G.; Scott, S.L.; Hayton, T.W. Ligand-Exchange-Induced Growth of an Atomically Precise Cu_{29} Nanocluster from a Smaller Cluster. *Chem. Mater.* **2016**, *28*, 8385–8390. [CrossRef]

183. Chakrahari, K.K.; Liao, J.-H.; Kahlal, S.; Liu, Y.-C.; Chiang, M.-H.; Saillard, J.-Y.; Liu, C.W. $[Cu_{13}\{S_2CN^nBu_2\}_6(acetylide)_4]^+$: A Two-Electron Superatom. *Angew. Chem. Int. Ed.* **2016**, *55*, 14704–14708. [CrossRef] [PubMed]

184. Negishi, Y.; Kurashige, W.; Niihori, Y.; Nobusada, K. Toward the Creation of Stable, Functionalized Metal Clusters. *Phys. Chem. Chem. Phys.* **2013**, *15*, 18736–18751. [CrossRef]

185. Puls, A.; Jerabek, P.; Kurashige, W.; Förster, M.; Molon, M.; Bollermann, T.; Winter, M.; Gemel, C.; Negishi, Y.; Frenking, G.; et al. A Novel Concept for the Synthesis of Multiply Doped Gold Clusters $[(M@Au_nM'_m)L_k]^{q+}$. *Angew. Chem. Int. Ed.* **2014**, *53*, 4327–4331. [CrossRef]

186. Kurashige, W.; Niihori, Y.; Sharma, S.; Negishi, Y. Recent Progress in the Functionalization Methods of Thiolate-Protected Gold Clusters. *J. Phys. Chem. Lett.* **2014**, *5*, 4134–4142. [CrossRef]

187. Kurashige, W.; Niihori, Y.; Sharma, S.; Negishi, Y. Precise Synthesis, Functionalization and Application of Thiolate-Protected Gold Clusters. *Coord. Chem. Rev.* **2016**, *320–321*, 238–250. [CrossRef]

188. Niihori, Y.; Hossain, S.; Kumar, B.; Nair, L.V.; Kurashige, W.; Negishi, Y. Perspective: Exchange Reactions in Thiolate-Protected Metal Clusters. *APL Mater.* **2017**, *5*, 053201. [CrossRef]

189. Niihori, Y.; Hossain, S.; Sharma, S.; Kumar, B.; Kurashige, W.; Negishi, Y. Understanding and Practical Use of Ligand and Metal Exchange Reactions in Thiolate-Protected Metal Clusters to Synthesize Controlled Metal Clusters. *Chem. Rec.* **2017**, *17*, 473–484. [CrossRef]

190. Kumar, B.; Kawawaki, T.; Shimizu, N.; Imai, Y.; Suzuki, D.; Hossain, S.; Nair, L.V.; Negishi, Y. Gold Nanoclusters as Electrocatalysts: Size, Ligands, Heteroatom Doping, and Charge Dependences. *Nanoscale* **2020**, *12*, 9969–9979. [CrossRef]

191. Zhang, H.; Watanabe, T.; Okumura, M.; Haruta, M.; Toshima, N. Catalytically Highly Active Top Gold Atom on Palladium Nanocluster. *Nat. Mater.* **2012**, *11*, 49–52. [CrossRef]

192. Kusada, K.; Yamauchi, M.; Kobayashi, H.; Kitagawa, H.; Kubota, Y. Hydrogen-Storage Properties of Solid-Solution Alloys of Immiscible Neighboring Elements with Pd. *J. Am. Chem. Soc.* **2010**, *132*, 15896–15898. [CrossRef] [PubMed]

193. Akutsu, M.; Koyasu, K.; Atobe, J.; Hosoya, N.; Miyajima, K.; Mitsui, M.; Nakajima, A. Experimental and Theoretical Characterization of Aluminum-Based Binary Superatoms of $Al_{12}X$ and Their Cluster Salts. *J. Phys. Chem. A* **2006**, *110*, 12073–12076. [CrossRef] [PubMed]

194. Negishi, Y.; Nakamura, Y.; Nakajima, A.; Kaya, K. Photoelectron Spectroscopy of Gold–Silver Binary Cluster Anions $(Au_nAg_m^-; 2 \leq n + m \leq 4)$. *J. Chem. Phys.* **2001**, *115*, 3657. [CrossRef]

nanomaterials

Article

PbS Quantum Dots Saturable Absorber for Dual-Wavelength Solitons Generation

Ling Yun [1] and Wei Zhao [2],*

[1] Advanced Photonic Technology Lab, Nanjing University of Posts and Telecommunications, Nanjing 210046, China; yunling@njupt.edu.cn
[2] State Key Laboratory of Transient Optics and Photonics, Xi'an Institute of Optics and Precision Mechanics, Chinese Academy of Sciences, Xi'an 710119, China
* Correspondence: weiz@opt.ac.cn

Abstract: PbS quantum dots (QDs), a representative zero-dimensional material, have attracted great interest due to their unique optical, electronic, and chemical characteristics. Compared to one- and two-dimensional materials, PbS QDs possess strong absorption and an adjustable bandgap, which are particularly fascinating in near-infrared applications. Here, fiber-based PbS QDs as a saturable absorber (SA) are studied for dual-wavelength ultrafast pulses generation for the first time to our knowledge. By introducing PbS QDs SA into an erbium-doped fiber laser, the laser can simultaneously generate dual-wavelength conventional solitons with central wavelengths of 1532 and 1559 nm and 3 dB bandwidths of 2.8 and 2.5 nm, respectively. The results show that PbS QDs as broadband SAs have potential application prospects for the generation of ultrafast lasers.

Keywords: fiber laser; mode locking; PbS quantum dots

Citation: Yun, L.; Zhao, W. PbS Quantum Dots Saturable Absorber for Dual-Wavelength Solitons Generation. *Nanomaterials* **2021**, *11*, 2561. https://doi.org/10.3390/nano11102561

Academic Editors: John Vakros, George Avgouropoulos and Efrat Lifshitz

Received: 16 July 2021
Accepted: 25 September 2021
Published: 29 September 2021

Publisher's Note: MDPI stays neutral with regard to jurisdictional claims in published maps and institutional affiliations.

1. Introduction

Low-dimensional materials have attracted extensive interest in applied physics due to their excellent optical, electronic, and chemical characteristics [1–3]. Two-dimensional (2D) graphene [4,5], black phosphorus [6], MXenes [7], antimonene [8], transition metal dichalcogenides [9], topological insulators [10], and one-dimensional (1D) carbon nanotubes [11,12] have been employed as saturable absorbers (SAs) to obtain ultrafast pulses in passively mode-locked fiber lasers. Among these SAs, the main problem is that it is difficult to achieve short carrier lifetime, high thermal damage, large modulation depth, and wide bandwidth in an individual material at the same time. Therefore, one of the ways to solve the problem is to find a new saturable absorption material that can effectively adjust the above parameters.

Semiconductor quantum dots (QDs) are particularly charming materials showing strong quantum confinement effects, as the size of QDs is close to the Bohr radius of the exciton [13]. Strong confinement not only produces interesting new effects but also strengthens the nonlinear optical characteristics [14]. Among numerous semiconductor QDs, PbS QDs possess smaller carrier effective masses and larger optical dielectric constant, leading to a large exciton Bohr radius (~18 nm), which makes relatively large QDs have strong quantum confinement effects [14]. Therefore, combined with the small energy bandgap (~0.41 eV) of PbS, the wavelength range of the excitonic absorption peak is from the visible to 3 μm via changing the size of PbS QDs [15]. This means that PbS QDs can promote the saturable absorption in a large spectral range by changing the size of QDs [16]. As a result, PbS QDs as SAs have been used for Q-switching or mode locking in various near-infrared lasers [13,16,17]. The most obvious advantages of PbS QDs-based SA are adjustable absorption peak, large third-order nonlinear susceptibility, fast response time, large modulation depth, and high damage threshold [18,19].

On the other hand, multi-wavelength passively mode-locked fiber lasers have been investigated extensively in the advancement of fascinating applications as optical fiber sensing, biomedical research, and wavelength division multiplexing (WDM) optical communication [20,21]. Several types of the saturable absorption materials that can realize multi-wavelength passive mode locking have been studied in depth [22–26]. Based on SESAM, Wu et al. realized the dual-wavelength (1553 and 1562 nm) dissipative solitons in Er-doped fiber laser operating at normal dispersion regime [22]. By virtue of carbon nanotube SA, dual-wavelength vector solitons centered at 1533 and 1557 nm were achieved by Zhao et al. [23]. Tunable dual- and triple-wavelength dissipative solitons were obtained from a Yb-doped fiber laser using graphene-oxide mode locker [24]. In a previous paper, we also reported the generation of a dual-wavelength polarization-locked vector solitons fiber laser using black phosphorus SA [25]. However, as far as we know, there is no report of multi-wavelength solitons operating in fiber-based PbS QDs mode-locked fiber lasers.

In this context, PbS QDs are fabricated via a modified hot-injection method. A dual-wavelength passively mode-locked Er-doped fiber laser is realized by using fiber-based PbS QDs as intracavity mode-locked devices. The stable dual-wavelength conventional solitons with central wavelengths of 1532 and 1559 nm, and 3 dB spectral bandwidths of 2.8 and 2.5 nm are obtained. By finely tuning the pump intensity and polarization state, the dual-wavelength mode locking can be switched into single-wavelength operation state.

2. Materials and Methods

The PbS QDs coated with oleic acid were prepared by a modified hot-injection meth-od by precisely controlling the mass of the precursor, reaction temperature, environment, and reaction time [18,27]. The preparation details are given in Figure 1a. Firstly, the Pb precursor was formed by putting lead oxide (Sinopharm Chemical Reagent Co., Ltd., Shanghai, China) (450 mg), octadecene (Sinopharm Chemical Reagent Co., Ltd., Shanghai, China) (30 mL), and oleic acid (Sinopharm Chemical Reagent Co., Ltd., Shanghai, China) (10 mL) into a three-necked flask and heating it at 120 °C for 1 h in vacuum. Secondly, the S precursor was prepared by mixing sulfur powder (Sinopharm Chemical Reagent Co., Ltd., Shanghai, China) (32 mg) with oleic amine (Sinopharm Chemical Reagent Co., Ltd., Shanghai, China) (10 mL) and heated to 120 °C until it was completely dissolved. Thirdly, the prepared S precursor was rapidly injected into a three-necked flask containing a Pb source under the protection of argon gas; then, it was cooled to room temperature with an ice water bath quickly. Finally, the sample was separated with ethanol and centrifuged at 12,000 r/min for 3 min. The obtained PbS QDs were dried in vacuum, dissolved in cyclohexane solution (Sinopharm Chemical Reagent Co., Ltd., Shanghai, China), and stored at a concentration of 10 mg/ mL. The transmission electron microscope (TEM) (Hitachi, Tokyo, Japan) image of PbS QDs is given in Figure 1b, which shows that PbS QDs are spherical and have good dispersivity, and the average particle size is ~5.7 nm.

The fiber-based PbS QDs mode-locked device was prepared by dropping PbS QDs solution on the end face of an optical fiber connector and evaporated slowly at room temperature and pressure. Compared with the SA prepared by other methods such as mechanical exfoliation or CVD growth, the PbS QDs SA prepared by this scheme avoids the complicated and time-consuming growth and post transfer processes, and it not only overcomes the mechanical damage but also improves the damage threshold of the laser. Based on the dual-path detection system [9,18], the nonlinear optical absorption characteristics of the PbS QDs have been studied. As illustrated in Figure 2, PbS QDs show strong saturable absorption behavior at 1550 nm. The experimental results show that the unsaturable loss, saturation intensity, and modulation depth of the PbS QDs SA are ~21%, ~0.22 MW/cm^2, and ~44% respectively. To the best of our knowledge, our SA exhibits high modulation depth compared with that reported to date. The corresponding digital photograph of the PbS QDs mode-locker is shown in the inset of Figure 2.

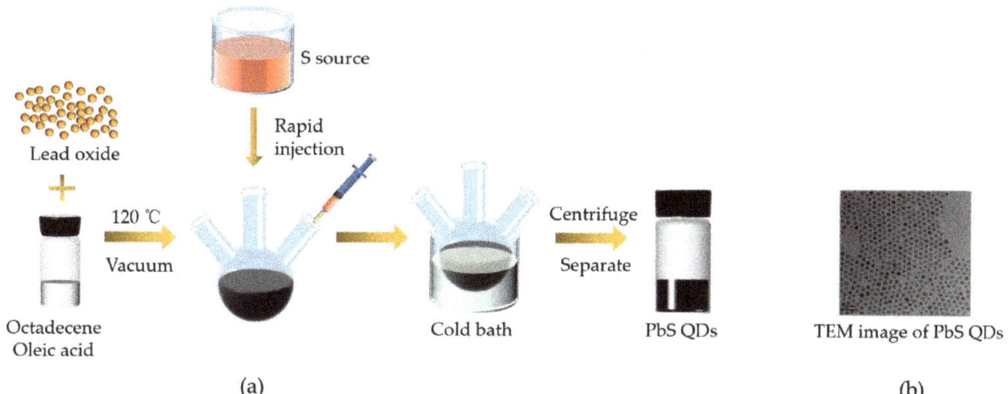

Figure 1. (**a**) The principle diagram of PbS quantum dots (PbS QDs) preparation. (**b**) The transmission electron microscope (TEM) image of PbS QDs.

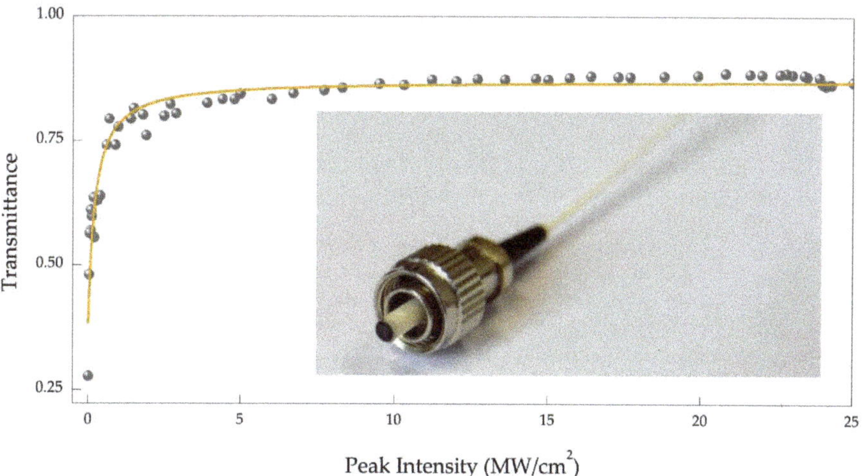

Figure 2. Nonlinear saturable absorption curve of PbS QDs saturable absorber (SA). Inset: PbS QDs mode locker.

3. Results and Discussion

The PbS QDs mode-locked Er-doped fiber laser operating in net anomalous dispersion regime is depicted in Figure 3. The ring cavity is composed of a 5.3 m erbium-doped fiber (EDF, $D = -16$ ps/nm/km) and 23.2 m single-mode fiber (SMF, $D = 17$ ps/nm/km). The net cavity dispersion is about -0.39 ps^2. The EDF served as a gain medium is pumped by a laser diode (LD, 980 nm) (Connet, Shanghai, China) through a WDM (980/1550 nm) (Connet, Shanghai, China). A polarization-insensitive isolator (PI-ISO) provides unidirectional operation. A polarization controller (PC) (Connet, Shanghai, China) is used to adjust the polarization and balance the gain distribution of the EDF by controlling the cavity loss. The pulses are extracted from the cavity with a 10% output coupler (OC). The PbS QDs SA device is assembled by sandwiching a fiber connector between two FC/PC fiber ferrules. The performance of the laser is recorded by an optical spectrum analyzer (Yokogawa AQ6370D) (Yokogawa, Tokyo, Japan), a second harmonic generation intensity autocorrelator (APE PulseCheck SM1600) (APE, Munich, Germany), an oscilloscope (RIGOL DS4050) (Tektronix, Johnston, OH, USA), and a radio-frequency analyzer (RS-FSV30) (Tektronix, Johnston, OH, USA).

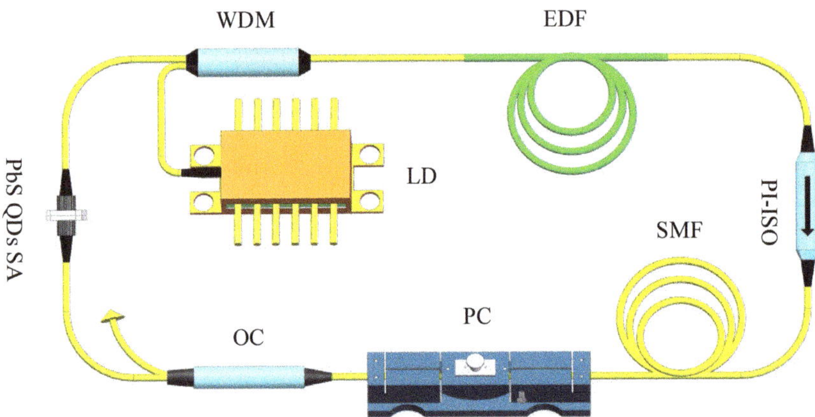

Figure 3. Laser setup. Laser diode (LD), wavelength division multiplexing (WDM), erbium-doped fiber (EDF), polarization-insensitive isolator (PI-ISO), single-mode fiber (SMF), polarization controller (PC), output coupler (OC), and Lead sulfide quantum dots saturable absorber (PbS QDs SA).

Based on the above experimental setup, the stable dual-wavelength mode-locked laser pulses are generated when the pump strength is scaled to 200 mW and the PC is finely tuned, as illustrated in Figure 4. The spectrum in Figure 4a appears to be obvious Kelly sidebands, which is a typical feature of the conventional solitons in the anomalous dispersion region [28]. The central wavelengths are 1532 and 1559 nm, and the corresponding 3 dB bandwidths are measured to be 2.8 and 2.5 nm, respectively. The spectrum of dual-wavelength solitons exhibits almost the same peak intensity, and the center wavelength spacing $\Delta\lambda$ is 27 nm. Figure 4b illustrates the oscilloscope trace, in which two pulse trains are formed simultaneously. There are two conventional solitons propagating in the laser cavity, and the pulse energy of each soliton varies slightly with the height. Under proper triggering, one pulse sequence stops, and the other moves on the oscilloscope screen. The results show that the two pulse sequences have different group velocities [24]. The corresponding radio-frequency spectrum is demonstrated in Figure 4c. Different from the single-wavelength soliton mode-locked, there are two fundamental frequencies that appear in the dual-wavelength mode-locked spectrum, corresponding to two mode-locked states. The fundamental frequencies are ~7.200695 MHz and ~7.201131 MHz, which are consistent with the mode-locked wavelengths of 1559 and 1532 nm, respectively. The formation of dual-wavelength conventional solitons may be due to the interaction of EDF gain spectrum and cavity-filtering effect [29]. Both of the signal-to-noise ratios are as high as 60 dB, which indicates good temporal stability of the PbS QDs-SA based dual-wavelength mode-locking operation. The frequency interval Δf is 436 Hz. Furthermore, the relationship between Δf and $\Delta\lambda$ is theoretically expressed as [30]:

$$\Delta f = \frac{c^2 D \Delta\lambda}{n^2(L + LD\Delta\lambda c/n)}$$

where L is the fiber length, n is the refractive index of fiber, D is the dispersion parameter, and c is the speed of light. Here, $L = 28.5$ m, $n = 1.46$, $D_{SMF+EDF} = 10.86$ ps/nm/km, and $c = 3 \times 10^8$ m/s. Therefore, the calculated $\Delta f = 434$ Hz, which is basically consistent with the experimental results.

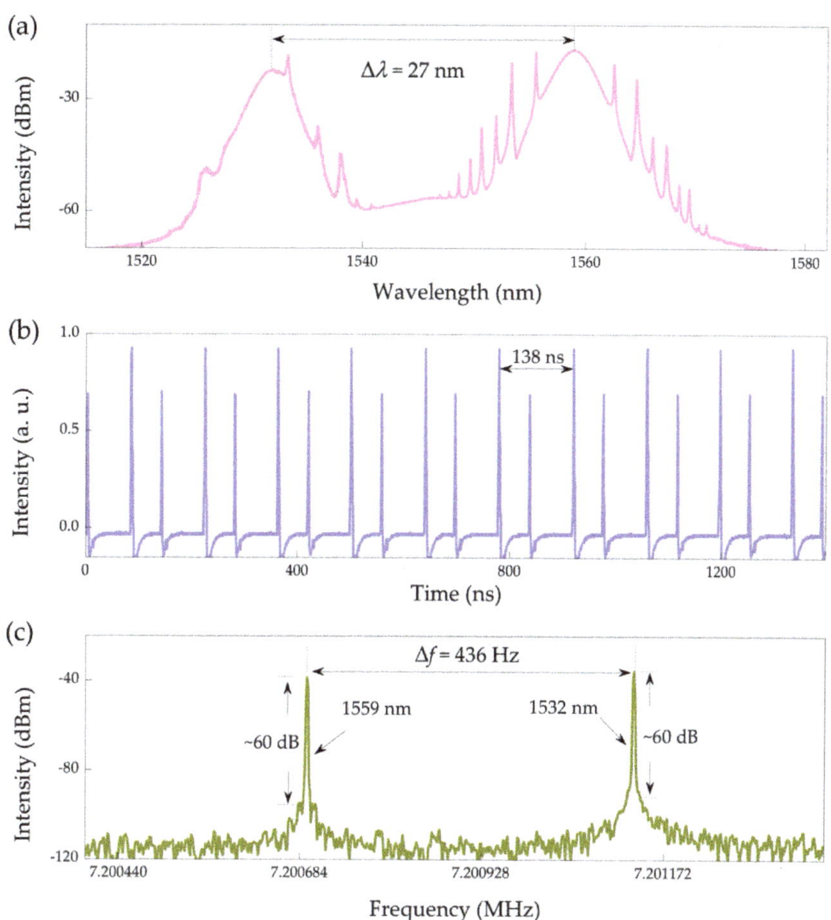

Figure 4. Dual-wavelength conventional solitons. (**a**) Optical spectrum at 1532 and 1559 nm, (**b**) pulse train, and (**c**) corresponding fundamental radio-frequency spectrum.

By decreasing the pump strength to 80 mW and carefully tuning the PC, dual-wavelength mode locking can be switched into single-wavelength mode locking. The characteristics of the proposed single-wavelength operation are presented in Figure 5. As shown in Figure 5a, the dual-wavelength conventional solitons at 1532 nm disappears, and the single-wavelength soliton remains at 1559 nm. The spectrum exhibits symmetric sid bands, and the 3 dB bandwidth is 2.4 nm. The autocorrelation trace is shown in Figure 5b. The pulse envelope is fitted with a Sech^2 function. The pulse duration is ~1.11 ps, so that the time-bandwidth product equals to 0.33. Therefore, the single-wavelength conventional soliton is nearly chirp-free. Figure 5c shows the fundamental frequency of ~7.200698 MHz, which is corresponding to a pulse interval of ~138 ns. The signal-to-noise ratio of the soliton is ~60 dB, indicating a stable single-wavelength mode locking. The average output power of the single pulse is 1.9 mW, corresponding to a pulse energy and peak power of ~0.26 nJ and 234 W, respectively.

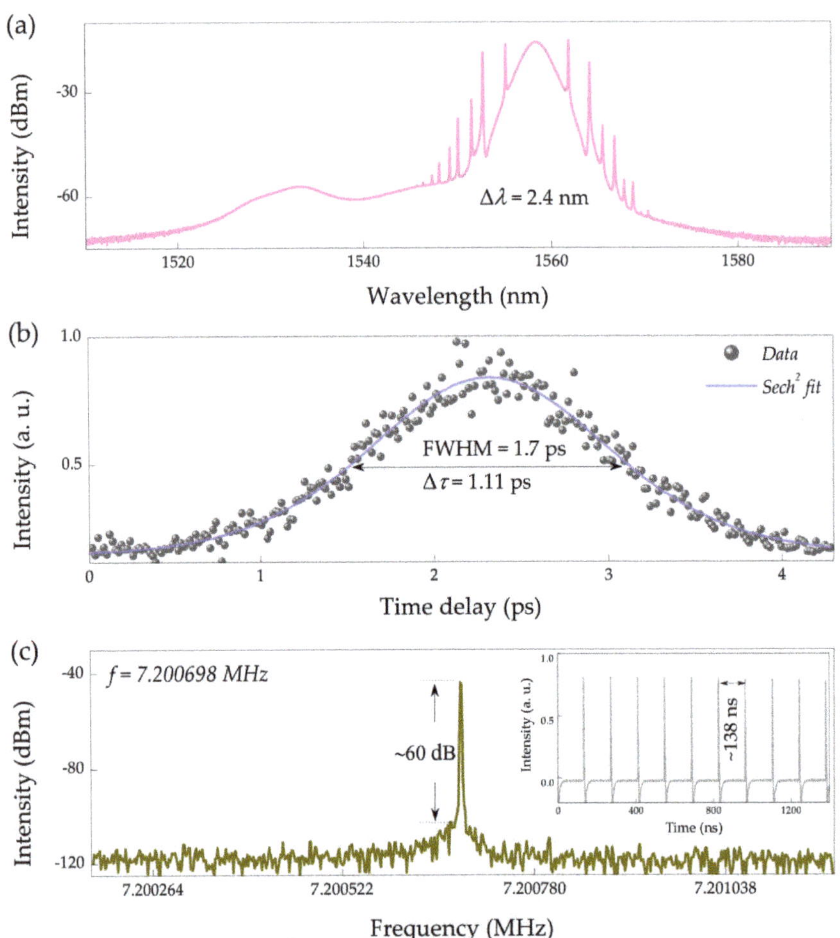

Figure 5. Single-wavelength conventional soliton. (**a**) Optical spectrum at 1559 nm, (**b**) autocorrelation trace, and (**c**) corresponding fundamental radio-frequency spectrum. Inset: pulse train.

When the pump strength increases from 80 to 120 mW, the stable single wavelength mode locking centered at 1532 nm with a 3 dB bandwidth of 3.5 nm is realized, as demonstrated in Figure 6. By Sech² fitting, the pulse duration of the conventional soliton is about 0.78 ps, the corresponding time bandwidth product is calculated to be 0.36, with slight chirp. The fundamental frequency is ~7.201135 MHz, which corresponds to the round-trip time of the cavity length of the fiber laser. The radio-frequency spectrum gives a signal-to-noise ratio of ~65 dB, indicating low-amplitude fluctuations and stable single-wavelength mode-locking state. When the pump power is 500 mW (maximum pump power available of LD in the experiment), the mode-locking operation of the fiber laser is still stable, which shows that the fiber-based PbS QDs SA fiber has good thermal damage. The average output power of the laser cavity is 12 mW, and the corresponding single pulse energy is 1.7 nJ. Therefore, the thermal damage threshold of PbS QDs SA is greater than 30 mJ/cm².

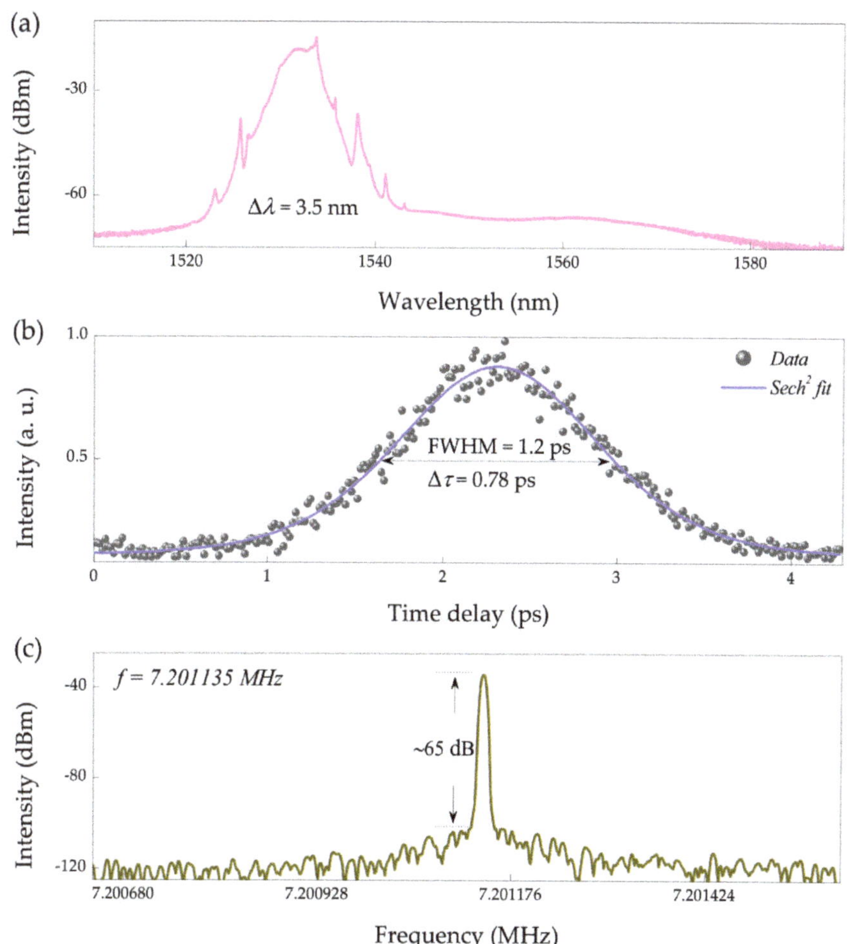

Figure 6. Single-wavelength conventional soliton. (**a**) Optical spectrum at 1532 nm, (**b**) autocorrelation trace, and (**c**) corresponding fundamental radio-frequency spectrum.

4. Conclusions

A passively mode-locked dual-wavelength Er-doped fiber laser is demonstrated with a fiber-based PbS QDs SA. Compared with other nanomaterials, PbS QDs prepared by a modified hot-injection method have the advantages of fast relaxation time, wide bandwidth, large modulation depth, and thermal damage. Based on this PbS QDs SA, the pulse laser can operate in a dual-wavelength conventional solitons region centered at 1532 and 1559 nm with 3 dB bandwidths of 2.8 and 2.5 nm, respectively. The experimental results reveal that our PbS QDs can be adopted as a broadband SA for application in pulse lasers.

Author Contributions: Conceptualization W.Z.; writing—original draft preparation, L.Y.; writing—review and editing, L.Y. and W.Z. All authors have read and agreed to the published version of the manuscript.

Funding: This research was funded by National Natural Science Foundation of China, grant number 61905118; Nanjing University of Posts and Telecommunications, grant number NY218023; and Research Center of Optical Communications Engineering & Technology, Jiangsu Province, grant number ZXF201905.

Institutional Review Board Statement: Not applicable.

Informed Consent Statement: Not applicable.

Data Availability Statement: Not applicable.

Conflicts of Interest: The authors declare no conflict of interest.

References

1. Martinez, A.; Sun, Z. Nanotube and graphene saturable absorbers for fibre lasers. *Nat. Photon.* **2013**, *7*, 842–845. [CrossRef]
2. Xu, N.; Wang, H.; Zhang, H.; Guo, L.; Shang, X.; Jiang, S.; Li, D. Palladium diselenide as a direct absorption saturable absorber for ultrafast mode-locked operations: From all anomalous dispersion to all normal dispersion. *Nanophotonics* **2020**, *9*, 4295–4306. [CrossRef]
3. Feng, J.; Li, X.; Shi, Z.; Zheng, C.; Li, X.; Leng, D.; Wang, Y.; Liu, J.; Zhu, L. 2D ductile transition metal chalcogenides (TMCs): A novel high-performance Ag_2S nanosheets for ultrafast photonics. *Adv. Opt. Mater.* **2019**, *8*, 1901762. [CrossRef]
4. Wu, X.; Yu, S.; Yang, H.; Li, W.; Liu, X.; Tong, L. Effective transfer of micron-size graphene to microfibers for photonic applications. *Carbon* **2016**, *96*, 1114–1119. [CrossRef]
5. Lee, E.; Choi, S.; Jeong, H.; Park, N.; Yim, W.; Kim, M.; Park, J.; Son, S.; Bae, S.; Kim, S.; et al. Active control of all-fibre graphene devices with electrical gating. *Nat. Commu.* **2015**, *6*, 6851. [CrossRef] [PubMed]
6. Wang, T.; Jin, X.; Yang, J.; Wu, J.; Yu, Q.; Pan, Z.; Shi, X.; Xu, Y.; Wu, H.; Wang, J.; et al. Oxidation-resistant black phosphorus enable highly ambient-stable ultrafast pulse generation at a 2 μm Tm/Ho-doped fiber laser. *ACS Appl. Mater. Interfaces* **2019**, *11*, 36854–36862. [CrossRef] [PubMed]
7. Fu, B.; Sun, J.; Wang, C.; Shang, C.; Xu, L.; Li, J.; Zhang, H. MXenes: Synthesis, optical properties, and applications in ultrafast photonics. *Small* **2021**, *17*, 2006054. [CrossRef] [PubMed]
8. Song, Y.; Liang, Z.; Jiang, X.; Chen, Y.; Li, Z.; Lu, L.; Ge, Y.; Wang, K.; Zheng, J.; Lu, S.; et al. Few-layer antimonene decorated microfiber: Ultra-short pulse generation and all-optical thresholding with enhanced long term stability. *2D Mater.* **2017**, *4*, 045010. [CrossRef]
9. Mao, D.; Du, B.; Yang, D.; Zhang, S.; Wang, Y.; Zhang, W.; She, X.; Cheng, H.; Zeng, H.; Zhao, J. Nonlinear saturable absorption of liquid-exfoliated molybdenum/tungsten ditelluride nanosheets. *Small* **2016**, *12*, 1489–1497. [CrossRef] [PubMed]
10. Luo, Z.; Liu, M.; Liu, H.; Zheng, X.; Luo, A.; Zhao, C.; Zhang, H.; Wen, S.; Xu, W. 2 GHz passively harmonic mode-locked fiber laser by a microfiber-based topological insulator saturable absorber. *Opt. Lett.* **2013**, *38*, 5212–5215. [CrossRef] [PubMed]
11. Wang, F.; Rozhin, A.; Scardaci, V.; Sun, Z.; Hennrich, F.; White, I.; Milne, W.; Ferrari, A. Wideband-tuneable, nanotube modelocked, fibre laser. *Nat. Nanotechnol.* **2008**, *3*, 738–742. [CrossRef]
12. Gladush, Y.; Mkrtchyan, A.; Kopylova, D.; Ivanenko, A.; Nyushkov, B.; Kobtsev, S.; Kokhanovskiy, A.; Khegai, A.; Melkumov, M.; Burdanova, M.; et al. Ionic liquid gated carbon nanotube saturable absorber for switchable pulse generation. *Nano Lett.* **2019**, *19*, 5836–5843. [CrossRef]
13. Rafailov, E.; Cataluna, M.; Sibbett, W. Mode-locked quantum-dot lasers. *Nat. Photon.* **2007**, *1*, 395–401. [CrossRef]
14. Guerreiro, P.; Ten, S.; Borrelli, N.; Butty, J.; Jabbour, G.; Peyghambarian, N. PbS quantum-dot doped glasses as saturable absorbers for mode locking of a Cr:Forsterite laser. *Appl. Phys. Lett.* **1997**, *71*, 1595–1597. [CrossRef]
15. Moreels, I.; Lambert, K.; Smeets, D. Size-dependent optical properties of colloidal PbS quantum dots. *ACS Nano* **2009**, *3*, 3023–3030. [CrossRef]
16. Lagatsky, A.; Malyarevich, A.; Savitski, V.; Gaponenko, M.; Yumashev, K.; Zhilin, A.; Brown, C.T. PbS quantum-dot-doped glass for efficient passive mode locking in a CW Yb: KYW laser. *IEEE Photon. Techol. Lett.* **2006**, *18*, 259–261. [CrossRef]
17. Lee, Y.; Chen, C.; Huang, C.; Chen, S.; Jiang, J. Passively Q-switched Er3+-doped fiber lasers using colloidal PbS quantum dot saturable absorber. *Opt. Express* **2016**, *24*, 10675–10681. [CrossRef] [PubMed]
18. Yun, L.; Qiu, Y.; Yang, C.; Xing, J.; Yu, K.; Xu, X.; Wei, W. PbS quantum dots as a saturable absorber for ultrafast laser. *Photon. Res.* **2018**, *6*, 1028–1032. [CrossRef]
19. Wundke, K.; Pötting, S.; Auxier, J.; Schülzgen, A.; Peyghambarian, N.; Borrelli, N. PbS quantum-dot-doped glasses for ultrashort pulse generation. *Appl. Phys. Lett.* **2000**, *76*, 10–12. [CrossRef]
20. Zhang, H.; Tang, D.; Wu, X.; Zhao, L. Multi-wavelength dissipative soliton operation of an erbium doped fiber laser. *Opt. Express* **2009**, *17*, 12692–12697. [CrossRef] [PubMed]
21. Liu, X.; Han, D.; Sun, Z.; Zeng, C.; Lu, H.; Mao, D.; Cui, Y.; Wang, F. Versatile multi-wavelength ultrafast fiber laser mode-locked by carbon nanotubes. *Sci. Rep.* **2013**, *3*, 2718. [CrossRef]
22. Wu, Z.; Fu, S.; Chen, C.; Tang, M.; Shum, P.; Liu, D. Dual-state dissipative solitons from an all-normal dispersion erbium-doped fiber laser: Continuous wavelength tuning and multi-wavelength emission. *Opt. Lett.* **2015**, *40*, 2684–2687. [CrossRef]
23. Zhao, C.; Dai, L.; Huang, Q.; Huang, Z.; Mou, C.; Araimi, M.; Rozhin, A.; Sergeyev, S.; Luo, Z. Dynamic polarization attractors of dissipative solitons from carbon nanotube mode-locked Er-doped laser. *Nanophotonics* **2020**, *9*, 2437–2443. [CrossRef]
24. Huang, S.; Wang, Y.; Yan, P.; Zhao, J.; Li, H.; Lin, R. Tunable and switchable multi-wavelength dissipative soliton generation in a graphene oxide mode-locked Yb-doped fiber laser. *Opt. Express* **2014**, *22*, 11417–11426. [CrossRef] [PubMed]
25. Yun, L. Black phosphorus saturable absorber for dual-wavelength polarization-locked vector soliton generation. *Opt. Express* **2017**, *25*, 32380–32385. [CrossRef]

26. Zhang, Y.; Li, X.; Qyyum, A.; Feng, T.; Guo, P.; Jiang, J. Lead sulfide nanoparticles for dual-wavelength ultrashort pulse generation. *Nanotechnology* **2020**, *31*, 085202. [CrossRef]
27. Hines, M.; Scholes, G. Colloidal PbS nanocrystals with size-tunable near-infrared emission: Observation of post-synthesis self-narrowing of the particle size distribution. *Adv. Mater.* **2003**, *15*, 1844–1849. [CrossRef]
28. Liu, X.; Yang, H.; Cui, Y.; Chen, G.; Yang, Y.; Wu, X.; Yao, X.; Han, D.; Han, X.; Zeng, C.; et al. Graphene-clad microfibre saturable absorber for ultrafast fibre lasers. *Sci. Rep.* **2016**, *6*, 26024. [CrossRef]
29. Shang, C.; Zhang, Y.; Qin, H.; He, B.; Zhang, C.; Sun, J.; Li, J.; Ma, J.; Ji, X.; Xu, L.; et al. Review on wavelength-tunable pulsed fiber lasers based on 2D materials. *Opt. Laser Technol.* **2020**, *131*, 106375. [CrossRef]
30. Agrawal, G. *Nonlinear Fiber Optics*, 4th ed.; Academic Press: Cambridge, MA, USA, 2007.

Article

Fabrication of Nanoyttria by Method of Solution Combustion Synthesis

Magdalena Gizowska [1,*], Milena Piątek [2], Krzysztof Perkowski [1], Gustaw Konopka [2] and Irena Witosławska [1]

[1] Department of Ceramics and Composites, Division of Ceramic and Concrete in Warsaw, Łukasiewicz Research Network—Institute of Ceramics and Building Materials, 9 Postępu Street, 02-676 Warsaw, Poland; k.perkowski@icimb.pl (K.P.); i.witoslawska@icimb.pl (I.W.)

[2] Research Laboratory, Division of Ceramic and Concrete in Warsaw, Łukasiewicz Research Network—Institute of Ceramics and Building Materials, 4 Kupiecka Street, 03-042 Warsaw, Poland; m.piatek@icimb.pl (M.P.); g.konopka@icimb.pl (G.K.)

* Correspondence: m.gizowska@icimb.pl

Received: 6 April 2020; Accepted: 24 April 2020; Published: 27 April 2020

Abstract: In the work the research on properties of an yttria nanopowder obtained by solution combustion synthesis (SCS) in terms of its application in ceramic technology is presented. In order to characterize the SCS reaction the decomposition of yttrium nitrate, glycine and their solution was investigated using differential thermal analysis coupled with FT-IR spectrometry of the gases emitted during the measurements. The product obtained in the SCS process was characterized in terms of its microstructure, particle size distribution and BET specific surface. Although the obtained powders showed nanoscaled structures, only after calcination at a temperature of 1100 °C nanosized particles were revealed. The calcined powder occurred in an agglomerated state (cumulants mean Z_{ave} = 1.3 μm). After milling particle size was successfully decreased to Z_{ave} = 0.28 μm. The deagglomerated powder was isostatically densified and tested for sintering ability. The obtained nanopowder showed very high sintering activity as the shrinkage onset was detected already at a temperature of about 1150 °C.

Keywords: DIL; DTA; DTG; FT-IR; SCS; yttria nanopowder

---

1. Introduction

A wide variety of applications of yttria is a driving force for the development of methods to fabricate pure powders. Doped or surface modified yttria nanoparticles find application in medicine and electronics [1–4].

Ceramic nanoparticles recently gained interest as an reinforcement of light weight alloys. Lately it was proven that the addition of 2.5 wt% ceramic nanoparticles caused an increase of the microhardness of Ti6Al4 V by 50%, manufactured by selective laser melting (SLM) [5]. Due to its high hardness and low reactivity with molten metals yttria can be an interesting material for metal matrix composites. Additionally, yttria has a relatively high thermal conductivity (8–12 W/m·K), which in case of the SLM technique can prove to be beneficial for the forming and densification of produced parts due to better heat transfer in the area of the laser's operation.

Yttria is also widely used in ceramic technology. Dense yttria ceramics find application as refractive ceramics (e.g., coatings and crucibles for molten reactive metals), optic devices (i.e., infrared missile domes) [6,7]. For ceramic technology nanopowder is desired due to the possibility of reducing the sintering temperature [8].

Solution combustion synthesis (SCS), which is based on the high energy reaction between metal nitrates and a reducing agent, is a promising method for the fabrication of nanopowders. Unlike sol-gel or precipitation technique [9,10] it is less time-consuming and requires fewer technological steps, as the synthesis byproducts undergo thermal decomposition. The principles of the SCS reaction are also used in technical scale and in continuous technology [11,12]. Recently, the mixture of an oxidizer-reducing agent solution has been used for the production of a powder in aerosol flame synthesis and spray pyrolysis [13–15]. The self-propagating high-temperature reaction is used in order to provide finer structuration of the powders produced in these methods.

A particularly interesting application of the SCS principle is a technique used to fabricate thin films for electronic devices such as solar cells, where the surface is coated with the solution containing the oxidizer–"fuel" mixture which is subsequently heated to produce a uniform oxide film [16].

The SCS method is based on the high-temperature self-propagating red-ox reaction between a metal nitrate and a reducing agent ("fuel"), which leads to the fabrication of metal oxide or its precursor, according to the overall SCS formula [12], non-toxic gas products. In the role of fuel, the following substances can be used: carbohydrazide, urea, amino acids (e.g., glycine, L-alanine) organic acids and saccharides [17–20].

Based on previous research, glycine is the most promising fuel for SCS reaction fuel for the fabrication of yttria [21]. The general formula of the reaction of yttrium nitrate and glycine is as follows (1):

$$6\,Y\,(NO_3)_3 + 10\,NH_2CH_2COOH \rightarrow 3\,Y_2O_3 + 20\,CO_2 + 25\,H_2O + 14\,N_2 \tag{1}$$

During the reaction vast amounts of gases are emitted which causes nanostructuration of the produced grains. Usually, the powders obtained by using this method show a high specific area and very complex particle morphology. Such microstructure is not beneficial for the fabrication of dense ceramics. Densification of nanopowders is a challenge in itself, as the smaller the particle, the more the particle-particle contact and the torque occurring between particles hinders the packing. Additionally, agglomerated particles need even higher pressures during compaction in order to destroy the inner structure of the particles. In the presented work we prove that with carefully designed technological steps the yttria powder obtained by using the SCS method has sintering ability and can be used for a ceramic application.

2. Materials and Methods

For the solution combustion synthesis of yttria yttrium nitrate hexahydrate (Sigma-Aldrich, St. Louis, MO, USA, purity 99.8%) and glycine (Sigma-Aldrich, purity ≥ 99%) was used. Glycine was used as the reducing agent and yttrium nitrate was both the precursor salt for yttria synthesis and the oxidizer in the redox reaction.

The solution combustion synthesis (SCS) was carried out in a quartz beaker. After the dissolution of the reagents in deionized water, the water was evaporated, and a gel was formed. The gel was then heated to the reaction initiation temperature. Once that temperature was reached, the high-temperature, self-propagating redox reaction took place [22]. The substrates were added in stoichiometric amounts, in each batch the aim was to obtain 5 g of yttria.

Thermogravimetric analysis was carried out in alumina crucibles using the thermal analyzer TG 449 F1 Jupiter (Netzsch Gerätebau GmbH, Selb, Germany). The gaseous products emitted were analyzed by FT-IR spectroscopy using a coupled FT-IR spectrometer (Tensor 27, Bruker, Billerica, MA, USA). The signals were identified based on the NIST database [23] and literature on glycine decomposition [24]. The curves of intensity of the characteristic absorbance wavenumber of a specific substance were subtracted and plotted as a function of temperature to investigate the reaction mechanism.

The powders' microstructure was characterized utilizing scanning electron microscopy (Nova NanoSEM 200, FEI Company, Hillsboro, OR, USA).

Obtained powders were deagglomerated in an attritor mill (Netzsch MiniCer, 1000 rpm) using zirconia balls of diameter of 0.4 mm. The milled powders mixed with binder and plasticizer were cryogranulated.

The particle size distribution of the powders was measured in aqueous suspensions by technique of dynamic light scattering (DLS), using zeta potential analyzer Zetasizer Nano ZS (Malvern Instruments Ltd., Worcestershire, UK). The analysis results are presented in terms of Z average (Z_{ave}) and polydispersity (Pd). These are supported by a median particle diameter (d_{V50}). Z average (also called the cumulants mean or harmonic intensity averaged particle diameter) is a mean value from the intensity distribution, which is the primary result obtained from the measurement thus the most stable result. Polydispersity derives from the polydispersity index (calculated from the cumulants analysis) and is the width of the estimated Gaussian distribution.

The specific surface area was measured by use of the BET technique (Gemini VII, Micromeritics Instrument Corp., Norcross, GA, USA). Based on the results of the BET, the equivalent spherical-particle diameter (d_{BET}) was calculated.

The measurement of linear changes of pressed granulates were conducted using a Netzsch high-temperature dilatometer (model Dil 402E) equipped with a graphite furnace. The measurement was carried out in a temperature range of RT to 1700 °C with a heating rate of 10 °C/min and the isothermal stage at the maximum temperature for a duration of 10 min.

Prior to measurement, calibration was carried out with a graphite standard of known properties and expansion. The measurement was carried out under the same conditions (temperature heating program, atmosphere, gas flow rate) to determine the signals related to the expansion of the dilatometer elements and to correct the results obtained during the proper measurement.

3. Results

Figure 1 shows the results of thermal analysis for yttrium nitrate hexahydrate. In Figure 1 in the top graph the curves corresponding to mass loss, mass loss derivative and thermal effects of thermal decomposition of yttrium nitrate hexahydrate are presented. The graphs below represent the absorbance intensity trends of selected wavenumbers as a function of temperature.

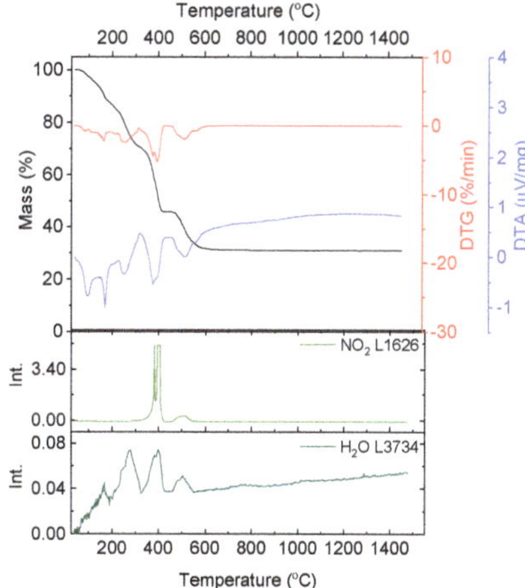

Figure 1. Thermal analysis results of the decomposition of yttrium nitrate hexahydrate in argon flow; absorbance intensities of traces for wavenumbers: 1626 and 3734 cm^{-1} corresponding with absorbance peaks of NO_2 and H_2O, respectively.

In the gaseous products resulting from yttrium nitrate decomposition water and nitrogen dioxide are detected. The first endothermic effect detected on the DTA curve (Figure 1) corresponds with the melting of the salt. Minor weight loss is then observed (3.44%) related to the evaporation of adsorbed water and the small signal on the DTG curve with a minimum at a temperature of 87 °C. At a temperature of about 108 °C dehydration begins and is followed in two stages (Figure 1):

- 108–193 °C with a maximum of mass loss rate at a temperature of 166.1 °C and endothermic peak at 170.8 °C, $\Delta m_{108-193\,°C} = 9.48\%$,
- 193–327 °C with a maximum of mass loss rate at a temperature of 267.5 °C and endothermic peak at 273.9 °C, $\Delta m_{193-327\,°C} = 18.74\%$.

In the temperature range of 108–327 °C the total mass loss is 28.12% which is close to the theoretical value of the complete dehydration of the salt (28.20%), which is confirmed by FT-IR data since exclusively the signal of water is visible (Figure 1).

Further mass loss occurs in two steps and corresponds to the degradation of the nitrate. The first distinctive mass loss ($\Delta m = 25.16\%$) occurs in the temperature range of 327–444 °C with a maximum mass loss rate at T = 397.7 °C and an endothermic peak at T = 398 °C (Figure 1). During the last decomposition stage, a mass loss of 13.32% occurs with a maximum mass loss rate at a temperature of 521.7 °C and an endothermic peak of 521.8 °C. Above the temperature of 641 °C the mass of the sample is stable. In both stages the signal indicating the presence of NO_2 is visible. Yttrium nitrate primarily decomposes to yttria and nitrogen pentaoxide, which is unstable and converts to nitrogen dioxide and oxygen. The two observed steps are a consequence of partial decomposition and the forming of cyclic oxynitrates [25].

In the temperature range of 327–641 °C the total mass loss equals 38.48%, which is in good consistency with the theoretical value of the decomposition reaction stoichiometry (42.30%) (2).

$$2\,Y(NO_3)_3 \xrightarrow{T} Y_2O_3 + 3\,N_2O_5 \tag{2}$$

Total mass loss observed during the decomposition of yttrium nitrate hexahydrate is 69.68% and corresponds well to the theoretical value of mass loss (70.52%).

In Figure 2 the results of the thermal analysis for glycine are presented.

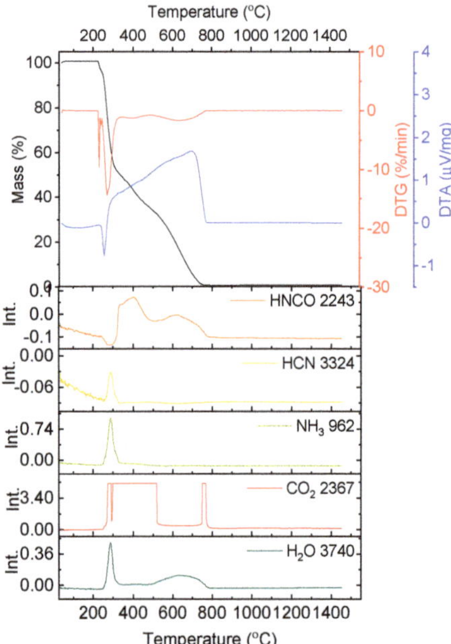

Figure 2. Thermal analysis results of the decomposition of glycine in synthetic air; absorbance intensities of traces for wavenumbers: 2243, 3324, 962, 2367 and 3740 cm^{-1} corresponding with absorbance peaks of HCNO, HCN, NH_3, CO_2 and H_2O, respectively.

The decomposition of glycine begins at a temperature of 228.6 °C. Up to a temperature of 303.8 °C the mass loss amounts to 46.26% with two overlapping maximum mass loss rates at T = 235.0 and 269.7 °C and with a sharp endothermic peak at 252.3 °C. In the exhaust gases HCNO, HCN, NH_3, CO_2 and H_2O were detected.

The next decomposition stage occurs at a temperature range of 269.7–473.6 °C with a minimum on DTG at T = 391.9 °C and no distinctive effect on the DTA curve. The mass loss equals 18.46% and is connected with the emission of CO_2 and HNCO.

The last stage of decomposition is the residual burnout (exothermic peak at T = 695.2 °C) and it ends at 835.8 °C (Δm = 35.34%).

According to literature [24] the decomposition begins with the emission of NH_3. Simultaneously, glycine can undergo condensation and cyclization reactions through dehydration reactions (Equations (3) and (4)). Subsequently, HCN, HNCO and CO is emitted due to selective cracking of cyclic amides. In air HNC and CO oxidize to HNCO and CO_2, respectively (5) [24].

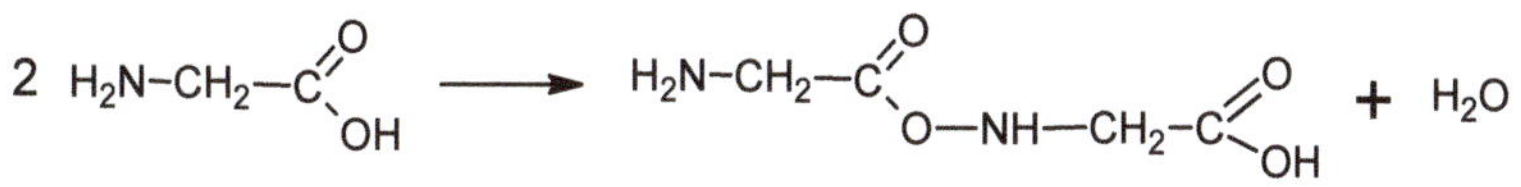

$$2\ H_2N\text{–}CH_2\text{–}COOH \longrightarrow H_2N\text{–}CH_2\text{–}CO\text{–}O\text{–}NH\text{–}CH_2\text{–}COOH + H_2O \tag{3}$$

$$2\ H_2N\text{–}CH_2\text{–}CO\text{–}O\text{–}NH\text{–}CH_2\text{–}COOH \longrightarrow \text{[diketopiperazine ring]} + H_2O \tag{4}$$

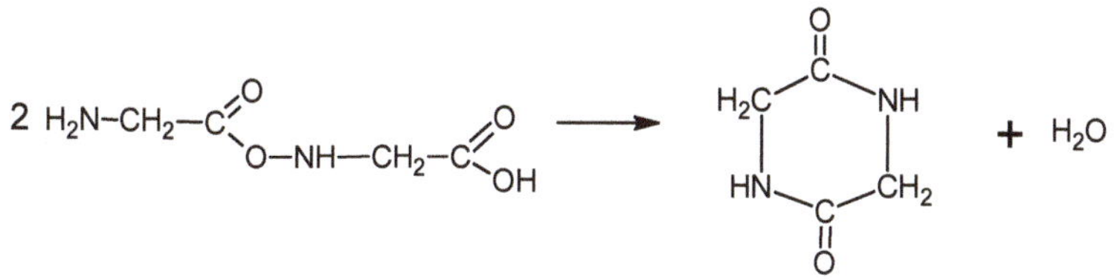

$$\text{[cyclic amide]} \longrightarrow HN{=}C{=}O \rightarrow HC{\equiv}N + CO \xrightarrow{[O_2]} HN{=}C{=}O + CO_2$$
$$\xrightarrow{-H_2} CO + H_2C{=}NH \rightarrow HC{\equiv}N \xrightarrow{[O_2]} HN{=}C{=}O + CO_2 \tag{5}$$

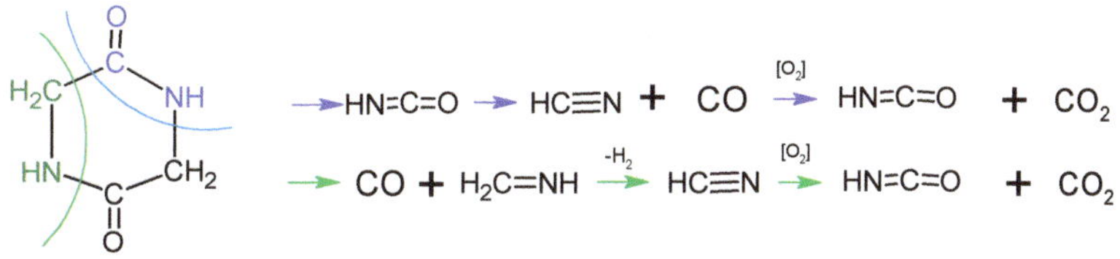

In Figure 3 the results of the thermal analysis for the solution containing stoichiometric amounts of yttrium nitrate hexahydrate and glycine are presented. The measurement was conducted in synthetic air flow, to best imitate the conditions of the synthesis, which takes place in an open quartz beaker.

The slope on the TG curve produced during the thermogravimetric measurement of the solution containing yttrium nitrate and glycine begins below 100 °C. The first endothermic effect (with the peak at T = 124.3 °C) continues up to a temperature of 182.0 °C and is connected with a mass loss of 71.79%. The mass loss is attributed mainly to the evaporation of water. However, on the FT-IR spectra signals resulting from the presence of HCN, NH_3 and NO_2 are also visible (Figure 3). This is surprising, as these compounds result from the degradation of glycine and yttrium nitrate, which according to the results presented in Figures 1 and 2 should be stable in this temperature range.

At a temperature of 238.2 °C the red-ox reaction between yttrium nitrate and glycine begins. It is distinguished by the exothermic peak on the DTA curve (T_{peak} = 244.1 °C) and an abrupt weight loss (Δm = 15.93%, Figure 3). At a temperature of 263.0 °C the process ends and with further temperature increase minor mass loss is observed (Δm = 3.96%). On FT-IR spectra not only CO_2 and H_2O, but also HCN, HCNO, NH_3 and NO_2 are detected. The yttria powders prepared by solution combustion synthesis using glycine and yttrium nitrate hexahydrate were investigated in terms of microstructure (Figure 4), particle size distribution and specific surface area (Table 1).

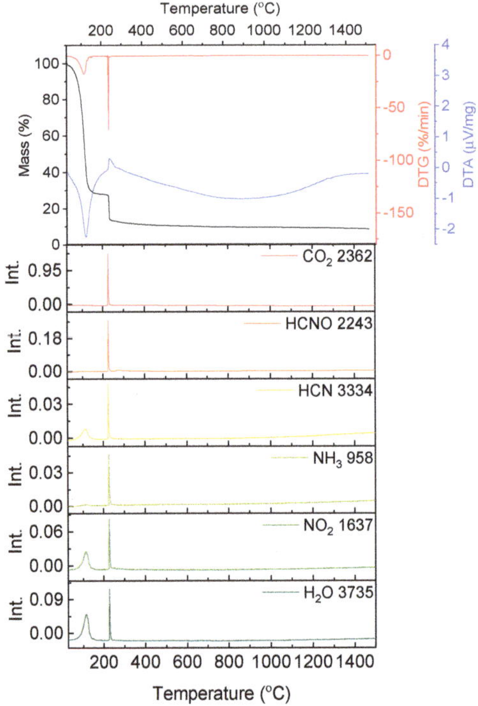

Figure 3. Results of the thermal analysis performed in synthetic air on the water solution containing yttrium nitrate and glycine; absorbance intensities of traces for wavenumbers: 2362, 2243, 3324, 958, 1637 and 3735 cm^{-1} corresponding to absorbance peaks of CO_2, HCNO, HCN, NH_3, NO_2 and H_2O, respectively.

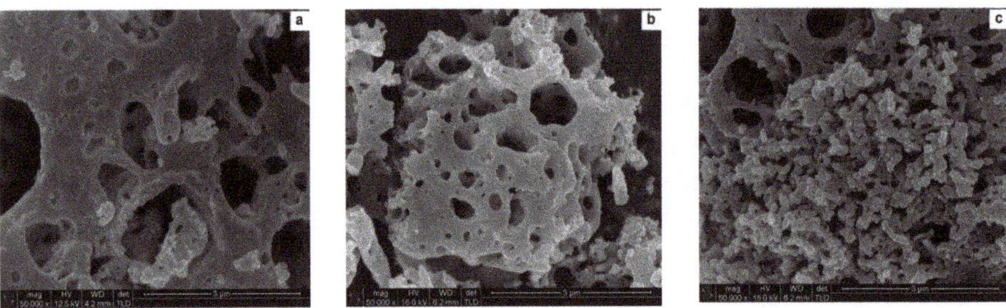

Figure 4. Micrographs of yttria nanopowder obtained by solution combustion synthesis (SCS) by the reaction of yttria nitrate with glycine: (**a**) not calcined, (**b**) calcined at a temperature of 800 °C, (**c**) calcined at a temperature of 1100 °C.

The synthesized powders are characterized by a highly porous microstructure (Figure 4a). The cumulants mean measured by technique of dynamic light scattering (DLS) is 2354 nm with a broad particle distribution (Pd = 1498 nm).

To burn out the substrates' residues the powders were calcined. After calcination at a temperature of 800 °C the powders sponge-like microstructure remained intact (Figure 4b). The DLS analysis provides information about average particle size expressed in the cumulants mean of 2649 nm while d_{V50} = 0.9 μm (Table 1). These strong discrepancies result from high polydispersity of particle size (expressed in a high value of polydispersity width—Pd), also visible in the SEM image (Figure 4b) where a micrometric sized particle is accompanied by some smaller grains at the image boundary. The BET surface is 19.7 m^2/g demonstrating a relatively high level of surface development.

Table 1. Particle size measured by method of dynamic light scattering (DLS) and calculated from BET specific surface area of yttria powders obtained by using the SCS method (Z_{ave}—cumulants mean, Pd—polydispersity width, d_{V50}—median diameter of the particle size distribution, S_{BET}—specific surface area, d_{BET}—BET equivalent spherical particle diameter).

	Calcination Temperature, °C	Z_{ave}, nm	Pd Width nm	d_{V50}, nm	S_{BET}, m^2/g	d_{BET}, nm
not milled	not calcined	2354	1498	1500	-	-
	800	2694	2319	859	19.7	61
	1100	1338	813	1190	11.5	104
15 min of milling in an attritor mill	800	570	329	747	-	-
	1100	275	137	352	-	-

Powders calcined at a temperature of 1100 °C reveal a finely grained microstructure. The microstructure transformation occurs without mass loss (Figure 3). The fine (about 100 nm in diameter) grains occur in agglomerates, with Z_{ave} = 1338 nm and d_{V50} = 1190 nm. Together with agglomerate size, BET surface is decreased as well and equals 11.5 m^2/g (Table 1). During calcination at a temperature of 1100 °C the disordered matter in nanostructures of high surface energy undergoes diffusion and reorganization into grains. The "sponge-like" structure undergoes conversion—its thin walls disappear and in its place uniform globular grains are produced. The specific surface decreases in the process of matter diffusion and grain formation, but the globular grains are connected by van der Waals forces or sintering necks. Such structure is more probable to disintegrate than the initial "sponge-like" aggregates which is portrayed in the DLS analysis. This kind of structure is still not beneficial for ceramic technology, as agglomerates are very difficult to densify during conventional pressing techniques and may cause fluctuations in density in the bulk of the pressed sample.

To deagglomerate the obtained powder high-energy milling was implemented. In Figure 5 the trend of cumulants mean vs milling time and in Figure 6 size distributions of the milled powders are presented.

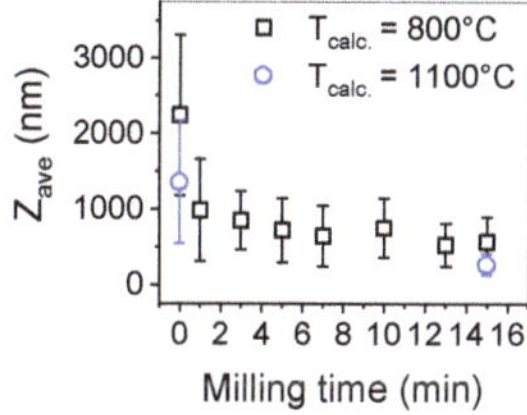

Figure 5. Particle diameter (expressed in cumulants mean—Z-ave) of yttria powder obtained by the SCS method.

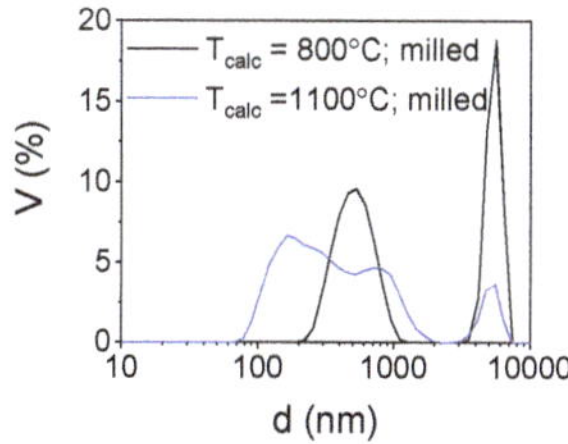

Figure 6. Particle size distribution of the milled powders.

In order to estimate the optimal time of milling the size distribution of yttria powder calcined at a temperature of 800 °C was measured after 1, 3, 5, 7, 10, 13 and 15 min (Figure 5). After 10 min, the particle size was about 500–600 nm and remained unchanged until maximum milling time which was 15 min. Basing on these observations and previous experience considering deagglomeration of yttria in water using an attritor mill [26], the milling time was set for 15 min and yttria powders were milled in these conditions. The results of particle size measurements of the milled powders are presented in Table 1 and in Figure 5. The size distribution of yttria powder calcined at the temperature of 800 °C revealed some agglomeration, as agglomerates or aggregates of 5 μm are visible. Agglomerates of powder calcined at a temperature of 1100 °C were disintegrated more effectively during milling (Z_{ave} = 275 nm and d_{V50} = 352 nm) which is consistent with the microstructural observations of the powder before milling. The agglomerates visible in Figure 4c underwent partial disintegration, the particle size distribution showed in Figure 6 indicates that the milled powder is trimodal with peaks at 0.275, 0.850 and 5 μm, which suggests that most of the agglomerates were disintegrated into particles with diameters of about 275 nm with some 3–4 particle agglomerates (d = 850 nm) and some bigger agglomerates left intact. In Figure 7 the SEM images of the milled powders are presented, where particles of about 100–150 nm with some bigger agglomerates are visible. This powder was cryo-granulated with the addition of binder and plasticizer, die pressed and densified in a cold isostatic press under a pressure of 150 MPa before taking dilatometric measurements (Figure 8). The suspension used for cryo-granulation was very diluted (c_{solids} ≈ 5 vol%), which is the reason why the powder does not appear in proper granules. Instead it occurs as separate particles and small agglomerates (Figure 6).

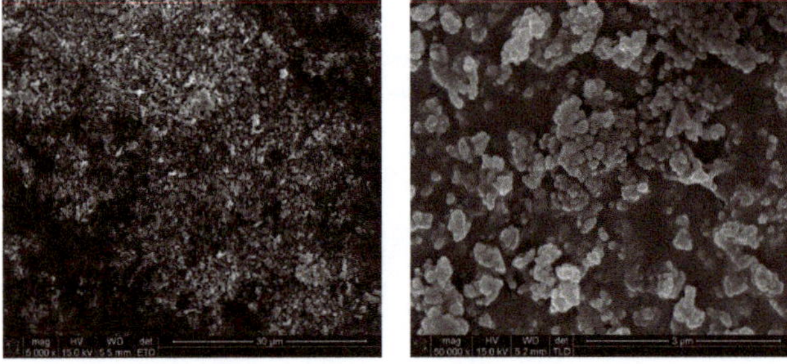

Figure 7. Micrographs of cryo-granulated yttria obtained by solution combustion synthesis through the reaction of yttria nitrate with glycine calcined at a temperature of 1100 °C after milling for 15 min in an attritor mill at a speed of 1000 rpm.

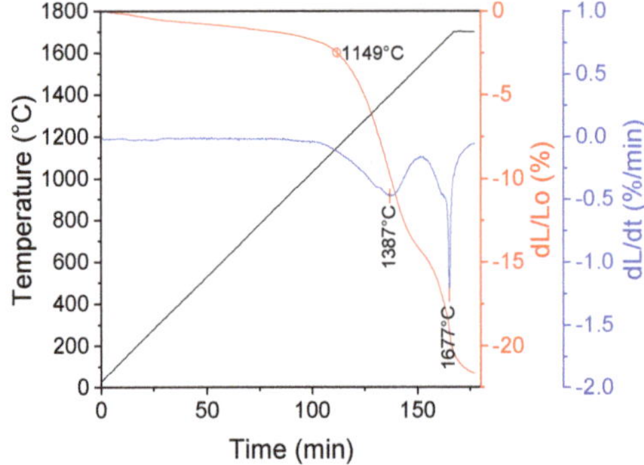

Figure 8. Dilatometric curve of a sample prepared from yttria powder obtained by method of the SCS with glycine.

The sintering starts at a temperature of 1149 °C and proceeds in two steps: the first—with the maximum sintering rate at a temperature of 1387 °C ($dL/L_0 = 14.33\%$) and the second with the maximum sintering rate at $T = 1677$ °C ($dL/L_0 = 7.33\%$).

4. Discussion

The general formula for the red-ox reaction of solution combustion synthesis (Equation (1)) suggests that the byproducts consist of the non-toxic gases: CO_2, H_2O and N_2. However, the FT-IR analysis of the gases emitted during the reaction showed also such specimens as HCN, HNCO, NH_3 (Figure 3). This indicates that the SCS reaction (1) conducted even in a very well controlled environment with homogenous temperature distribution (small sample in crucible and TGA chamber) is accompanied by decomposition reactions of the substrates: yttrium nitrate and glycine.

What is more, it was observed that below the redox reaction ignition point (238 °C) the presence of HCN, NH_3 and NO_2 was detected on FT-IR spectra (Figure 3). The specimens occur already at a temperature of about 124 °C.

Such observations have been also made by Biamino et al. [27] in the investigation of SCS with urea as fuel. In the aforementioned work [27] it was suggested that the emission of nitrate oxides derives from the direct reaction of nitrate with urea, which occur at a temperature below the reaction ignition point. In case of the reaction of yttrium nitrite with glycine signals deriving from NO_2 and HCN are visible in FT-IR spectra already at a temperature of about 124 °C (Figure 3). According to [27] the corresponding reaction of glycine and yttrium nitrite can be described as follows (6, 7):

$$26\ Y(NO_3)_3 + 6\ NH_2CH_2COOH \rightarrow 13\ Y_2O_3 + 84\ NO_2 + 12\ CO_2 + 15\ H_2O \tag{7}$$

$$2\ Y(NO_3)_3 + 24\ NH_2CH_2COOH \rightarrow Y_2O_3 + 30\ HCN + 18\ CO + 45\ H_2O \tag{8}$$

In both cases the reaction proceeds with the formation of carbon oxides which were not detected during the measurement below the ignition point (238 °C). Reactions 6 and 7 may take place during the exothermic reaction after reaching the ignition point, as HCN and NO_2 were detected then, together with carbon dioxide (the measurement was carried out in air which can cause the oxidation of carbon monoxide and cyanic acid to isocyanic acid). This suggests that the reduction of the nitrate does not cause a complete degradation of the glycine carbon chain. Presumably, the presence of nitrate in the solution can trigger the first step of glycine degradation i.e., removal of ammonia, which is visible in FT-IR spectra and can cause a degradation of the cyclic amides (5). Ammonia and intermediate compounds derived from the decomposition of cyclic peptides may react with nitrite in accordance with the following general formulas (8, 9):

$$14\ Y(NO_3)_3 + 6\ NH_3 \rightarrow 7\ Y_2O_3 + 48\ NO_2 + 9\ H_2O \tag{9}$$

$$4\ Y(NO_3)_3 + 6\ H_2CNH \rightarrow 2\ Y_2O_3 + 6\ HCN + 12\ NO_2 + 6\ H_2O \tag{10}$$

The assumption is consistent with macroscopic observations as precipitation was observed when the solutions containing the reagents were left to age for a month at room temperature. This suggests that a reaction between reagents took place.

Combustion synthesis is a widely used method for the production of nanopowders in both, laboratory and semi-technical scale. The presented results indicate that for a scaling-up of the process special precautions must be undertaken, which will provide for the neutralization of the hazardous nitrogen derived compounds.

The reaction of yttrium nitrate with glycine leads to the fabrication of nanostructured powder. The SEM observation of the powders showed an agglomerated, "sponge-like" microstructure of the particles. The structure remains stable at 800 °C as the microstructure of the powder remains intact after calcination at this temperature. Such morphology may be beneficial in some applications [1–4]. However, for ceramic technology the agglomeration is undesired as the densification of the nanopowder is severely hindered. After calcination at a temperature of 1100 °C the microstructure of the powder underwent modification and the "sponge-like" structure transformed into agglomerates consisting of globular grains with a diameter of about 100 nm. Milling of the powder calcined at a higher temperature was more effective as the particle diameter measured by method of DLS was decreased to 275 nm. Despite the milling, some agglomeration was observed in the size distribution curve (Figure 6). Further studies will be focused on optimization of milling conditions.

Sintering of calcined powder starts at a temperature of 1149 °C and proceeds in two distinctive stages. Such behavior indicates that the sintering process is divided into two stages: densification of spherical particles (reorganization of particles without particle growth) and grain growth, which was also observed by other researchers [28]. Another explanation of this phenomenon is that the first densification stage corresponds to the sintering of grains within the agglomerates, afterwards the sintering of the agglomerate domains and the separate particles takes place [29,30].

5. Conclusions

Solution combustion synthesis is an effective method for yttria nanopowder fabrication. The high-temperature self-propagating red-ox reaction between yttrium nitrate and glycine lowers the temperature of the nitrate decomposition and yttria formation from 642 to 263 °C. The vast amounts of gases emitted during decomposition implies the formation of nanoparticles. The yttria powders obtained by method of the SCS have a high sintering ability an can be applied in ceramic technology.

Author Contributions: Conceptualization, M.G. and K.P.; investigation, M.G., M.P., K.P. and G.K.; project administration, M.G. and K.P.; supervision, K.P. and I.W.; validation, M.P.; Visualization, M.G.; writing—original draft, M.G.; writing—review & editing, I.W. All authors have read and agreed to the published version of the manuscript.

Funding: This research was funded by the Polish Ministry of Science and Higher Education.

Acknowledgments: This work was funded by the Polish Ministry of Science and Higher Education.

Conflicts of Interest: The authors declare no conflict of interest.

References

1. Kumar, S.; Panwar, S.; Kumar, S.; Augustine, S.; Malhotra, B.D. Biofunctionalized Nanostructured Yttria Modified Non-Invasive Impedometric Biosensor for Efficient Detection of Oral Cancer. *Nanomaterials* **2019**, *9*, 1190. [CrossRef] [PubMed]

2. Kang, Y.C.; Lenggoro, I.W.; Park, S.B.; Okuyama, K. YAG: Ce Phosphor Particles Prepared by Ultrasonic Spray Pyrolysis. *Mater. Res. Bull.* **2000**, *35*, 789–798. [CrossRef]

3. Chang, H.; Lenggoro, I.W.; Okuyama, K.; Kim, T.O. Continuous Single-Step Fabrication of Nonaggregated, Size-Controlled and Cubic Nanocrystalline Y_2O_3:Eu^{3+} Phosphors Using Flame Spray Pyrolysis. *Jpn. J. Appl. Phys.* **2004**, *43*, 3535. [CrossRef]

4. Lee, S.K.; Yoon, H.H.; Park, S.J.; Kim, K.H.; Choi, H.W. Photoluminescence Characteristics of $Y_3Al_5O_{12}$:Ce^{3+} Phosphors Synthesized Using the Combustion Method with Various Reagents. *Jpn. J. Appl. Phys.* **2007**, *46*, 7983–7986. [CrossRef]

5. Hattal, A.; Chauveau, T.; Djemai, M.; Fouchet, J.J.; Bacroix, B.; Dirras, G. Effect of nano-yttria stabilized zirconia addition on the microstructureand mechanical properties of Ti_6Al_4V parts manufactured by selectivelaser melting. *Mat. Des.* **2019**, *180*, 107909.

6. Micheli, A.L.; Dungan, D.F.; Mantese, J.V. High-density yttria for practical ceramic applications. *J. Am. Ceram. Soc.* **1992**, *75*, 709–711. [CrossRef]

7. Eilers, H. Fabrication, optical transmittance, and hardness of IR-transparent ceramics made from nanophase yttria. *J. Eur. Ceram Soc.* **2007**, *27*, 4711–4717. [CrossRef]

8. Bernard-Granger, G.; Guizard, C.; San-Miguel, L. Sintering Behavior and Optical Properties of Yttria. *J. Am. Ceram. Soc.* **2007**, *90*, 2698–2702. [CrossRef]

9. Serivalsatit, K.; Kokuoz, B.; Yazgan-Kokuoz, B.; Kennedy, M.; Ballato, J. Submicrometer grain-sized transparent erbium-doped scandia ceramics. *J. Am. Ceram. Soc.* **2010**, *93*, 1320–1325. [CrossRef]

10. Mellado-Vázquez, R.; García-Hernández, M.; López-Marure, A.; López-Camacho, P.Y.; Morales-Ramírez Á, J.; Beltrán-Conde, H.I. Sol-gel synthesis and antioxidant properties of yttrium oxide naocrystallites incorporating P-123. *Materials* **2014**, *7*, 6768–6778. [CrossRef]

11. Levashov, E.A.; Mukasyan, A.S.; Rogachev, A.S.; Shtansky, D.V. Self-propagating high-temperature synthesis of advanced materials and coatings. *Int. Mater. Rev.* **2017**, *62*, 203–239. [CrossRef]

12. Buesser, B.; Pratsinis, S.E. Design of Nanomaterial Synthesis by Aerosol Processes. *Annu. Rev. Chem. Biomol. Eng.* **2012**, *3*, 103–127. [CrossRef] [PubMed]

13. Cho, J.S.; Jung, K.Y.; Kang, Y.C. Yolk-shell Structured Gd_2O_3:Eu^{3+} Phosphor Prepared by Spray Pyrolysis: The Effect of Preparation Conditions on Microstructure and Luminescence Properties. *Phys. Chem. Chem. Phys.* **2015**, *17*, 1325–1331. [CrossRef] [PubMed]

14. Jian, G.; Xu, Y.; Lai, L.-C.; Wang, C.; Zachariah, M.R. Mn_3O_4 Hollow Spheres for Lithium-Ion Batteries with High Rate and Capacity. *J. Mater. Chem. A* **2014**, *2*, 4627–4632. [CrossRef]

15. Jung, J.W.; Chueh, C.C.; Jen, A.K.Y. A Low-Temperature, Solution-Processable, Cu-Doped Nickel Oxide Hole-Transporting Layer via the Combustion Method for High-Performance Thin-Film Perovskite Solar Cells. *Adv. Mater.* **2015**, *27*, 7874–7880. [CrossRef]

16. Varma, A.; Mukasyan, A.S.; Rogachev, A.S.; Manukyan, K.V. Solution Combustion Synthesis of Nanoscale Materials. *Chem. Rev.* **2016**, *116*, 14493–14586. [CrossRef]

17. Kingsley, J.J.; Patil, K.C. A novel combustion process for the synthesis of fine particle alpha-alumina and related oxide materials. *Mater. Lett.* **1988**, *6*, 427–432. [CrossRef]

18. Mangalaraja, R.V.; Mouzon, J.; Hedström, P.; Kero, I.; Ramam, K.V.S.; Camurri, C.P.; Od´en, M. Combustion synthesis of Y_2O_3 and Yb-Y_2O_3 Part I. Nanopowders and their characterization. *J. Mater. Process Technol.* **2008**, *208*, 415–422. [CrossRef]

19. Yen, W.M.; Weber, M.J. *Inorganic Phosphors: Compositions, Preparation and Optical Properties*; CRC Press: Boca Raton, FL, USA, 22 June 2004.

20. Dupont, A.; Parent, C.; Le Garrec, B.; Heintz, J.M. Size and morphology control of Y_2O_3 nanopowders via a sol-gel route. *J. Solid State Chem.* **2003**, *171*, 152–160. [CrossRef]

21. Gizowska MKobus, I.; Perkowski, K.; Piątek MKonopka, G.; Witosławska, I.; Osuchowski, M. Size and morphology of yttria nanopowders obtained by Solution Combustion Synthesis. *Arch. Metall. Mater.* **2018**, *63*, 743–748.

22. Gizowska, M.; Kobus, I.; Perkowski, K.; Zalewska, M.; Konopka, G.; Witosławska, I.; Osuchowski, M.; Jakubiuk, K.; Witek, A. Otrzymywanie i charakterystyka nanoproszków $Y_3Al_5O_{12}$ syntezowanych na drodze reakcji spaleniowej inicjowanej promieniowaniem mikrofalowym. *Szkło i Ceramika* **2016**, *5*, 23–26.

23. NIST Standard Reference Database 69: NIST Chemistry WebBook. Available online: http://webbook.nist.gov/chemistry (accessed on 6 April 2017).

24. Li, J.; Wang, Z.; Yang, X.; Hu, L.; Liu, Y.; Wang, C. Evaluate the pyrolysis pathway of glycine and glycylglycine by TG–FT-IR. *J. Anal. Appl. Pyrolysis* **2007**, *80*, 247–253. [CrossRef]

25. Melnikov, P.; Nascimento, V.A.; Consolo LZ, Z.; Silva, A.F. Mechanism of thermal decomposition of yttrium nitrate hexahydrate, $Y(NO_3)_3 \cdot 6H_2O$ and modeling of intermediate oxynitrates. *J. Therm. Anal. Calorim.* **2013**, *111*, 115–119. [CrossRef]

26. Zalewska, M. *Porous Composite Materials in the Process of Removing Virus-Like Particles from Water*; Warsaw University of Technology, Faculty of Chemistry: Warsaw, Poland, 2015.

27. Biamino, S.; Badini, C. Combustion synthesis of lanthanum chromite starting from water solutions: Investigation of process mechanism by DTA–TGA–MS. *J. Eur. Ceram. Soc.* **2004**, *24*, 3021–3034. [CrossRef]

28. Dupont, A.; Largeteau, A.; Parent, C.; Le Garrec, B.; Heintz, J.M. Influence of the yttria powder morphology on its densification ability. *J. Eur. Ceram. Soc.* **2005**, *25*, 2097–2103. [CrossRef]

29. Falkowski, P. Effect of Selected Monosaccharides on the Liquefaction Process of Ceramic Nanopowders. Ph.D. Thesis, Warsaw University of Technology, Faculty of Chemistry, Warsaw, Poland, 2009.

30. Mayo, M.J.; Hague, D.C. Porosity-Grain Growth Relationships in the Sintering of Nanocrystalline Ceramics. *Nanostr. Mater.* **1993**, *3*, 43–52. [CrossRef]

Article

New Insights on the Nickel State Deposited by Hydrazine Wet-Chemical Synthesis Route in the Ni/BCY15 Proton-Conducting SOFC Anode

Dimitrinka Nikolova [1,*], Margarita Gabrovska [1], Gergana Raikova [2], Emiliya Mladenova [2], Daria Vladikova [2], Krassimir L. Kostov [3] and Yordanka Karakirova [1]

[1] Institute of Catalysis, Bulgarian Academy of Sciences, 1113 Sofia, Bulgaria; margarita.gabrovska@abv.bg (M.G.); daniepr@ic.bas.bg (Y.K.)
[2] Acad. Evgeni Budevski Institute of Electrochemistry and Energy Systems, Bulgarian Academy of Sciences, 1113 Sofia, Bulgaria; graikova@iees.bas.bg (G.R.); e_mladenova@iees.bas.bg (E.M.); d.vladikova@iees.bas.bg (D.V.)
[3] Institute of General and Inorganic Chemistry, Bulgarian Academy of Sciences, 1113 Sofia, Bulgaria; klkostov@svr.igic.bas.bg
* Correspondence: dimi_nik@abv.bg; Tel.: +359-2979-3578

Citation: Nikolova, D.; Gabrovska, M.; Raikova, G.; Mladenova, E.; Vladikova, D.; Kostov, K.L.; Karakirova, Y. New Insights on the Nickel State Deposited by Hydrazine Wet-Chemical Synthesis Route in the Ni/BCY15 Proton-Conducting SOFC Anode. *Nanomaterials* **2021**, *11*, 3224. https://doi.org/10.3390/nano11123224

Academic Editors: Christian M. Julien and Carlos M. Costa

Received: 25 October 2021
Accepted: 25 November 2021
Published: 27 November 2021

Publisher's Note: MDPI stays neutral with regard to jurisdictional claims in published maps and institutional affiliations.

Abstract: Yttrium-doped barium cerate (BCY15) was used as an anode ceramic matrix for synthesis of the Ni-based cermet anode with application in proton-conducting solid oxide fuel cells (pSOFC). The hydrazine wet-chemical synthesis was developed as an alternative low-cost energy-efficient route that promotes 'in situ' introduction of metallic Ni particles in the BCY15 matrix. The focus of this study is a detailed comparative characterization of the nickel state in the Ni/BCY15 cermets obtained in two types of medium, aqueous and anhydrous ethylene glycol environment, performed by a combination of XRD, N_2 physisorption, SEM, EPR, XPS, and electrochemical impedance spectroscopy. Obtained results on the effect of the working medium show that ethylene glycol ensures active Ni cermet preparation with well-dispersed nanoscale metal Ni particles and provides a strong interaction between hydrazine-originating metallic Ni and cerium from the BCY15 matrix. The metallic Ni phase in the pSOFC anode is more stable during reoxidation compared to the Ni cermet prepared by the commercial mechanical mixing procedure. These factors contribute toward improvement of the anode's electrochemical performance in pSOFC, enhanced stability, and a lower degradation rate during operation.

Keywords: Ni/BCY15 proton-conducting anode; hydrazine reduction; Ni–Ce interaction; SOFC; electrochemical impedance spectroscopy; XPS; EPR

1. Introduction

In July 2020, the European Commission launched the European Clean Hydrogen Strategy alongside the Strategy for Energy System Integration. Investment in hydrogen will be a critical growth engine in the context of recovery from the COVID-19 crisis and in the longer term, an important component of Europe's industrial competitiveness.

The production and energy-related consumption of hydrogen by 2050–2100 is expected to exceed the current level by tens or even hundreds of times [1]; accordingly, its multiple utilization is expected to grow and evolve.

The fuel cell industry uses hydrogen as feedstock, which is considered the most environmentally friendly fuel. Nowadays, hydrogen and fuel cell technologies offer greater personal choice in the transition to a low-carbon economy. Hydrogen fuel cells have become a key part since they are seen as reliable emission-free generating systems, alternative to polluting processes based on conventional processes of fossil fuel combustion. Hydrogen and fuel cells are seeing resurgence in interest: large-scale production of fuel cell vehicles has begun, and hundreds of thousands of homes are now heated and powered by fuel cells [2]. Thus, the development of hydrogen fuel cells, producing clean electricity by electrochemical reaction, has an important role in the improvement of human living conditions.

Solid oxide fuel cells (SOFCs) are a promising technology that can provide efficient and clean energy production, generating power from hydrogen, natural gas, and other renewable fuel. They have a number of advantages, such as flexibility towards the type of fuel, ability to tolerate the presence of impurities, higher efficiency, and application of non-noble metal catalysts. For their commercialization, however, lower cost and better durability of the performance are needed. The main pathway to realize these objectives is the reduction of the operating temperature. The conventional SOFCs operate at very high temperatures within 800–1000 °C [3–8].

In the moment, the commercial goal is 600–700 °C. These conditions can increase cell stability, improve materials' compatibility, and ensure cheaper metallic alloys. Since the limitations of the temperature decrease come mainly from the reduced performance of the active materials, approaches for increasing the electrocatalytic activity and ionic conductivity are highly appreciable. The lower activation energy of proton-conducting oxide materials has made them attractive candidates for electrolyte materials operating in the intermediate temperature range (500–700 °C), which opened the direction of the proton-conducting solid oxide fuel cells' (pSOFCs) development [4].

The principal disadvantage of classical SOFC and of its proton-conducting modification (pSOFC) is the formation of water at the electrodes, where it mixes with the fuel/oxidizer, resulting in electromotive force losses and decreased catalytic activity and durability. A successful attempt has been reported to eliminate water production at the electrodes with the so-called dual-membrane fuel cell (dmFC) design. The latter introduces a separate compartment (central membrane, CM) for the production and evacuation of water. It has mixed ionic (proton and oxide ion) conductivity, which ensures penetration of the protons and oxide ions produced at the electrode/electrolyte interface. They react in the CM to form water that is evacuated through the porous microstructure. This original approach opens a new niche for the development of reversible flexible fuel cells operating at 600–700 °C due to the application of proton-conducting oxide materials [9–14].

The type of the electrolyte has a large impact on the optimum cell performance through its contribution to the ohmic internal resistance. For many years, barium cerate-based materials have been known as the electrolyte material with high performance in terms of proton conductivity [4], in particular the yttrium-doped barium cerate, $BaCe_{0.85}Y_{0.15}O_{2.925}$ (BCY15). Since BCY15 exhibits a high level of oxide ion and proton conduction at medium temperatures (600–700 °C), the central membrane in the dmFC has been simplified by using a single material with sufficient mixed ionic conductivity as BCY15 [14–18].

In responding to the needs of modern society for the replacement of critical components in electrocatalysts from platinum group metals [19], nickel continues to be a non-precious metal of particular interest for the development of the cermet anodes, not only for its catalytic properties, but also for the effective cost reduction for the production of inexpensive, environmentally friendly energy and storage systems.

By analogy with classical cermet anodes, a composite Ni/BCY15 anode has been applied for dmFC construction, and thus the difference in the thermal expansion coefficient between the electrolyte and anode was avoided. Generally, Ni-based cermet production presents incorporation of commercial NiO powder in BCY electrolytes by the traditional oxide powder mixture reaction, consisting of mixing, cold pressing, and sintering at high temperatures (1100–1400 °C) [20,21]. During the solid-state reaction, cationic $Ni^{2+} \rightarrow Ce^{4+}$ partial substitution occurs in the BCY structure [22,23].

Reduction treatment of the anode before the operation of the cell under H_2/Ar atmosphere at high temperatures (700–800 °C) is then performed to produce metallic Ni, which leads to a morphology change of the composite, bringing about an increase of the anode porosity [22,24–26].

Our investigations on electrolyte-supported half-cell BCY15-NiO/BCY15/BCY15-NiO with cermet deposition by tape casting and sintering at temperatures below 1250 °C, which ensure the required porosity, showed a drastic increase of both the resistivity of the electrolyte and the area-specific resistance (ASR) of the anode. TEM/STEM analysis registered an extensive precipitation of nanosized Y_2O_3 particles along the NiO/BCY15 and the anode/BCY15 electrolyte interface. Their concentration decreases towards the depth of the electrolyte. Thus, an efficient electrical barrier between the protonic (BCY15) and the metallic (Ni) phase is produced. This phenomenon may be initiated by slight diffusion of Ba toward NiO. For sintering temperatures above 1350 °C, an improvement of both the ASR and resistivity of the electrolyte was observed. X-ray diffraction analysis shows that the system is stabilized towards segregation of yttria, however, the density is under the required limits and a small quantity of a new cerium yttrium oxide phase is registered [26,27].

To overcome the foregoing problems of the commercial mechanical mixing procedure, a new and original approach to metallic Ni introduction in the anode ceramic matrix of BCY15 is introduced based on wet-chemical synthesis of Ni/BCY15 cermet. It is cost-effective and energy-efficient, offering better structural control over the Ni metal particles in the BCY15 matrix for the production of anodes with good connectivity in both metallic and electrolyte phases, satisfactory porosity for the gas transport, and high electrical conductivity. In addition, it is expected that the sintering procedure based on the newly developed Ni/BCY cermet, which will induce partial or full oxidation of the Ni particles and possible ejection out of the Ni-YSZ/air surface, will lead to relief of the internal stress and thus to increased durability [28], which also needs to be checked.

Our expectations were to reach higher Ni^0 dispersion based on the investigations of Wojcieszak et al. The authors have found that the hydrazine reduction method is better to prepare a Ni/CeO_2 catalyst with higher dispersion and smaller particle size compared to the conventional H_2 reduction [29]. The synthesis procedure is based on concomitant nickel introduction in the anode ceramic matrix and gaining metallic Ni particles on the electrolyte surface via wet-reduction by hydrazine reducing agent [30]. The formation of finely divided metal particles proceeds in alkaline solution [31–33] and can be summarized by the reaction: $2Ni^{2+}+N_2H_4+4OH^- \rightarrow 2Ni^0+N_2+4H_2O$. Fundamentally, the severe problem regarding the ceramic BCY structure was the low chemical stability of barium cerate to water [34]. Our challenge was to protect the BCY15 matrix during hydrazine wet-reduction by application of non-aqueous medium [35]. We used ethylene glycol as a protective agent and solvent, but also because of its reduction properties. Ethylene glycol is known as a reducing agent used to obtain nanocrystalline metallic powders by employing the polyol method [36].

This work is focused on the detailed comparison between Ni/BCY15 cermets obtained by hydrazine wet-reduction using two types of media: aqueous medium and anhydrous ethylene glycol environment. The target of this evaluation is to explain the impact of the applied medium on the nickel state as a catalytic active component in the synthesized Ni/BCY15 cermets as an additional beneficial factor for the electrochemical activity of pSOFC anodes. For this purpose, a comparison was also performed between hydrazine-prepared Ni cermet and Ni cermet prepared by the commercial mechanical mixing procedure. The employed methodology involved combining the results from several methods, including X-ray powder diffraction, N_2 physisorption measurements, scanning electron microscopy, electron paramagnetic resonance, X-ray photoelectron spectroscopy, and electrochemical impedance spectroscopy. To our knowledge, the analysis of the bare BCY15 ceramic matrix by electron paramagnetic resonance and X-ray photoelectron spectroscopy, as well as the detailed examination of the surface of Ni/BCY15 cermets by X-ray photoelectron spectroscopy, are novelties in the development of pSOFC anodes.

2. Materials and Methods

2.1. Materials

$BaCe_{0.85}Y_{0.15}O_{2.925}$ (BCY15) powder (Marion Technology, Verniolle, France), used as the anode ceramic matrix, was fabricated by the auto-combustion process starting from metal nitrates and applying urea as a reducing agent. Sintering of the precursor at 1100–1150 °C in a carrier gas (helium or argon) for complete CO_2 elimination ensured the production of single-phase powder with 48% porosity and a dominating particle size around 200 nm, and a minor degree of agglomeration. Before synthesis, the BCY15 powder was thermally pretreated at 1100 °C for 2 h.

Nickel chloride hexahydrate ($NiCl_2 \cdot 6H_2O$), hydrazine monohydrate (99+% $N_2H_4 \cdot H_2O$), sodium hydroxide (NaOH), and anhydrous sodium carbonate (Na_2CO_3), of analytical grade, all procured by Alfa Aesar (Ward Hill, MA, USA), were used.

2.2. Synthesis

Two Ni/BCY15 cermets of 50 g each were synthesized by wet-reduction with hydrazine using different media, such as deionized water and ethylene glycol (EG). These media were used for the preparation of the initial $NiCl_2 \cdot 6H_2O$ solution and as environment during the reduction reaction. The synthesis was performed according to the reported wet-reduction procedure [30,35]. A NiO to BCY15 volume ratio of 44.4:55.6 was determined with both Ni/BCY15 cermets to match the Ni metal content of 32 wt.%. A pre-set amount of $N_2H_4 \cdot H_2O$ solution was used to provide a N_2H_4 to Ni^0 weight ratio of 6:1. This ratio was chosen based on studies of Huang et al. [37]. These authors have established that the optimal N_2H_4 to Ni^0 molar ratio is 4, which ensured complete reduction of the Ni^{2+} ions to Ni^0 in the solution free of $Ni(OH)_2$. The reduction of hydrazine complex precursors proceeds via formation of $Ni(OH)_2$, which is further decomposed by the equation: $2Ni(OH)_2 + N_2H_4 \rightarrow 2Ni + N_2 + 4H_2O$. The reason for using a higher amount of $N_2H_4 \cdot H_2O$ ($N_2H_4:Ni^0 = 6:1$) was a possible reduction of Ce^{4+} from the BCY15 structure to Ce^{3+} via treatment with hydrazine, which would take place in a thin surface layer by analogy with the known ceria reduction by hydrazine [29,38]. The installation equipment consisted of a five-liter glass reactor with a stirrer and a steam jacket equipped with pH electrodes, a thermocouple, a reflux condenser, and peristaltic pumps. The mixture of nickel hydrazine complex and BCY15 powder was reduced to nickel black deposited on the BCY15 surface by addition of an appropriate amount of alkaline solution (a mixture of NaOH and Na_2CO_3) to keep a constant alkaline pH value upon heating to 95 °C. For the purpose of complete reduction, these reaction conditions were retained for 1 h under vigorous stirring. The suspension was washed with deionized water several times up to neutral pH and absence of Cl^- ions. The samples were dried in the air at 100 °C for 20 h. Both obtained samples were further denoted by an extension to indicate the applied medium, namely Ni/BCY15-W and Ni/BCY15-EG.

2.3. Catalyst Characterization

2.3.1. Standard Characterization

N_2 physisorption measurements were carried out on a Quantachrome Instruments NOVA 1200 e (Boynton Beach, FL, USA) apparatus by low-temperature (77.4 K) nitrogen adsorption after sample outgassing in a vacuum at 80 °C for 16 h. The nitrogen adsorption-desorption isotherms were analyzed through the linear part of the curves to evaluate the specific surface area calculated by means of the Brunauer–Emmett–Teller (BET) equation. Pore size distribution (PSD) was obtained by the Barrett–Joyner–Halenda method using the desorption branch of the isotherms.

X-ray powder diffraction (XRD) data were collected on a Bruker D8 Advance (Bruker-AXS, Karlsruhe, Germany) diffractometer employing $CuK\alpha$ radiation (λ = 0.15406 nm), operated at U = 40 kV and I = 40 mA. The crystalline phases were identified by means of International Centre for Diffraction Data (ICDD) powder diffraction files. The semi-quantitative analysis (as wt.%) was performed with the program Diffrac.Eva V4 (Diffrac.Eva Version 4, Bruker AXS GmbH, Karlsruhe, Germany) using the ICDD PDF-2 2021 database.

2.3.2. Scanning Electron Microscopy (SEM)

The morphological studies were performed using a JEM 200 CX (Instruments, München, Germany) transmission scanning microscope equipped with ASID 3D.

2.3.3. Electron Paramagnetic Resonance (EPR) Spectroscopy

Electron paramagnetic resonance measurements were performed at a temperature of 123 K using a JEOL JES-FA 100 EPR spectrometer (JEOL, Tokyo, Japan) operating at the X band (~9.8 GHz). The magnetic field was modulated at 100 kHz and the g values were determined from precise frequency and magnetic field values. The samples were placed in a quartz tube and fixed in the center of a standard TE011 cylindrical resonator. The EPR spectra were recorded under the following conditions: microwave power 5 mW, modulation amplitude 0.5 mT, sweep 500 mT, time constant 0.03 s, and sweep time 2 min.

2.3.4. X-ray Photoelectron Spectroscopy

X-ray photoelectron spectroscopy (XPS) measurements were carried out on an AXIS Supra electron spectrometer (Kratos Analyical Ltd., Trafford Park, Stretford, United Kingdom) with a base vacuum in the analysis chamber of around 10^{-8} Pa. The samples were irradiated with Mg Kα photons with energy of 1253.6 eV. According to their kinetic energy, the photoemitted electrons were separated by a 180°-hemispherical analyzer with total instrumental resolution of ~1.0 eV (as measured by the FWHM (full width at half maximum) of the $Ag3d_{5/2}$ line) at a pass energy of 20 eV. Due to the charging effect, a resolution of ~1.3 eV was measured on the isolated samples. Energy calibration was performed by normalizing the C1s line of adventitious adsorbed hydrocarbons to 285.0 eV. The analysis area was $300 \times 700\ \mu^2$. The concentrations (as at.%) of the observed chemical elements were calculated by normalizing the areas of their most intense photoelectron peaks to their relative sensitivity factors using commercial spectrometer software (SpecsLab2 CasaXPS software, Casa Software Ltd, Bay House 5 Grosvenor Terrace Teignmouth, United Kingdom). The accuracy of the binding energy determination was within ±0.1 eV.

'Bare' anode samples were prepared using a standard ceramic technology by cold pressing (3 t/5 min) to obtain pressed tablets followed by sintering in air at 1200 °C for 5 h. Volumetric shrinkage of each tablet did not exceed 6%. The sintering procedure was used for real electrodes, leading to obtaining dense cermet tablets with definite porosity and grain size.

2.4. Electrochemical Measurements

The electrochemical impedance spectroscopy (EIS) measurements were performed on an IVIUM—CompactStat e10030 (Alvatek Ltd, Tetbury, England) in the temperature interval of 100–750 °C and frequency range of 1 MHz–0.01 Hz, with a density of 5 points/decade and amplitude of the AC signal of 1 mA in the reduction atmosphere. In situ redox cycling analysis was performed on two types of symmetrical half cells fabricated by tape casting of the cermet slurry on the BCY15 dense electrolyte support and sintered at 1350 °C for 4 h using the following compositions: (i) Ni/BCY15-Mech obtained by a ball-milled mixture of 30 vol.% BCY15 and 70 vol.% NiO (NOVAMET, HP green NiO-Type A) powders, and (ii) Ni/BCY15-EG powders prepared by wet-reduction with hydrazine in ethylene glycol.

The impedance measurements were preceded by a specially developed reduction procedure [39] performed at 800 °C in the reducing atmosphere, starting with the H_2+N_2 mixture and followed by reduction in pure hydrogen. This treatment transforms the NiO to Ni. Each sample was subjected to six redox cycles following the regime presented in Table 1.

Table 1. Redox cycling regime.

Oxidation			Reduction		
Duration	N_2	Air	Duration	N_2	H_2
(Min)	(mL/Min/cm^2)	(mL/Min/cm^2)	(Min)	(mL/Min/cm^2)	(mL/Min/cm^2)
2	3.97	3.97	6	3.97	3.97
6	3.97	0	10	22.28	0
			5	3.97	0

3. Results and Discussion

3.1. X-ray Powder Diffraction

The basic analysis of the impact of the wet-chemical synthesis route on the structural characteristics of as-prepared Ni/BCY15 cermets was performed by XRD. The patterns of bare BCY15 and as-prepared Ni/BCY15 cermets were graphically presented in our previous study [30,35]. They exhibit the characteristic diffraction lines of single-phase isostructural $BaCeO_3$ (ICDD-PDF file 00-022-0074). After Ni^0 deposition on a BCY15 ceramic matrix, typical reflections of metallic Ni (ICDD-PDF file 00-004-0850) were observed and parasite $BaCO_3$ phase (ICDD-PDF file 00-045-1471) was also registered, however, pattern intensities in this case of Ni/BCY15-EG were negligible [35] if compared with very well-organized and intense reflections of the $BaCO_3$ phase observed in the aqueous environment [30]. However, at Ni/BCY15-W, only two phases were detected, namely metallic Ni and parasite $BaCO_3$. There were no reflections for $BaCeO_3$ structure (Table 2). As it was mentioned above, during synthesis, the deionized water solvent and medium in the reactor causes decomposition of the BCY15 perovskite structure and readily transforms part of the $BaCe_{0.85}Y_{0.15}O_{2.925}$ material to barium oxide. Then, the latter reacts with carbon dioxide and water to yield barium carbonate [40]. Barium carbonate formation in ethylene glycol anhydrous synthesis is explained by partial decomposition of the BCY15 structure due to surface reduction of Ce^{4+} to Ce^{3+} by hydrazine, thus liberating some barium, which reacts with CO_2 to $BaCO_3$.

Table 2. XRD and N_2 physisorption parameters of as-prepared Ni/BCY15 cermets.

Sample	Phase	Phase Quantity, wt.%	$L_{Ni(111)}$ nm	S_{BET} m^2/g	d_{av} nm	Constant C
BCY15	$BaCeO_3$	–		3	14	90
Ni/BCY15-W	Ni^0	40	15.7			
	–	–		8	11.5	97
	$BaCO_3$	60				
Ni/BCY15-EG	Ni^0	80	13.0			
	$BaCeO_3$	16		11	9.1	82
	$BaCO_3$	4				

In this study, the semi-quantitative analysis was performed by XRD spectra to estimate the amount of parasite $BaCO_3$ phase at both as-prepared Ni/BCY15 cermets listed in Table 2. It is evident that the amount of $BaCO_3$ phase at Ni/BCY15-W is too much (60 wt.%), while the quantity is very low at Ni/BCY15-EG (4 wt.%). Indubitably, the application of anhydrous ethylene glycol medium leads only to a low extent of BCY15 structure decomposition. In addition to these comments, the great amount of $BaCO_3$ is the reason for no registration of the main $BaCeO_3$ perovskite structure.

It is also evident from the data in the Table 2 that Ni^0 phase quantity at as-prepared Ni/BCY15-EG is higher (80 wt.%) than the theoretical Ni metal content of 32 wt.%. The explanation is the high mass absorption index of the nickel originated from its a large atomic number, thus leading to very good X-ray absorption. Another observation was that this Ni^0 phase quantity is higher at Ni/BCY15-W (40 wt.%), due to the better covering of the BCY15 with metallic Ni particles and a lesser extent of decomposition of the BCY15 structure.

Our previous XRD examination [30,35] revealed that the sintering in air at 1350 °C, which is an obligatory stage in the technological cycle for anode tablets' preparation and cannot be avoided, generates complete decomposition of parasitic $BaCO_3$ phase at both Ni cermets. NiO and restored $BaCeO_3$ perovskite phase as prevailing phases and formation of impurity phases were noticed, with a total amount of 6 wt.% obtained in sintered Ni/BCY15-W, as presented in Table 3. In contrast, XRD patterns of sintered BCY15/Ni-EG exhibited reflections of NiO, preserved $BaCeO_3$ perovskite structure, and traces of 2% Y_2BaNiO_5 oxide.

Table 3. XRD parameters of sintered and reduced Ni/BCY15 anode tablets.

Stage	Ni/BCY15-W		Ni/BCY15-EG	
	Phase	Phase Quantity, wt.%	Phase	Phase Quantity, wt.%
Sintered anode tablets	NiO	40	NiO	68
	$BaCeO_3$	54	$BaCeO_3$	30
	$BaNiO_{2.36}$	3	Y_2BaNiO_5	2
	$Y_{0.10}Ce_{0.90}O_{1.95}$	3		
Reduced anode tablets	Ni^0	80	Ni^0	92
	$BaCeO_3$	17	$BaCeO_3$	8
	$BaNiO_{2.36}$	2		
	$Y_{0.10}Ce_{0.90}O_{1.95}$	1		

The Ni/BCY15 anodes were processed in the reducing atmosphere before the impedance measurements to transform NiO to metallic Ni, noted in Table 3 as reduced anode tablets. The existence of $BaNiO_{2.36}$ phase in the reduced Ni/BCY15-W anode indicated that the Ni ions are strongly included in this structure (2 wt.% amount), which is stable at exposure in the reducing atmosphere at 800 °C. The important finding is that after reduction treatment, only metallic Ni and the BCY15 structure existed in the Ni/BCY15-EG anode. The phase quantity of 92 wt.% Ni^0 is due to the very intensive peaks of better-crystalized metallic nickel phase compared to as-prepared Ni/BCY15-EG cermet.

Data of the mean crystallite sizes (L_{Ni}) of the formed metallic nickel in Table 2 clearly show a smaller size of the nickel in Ni/BCY15-EG, with 17%, compared to Ni/BCY15-W. This observation confirms ethylene glycol's contribution as a reducing agent to decrease Ni^0 size.

The XRD data show that no significant structural degradation was observed for the BCY15 ceramic matrix during wet-chemical synthesis of Ni/BCY15-EG as well as Ni/BCY15 anode tablets. Thus, the bulk structural analysis convincingly supports the advantage of the hydrazine wet approach via ethylene glycol assistance.

3.2. N_2 Physisorption

The first common method for surface analysis of both the as-prepared cermets is N_2 physisorption, which is known as a technique to study pore characteristics of solid materials. Pore size distribution (PSD) provides information about mesopores, which may facilitate the transport of reagent molecules to and from active sites, being an essential factor for catalyst design [41,42].

BCY15 support is a typical macroporous material, which is confirmed by the Type II isotherm characteristic of aggregated powders, such as clays, cements, etc., having a small H3-type hysteresis loop at the highest pressures [30]. Such an isotherm shape possessing a loop is referred to as a pseudo-Type II isotherm, adopted as Type II(b) [43,44]. Nickel deposition did not change the isotherm type of both Ni/BCY15 cermets, however, the hysteresis loops were increased in size relative to bare BCY15 [30,35]. The hysteresis loop type is also H3 due to mesopores being formed from non-rigid aggregates of plate-like particles [43,44]. The formation of clearly manifested hysteresis may be due to metallic Ni particles forming their own mesopore structure. The specific surface area increases more visibly with Ni/BCY15-EG (S_{BET} = 11 m^2/g) in relation to that of BCY15 (Table 2).

Pore size distribution of BCY15 and both nickel-containing samples is displayed in Figure 1. A uniform pore system of the BCY15 ceramic matrix is illustrated and the presence of some macropores above 50 nm indicates a macroporous material. The PSD curve of Ni/BCY15-W shows a polydisperse character in the 3–50 nm range. Compared with BCY support, new mesopores can be observed due to the formation of metallic Ni phase, with an average pore diameter (d_{av}) of 11.5 nm (Table 2). In principle, this polydisperse character is the same after synthesis in EG medium, however, with intense maxima in the 3–14 nm range and less mesopores within 14–50 nm, with a lower d_{av} value of 9.1 nm (Table 2). Obviously, the formation of metallic nickel phase is accompanied by the creation of more mesopores. This observation could be explained by the presence of smaller Ni^0 particles that expose smaller mesopores of 3–6 nm size, in comparison with Ni/BCY15-W PSD data. PSD intensity and shape within this range (Figure 1) indicate a relatively narrow distribution. The presence of larger amounts of pores in the Ni/BCY15-EG sample synthesized in ethylene glycol leads to a higher S_{BET} value (Table 2), as clearly seen in Figure 1.

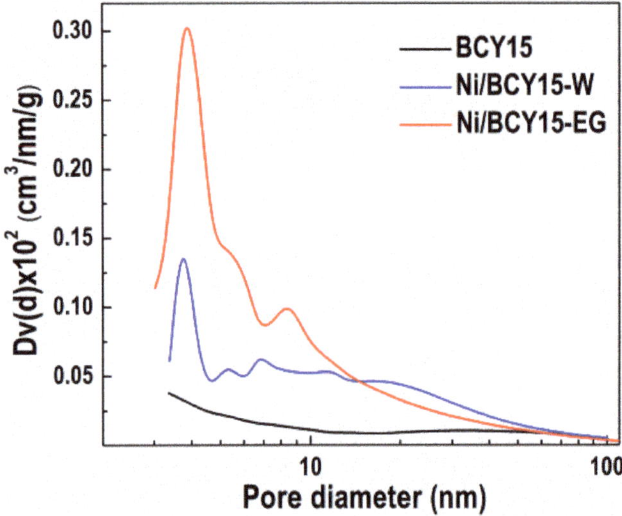

Figure 1. Pore size distribution of BCY15, Ni/BCY15-W, and Ni/BCY15-EG samples.

Constant C in the BET equation is a fundamental parameter in the analysis of adsorption isotherms, which provides information about the magnitude of the adsorbent–adsorbate interaction force at the surface of the solids. The C constant values presented in Table 2 follow the range of 82–97. In general, Type II provides C values between 20 and 200 [45,46]. Changes in surface polarity may reflect the interaction of a quadrupole N$_2$ molecule with a surface, leading to changes of the C parameter value. Based on the above-mentioned statements, the difference between BCY15 and Ni/BCY15-W values of C indicates an increase of the polarity due to the abundance of OH groups, which is a result of the interaction between the hydrophilic BCY15 and deionized water, leading to a partial decomposition of the initial BCY15 structure, as shown by XRD. A lower C value of Ni/BCY15-EG in comparison with Ni/BCY15-W presumes a decrease in surface polarity due to the effect of ethylene glycol medium and reflection to preserve the BCY15 structure. It is worth mentioning that the latter C value is also lower than that of the bare BCY15. The reduced surface polarity at Ni/BCY15-EG is consistent with more uniform coating of the BCY15 surface with metallic Ni particles and negligible decomposed BCY15 structure relative to Ni/BCY15-W. This fact favors the EG route synthesis. In addition, a larger amount of mesopores (higher S_{BET}) in the Ni/BCY15-EG cermet facilitates the transport of reagent molecules to and from active sites.

3.3. Structural and Morphological Characterization

The SEM technique was used to characterize the surface microstructure of deposited Ni^0 on the surface of the BCY15 matrix for the same magnification. Image analysis shows that a different morphology was observed for as-prepared Ni/BCY15-W (Figure 2a) and Ni/BCY15-EG (Figure 2b) cermets.

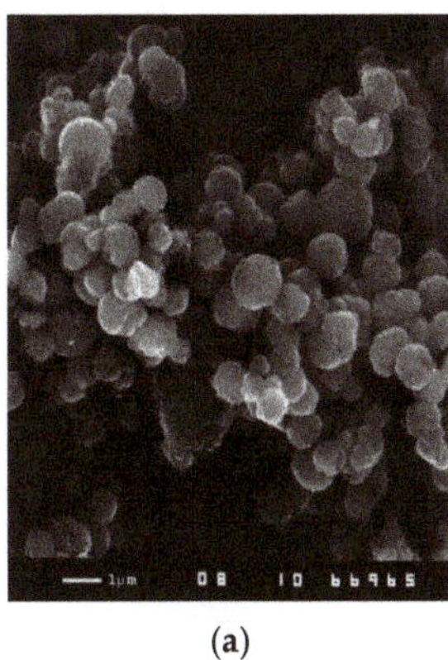
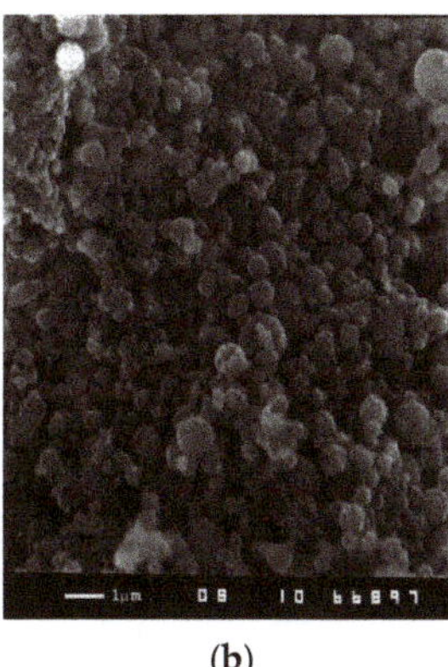

(a)   (b)

Figure 2. Micrographs of (**a**) Ni/BCY15-W and (**b**) Ni/BCY15-EG samples at 10,000× magnification.

The Ni^0 particles in the Ni/BCY15-W cermet are characterized by different size (Figure 2a). The presence of agglomerated Ni particles forming bigger and smaller agglomerates during the wet-reduction synthesis in aqueous medium is clearly observed. Metallic nickel is unevenly distributed on the BCY15 surface. The possibility that part of the surface Ni^0 particles will be covered with $BaCO_3$ phase, formed after the partial destruction of the BCY15 perovskite structure, cannot be excluded either, as is shown by XRD (Table 2). XRD evidence further indicates (Table 2) that after partial destruction of the BCY15 perovskite structure, the surface Ni^0 particles may be covered by $BaCO_3$ phase to a certain extent.

A SEM image of the Ni/BCY15-EG cermet surface reveals that the microstructure seems homogeneous, exposing spherical particles of similar size and no sign of agglomeration (Figure 2b). Undoubtedly, ethylene glycol anhydrous synthesis provides uniformly distributed Ni^0 particles of smaller particle size.

It is well-known that high metal loadings affect metal dispersion on the support. However, the use of ethylene glycol as a protective solvent saves not only the BCY15 structure but also, owing to its properties as a reducing and dispersive agent, leads to a finely divided metal Ni phase of smaller particle sizes.

3.4. Electron Paramagnetic Resonance

EPR spectroscopy characterized by high selectivity and sensitivity (10^{-11}–10^{-12} mol/L) was used to gain further information about the bulk nature of both the bare BCY15 matrix and active metallic Ni phase to find detailed explanations for differences in the electrochemical activity of Ni/BCY15-W and Ni/BCY15-EG anodes. So far, EPR analysis of the yttrium-doped barium cerate ($BaCe_{0.85}Y_{0.15}O_{2.925}$; BCY15) structure is still lacking in the current literature.

Several signals have been registered in the EPR spectrum of bare BCY15 (Figure 3a). One of them, with $g_{\perp} = 1.9436$ and $g_{||} = 1.9321$, is associated with the presence of Ce^{3+} sites, narrow lines marked with a circle in Figure 3 due to O^{2-} species on the cerium surface ($Ce^{4+} - O^{2-}$), and a signal with $g = 2.2866$, which could be ascribed to Ce^{3+} in a distortion polyhedron of defect associations {Ce$_{Ce}$'VÖ} [47]. The g-factor values have small deviations compared to the typical EPR signals of Ce^{3+} ion and oxygen vacancies reported in the literature [38,48,49], which can be attributed to the incorporated Y^{3+} ions which partially replaced Ce^{4+} host cations into the BCY15 structure by creation of oxygen anion vacancies and Ce^{3+} sites, in agreement with a study of Y-doped CeO$_2$ [50].

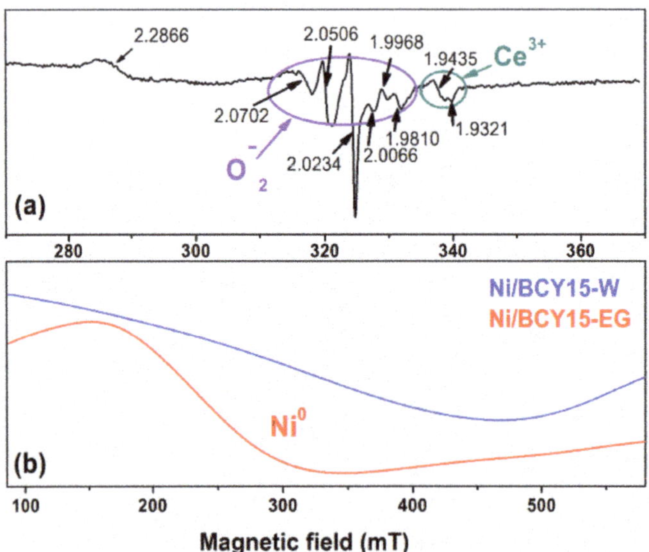

Figure 3. EPR spectra recorded at 123 K: (**a**) bare BCY15 and (**b**) Ni/BCY15-W and Ni/BCY15-EG cermets.

After metallic Ni deposition, broad lines were only detected in the EPR spectra of as-prepared Ni/BCY15-W and Ni/BCY15-EG cermets (Figure 3b). It is good to mention that unlike a simple EPR signal of ceria reduced in H$_2$ at 500 °C, hydrazine-treated ceria displays a complicated EPR spectrum with signals at g-values of 1.979, 1.998, and 2.013 [38]. However, the high amount of nickel (32 wt.% metallic Ni) makes it impossible to record signals corresponding to paramagnetic Ni^0 and Ni^{2+}, as well as Ce^{3+} ions. However, because of the broad EPR spectrum of freshly prepared Ni/BCY15 cermets, according to literature data, the Ni^{2+} ions may originate from probable superficial reoxidation of the metallic Ni particles, in part after the hydrazine reduction during subsequent washing and drying stages [29]. It has been established that the interaction between metallic nickel phase and ceria support in hydrazine Ni/CeO$_2$ catalysts highly preserves the nickel from partial oxidation in air [29]. The EPR data show that the Ni/BCY15-EG spectrum is better resolved. This narrowing of the signal supposes that the interaction between hydrazine-originating metallic Ni and cerium from the BCY15 matrix in the anhydrous ethylene glycol environment is stronger than in aqueous medium.

Apparently, the magnitude of the nickel–cerium interaction would be an additional factor for better electrochemical performance of the Ni/BCY15-EG anode, combined with smaller and more uniform size of better-distributed metallic Ni particles.

3.5. X-ray Photoelectron Spectroscopy

Surface structure properties of Ni/BCY15-W and Ni/BCY15-EG cermets are an important factor that controls the performance of these materials as anodes. X-ray photoelectron spectroscopy is a powerful and sensitive probe for surface examination and ensures reliable identification of the surface composition, oxidation states, and the relative dispersion of the components. Determination of the nickel chemical state on the BCY15 surface in as-prepared Ni/BCY15-W and Ni/BCY15-EG cermets is important because nickel is the key to adsorption, dissociation, and oxidation of the hydrogen, and an electronic conductor that provides the electronic conductivity of the anode.

The main Ni2p$_{3/2}$ peaks with asymmetry toward a lower binging energy (BE), accompanied by shake-up satellite lines, were registered at 856.0 eV in the spectra of both Ni/BCY cermets (Figure 4, Table 4). The position of the peaks clearly indicates the presence of Ni^{2+} ions on the surface, which can be attributed to Ni(OH)$_2$ (855.3–856.6 eV) [51–53]. It should be noted that the XRD data show only the presence of Ni^0, while the Ni^{2+} oxidation state originates from surface oxidation of metallic Ni after the hydrazine reduction during subsequent water washing and air-drying stages, as already mentioned [29].

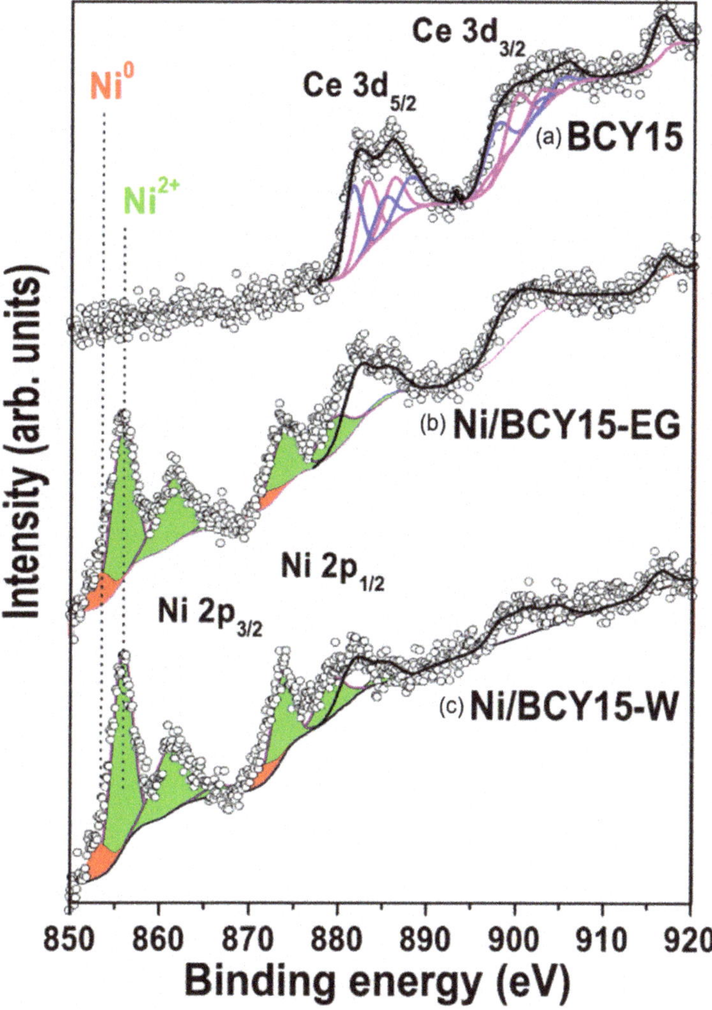

Figure 4. Ni2p and Ce3d photoelectron regions of the studied samples: (**a**) BCY15, (**b**) Ni/BCY15-EG, and (**c**) Ni/BCY15-W. Ni^0 and Ni^{2+} species are colored in red and green, respectively. Black contour lines outline the total Ce3d signals from the studied samples. Contributions of the Ce^{3+} and Ce^{4+} oxidation states in spectrum (**a**) are marked in purple and blue, respectively.

Table 4. Binding energies of the peaks of the BCY15 surface in as-prepared Ni/BCY15-W and Ni/BCY15-EG cermets, referred to as Cls, with BE = 285.0 eV.

Sample	Binding Energy (eV)				
	Ni2p		Ce3d	O1s	
	Ni^0	Ni^{2+}	Ce^{4+} Satellite	I	II
BCY15	–	–	916.1	528.8	531.5
Ni/BCY15-W	853.4	856.0	916.3	529.4	531.7
Ni/BCY15-EG	853.4	856.0	916.7	529.4	531.5

Curve fitting of the Ni2p level shows that the low-energy peak registered at 853.4 eV (Table 4) is present in the spectra of both samples (Figure 4). According to literature data, the Ni $2p_{3/2}$ spectrum of pure nickel metal is registered in the range of 852.7–853 eV [51–54]. Following the above results, the BE value corresponds to the presence of Ni^0 on the surface. A small shift to higher binding energies is a result of the influence of the nearest neighbors around the metallic nickel particles [53]. Thus, this shift can be attributed to the established interaction of the metallic Ni particles with Ce^{3+} sites [50,55]. The Ce^{3+} sites are additionally formed in the structure of the BCY15 support by thin surface layer reduction of Ce^{4+} during hydrazine wet-reduction synthesis of both as-prepared Ni/BCY15-W and Ni/BCY15-EG cermets, similar to the studies of ceria reduction by the hydrazine reduction technique [29,38]. Contributions of metallic Ni and the Ni^{2+}/Ni^0 ratio are presented in Table 4. The data reveal that a larger amount of Ni^0 is present on the surface of the Ni cermet synthesized in ethylene glycol anhydrous medium (Ni/BCY15-EG).

The claim stated above is based not only on the EPR data, but also on determining the Ce chemical state through analysis of the Ce3d photoelectron region. For this purpose, the surface analysis of bare BCY15 was performed as a reference point (Figure 4a). So far, no Ce3d spectrum of BCY15 has been reported in the literature.

Reference data using appropriate standards for the various chemical states of Ce [56,57] indicate that the BCY15 spectrum can be fitted with the characteristic spectra of the Ce^{4+} and Ce^{3+} oxidation states. Their peak contributions are marked in blue and purple, respectively. Deconvoluted XPS spectra contain many satellite peaks, which can be separated in two multiplet groups with spin-orbital splitting of 18.4 and 18.7 eV for Ce^{4+} and Ce^{3+}, respectively. According to the literature [56], the origin of Ce^{4+} and Ce^{3+} is due to the final 4f occupied state connected with O2p-Ce4f charge transfer during photoemission. The highest binding energy peak at 916.1 eV is characteristic of the Ce^{4+} oxidation state without any contribution by Ce^{3+} ions. The detection of this peak in the spectrum unambiguously indicates the presence of the Ce^{4+} oxidation state in the Ce-containing complex compounds. Wang and Meng [57] and Shyu et al. [58] have proposed a methodology for determining the relative Ce^{4+} concentration compared with that of Ce^{3+} by using the relative area of the highest binding-energy satellite peak to the total Ce3d area. Results of the studied samples are presented in Table 5. As can be seen, the relative concentrations of Ce^{4+} and Ce^{3+} in BCY15 and Ni/BCY15-EG are similar, 62% and 65% respectively, and increase significantly to 84% for Ni/BCY15-W. Therefore, analysis of the Ce3d spectra by deconvolution shows coexistence of the Ce^{4+} and Ce^{3+} oxidation states, in agreement with literature data [59–63]. The presence of Ce^{3+} sites in BCY15 was also found by EPR, caused by Y^{3+} ion-doping in the matrix.

Table 5. XPS data on the BCY15 surface of as-prepared Ni/BCY15-W and Ni/BCY15-EG cermets.

Sample	Ni^{2+}/Ni^0 Ratio	Contribution, %		Ni/(Ba+Ce+Y)	Ba/(Ce+Y)
		Ni^0	Ce^{4+}		
BCY15	–	–	62	–	0.63
Ni/BCY15-W	10.4	8.8	84	1.82	1.52
Ni/BCY15-EG	9.9	9.2	65	1.94	0.35

The position of the Ce^{4+} satellite registered at 916.1 eV in BCY15 (Figure 4, Table 4) is shifted toward higher BEs after metallic Ni deposition at 916.3 and 916.7 eV for Ni/BCY15-W and Ni/BCY15-EG, respectively. The shift of the Ce3d line is larger than ± 0.1 eV experimental accuracy and it can be explained by different electron densities (different bond lengths) around the cerium atoms after introduction of metallic Ni particles in the BCY15 matrix. Bearing in mind the above-mentioned results of Ni^0 and Ni^{2+} BEs, namely that metallic Ni is situated at a BE of 853.4 eV (Table 4) and the Ni^{2+} oxidation state arose from partial surface reoxidation of metallic Ni particles to $Ni(OH)_2$, the observed Ce^{4+} satellite shift could be indicative of the interaction strength between metallic Ni and Ce^{3+} species. The satellite position is higher by 0.4 eV in the Ni/BCY15-EG spectrum compared with that of Ni/BCY15-W, and by 0.6 eV relative to BCY15. It is known that the chemical shift of the XPS peak (ΔBE) serves to evaluate the degree of interaction between the individual components in the studied system. Due to the higher BE value of 916.7 eV of the Ni/BCY15-EG cermet, the kinetic energy is lower ($E_{kinetic} = h\nu - E_{binding}$). This affords a possibility to claim that upon reduction, the formed Ni^0–Ce^{3+} bond is stronger after synthesis using ethylene glycol as a medium and reducing agent. The observation is also in agreement with better-resolved EPR spectra (Figure 3).

Considering the data in Table 5, it follows that the relative concentration of the Ce^{3+} ions with respect to all Ce ions is 38% in the surface layers of the BCY15 sample. It decreases slightly to 35% with Ni/BCY15-EG and reaches 16% in the BCY15/Ni-W sample, showing that the Ce^{3+} sites are twice as many on the surface of BCY15/Ni-EG cermet after ethylene glycol anhydrous application, relative to aqueous medium in N_2H_4-reduction synthesis. Since Ce^{3+} ions in hydrazine-treated ceria are stable only in the presence of N_2H_4 and pure CeO_2 is restored upon drying of the reduced powder in air, the data on the relative Ce^{3+} concentration in both Ni/BCY15 samples points to a better-stabilized Ce^{3+} oxidation state on the BCY15/Ni-EG cermet surface due to the bond with metallic Ni.

The O1s spectrum consists of two peaks, namely a less intensive low-energy peak at 528.8 eV and a most intense higher energy peak at 531.5 eV, henceforth noted for convenience as peak I and peak II (Figure 5, Table 4). References for analysis of the O1s level of BCY15 are lacking in the current literature.

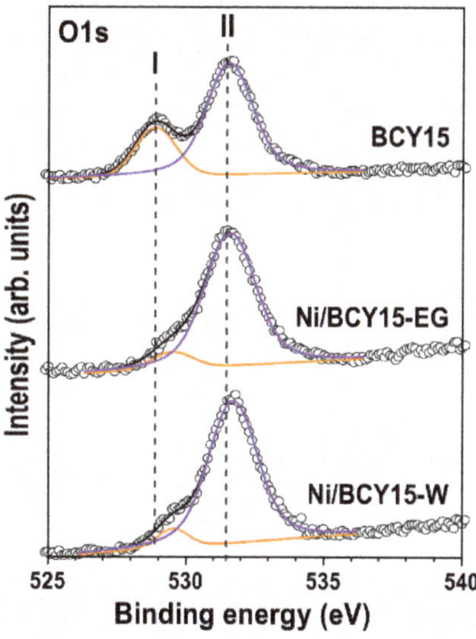

Figure 5. O1s level of the BCY15 surface in as-prepared Ni/BCY15-W and Ni/BCY15-EG cermets.

It is known that a low binding energy peak at 528.7–529.2 eV in the O1s spectrum of pure CeO_2 originates from lattice oxygen atoms (O^{2-}), and a high binding energy peak at 531.5–532.0 eV is formed from hydroxyl groups [59,60,63,64]. On this basis, peak I can be attributed to oxygen atoms in the lattice of BCY15 ($BaCe_{0.85}Y_{0.15}O_{2.925}$), whereas peak II can be recognized as due to physisorbed hydroxyl groups coming from atmospheric moisture uptake as a result of yttrium-doped barium cerate matrix affinity to H_2O [40].

A considerably broader main O1s peak, being asymmetrical on the low binding energy side, characterizes the same spectra of as-prepared Ni/BCY15-W and Ni/BCY15-EG cermets. It is obvious that the low-energy peak I with BCY15 is transformed into a shoulder (Figure 5), as better outlined for Ni/BCY15-W. Peak I in the Ni/BCY15-W and Ni/BCY15-EG spectra can be associated with O^{2-} in Ce–O bonding. Compared with bare BCY15, there is a noticeable decrease in intensity of these peaks, which could be related to Ce^{4+} reduction. The same observation has been reported for the Ni/CeO_2 system [65]. Besides, as deconvolution results show, the peaks are shifted to higher BE at 529.4 eV for both Ni/BCY15 cermets (Figure 5, Table 4). These observations are because of two simultaneous processes: surface reduction of Ce(IV) to Ce(III) by loss of lattice oxygen and Ni interaction with cerium during the hydrazine reduction process. The aforementioned statement is supported by literature data on binding energies of oxygen ions in oxygen-deficient regions caused by oxygen vacancies (O_x^-) in a matrix of metal oxides registered at 529.9–531.1 eV [65–69]. The O1s peak for the reduced state of cerium, oxygen in the Ce_2O_3 lattice, is also registered at higher BE values, namely 529.8–530 eV [60,63]. Evidently, a lower peak intensity in combination with peak position in the spectrum of Ni/BCY15-EG cermet can be associated with a deeper reduction of cerium, and it is a further indication for a stronger Ni^0–Ce^{3+} interaction in ethylene glycol-assisted synthesis.

The spectra exhibit a second symmetrical high binding energy peak. This peak, denoted as peak II, is centered at the same BE of 531.5 eV for Ni/BCY15-EG and shifted to 531.7 eV for Ni/BCY15-W, i.e., more than an instrument accuracy of ± 0.1 eV. Both peak II values are more intense related to the BCY15 spectrum. They originate from two types of oxygen: OH groups from $Ni(OH)_2$ and physisorbed humidity, and oxygen in CO_3 groups due to partial decomposition of the BCY structure to $BaCO_3$, as evidenced through the amount of $BaCO_3$ phase at Ni/BCY15-W of 60 wt.% calculated by the semi-quantitative XRD analysis (Table 2). The O1s of pure $BaCO_3$ has been registered at 531.1 eV [70]. Therefore, it is not possible to determine the contribution of each type of oxygen to the position of peak II.

Attention was also paid to another important factor—relative dispersion of the components on the surface. An estimation of nickel atoms' dispersion was performed by calculation of the Ni/(Ba+Ce+Y) ratio and is summarized in Table 5. Thus, Ni dispersion on the BCY15 surface is higher for the Ni/BCY15-EG cermet, as already shown by PSD (Figure 1), while the SEM image (Figure 2b) indicates that the BCY15 surface is uniformly covered by deposited smaller metallic Ni particles.

Another intention was to evaluate the BCY15 surface after the synthesis of both Ni cermets. A Ba/(Ce+Y) ratio was calculated (Table 5) to consider partial decomposition of the BCY matrix during synthesis and liberation of barium. A value of 1.52 for Ni/BCY15-W relative to 0.63 for bare BCY15 proved that there was more of a presence of barium on the surface due to the formation of more $BaCO_3$ using aqueous medium. However, Ni/BCY15-EG manifested a lower value (0.35) by comparison with BCY15, which undoubtedly confirms that the BCY15 surface is more uniformly covered with nickel atoms, resulting in shielding of the signal from the other chemical elements, thus being indirect proof for better deposition.

In summary, the surface science study showed that nickel dispersion is increased by using the ethylene glycol environment and that metallic Ni is better stabilized over a partially reduced cerium surface due to a stronger interaction. As a result, further evaluation proceeded with the BCY15/Ni-EG anode.

3.6. Electrochemical Performance

The electrical properties of the anodes prepared with EG were evaluated using comparative electrochemical impedance analysis in respect of symmetrical half-cells prepared from Ni/BCY-Mech. Each half-cell was subjected to six redox cycles following the standardized procedure presented in Table 1. Samples of Ni/BCY15-W were excluded from the tests since their composition does not correspond to the required one due to the strong decomposition of the BCY15 structure in accordance with XRD data.

As already mentioned, during sintering in air, which is an obligatory stage in the technological cycle that cannot be avoided, the initial metallic Ni used for synthesis of Ni/BCY15-EG oxidizes partially or fully to NiO. In principle, the process is associated with volume expansion and should cause stresses on the electrolyte matrix and the appearance of cracks or delamination [71]. However, the application of the EG technological approach ensures production of well-dispersed nanosized metallic nickel particles on the BCY15 matrix and provides a strong metal–partially reduced cerium interaction through Ce^{3+} sites, determining the electrochemical activity of nickel. Both factors suppose improved electrochemical characteristics and increased stability, which is a prerequisite for decreased anode degradation under operating conditions. In addition, the nanometric Ni phase would ensure easier spills during oxidation, and instead of forming cracks, this will increase the voids, thus improving the tolerance with respect to Ni agglomeration and redox cycling [28].

In principle, redox cycling can be considered an accelerated stress test regarding Ni cermet degradation, since it leads to similar degradation mechanisms, namely Ni coarsening and Ni depletion, which decrease the triple-phase boundary points and thus the catalytic activity and conductivity.

Figure 6 displays typical impedance diagrams of reduced and oxidized states, which are used for evaluation of the electrolyte resistance and the polarization resistance of the Ni/BCY15-Mech anode, presented as area-specific resistance (ASR). A summary of the results from the impedance measurements is provided in the form of Arrhenius plots (Figure 7).

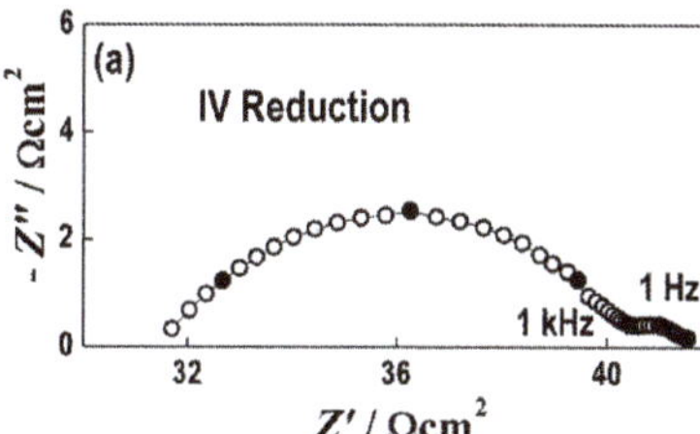

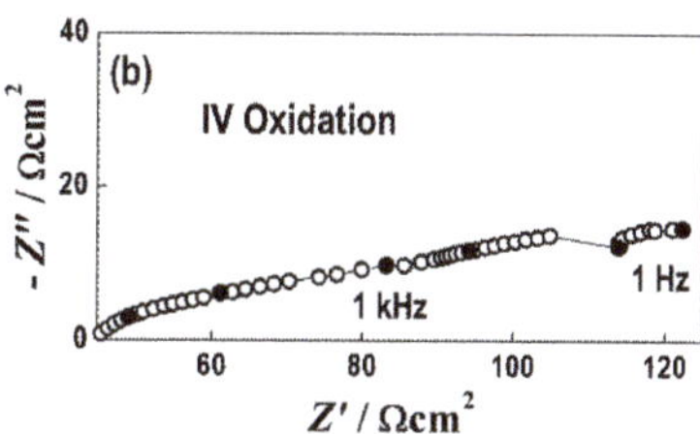

Figure 6. Impedance diagrams of reduced (**a**) and oxidized (**b**) states (4th cycle) of symmetrical half-cells with the Ni/BCY-Mech anode.

The results of the electrolyte resistance show that during reoxidation, it increases in a similar way for both samples, Ni/BCY15-EG and Ni/BCY15-Mech (Figure 7a). The different values of the electrolyte resistance, 88% and 90% for Mech and EG samples respectively, are probably due to the different porosity of the electrolytes.

Bearing in mind the electrode polarization resistance, however, a drastic increase with the Ni/BCY15-Mech sample was observed. Although the initial resistance was lower for the pristine sample, on the 6th cycle, it was already two times higher than that of the Ni/BCY15-EG sample, i.e., a very fast degradation rate was observed (Figure 7b). This result confirms the higher stability of the microstructure obtained by the EG method in relation to Ni coarsening and migration. The smaller size of the Ni particles and thus bigger pore sizes also ensure higher mechanical stability.

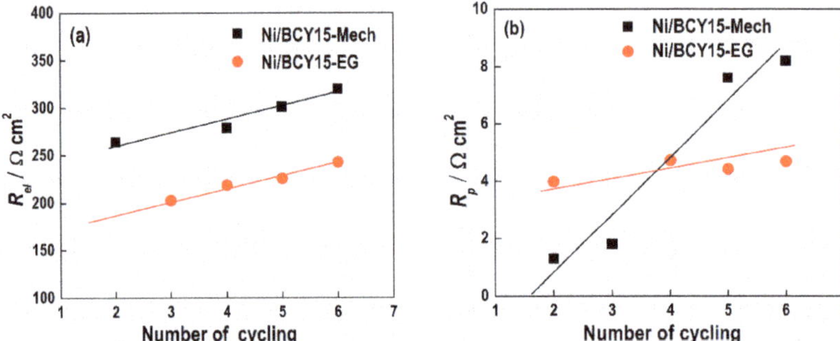

Figure 7. Dependences of electrolyte resistance (**a**) and anode polarization resistance (**b**) on the number of cycles for Ni/BCY15-EG (●) and Ni/BCY15-Mech samples (■).

Thus, the observed electrochemical behavior of Ni/BCY15-EG anodes confirms that the method of wet-reduction by the hydrazine reducing agent guarantees better dispersion of the Ni metal particles in the BCY15 matrix, minimizing the microstructural changes during its reoxidation, and accordingly, strongly influences the degradation processes and whole cell performance.

4. Discussion and Summary

Analysis of the nickel state in the bulk and on the surface of the BCY15 matrix provided an opportunity to discover the origin of the differences between two cermets (Ni/BCY15-W and Ni/BCY15-EG) prepared via aqueous and ethylene glycol environments.

XRD analysis showed that the hydrazine wet-reduction methodology ensured complete reduction of the formed Ni^{2+} hydrazine complex ($[Ni(N_2H_4)_x]Cl_y$) to Ni^0 particles. An advantage of this route is the easy preparation of fine nickel powders at a low reaction temperature (80–95 °C) and a simple procedure compared with the classical synthesis method, which involves a solid-state reaction between NiO powder and BCY electrolytes at high temperatures of 1100–1400 °C, followed by hydrogen reduction at a temperature of 700–800 °C.

Our findings showed that the hydrazine wet-reduction approach to synthesize Ni/BCY15-EG cermet in ethylene glycol medium provides a metallic Ni phase in the pSOFC anode that is more stable during reoxidation compared to Ni/BCY15-Mech cermet prepared by the commercial mechanical mixing procedure. This fact clearly outlines another advantage of this method for metallic Ni incorporation into the electrolyte matrix. Ethylene glycol, used not only as anhydrous medium but also as an additional reducing agent, provided nano-scaled Ni^0 particles of narrow size distribution, a higher specific surface, and improved dispersion on Ni/BCY15-EG cermet (XRD, N_2 sorption, SEM, XPS). These features facilitated fuel molecules' access to the nickel sites. As a result, more nickel in the Ni/BCY15-EG anode responds faster to the hydrogen, and thus the percolation process is more rapidly achieved, and full connectivity among the metallic Ni particles during creation of the conductive network in the electrode is attained [30,35].

Another determining factor for better electrochemical performance of the Ni/BCY15-EG anode if compared with Ni/BCY15-W (XPS, EPR) is the interaction magnitude between nickel and cerium in the as-prepared cermet. The suggestion is that the nickel–cerium interaction has arisen during decomposition of the Ni^{2+} hydrazine complex, followed by transformation of the Ni^{2+} ions into metallic Ni. It is useful to note that a mutual influence of nickel and ceria exists during the N_2H_4-reduction process [29]. A stronger interaction in as-prepared Ni/BCY15-EG definitely takes place in the bulk BCY15 structure (EPR), not only on the surface (XPS). It is assumed that Ni^0 stabilization occurs through bonding to Ce^{3+} sites formed by lattice oxygen transfer. The interaction affinity is kept after reoxidation during electrode preparation, and it is the reason for a completely formed conductive Ni^0 network characterized by low resistance after 24 h exposure of the cell to ambient conditions. Stabilization of the metallic nickel state on the surface determines its resistance to oxidation and offers an increased number of interacting metal atoms, leading to better electronic conductivity. Consequently, this is a precondition for better durability of the Ni/BCY15-EG anode.

Our finding is in agreement with a fundamental study of the Ni/CeO_2 system [62]. The authors have detected uniform distribution of Ni particles over ceria surface, and this has been explained as a result of surface defects (oxygen vacancies) being the nucleation sites for nickel. It is stated that the Ce^{3+} sites do not only exist in the regions near the metal, i.e., at the metal–ceria interfaces [50]. In a theoretical study with density functional theory (DFT) calculations applied for estimation of Ni deposition on stoichiometric (CeO_2) and reduced (Ce_2O_3) cerium oxide surfaces, Liu et al. have claimed that the nickel oxidation state is 2+ on CeO_2, and it remains metallic over reduced Ce_2O_3 [55]. The presented model surface shows that Ni^0 interacts with Ce^{3+} sites. In addition, Lustemberg et al. have shown a change occurring under reduction conditions: $Ni^{2+}-CeO_2 \rightarrow Ni^0-CeO_{2-x}$. Upon increasing Ce^{4+} reduction by oxygen removal, the oxidation state of nickel was transformed from +2 to 0 [72]. The authors also revealed that the most stable Ni^0 particles are located on O-bridge sites, whereas O-top and O-hollow sites are less stable. Hence, Ni^0 is stabilized by bonding to Ce^{3+} sites that arise during hydrazine reduction.

By the same analogy, uniform distribution of nickel nanoparticles in Ni/BCY15-EG cermet is suggested by an increase of the oxygen vacancies on the BCY15 surface and the accompanying stronger interaction between metallic Ni and Ce^{3+} sites in the BCY15 matrix due to deeper reduction by ethylene glycol. Furthermore, the interaction between cerium and Ni could also change the electronic properties of nickel [62], thus affecting its electrochemical reactivity.

The application of the wet-chemical synthesis route in an anhydrous ethylene glycol environment provided not only anode ceramic matrix structure preservation, but also a new microstructure of the Ni/BCY15 pSOFC anode more tolerant to redox cycling, i.e., the microstructure is more stable to changes causing degradation. The advantage of the hydrazine wet-reduction methodology for improvement of the electrochemical behavior of the Ni/BCY15 pSOFC anode is indubitable.

Author Contributions: Conceptualization, D.N. and M.G.; methodology, D.N. and M.G.; data analysis and interpretation, D.N.; electrochemical investigation, G.R. and E.M.; electrochemical data analysis and interpretation, G.R. and D.V.; XPS investigation, K.L.K.; EPR investigation, Y.K.; writing—original draft preparation, D.N.; writing—review and editing, M.G., G.R., D.V., K.L.K. and Y.K.; visualization, D.N.; supervision, D.V.; project administration, M.G. and G.R.; funding acquisition, D.V. All authors have read and agreed to the published version of the manuscript.

Funding: The research leading to these results received funding from the Bulgarian Ministry of Education and Science under the National Research Program "Low-Carbon Energy for the Transport and Households, EPLUS", approved by DCM No. 577/17.08.2018, and the National Roadmap for Re-search Infrastructure 2017–2023 "Energy storage and hydrogen energetics (ESHER)", approved by DCM No. 354/29.08.2017, which provided the equipment used.

Data Availability Statement: The data presented in this study are available in the article.

Acknowledgments: This scientific direction of development was inspired by the late Zdravko Stoynov.

Conflicts of Interest: The authors declare no conflict of interest.

References

1. Marchenko, O.V.; Solomin, S.V. The future energy: Hydrogen versus electricity. *Int. J. Hydrogen Energ.* **2015**, *40*, 3801–3805. [CrossRef]
2. Staffell, I.; Scamman, D.; Abad, A.V.; Balcombe, P.; Dodds, P.E.; Ekins, P.; Shah, N.; Ward, K.R. The role of hydrogen and fuel cells in the global energy system. *Energy Environ. Sci.* **2019**, *12*, 463–491. [CrossRef]
3. Stambouli, A.; Traversa, E. Solid oxide fuel cells (SOFCs): A review of an environmentally clean and efficient source of energy. *Renew. Sust. Energ. Rev.* **2002**, *6*, 433–455. [CrossRef]

4. Fabbri, E.; Pergolesi, D.; Traversa, E. Materials challenges toward proton-conducting oxide fuel cells: A critical review. *E. Chem. Soc. Rev.* **2010**, *39*, 4355–4369. [CrossRef]
5. Tong, J.; Clark, D.; Bernau, L.; Subramaniyan, A.; O'Hayre, R. Proton-conducting yttrium-doped barium cerate ceramics synthesized by a cost-effective solid-state reactive sintering method. *Solid State Ionics* **2010**, *181*, 1486–1498. [CrossRef]
6. Sawant, P.; Varma, S.; Gonal, M.R.; Wani, B.N.; Prakash, D.; Bharadwaj, S.R. Effect of Ni Concentration on Phase Stability, Microstructure and Electrical Properties of $BaCe_{0.8}Y_{0.2}O_3$-Ni Cermet SOFC Anode and its application in proton conducting ITSOFC. *Electrochim. Acta* **2014**, *120*, 80–85. [CrossRef]
7. Feng, W.; Wu, W.; Jin, C.; Zhou, M.; Bian, W.; Tang, W.; Gomez, J.; Boardman, R.; Ding, D. Exploring the structural uniformity and integrity of protonic ceramic thin film electrolyte using wet powder spraying. *J. Power Sources Adv.* **2021**, *11*, 100067. [CrossRef]
8. Hussain, S.; Yangping, L. Review of solid oxide fuel cell materials: Cathode, anode, and electrolyte. *Energy Transit.* **2020**, *4*, 113–126. [CrossRef]
9. Cellule de Pile a Combustible Haute Temperature a Conduction Mixte Anionique et Protonique. Patent No. 0550696000, 7 March 2005.
10. Thorel, A.; Chesnaud, A.; Viviani, M.; Barbucci, A.; Presto, S.; Piccardo, P.; Ilhan, Z.; Vladikova, D.; Stoynov, Z. IDEAL-Cell, a high temperature innovative dual membrane fuel-cell. Chapter 3, Cell designs, processing and performance. In Proceedings of the Solid Oxide Fuel Cells 11 (SOFCXI), Vienna, Austria, 4–9 October 2009; Singhal, S.C., Yokokawa, H., Eds.; The Electrochemical Society, ECS Transactions: Pennington, NJ, USA, 2009; Volume 25, pp. 753–762.
11. Vladikova, D.; Stoynov, Z.; Raikova, G.; Thorel, A.; Chesnaud, A.; Abreu, J.; Viviani, M.; Barbucci, A.; Presto, S.; Carpanese, P. Impedance spectroscopy studies of dual membrane fuel cell. *Electrochim. Acta* **2011**, *56*, 7955–7962. [CrossRef]
12. Vladikova, D.; Stoynov, Z.; Raikova, G.; Krapchanska, M.; Thorel, A.; Chesnaud, A. Dual membrane fuel cell—Impedance approach for proof of concept. *Bulg. Chem. Commun.* **2012**, *44*, 364–370.
13. Thorel, A.S.; Abreu, J.; Ansar, S.A.; Barbucci, A.; Brylewski, T.; Chesnaud, A.; Ilhan, Z.; Piccardo, P.; Prazuch, J.; Presto, S.; et al. Proof of Concept for the Dual Membrane Cell: I. Fabrication and Electrochemical Testing of First Prototypes. *J. Electrochem. Soc.* **2013**, *160*, F360–F366. [CrossRef]
14. Masson, D.; Perrozzi, F.; Piccardo, P.; Viviani, M.; Pilot, C.; Stoynov, Z.; Vladikova, D.; Chesnaud, A.; Thorel, A. Shaping of a dual membrane SOFC and first electrochemical tests in a dedicated 3-chamber set-up. Chapter 7, Cell designs, fabrication, performance and durability. In Proceedings of the Solid Oxide Fuel Cells 14 (SOFC-XIV), Glasgow, Scotland, 26–31 July 2015; Singhal, S.C., Eguchi, K., Eds.; The Electrochemical Society, ECS Transactions: Pennington, NJ, USA, 2015; Volume 68, pp. 1969–1978.
15. Raikova, G.; Krapchanska, M.; Genov, I.; Caboche, G.; Combemale, L.; Thorel, A.; Chesnaud, A.; Vladikova, D.; Stoynov, Z. Impedance investigation of $BaCe_{0.85}Y_{0.15}O_{3-\delta}$ properties for hydrogen conductor in fuel cells. *Bulg. Chem. Commun.* **2012**, *44*, 389–394.
16. Vladikova, D.; Stoynov, Z.; Chesnaud, A.; Thorel, A.; Viviani, M.; Barbucci, A.; Raikova, G.; Carpanese, P.; Krapchanska, M.; Mladenova, E. Application of yttrium doped barium cerate for improvement of the dual membrane SOFC design. *Int. J. Hydrogen Energ.* **2014**, *39*, 21561–21568. [CrossRef]
17. Krezhov, K.; Vladikova, D.; Raikova, G.; Malakova, T.; Genov, I.; Nonova, T.; Svab, E.; Fabian, M. $BaCe_{0.85}Y_{0.15}O_{3-\Delta}$ based materials for solid oxide fuel cells: Room-temperature neutron diffraction study. Neutron and Heavy Ion Radiations. In Proceedings of the International Conference on Radiation and Applications in Various Fields of Research, Niš, Serbia, 23–27 May 2016; Ristić, G.S., Ed.; RAD Centre: Niš, Serbia, 2016; Volume 1, pp. 117–123.
18. Raikova, G.; Krezhov, K.; Genov, I.; Thorel, A.; Chesnaud, A.; Malakova, T.; Vladikova, D.; Stoynov, Z. Structural and electrochemical characterization of yttrium doped barium cerate $BaCe_{0.85}Y_{0.15}O_{3-\alpha}$ for applications in solid oxide fuel cells. *Bulg. Chem. Commun.* **2017**, *49*, 162–170.
19. Miller, H.A.; Bouzek, K.; Hnat, J.; Loos, S.; Bernaecker, C.; Weissgaerber, T.; Rontzsch, L.; Meier-Haack, J. Green hydrogen from anion exchange membrane water electrolysis: A review of recent developments in critical materials and operating conditions. *Sustain. Energ. Fuels* **2020**, *4*, 2114–2133. [CrossRef]
20. Safari, A.; Panda, R.K.; Janas, V.F. Ferroelectricity. Materials, characteristics, and applications. In *Key Engineering Materials Advanced Ceramic Materials*; Mostaghaci, H., Ed.; Trans Tech Publications, Ltd.: Bäch, Switzerland, 1996; Volumes 122–124, pp. 35–70.
21. Liu, Y.; Shao, Z.; Mori, T.; Jiang, S.P. Development of nickel based cermet anode materials in solid oxide fuel cells—Now and future. *MRE* **2021**, *1*, 100003. [CrossRef]
22. Caldes, M.T.; Kravchyk, K.V.; Benamira, M.; Besnard, N.; Gunes, V.; Bohnke, O.; Joubert, O. Metallic Nanoparticles and Proton Conductivity: Improving Proton Conductivity of $BaCe_{0.9}Y_{0.1}O_{3-\delta}$ Using a Catalytic Approach. *Chem. Mater.* **2012**, *24*, 4641–4646. [CrossRef]
23. Costa, R.; Grünbaum, N.; Berger, M.-H.; Dessemond, L.; Thorel, A. On the use of NiO as sintering additive for $BaCe_{0.9}Y_{0.1}O_{3-\alpha}$. *Solid State Ionics* **2009**, *180*, 891–895. [CrossRef]
24. Fua, Q.; Tietz, F.; Sebold, D.; Tao, S.; Irvine, J.T.S. An efficient ceramic-based anode for solid oxide fuel cells. *J. Power Sources* **2007**, *171*, 663–669. [CrossRef]
25. Essoumhi, A.; Taillades, G.; Taillades-Jacquin, M.; Jones, D.J.; Rozière, J. Synthesis and characterization of Ni-cermet/proton conducting thin film electrolyte symmetrical assemblies. *Solid State Ionics* **2008**, *179*, 2155–2159. [CrossRef]

26. Chesnaud, A.; Thorel, A.; Valy, J.; Vladikova, D.; Stoynov, Z.; Raikova, G. Optimisation of a BCY15/Ni composite as an anode for a dual membrane fuel cell. Preparation and DC resistivity measuremen. In Proceedings of the Internat. Workshop "Advances and Innovations in SOFCs 2. From Materials to the Systems", Katarino, Bulgaria, 11–16 September 2011; Vladikova, D., Stoynov, Z., Eds.; Acad. Evgeni Budevski Institute of Electrochemistry and Energy Systems: Sofia, Bulgaria; Bulgarian Academy of Sciences: Sofia, Bulgaria, 2011; pp. 45–52.

27. Raikova, G.; Caboche, G.; Combemale, L.; Thorel, A.; Chesnaud, A.; Abreu, J.; Vladikova, D.; Stoynov, Z. Optimization of the anode compartment of a dual membrane fuel cell. In Proceedings of the Internat. Workshop "Advances and Innovations in SOFCs 2. From Materials to the Systems", Katarino, Bulgaria, 11–16 September 2011; Vladikova, D., Stoynov, Z., Eds.; Acad. Evgeni Budevski Institute of Electrochemistry and Energy Systems: Sofia, Bulgaria; Bulgarian Academy of Sciences: Sofia, Bulgaria, 2011; pp. 37–44.

28. Young, J.L.; Molero, H.; Birss, V.I. The effect of pre-oxidation treatments on the oxidation tolerance of Ni-yttria-stabilized zirconia anodes in solid oxide fuel cells. *J. Power Sources* **2014**, *271*, 538–547. [CrossRef]

29. Wojcieszak, R.; Monteverdi, S.; Bettahar, M.M. Ni/CeO$_2$ catalysts prepared by aqueous hydrazine reduction. *Colloids Surf. A Physicochem. Eng. Asp.* **2008**, *317*, 116–122. [CrossRef]

30. Gabrovska, M.; Nikolova, D.; Mladenova, E.; Vladikova, D.; Rakovsky, S.; Stoynov, Z. Ni incorporation in pSOFC anode ceramic matrix: Part I. Wet chemical reduction in an aqueous medium. *Bulg. Chem. Commun.* **2017**, *49*, 171–178.

31. Goia, D.V. Preparation and formation mechanisms of uniform metallic particles in homogeneous solutions. *J. Mater. Chem.* **2004**, *14*, 451–458. [CrossRef]

32. Park, J.; Chae, E.; Kim, S.; Lee, J.; Kim, J.; Yoon, S.; Choi, J.-Y. Preparation of fine Ni powders from nickel hydrazine complex. *Mater. Chem. Phys.* **2006**, *97*, 371–378. [CrossRef]

33. Wang, D.P.; Sun, D.-B.; Yu, H.-Y.; Meng, H.-M. Morphology controllable synthesis of nickel nanopowders by chemical reduction process. *J. Cryst. Growth* **2008**, *310*, 1195–1201. [CrossRef]

34. Tanner, C.W.; Virkar, A.V. Instabiliiy of BaCeO$_3$ in H$_2$O-Containing Atmospheres. *J. Electrochem. Soc.* **1996**, *143*, 1386–1389. [CrossRef]

35. Gabrovska, M.; Nikolova, D.; Mladenova, E.; Vladikova, D.; Rakovsky, S.; Stoynov, Z. Ni incorporation in pSOFC anode ceramic matrix: Part II. Wet chemical reduction in an anhydrous medium. *Bulg. Chem. Commun.* **2018**, *50*, 119–126.

36. Kurihara, L.K.; Chow, G.M.; Schoen, P.E. Nanocrystalline metallic powders and films produced by the polyol method. *Nanostruct. Mater.* **1995**, *5*, 607–613. [CrossRef]

37. Huang, G.-Y.; Xu, S.-M.; Xu, G.; Li, L.-Y.; Zhang, L.-F. Preparation of fine nickel powders via reduction of nickel hydrazine complex precursors. *Trans. Nonferrous Met. Soc. China* **2009**, *19*, 389–393. [CrossRef]

38. Filotti, L.; Bensalem, A.; Bozon-Verduraz, F.; Shafeev, G.A.; Voronov, V.V. A comparative study of partial reduction of ceria via laser ablation in air and soft chemical route. *Appl. Surf. Sci.* **1997**, *109–110*, 249–252. [CrossRef]

39. Vladikova, D.E.; Stoynov, Z.B.; Wuillemin, Z.; Montinaro, D.; Piccardo, P.; Genova, I.; Rolland, M. Impedance studies of the reduction process in NiO-YSZ SOFC anodes. Chapter 7, Cell designs, fabrication, performance and durability. In Proceedings of the Solid Oxide Fuel Cells 14 (SOFC-XIV), Glasgow, Scotland, 26–31 July 2015; Singhal, S.C., Eguchi, K., Eds.; The Electrochemical Society, ECS Transactions: Pennington, NJ, USA, 2015; Volume 68, pp. 1161–1168.

40. Zakowsky, N.; Williamson, S.; Irwine, J.T.S. Elaboration of CO$_2$ tolerance limits of BaCe$_{0.9}$Y$_{0.1}$O$_{3-d}$ electrolytes for fuel cells and other applications. *Solid State Ionics* **2005**, *176*, 3019–3026. [CrossRef]

41. Groen, J.C.; Peffer, L.A.A.; Pérez-Ramırez, J. Pore size determination in modified micro- and mesoporous materials. Pitfalls and limitations in gas adsorption data analysis. *Microporous Mesoporous Mater.* **2003**, *60*, 1–17. [CrossRef]

42. Groen, J.C.; Pérez-Ramırez, J. Critical appraisal of mesopore characterization by adsorption analysis. *J. Appl. Catal. A Gen.* **2004**, *268*, 121–125. [CrossRef]

43. Rouquerol, F.; Rouquerol, J.; Sing, K.S.W.; Llewellyn, P.; Maurin, G. *Adsorption by Powders and Porous Solids Principles, Methodology and Applications*, 2nd ed.; Academic Press: Cambridge, MA, USA; Elsevier Ltd.: Amsterdam, The Netherlands, 2014.

44. Thommes, M.M.; Kaneko, K.; Niemark, A.V.; Olivier, J.P.; Rodriguez-Reinoso, F.; Rouquerol, J.; Sing, K.S.W. Physisorption of gases, with special reference to the evaluation of surface area and pore size distribution (IUPAC Technical Report). *Pure Appl. Chem.* **2015**, *87*, 1051–1069. [CrossRef]

45. Lecloux, A.; Pirard, J.P. The Importance of Standard Isotherms in the Analysis of Adsorption Isotherms for Determining the Porous Texture of Solids. *J. Colloid Interface Sci.* **1979**, *70*, 265–281. [CrossRef]

46. Serwicka, E.M. Surface area and porosity, X-ray diffraction and chemical analyses. *Catal. Today* **2000**, *56*, 335–346. [CrossRef]

47. Lia, L.; Lin, X. Solid solubility and transport properties of Ce$_{1-x}$Nd$_x$O$_{2-d}$ nanocrystalline solid solutions by a sol-gel route. *J. Mater. Res.* **2001**, *16*, 3207–3213. [CrossRef]

48. Bensalem, A.; Shafeev, G.; Bozon-Verduraz, F. Application of electroless procedures to the preparation of palladium catalysts. *Catal. Lett.* **1993**, *18*, 165–171. [CrossRef]

49. Abi-Aad, E.; Bennani, A.; Bonnelle, J.-P.; Aboukaïs, A. Transition-metal ion dimers formed in CeO$_2$: An EPR study. *J. Chem. Soc. Faraday Trans.* **1995**, *91*, 99–104. [CrossRef]

50. Wang, J.B.; Tai, Y.-L.; Dow, W.-P.; Huang, T.-J. Study of ceria-supported nickel catalyst and effect of yttria doping on carbon dioxide reforming of methane. *Appl. Catal. A Gen.* **2001**, *218*, 69–79. [CrossRef]

51. Kim, K.S.; Winograd, N. X-ray photoelectron spectroscopic studies of nickel-oxygen surfaces using oxygen and argon ion-bombardment. *Surf. Sci.* **1974**, *43*, 625–643. [CrossRef]
52. Löchel, B.P.; Strehblow, H.H. Breakdown of Passivity of Nickel by Fluoride: II. Surface Analytical Studies. *J. Electrochem. Soc.* **1884**, *131*, 713–723. [CrossRef]
53. Grosvenor, A.P.; Biesinger, M.C.; Smart, R.S.C.; McIntyre, N.S. New interpretations of XPS spectra of nickel metal and oxides. *Surf. Sci.* **2006**, *600*, 1771–1779. [CrossRef]
54. Carley, A.F.; Jackson, S.D.; O'Shea, J.N.; Roberts, M.W. The formation and characterisation of Ni^{3+}—An X-ray photoelectron spectroscopic investigation of potassium-doped Ni(110)–O. *Surf. Sci.* **1999**, *440*, L868–L874. [CrossRef]
55. Liu, Z.; Grinter, D.C.; Lustemberg, P.G.; Nguyen-Phan, T.-D.; Zhou, Y.; Luo, S.; Waluyo, I.; Crumlin, E.J.; Stacchiola, J.D.; Zhou, J.; et al. Dry Reforming of Methane on a Highly-Active $Ni-CeO_2$ Catalyst: Effects of Metal-Support Interactions on C–H Bond Breaking. *Angew. Chem. Int. Ed.* **2016**, *55*, 7455–7459. [CrossRef]
56. Beche, E.; Charvin, P.; Perarnau, D.; Abanades, S.; Flamant, G. Ce 3D XPS investigation of cerium oxides and mixed cerium oxide $(Ce_xTi_yO_z)$. *Surf. Interface Anal.* **2008**, *40*, 264–267. [CrossRef]
57. Wang, L.; Meng, F. Oxygen vacancy and Ce^{3+} ion dependent magnetism of monocrystal CeO_2 nanopoles synthesized by a facile hydrothermal method. *Mater. Res. Bull.* **2013**, *48*, 3492–3498. [CrossRef]
58. Shyu, J.Z.; Otto, K.; Watkins, W.L.H.; Graham, G.W. Characterization of Pd/γ-alumina catalysts containing ceria. *J. Catal.* **1988**, *114*, 23–33. [CrossRef]
59. Ingo, G.M.; Paparazzo, E.; Bagnarelli, O.; Zacchetti, N. XPS studies on cerium, zirconium and yttrium valence states in plasma-sprayed coatings. *Surf. Interface Anal.* **1990**, *16*, 515–519. [CrossRef]
60. Laachir, A.; Perrichon, V.; Badri, A.; Lamotte, J.; Catherine, E.; Lavalley, J.C.; El Fallah, J.; Hilaire, L.; Le Normand, F.; Quéméré, E.; et al. Reduction of CeO_2 by hydrogen. Magnetic susceptibility and Fourier-transform infrared, ultraviolet and X-ray photoelectron spectroscopy measurements. *J. Chem. Soc. Faraday Trans.* **1991**, *87*, 1601–1609. [CrossRef]
61. Deshpande, S.; Patil, S.; Kuchibhatla, S.V.N.T.; Seal, S. Size dependency variation in lattice parameter and valency states in nanocrystalline cerium oxide. *Appl. Phys. Lett.* **2005**, *87*, 133113. [CrossRef]
62. Zhou, Y.; Perket, J.M.; Crooks, A.B.; Zhou, J. Effect of Ceria Support on the Structure of Ni Nanoparticles. *J. Phys. Chem. Lett.* **2010**, *1*, 1447–1453. [CrossRef]
63. Praline, G.; Koel, B.E.; Hance, R.L.; Lee, H.-I.; White, J.M. X-ray photoelectron study of the reaction of oxygen with cerium. *J. Electron Spectrosc. Relat. Phenom.* **1980**, *21*, 17–30. [CrossRef]
64. Pereira, A.; Blouin, M.; Pillonnet, A.; Guay, D. Structure and valence properties of ceria films synthesized by laser ablation under reducing atmosphere. *Mater. Res. Express* **2014**, *1*, 015704. [CrossRef]
65. Caballero, A.; Holgado, J.P.; Gonzalez-delaCruz, V.M.; Habas, S.E.; Herranz, T.; Salmeron, M. In situ spectroscopic detection of SMSI effect in a Ni/CeO_2 system: Hydrogen-induced burial and dig out of metallic nickel. *Chem. Commun.* **2010**, *46*, 1097–1099. [CrossRef]
66. Wang, X.; Wang, X.; Di, Q.; Zhao, H.; Liang, B.; Yang, J. Mutual Effects of Fluorine Dopant and Oxygen Vacancies on Structural and Luminescence Characteristics of F Doped SnO_2 Nanoparticles. *J. Mater.* **2017**, *10*, 1398. [CrossRef] [PubMed]
67. Tu, Y.; Chen, S.; Li, X.; Gorbaciova, J.; Gillin, W.P.; Krause, S.; Briscoe, J. Control of oxygen vacancies in ZnO nanorods by annealing and their influence on ZnO/PEDOT: PSS diode behavior. *J. Mater. Chem. C* **2018**, *6*, 1815–1821. [CrossRef]
68. Saini, S.; Shah, J.; Kotnala, R.K.; Yadav, K.L. Nickel substituted oxygen deficient nanoporous lithium ferrite based green energy device hydroelectric cell. *J. Alloys Compd.* **2020**, *827*, 154334. [CrossRef]
69. Choi, M.; Ibrahim, I.A.M.; Kim, K.; Koo, J.Y.; Kim, S.J.; Son, J.-W.; Han, J.W.; Lee, W. Engineering of Charged Defects at Perovskite Oxide Surfaces for Exceptionally Stable Solid Oxide Fuel Cell Electrodes. *ACS Appl. Mater. Interfaces* **2020**, *12*, 21494–21504. [CrossRef] [PubMed]
70. Gauzzi, A.; Mathieu, H.J.; James, J.H.; Kellett, B. AES, XPS and SIMS characterization of $YBa_2Cu_3O_7$ superconducting high T_c thin films. *Vacuum* **1990**, *41*, 870–874. [CrossRef]
71. Liu, B.; Zhang, Y.; Tu, B.; Dong, Y.; Cheng, M. Electrochemical impedance investigation of the redox behaviour of a Ni–YSZ anode. *J. Power Sources* **2007**, *165*, 114–119. [CrossRef]
72. Lustemberg, P.G.; Ramírez, P.J.; Liu, Z.; Gutiérrez, R.A.; Grinter, D.G.; Carrasco, J.; Senanayake, S.D.; Rodriguez, J.A.; Ganduglia-Pirovano, M.V. Room-Temperature Activation of Methane and Dry Re-forming with CO_2 on $Ni-CeO_2$(111) Surfaces: Effect of Ce^{3+} Sites and Metal–Support Interactions on C–H Bond Cleavage. *ACS Catal.* **2016**, *6*, 8184–8191.

 *nanomaterials*

Review

How the Physicochemical Properties of Manufactured Nanomaterials Affect Their Performance in Dispersion and Their Applications in Biomedicine: A Review

Spiros H. Anastasiadis [1,2,*], **Kiriaki Chrissopoulou** [1], **Emmanuel Stratakis** [1,3], **Paraskevi Kavatzikidou** [1], **Georgia Kaklamani** [1] and **Anthi Ranella** [1]

1   Institute of Electronic Structure and Laser, Foundation for Research and Technology-Hellas, N. Plastira 100, 700 13 Heraklion, Crete, Greece; kiki@iesl.forth.gr (K.C.); stratak@iesl.forth.gr (E.S.); ekavatzi@iesl.forth.gr (P.K.); georgina@iesl.forth.gr (G.K.); ranthi@iesl.forth.gr (A.R.)
2   Department of Chemistry, University of Crete, 700 13 Heraklion, Crete, Greece
3   Department of Physics, University of Crete, 700 13 Heraklion, Crete, Greece
*   Correspondence: spiros@iesl.forth.gr; Tel.: +30-2810-391466

Citation: Anastasiadis, S.H.; Chrissopoulou, K.; Stratakis, E.; Kavatzikidou, P.; Kaklamani, G.; Ranella, A. How the Physicochemical Properties of Manufactured Nanomaterials Affect Their Performance in Dispersion and Their Applications in Biomedicine: A Review. *Nanomaterials* **2022**, *12*, 552. https://doi.org/10.3390/nano12030552

Academic Editors: John Vakros and George Avgouropoulos

Received: 17 November 2021
Accepted: 1 February 2022
Published: 6 February 2022

Publisher's Note: MDPI stays neutral with regard to jurisdictional claims in published maps and institutional affiliations.

Abstract: The growth in novel synthesis methods and in the range of possible applications has led to the development of a large variety of manufactured nanomaterials (MNMs), which can, in principle, come into close contact with humans and be dispersed in the environment. The nanomaterials interact with the surrounding environment, this being either the proteins and/or cells in a biological medium or the matrix constituent in a dispersion or composite, and an interface is formed whose properties depend on the physicochemical interactions and on colloidal forces. The development of predictive relationships between the characteristics of individual MNMs and their potential practical use critically depends on how the key parameters of MNMs, such as the size, shape, surface chemistry, surface charge, surface coating, etc., affect the behavior in a test medium. This relationship between the biophysicochemical properties of the MNMs and their practical use is defined as their functionality; understanding this relationship is very important for the safe use of these nanomaterials. In this mini review, we attempt to identify the key parameters of nanomaterials and establish a relationship between these and the main MNM functionalities, which would play an important role in the safe design of MNMs; thus, reducing the possible health and environmental risks early on in the innovation process, when the functionality of a nanomaterial and its toxicity/safety will be taken into account in an integrated way. This review aims to contribute to a decision tree strategy for the optimum design of safe nanomaterials, by going beyond the compromise between functionality and safety.

Keywords: physical/chemical characteristics; functionality; nanoparticles; nanomaterials

1. Introduction

The rapid expansion of nanotechnology and of the related synthesis and analysis tools has led to a significant increase of the variety of manufactured nanomaterials (MNMs) and of their range of applications. The term MNMs signifies *intentionally manufactured* materials 'containing particles, in an unbound state or as an aggregate or as an agglomerate and where, for 50% or more of the particles in the number size distribution, one or more external dimensions is in the size range 1–100 nm'. Moreover, fullerene, graphene, and carbon nanotubes with minimum diameters below 1 nm are included as well. The definition of 'nanomaterial' was given in 2011 in a European Commission recommendation [1], where nanomaterials were also categorized as natural, incidental, or manufactured. This expansion in the application of MNMs has significantly increased the probability of them coming in contact with humans, the environment, and, in general, the Earth system [2]. It is, therefore, of great importance to identify all probable deleterious effects that MNMs may have on both human health and the environment, early on in the innovation process. A first step towards achieving this objective is to be able to link the physicochemical characteristics of the manufactured nanomaterials to their functionality. At the same time, much research work is still required to both, advance our knowledge on the physicochemical characterization of MNMs, and to explore on how these characteristics and the resulting properties affect their potential to induce toxicity in different receptors, as well as determine their ultimate fate [3]. The importance of lacking the right correlations regarding how physicochemical characteristics influence the fate of manufactured nanomaterials has been emphasized in reports on the life-cycle assessment of these MNMs [4]. Moreover, correlating the physico-chemical characteristics of MNMs and their extensive (eco)toxicological assessment would allow the application of grouping and read-across methodological approaches, which have been extensively used for chemicals in general and, based on the similarity between substances and their behavior, could be used to fill data gaps for other MNMs, without performing additional effort, and time, consuming testing [5]. One should also refer here to a classic book by Otterstedt and Brandreth [6], which deals with the chemical and physical principles of methods for the preparation of MNMs, as well as with the description of their surface and of the methods of its modification. The applications of small particle technology are also demonstrated, considering how to make technically important materials.

When any type of a nanomaterial interacts with a biological medium, which can consist of proteins, membranes, cells, organelles, and nucleic acids, various kinds of nanoparticle/biological interfaces are established, where the behavior is governed by the relevant biophysicochemical interactions, as well as by colloidal forces. These kinds of interactions can lead to the formation of protein coronas on the surface of the nanomaterials, wrapping of nanoparticles by membranes, intracellular uptake, and biocatalytic processes that could potentially have biocompatible or bio-antagonistic outcomes. At the same time, the nanomaterial surface may suffer phase transformations, restructuring, and/or dissolution, due to the presence of the biomolecules and the dispersing liquid medium. Being able to understand the structure and the behavior at such interfaces would allow predictive relationships between structure and activity to be developed, which will be governed by the nanomaterial characteristics, such as size, shape, roughness, surface chemistry, and surface coatings. Such knowledge will be imperative for the safe use of the nanomaterials [7].

Our main objective has been to identify, classify, and prioritize the physicochemical characteristics of nanomaterials in relationship to their functionalities, in order to demonstrate the interrelationship between these different functionalities and to illustrate the effect of the physicochemical properties on the MNM performance. The number of different nanoparticles, their properties, and their practical uses are vast, as are their different physicochemical properties and the resulting biophysicochemical interactions at the respective interfaces. Thus, it is not possible to discuss all of them in sufficient detail. In this work, we present a short review of how specific key parameters of manufactured nanomaterials affect some of these functionalities, except toxicity, which is, by itself, a huge field of research. Key parameters relative to geometry (particle size, particle shape, and aspect ratio), chemistry (composition, surface groups, surface charge), crystallinity, morphology (topology, roughness, porosity, and surface area), surface functionalization (surface coatings, reactivity, and stability), and test media (mostly aqueous) are discussed in relation to MNM functionalities. These functionalities are discussed in terms of two groups: performance or properties, on the one hand, and applications, on the other. In the properties/performance functionalities we have included dispersion ability in aqueous media, solubility/dissolution characteristics, and hydrophobicity/hydrophilicity, which are directly affected by the physicochemical characteristics of the prepared nanomaterials, but, at the same time, they can have an effect on the activity and the practical uses of the MNMs. In the second functionality group, we have included applications such as the cellular uptake of the MNMs, as well as their optical, electronic, magnetic, and catalytic properties. Since the number of MNMs is vast, we tried to focus our report mainly on certain MNMs that are more frequently encountered in contact with humans, such as titania (TiO_2), silica (SiO_2), zinc oxide (ZnO), cerium oxide (CeO_2), iron oxide (Fe_3O_4), barium sulfate ($BaSO_4$), cadmium selenide (CdSe) quantum dots, gold (Au), silver (Ag), and various carbon nanomaterials such as carbon nanotubes (CNTs), graphene, graphene oxide, and reduced graphene oxide. It is noted that carbon black (nano)materials, which are broadly used in tires, are not discussed in this mini review, mostly because this is a very broad area, where various grades of carbon black are used, often with a non-disclosed primary particle size distribution, as well as different sizes and structures of aggregates [8].

One should note that being able to understand such interrelationships will allow engineering the MNMs so that one can maximize the benefits for functionality, while reducing the risks to human health and/or the environment and, moreover, being able to achieve this at an early phase of the innovation process. This would enable the consideration of safety aspects for humans and the environment early on in the process of designing a new product, so as to minimize or, even, eliminate the risks of adverse effects during its life cycle, which includes synthesis, storage, use, maintenance, and decommission.

2. How the Key Parameters Affect Functionalities with Respect to Performance

2.1. Dispersion Ability

The state of dispersion of nanomaterials in the different dispersing media is a very important characteristic of nanoparticulates; yet this state is very challenging to quantify, since dispersion is a very complicated (and little understood) process [9,10]. Controlling the dispersion of fine particles and preventing the formation of uncontrollable aggregates can lead to materials with improved properties [11]. The aggregation of nanomaterials depends both on the particle characteristics (e.g., size, shape, concentration, surface charge, and surface roughness) and on the physicochemical properties of the media (e.g., ionic strength, pH, and/or presence of organic macromolecules) [12]. In the absence of a surface coating, the aggregation/disaggregation of nanomaterials is mainly controlled by the intrinsic properties of the particles, such as size and zeta (ζ)-potential, as well as by the ionic strength of the solutions, as described by the DLVO theory proposed by Derjaguin, Landau, Verwey, and Overbeek [13,14].

Nanoparticles tend to agglomerate immediately in cell culture media. Thus, the effects of the various biological dispersion media on the state of aggregation of the nanoparticles has been extensively investigated in the literature, especially since these are critical in evaluating and interpreting the toxicological assay results [15]. At the same time, when natural organic matter (NOM) is present, it usually increases the stability of the nanoparticles in water [12,16], whereas chemical surfactants, serum, and/or proteins are frequently used to improve the dispersion and stabilization of nanoparticles [17].

2.1.1. Dispersibility of Metal and Metal Oxide Nanomaterials

Titanium oxide (TiO_2) nanoparticles are widely utilized in many different applications, for example, in cosmetics and sunscreen products; nevertheless, they may be toxic in certain cases and/or certain environments or aggregate in different culture media and, thus, the investigation of the degree of their dispersion is critical. Ultrapure water was found to disperse TiO_2 better than freshwater microalgae and daphnia aquatic culture media (Figure 1). The hydrodynamic size of the nanoparticles was found to slightly depend on concentration in the former case; whereas, the effect was significantly larger for the latter [18].

In contrast, attempts to disperse TiO_2 nanoparticles in water, even under strong sonication, led to sizes bigger than the hydrodynamic radius of the primary nanoparticles, indicating that the TiO_2 sample consists of a certain number of strong aggregates that cannot be broken down easily, even when ultra-sonication is utilized [19]; the dispersion state was much poorer when different cell culture media were used in the absence of any dispersing agents. Bovine serum albumin (BSA) greatly improved the dispersion of nanoparticles in many culture media, with the observed differences attributed to the different protein–nanoparticle interactions in the media. On the other hand, fetal bovine serum (FBS) was found to be the best agent for dispersing and stabilizing TiO_2 nanoparticles, due to the various proteins it comprises, which function in a synergistic manner. When rat and mouse bronchoalveolar lavage fluid (BALF) was used as a suspension medium, it was found to considerably reduce the aggregation of TiO_2 (as well as ultrafine and fine carbon black); whereas, the use of phosphate buffered saline (PBS) containing protein or DPPC alone, in similar concentrations to those found in BALF, was not successful in satisfactorily dispersing the particles [20]. In another study, similar nanoparticle size distributions were measured in water without and with bovine serum; whereby, further dilution in Roswell Park Memorial Institute (RPMI) cell culture medium resulted in significant aggregation [21].

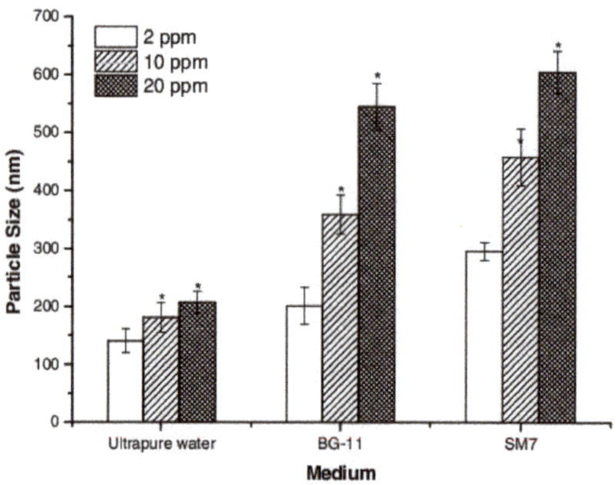

Figure 1. Dynamic light scattering (DLS) results for the size of TiO_2 agglomerates as a function of TiO_2 concentration in water, in freshwater microalgae cultured in Blue-Green medium (BG-11), and in daphnia magna cultured in simplified Elendt M7 medium (SM7). * denotes statistical differences from the control [18].

The type of biological medium in the presence of serum, together with the size of the nanoparticles, were found to affect the aggregation behavior of SiO_2 nanoparticles; their primary size was measured when dispersed in water or media without serum [15]. In contrast to SiO_2 nanoparticles, which showed a significant dependence of their measured size on the dispersion medium and/or on the presence of a protein, the size of poly(acrylic acid)-coated cobalt ferrite nanoparticles was found to be insensitive to the medium [22]. Moreover, the size of magnetic iron oxide nanoaggregates can be kept low, due to their stabilization via adsorption of FBS proteins [23], whereas the same protein reduces the agglomeration of zinc oxide nanoparticles [24], similarly to its effect on the dispersion of TiO_2 nanoparticles mentioned above [19]. For hydroxyapatite nanomaterials, the nanoparticle size decreased with increasing FBS concentration in conjunction with stirring, which provides the necessary steric and electrostatic repulsion to overcome the attractive van der Waals forces and preserve the dispersion stability for a long period [25]. Fetal calf serum (FCS) was not successful in supplementing the dispersion of Au nanoparticles of different sizes in deionized water (DI); whereas, when it was used in Dulbecco's modified eagle's medium (DMEM), it led to the formation of complexes [26].

Temporarily stable small aggregates were formed when Al_2O_3 nanoparticles were dispersed either in deionized water (DI) or in ethylene glycol [27], whereas CeO_2 nanoparticles formed a more stable dispersion only in water, in comparison to a fish medium in which sedimentation was clearly observed [28]. However, in both cases dispersions of small aggregates and not of primary particles were obtained. Moreover, citrate capped silver (Ag) nanoparticles in aqueous matrices were found to aggregate more pronouncedly in salty sea water compared to lake fresh water, due to the presence of natural organic matter (NOM), i.e., alginate humic and fulvic acids, and the low ionic strength of fresh water when compared to sea water [29]. The measured hydrodynamic radii were also found to decrease with increasing pH.

The dispersion of Ag nanoparticles, their aggregation, as well as the size of these aggregates and their stability were found to be very different in different organic solvents [30]. Ag (80 nm), hydrocarbon-coated Ag (15 nm and 25 nm), and polysaccharide-coated Ag (10, 25–30 and 80 nm) showed a similar tendency since they agglomerate at almost the same size when they are dispersed in water or media with serum; when media without serum were utilized, higher agglomeration sizes were obtained [31]. At the same time, the dispersion of metal and metal oxide nanoparticles, such as Al, Al_2O_3, Cu, SiO_2, TiO_2, and Ag, was investigated in water, cell culture media (RPMI-1640) only, and/or cell culture media with serum [31]. In the majority of cases, media without serum exhibited the worst dispersing ability, irrespectively of the kind of nanoparticles, their size, and/or their coating; whereas, in general, the media with serum were the best, differences in the final sizes were observed depending on the kind of nanoparticles in water. Moreover, the effect of the particle primary size on the agglomeration was very weak, if not absent. TiO_2 nanoparticles exhibited high agglomeration, whereas SiO_2 particles and SiO_2-coated fluorophores (35, 51, and 110 nm) were the only nanoparticles that were dispersed in a way whereby the size of the primary particles could be measured. The dispersibility of CuO and ZnO nanoparticles was tested in different mineral and complex test environments, as well as its relationship with toxicity towards selected environmentally relevant test organisms and mammalian cells in vitro [32]. Both, CuO and ZnO nanoparticles were very unstable and sedimentation was observed. A considerably high degree of agglomeration/sedimentation was observed in the mineral media that are used for key regulatory ecotoxicological assays (crustaceans, algae). On the contrary, the components of the complex test media (test environment with organic components) were found to be critical in dispersing the nanoparticles and preventing their sedimentation.

The crystallinity and the primary size of nanoparticles are also factors that influence their dispersibility. In the case of TiO_2, 100% anatase, 61–39% rutile-to-anatase, 40–60% rutile-to-anatase, as well as completely amorphous TiO_2 nanoparticles were evaluated in water and in media with and without serum [31]. The amorphous TiO_2 showed a high degree of agglomeration in all three suspending media, whereas the other TiO_2 particles showed slightly smaller aggregates in water, and only the 61% rutile TiO_2 showed a significant decrease in media with serum. The 61% rutile titania also exhibited the highest values of zeta-potential. Moreover, when the size of the TiO_2 nanoparticles was studied utilizing nominally 5, 10, 16, 50, and 100 nm nanoparticles, a high agglomeration was obtained in all three media, except the 10 nm TiO_2 in water. The effect of nanoparticle size on dispersibility has also been investigated with Au nanoparticles of 10, 50, 100, and 250 nm in aqueous suspensions diluted in phosphate buffered saline (PBS), to obtain a physiological solution [33]. The coexistence of agglomerates consisting of loosely arranged nanoparticles with individual ones was observed in all dispersions, except for the one of the largest nanoparticles, where there was not any obvious clustering. Particle shape also influences the electrostatic and steric repulsive forces, which are much stronger between two plate-like particles than between two spherical particles of the same volume, due to the much larger interaction surface between the plate-like particles [34].

2.1.2. Dispersibility of Carbon Nanomaterials

More so than the dispersion of inorganic, metallic, or metal oxide nanoparticles, the prevention of aggregation in carbon nanomaterials is of utmost importance, since their agglomeration may hinder the realization of their excellent properties. Enhanced dispersion and stabilization of carbon nanomaterials (CNMs), such as graphene oxide, graphene, carbon nanotubes, and fullerenes, especially in water, is a critical challenge, because of their tendency to aggregate, particularly in aqueous systems, due to significant van der Waals attractions and their specific hydrophobic interactions [35]. It is both the physicochemical properties of the carbon nanomaterials and the properties of the dispersion medium that influence the dispersion stability, which is further enhanced in aqueous media with NOM, due to the enhanced interactions assisted by the CNMs hydrophobic surfaces. Both single- and multi-wall carbon nanotubes (SWCNTs and MWCNTs) were found to disperse better in media with NOM than in natural water (Figure 2); nevertheless, functionalization of the MWCNTs can improve the dispersion and lead to differences among the different media [16]. The presence of proteins, lipids, or protein/lipid components is crucial for the dispersion of carbon nanomaterials such as fullerenes and single- and multi-wall carbon nanotubes in various media as well [36], whereas vehicles lacking lipids or proteins lead to the formation of the largest agglomerates.

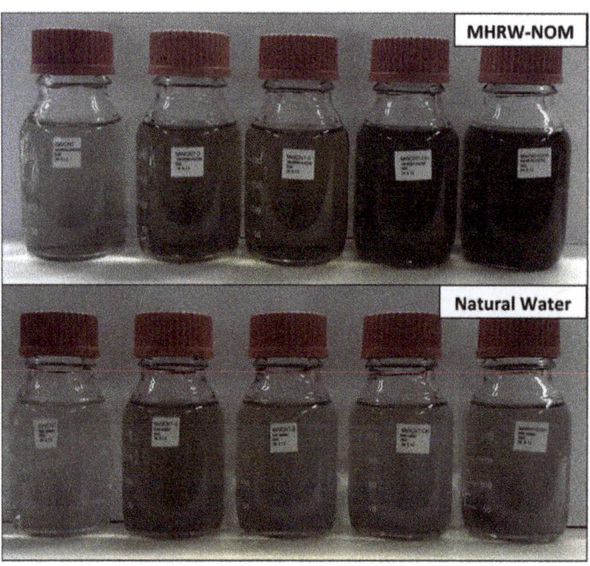

Figure 2. Different dispersibilities among CNT types and between different media, illustrated by CNT dispersions in MHRW-NOM (**top**) and natural water (**bottom**). From the left: SWCNT, MWCNT-15, MWCNT-30, MWCNT-OH, and MWCNT-COOH. (Reprinted with permission from ref. [16]. Copyright 2018 Elsevier).

Aqueous suspensions of nanosilver, nanocopper, and fullerenes (C60) [37] were prepared in deionized water and in filtered natural river water to examine the effect of different concentrations of dissolved organic carbon (DOC) and different ionic strengths of the solutions; it was found that water chemistry influences both the suspension/solubility of the nanomaterials, as well as their particle size distributions. The dispersion of carbon nanotubes and carbon black was studied in water, in cell culture media (RPMI-1640), and/or in cell culture media in the presence of serum [31]. SWCNTs, MWCNT-COOHs, and CNTs formed aggregates in deionized water, whereas carbon black showed a large range of agglomeration sizes (the smaller found in water) depending on the solvent used. Stable aqueous dispersions of fullerenes, C60 and C70, were prepared in a different study by injecting a saturated suspension of fullerenes in tetrahydofuran (THF) into water and subsequently removing the THF by purging with nitrogen gas [38]. Fullerenes were dispersed as monodisperse clusters in water, and the obtained dispersions exhibited excellent colloidal stability, despite the absence of any stabilizing agent. This was attributed to the negatively charged surfaces that led to significant electrostatic repulsion and, thus, caused the stability of the dispersions.

2.1.3. Surface Modification and Dispersibility

One of the most widely used methods to improve the dispersion stability of nanoparticles is their surface modification [39]. This necessitates a different designing of the surface structure, depending on the type of nanoparticle, as well as of the dispersing liquid media. Colloidal stability can be achieved by the adsorption, grafting, and/or coating of polymers, surfactants, and charged or biological molecules [34,40,41] that will provide electrostatic or steric repulsion between nanoparticles, thus, avoiding their agglomeration. In certain media, in order for a good dispersion of nanoparticles to be achieved, either a formulation with dispersants (usually amphiphilic molecules) or surface modification is requisite. For the latter case, the best functioning grafting molecules depend strongly on the size of the nanoparticle, with surfactants working better for small nanoparticles (<10–50 nm), whereas alkoxysilanes work better for larger ones (>50 nm) [42].

One of the simplest surface modification methods for improving dispersion stability is the adsorption of a polymeric dispersant on the surface of the nanoparticles; this methodology was presented in a comprehensive review [39]. Cationic or anionic polymer dispersants are commonly utilized to disperse nanoparticles, in either aqueous media or in organic solvents with high polarity; the polymer chains generate the steric repulsive force and increase the surface charge. Poly(acrylic acid) (PAA), sodium salts of PAA, as well as copolymers of acrylic acid and maleic acid are common anionic polymeric surfactants utilized to disperse oxide nanoparticles, such as TiO_2, $BaTiO_3$, Fe_2O_3, MgO, and Al_2O_3, whereas polyethyleneimine, PEI, is a commonly used cationic surfactant. The adsorption of the surfactants on the nanoparticles and the resulting range and magnitude of the repulsive force are influenced by a combination of various parameters, such as the suspension pH and solid fraction, the molecular weight of the polymer and its degree of dissociation, as well as the nanoparticle surface charge and its particle size. It was found that polymeric surfactants with a high molecular weight diffuse more difficultly around small nanoparticles and, thus, they cannot efficiently adsorb on their surface, failing to improve the dispersion stability of the suspension. Moreover, the dispersion stability can be affected by the surfactant structure. For example, for a polymer dispersant with a hydrophilic and a hydrophobic group, the ratio of the hydrophilic and hydrophobic sites controls the loop-train structure of the polymer adsorbed onto the particle surface, thus, affecting the dispersant ability. Copolymers possessing hydrophilic and hydrophobic segments are often utilized as anionic surfactants, to assist the dispersion of hydrophobic nanoparticles, such as SiC, CNTs, and coal, in aqueous media, since they can adsorb on the surface via their hydrophobic segments. Moreover, an aromatic monomer, such as styrene, can further improve the adsorption via both hydrophobic and π-π interactions. At the same time, the hydrophilic parts provide the necessary compatibility with the aqueous dispersing media and create an effective repulsive steric force. Cationic polymers, such as PEI, can also be utilized to enhance the dispersion of hydrophobic particles, such as SiC and CNTs, in aqueous media. Another method to improve the degree of dispersion of nanoparticles in various liquids is chemical modification of their surface. Silane coupling agents are utilized to alter the surfaces of oxide nanoparticles via the introduction of various reactive groups, such as epoxides, amines, and vinyls, on the particle surface and the subsequent grafting-from or grafting-to of polymers onto the surface. It is noted that neutral polymers, such as poly(ethylene oxide) or dextran, can also be employed as stealth coating agents to improve the colloidal stability and pass through physiological barriers; the most common cell targeting agents are proteins, enzymes, antibodies, or nucleotides [43].

Adsorption of certain surfactants on the outer or the inner surface of halloysite nanotubes has been utilized to increase their dispersibility, either in water or in organic solvents. At the same time, covalent or non-covalent functionalization of boron nitride nanotubes creates homogeneous dispersions in aqueous and organic media [44]. The dispersion stability of copper oxide (CuO) was investigated in different media, in their pristine form and when modified by four different stabilizing agents that gave them a negative (sodium ascorbate, ASC, and sodium citrate, CIT), a positive (polyethylenimine, PEI), or a neutral (polyvinylpyrrolidone, PVP) surface charge. The results showed that, in media with low ionic strength, the first two materials improved the dispersion by improving the repulsive potential, due to the negative charge, where PEI had the most significant effect, since it provides both electrostatic and steric stabilization, due to the positive charge and its polymeric nature, respectively. Amino acid and protein-rich media, however, control the stability irrespectively of the coating molecule [45].

An optimal concentration of sodium dodecylbenzene sulfonate (SDBS) was attained in the case of CuO and Al_2O_3 particles in deionized water, based on the reduction of their hydrodynamic radii that led to a concurrent decrease of viscosity and increase of thermal conductivity [46]. At the same time, SDBS and cetyltrimethylammonium bromide (CTAB) were utilized at low concentration and at exactly the critical micelle concentration (CMC) to assist the Al_2O_3 nanoparticle dispersion [47]. SDBS at CMC showed the best dispersion, because of the positive surface charge of alumina in the aqueous medium and its strong affinity for anionic groups, in contrast to CTAB, which, being a cationic surfactant, is repelled by the positively charged alumina surfaces. Similarly, SDBS was found to provide a better stability of Al_2O_3 nanoparticles than CTAB or SDS, whose performance was rather poor. In the former case, the measured hydrodynamic radius of the nanoparticles was approximately that of the primary ones, taking into account the size of the additional surfactant layer [48]. Beyond the stabilization in a simple nanofluid, SDBS shows a better and longer stabilization, lower hydrodynamic size, and narrower polydispersity than SDS, even for nanohybrid TiO_2-Ag nanoparticles [49]. In a similar way, a certain concentration of PVP surfactants in a Al_2O_3/ethylene glycol nanofluid provides the most stable dispersions for long durations, due to the polymeric chain interactions, in contrast to the case when SDS is used, where a fast sedimentation is observed [50].

In the case of titanium dioxide/distilled water nanofluids, the more stable dispersions were obtained when PVP was utilized as a stabilizer, whereas the use of the non-ionic surfactant polyoxyethylenesorbitan monolaurate (Tween 20) led to systems with lower viscosity; heat transfer is improved by both additives [51]. SDS also significantly influences the stability of TiO_2 nanoparticles, via different processes, which include surface adsorption and agglomeration (Figure 3).

These processes are reversible (desorption, disagglomeration) when the pH or the SDS concentration changes, whereas the concentration of the surfactants, the presence of divalent electrolytes, and the mixing procedure (successive or punctual addition) are of significant importance, because of the complex interplay among the adsorption/desorption of the surfactant, specific adsorption, hydrophobic effects, charge cation bridging, inversion, agglomeration, and disagglomeration [52].

The anionic surfactant SDS was found to be the best among non-ionic (TritonX 100, PEG), anionic (SDS), and cationic surfactants (CTAB) in stabilizing ZnO in aqueous media, as its utilization resulted in particles with a smaller size distribution and longer resistance to sedimentation, especially following sonication [53]. In contrast, the non-ionic surfactant PVP resulted in smaller hydrodynamic radii of zirgonium oxide, ZrO_2, compared to the ionic SDBS and to CTAB. PVP was found to create stable aqueous dispersions over a long period of time, with its concentration not playing a significant role [54]. Different concentrations of TiO_2 were better dispersed when FBS was used as the surfactant in the conventional F-12K plus FBS cell culture medium, in comparison with cases where the non-ionic block copolymer pluronic F68 or the semi-synthetic plant-derived DPPC were used as anti-agglomerating agents [17]. In all cases, the size of the particles increased as a function of their concentration. Similar results were observed when nickel oxide (NiO) nanoparticles were investigated in the same media. Covalently bound dextran on the surface of permanently magnetic nanoplatelets ensured robust steric stabilization in different physiological buffers and in complex biological media. These kinds of nanoparticles are keen to agglomerate, not only because of the van der Waals attraction, but due to dipole–dipole interactions as well [34]. The presence of humic acid (HA) as the natural organic matter in conjunction with ultra-sonication (and, more specifically, the addition of the dispersant before the sonication) were critical for achieving a stable dispersion of TiO$_2$ nanoparticles, together with the concentration of HA and the pH. At the same time, the optimum values of these parameters depend on the anatase or rutile crystalline phases of the nanoparticles [55].

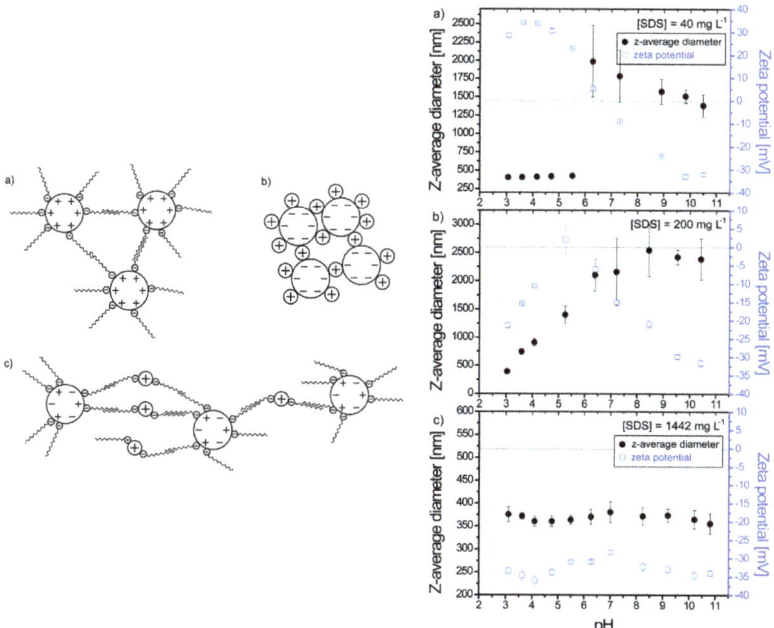

Figure 3. (**Left**) Schematic representations of TiO$_2$ and SDS interactions and agglomerate formation. (**a**): TiO$_2$–SDS agglomerates at pH 3.1. Hydrophobic interactions promote the formation of large agglomerates. (**b**): TiO$_2$ agglomerate formation at pH 8.2 in the presence of divalent cations ($\oplus$). Cation bridging between TiO$_2$ promotes agglomeration. (**c**): TiO$_2$–SDS agglomeration in the presence of divalent cations at pH 8.2. Cation bridging between SDS tails destabilizes the complexes. (**Right**) Z-average diameters and ζ-potential as a function of pH for (**a**): [SDS] = 40 mg L^{-1}: charge neutralization and inversion is observed. SDS–TiO$_2$ complex properties are mainly controlled by the TiO$_2$ surface properties. (**b**): [SDS] = 200 mg L^{-1}: the impact of SDS properties on the behavior of the TiO$_2$–SDS complexes is more pronounced. Charge neutralization occurs and the isoelectric point is obtained at pH 5.2; by further increasing the pH, negative values are obtained, due to surface deprotonation. (**c**): [SDS] = 1442 mg L^{-1}: the SDS–TiO$_2$ complexes exhibit stable Z-average diameter and ζ-potential in the full pH range. [TiO$_2$] = 50 mg L^{-1} (Reprinted from ref. [52]. Copyright 2017 The Royal Society of Chemistry).

Magnetic iron oxide nanoparticles were also functionalized by the acidic form of sophorolipids [56]. No stable dispersions were achieved in the absence of sophorolipids, whereas when sophorolipids were employed, a stable colloidal suspension of maghemite Fe$_2$O$_3$ nanoparticles, in coexistence with a black/brown precipitate, was obtained; the presence of the precipitate was attributed to the nanoparticle aggregation before the addition of the sophorolipids and/or the insufficient complexation by the sophorolipids. An increase in temperature further assisted the dispersion. Different organic ligands have been utilized to influence the colloidal stability of TiO$_2$ nanoparticles as a function of pH, electrolyte concentration, and dispersing medium, where different behaviors were observed depending on their functional group (Figure 4). It was shown that, in certain cases, the behavior was more influenced by the electrolyte concentration than by the pH, in contrast to other cases where, not only was the pH the main parameter, but it showed opposite effects for different modifiers. There were cases where none of these parameters were found to significantly influence the behavior or the final hydrodynamic radii measured in the dispersions [57].

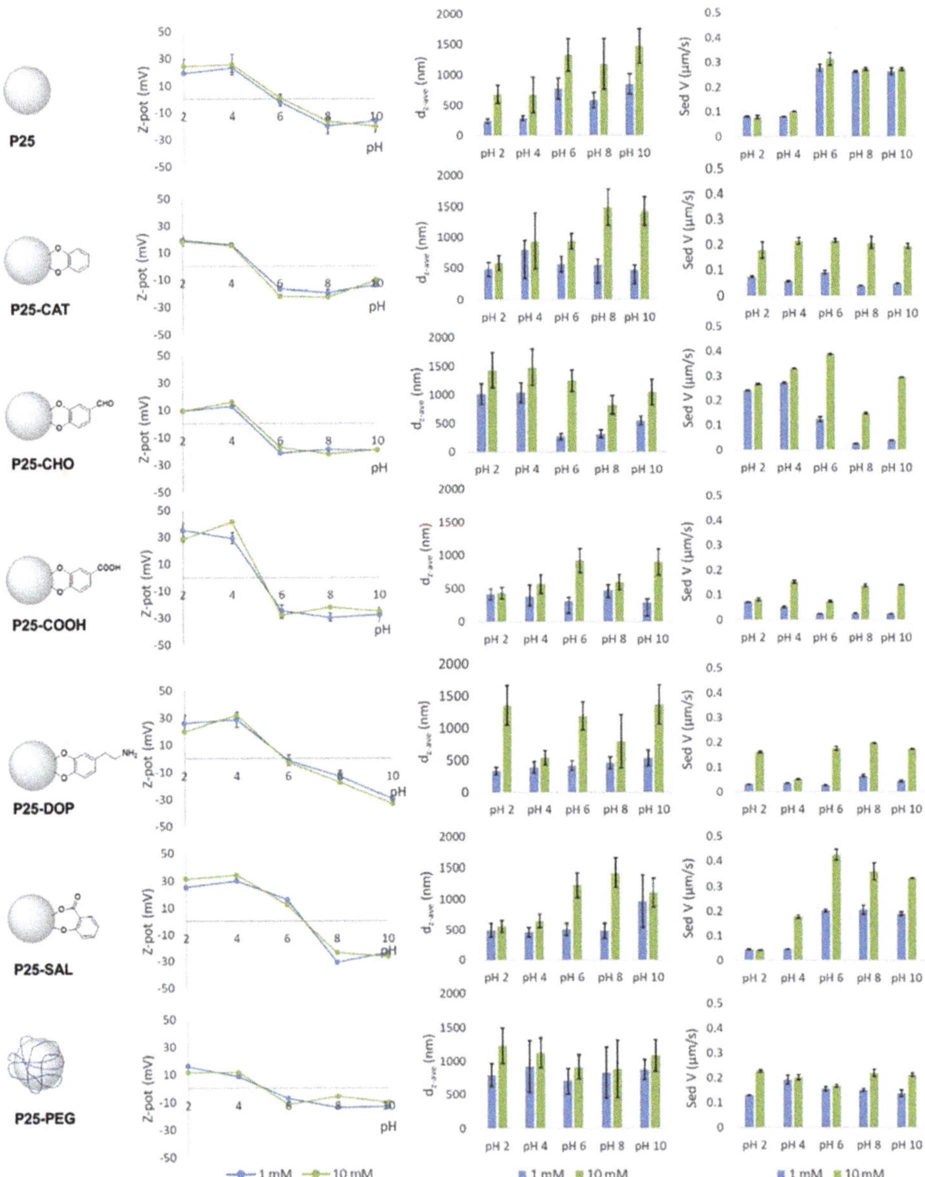

Figure 4. Zeta potential (Z-pot), hydrodynamic diameter ($d_{z\text{-ave}}$), and sedimentation velocity (sed V) of pristine and functionalized Aeroxide® P25 TiO_2 nanoparticles (declared average particle size: 21 nm) dispersed in 1 and 10 mM NaCl solution for pH values from 2 to 10. Catechol (CAT), 3,4-dihydroxybenzaldehyde (CHO), 3,4-dihydroxybenzoic acid (COOH), dopaminehydrochloride (DOP), salicylic acid (SAL), and polyethylene glycol (PEG, M_v 100,000) were utilized for the functionalization. (Reprinted with permission from ref. [57]. Copyright 2018 Elsevier).

Various mineral and complex test environments were used to examine the dispersibility of Ag nanoparticles [32]. In all liquid media, coated silver nanoparticles were significantly more stable compared to the uncoated ones. This was in agreement with the results of an independent study [58], which showed that uncoated Ag nanoparticles tend to precipitate in high ionic strength suspensions and sediment within a few hours. Furthermore, the dispersibility of both bare and surface-coated Ag nanoparticles with either poly(vinyl pyrrolidone) (PVP) or oleic acid (OA) was investigated, as well as its relation to bioaccumulation and reproductive toxicity in earthworms versus that of Ag ions [59]. Nanoparticles coated with PVP are hydrophilic and they usually form stable suspensions in polar solvents [60], whereas ones coated with OA are amphiphilic and form stable suspensions in both polar and non-polar solvents, as well as in polar/non-polar interface layers, depending on the pH of the suspension [61,62]. The primary particle diameters were determined by TEM, which showed that the OA-coated particles had a slightly smaller mean diameter than the PVP-coated ones. Dynamic light scattering measurements in DI water were in agreement with TEM concerning the size distributions of the PVP-coated nanoparticles, whereas they showed a greater ratio of larger aggregates for the OA-coated ones.

Surfactants also improve the stability of carbon nanomaterials (CNMs) in water, because of their adsorption via hydrophobic and π–π interactions. Ionic surfactants lead to stabilization of CNMs dispersions via the electrostatic repulsion between the charged hydrophilic head groups; a similar dispersion ability is obtained for both anionic and cationic types. Additionally, the purification process, as well as the surface-functionalization that defines the nanomaterial surface charge, influence the mechanism by which ionic surfactants can adsorb on the CNM surface. The phase behavior of carbon nanotubes (CNTs) in suspension depends strongly on the kind of surfactant used, its concentration, and on the type of interaction. Understanding the adsorption mechanism of ionic surfactants and the prediction of the colloidal stability of CNTs in different media requires knowledge of their surface charge. CNTs can be dispersed in water when coated by surfactants adsorbed on their surfaces, preferentially with those that have a relatively high hydrophilic–lipophilic balance [63]. The stability of aqueous dispersions of CNTs usually increases when sodium dodecyl sulfate (SDS) is utilized [64]. UV–vis spectroscopy has shown that the CNT/SDS dispersions exhibit very high stability; the amount of nanotubes in the supernatant liquid above the sediment decreased by only 15%, whereas the corresponding decrease in the case of bare CNTs was ~50% after 500 h was allowed for sedimentation. The interaction between CNTs and SDS via the hydrophobic segment results in a higher negative surface charge and steric repulsion, which enhances the stability of the CNT/SDS dispersion. It was, thus, concluded that a surfactant comprising of a single, long, straight-chain hydrophobic segment and a terminal hydrophilic group can be a suitable dispersant for stable CNT dispersions. Moreover, Tween 80 (T80), which is a non-ionic surfactant, was found to enhance the dispersion of multi-walled CNTs in aqueous media, whereas the presence of biological media, such as RPMI and DMEM cell culture media, improved the dispersion even further [65]. In that case, the stabilization was ascribed to steric effects, as there was no change in the zeta potential measurements.

2.1.4. Dispersion Medium and Dispersibility

The effect of ionic strength (IS) and solution pH on nanoparticle dispersion has also been extensively studied, for example for anatase TiO_2 nanoparticles with a primary particle size of 15 nm; the authors studied their influence on the hydrodynamic size and on the surface charge of the resulting 'particles' [66]. In one case, the nanoparticles were dispersed in NaCl solution with different concentrations to investigate the effect of the IS at constant pH and, in another, in solutions with the same ionic strength, but different pH adjusted by using HCl, NaOH and NaCl, and their combination. A large increase in the average size was found with increasing solution IS, since, at low IS, the electrostatic repulsive forces are dominant, whereas, when IS increases, the attractive forces dominate, resulting in a highly-agglomerated dispersion. Measurements of the average diameter of the TiO_2

dispersions and of the zeta potential as a function of pH at constant ionic strength were also performed. For pH values far from the isoelectric point (IEP), a high value of zeta potential was measured, and the electrostatic repulsion prevailed over the van der Waals attraction and agglomeration was suppressed. For pH approaching the IEP, the low surface charge leads to a reduction of the repulsive forces, which results in an increase of the hydrodynamic size and in the formation of large flocs that sediment due to gravitational forces in a short time. Analogous conclusions were obtained when the aggregation of TiO_2 was investigated for different concentrations of Suwannee river fulvic acid (SRFA) and various values of pH and ionic strengths [67]. The aggregation of bare TiO_2 nanoparticles increased for pHs close to the zero point of charge, whereas at constant pH, aggregation increased with ionic strength. Furthermore, adsorption of SRFA resulted in a smaller degree of aggregation of the TiO_2 nanoparticles, presumably due to enhanced steric repulsion. Dynamic light scattering showed that the TiO_2 particles readily form stable aggregates at pH ~4.5 in a NaCl solution adjusted to an ionic strength of 0.0045 M [68]. At the same pH, when the ionic strength increased to 0.0165 M, micron-sized aggregates were formed within 15 min. At all other pH values, micron-sized aggregates were found to form faster than the minimum detection time of 5 min, even at low ionic strengths when NaCl was used. However, micron-sized aggregates form much faster in an aqueous suspension in the presence of $CaCl_2$ than in respective suspensions in NaCl, showing that divalent cations may enhance the aggregation of titania.

Similar observations were made when the agglomeration of SiO_2 nanoparticles in aqueous media was studied for different ionic strengths and pH values [69]. Addition of different salts (NaCl, $MgCl_2$, $BaCl_2$ and $CaCl_2$) caused aggregation of the SiO_2 nanoparticles, whereas a change of the pH within the range investigated did not influence the degree of aggregation in the absence of an electrolyte. The type of cation significantly affected the aggregation, with divalent cations (Mg^{2+}, Ba^{2+} and Ca^{2+}) being more efficient in destabilizing the nanoparticle suspension than the monovalent Na^+ cations.

The effect of natural organic matter (NOM) on the aggregation of anatase TiO_2 nanoparticles was also evaluated [70]. Changes in the particle size were measured as a function of the concentration of three different electrolytes (NaCl, Na_2SO_4, and $CaCl_2$) and of the suspension pH. In general, the influence of the addition of an electrolyte in the absence of NOM followed DLVO theory. When the level of NOM adsorption on the titania surface was low, aggregation was induced, whereas an increase of the surface coverage could reduce the particle aggregation, even at high ionic strengths. The surface coverage was determined by the ratio of the concentration of NOM to that of the nanoparticles, whereas the mixing procedure was proven to be important, since it led to different final aggregation states. Ionic strength strongly influenced the aggregation behavior, whereas divalent cations and anions led to stronger destabilization of negatively or positively charged titania particles, respectively. Nanoparticles that were positively charged at low pH were more easily destabilized by $SO_4{}^{2-}$ compared to Cl^-, whereas the opposite was observed for Ca^{2+} compared to Na^+ for negatively charged nanoparticles at high pH. The addition of NOM at concentrations that create stable dispersions increased the stability of the suspensions with respect to Na_2SO_4 and NaCl but did not have much influence when $CaCl_2$ was used.

The effect of concentration of sodium dodecylbenzene sulfonate (SDBS) surfactant and of pH on the size of 'nanoparticles' of alumina (Al_2O_3) and copper in water was investigated [71]. Optimal values of SDBS concentration (0.10% for alumina and 0.07% for copper) and pH (pH ~8.0 for alumina and pH ~9.5 for copper) were found, at which the effective particle diameters exhibited minimum values. Hexadecyl trimethyl ammonium bromide assisted in obtaining Cu nanoparticles with more than one order of magnitude smaller sizes in aqueous suspensions [72].

The degree of aggregation of CNMs increases at low pH, mainly due to the relatively smaller negative charge, although the degree of dispersion generally depends on the dispersing agent [35]. The dispersion of CNMs is significantly influenced by the presence of dissolved ions in water as well, where the aggregation of CNMs increases as the ionic strength increases, as expected. However, beyond a certain value of the ionic strength, there is no additional increase in the degree of aggregation, signifying that the electrostatic repulsive forces are already shielded. Moreover, increasing temperature results in an increase of the stability of CNM suspensions, most probably because of the disruption of weak interaction forces, increased Brownian motion (and, thus, collisions), and reduced zeta potential. Cellulose nanocrystals suspended in water also show pH-dependent size and viscosity; both quantities increase in acidic or alkaline conditions, whereas they obtain their lowest values at neutral pH [73].

Synthesized core-shell ZnS-coated CdSe nanocrystal quantum dots (QDs) were further coated to possess single $-NH_2$, -COOH, -OH, or dual $-NH_2$/OH and -OH/COOH functional groups [74]. The surface charge, as measured by zeta-potential measurements, varied depending on the functional group; it was found that QD-COOH and QD-OH/COOH were highly negatively charged, whereas $QD-NH_2$ and $QD-NH_2$/OH were positively charged. QD with hydroxyl groups were less negatively charged than the QDs with carboxylic acid groups, whereas QDs with both -OH and -COOH or $-NH_2$ groups had median charge. $QD-NH_2$ showed a broad particle distribution in contrast to QDs with -COOH groups that exhibited a much narrower distribution, while functionalization of the QD surface with -OH groups led to improved dispersion and stability under hypertonic conditions. In contrast, all QDs were stable in nonelectrolyte solutions. Moreover, all functionalized QDs were stable under weak alkaline conditions, whereas only $QD-NH_2$ was stable under acidic conditions.

In conclusion, the investigation of the dispersibility of nanoparticles is a complicated process, since nanomaterials constitute dynamic entities that undergo physical and chemical transformations when mixed with environmental, synthetic, or biological media of different complexities, the characteristics of which affect the behavior to a large extend.

2.2. Solubility and Dissolution of Nanoparticles

The possibility of nanoparticles dissolving within the suspending medium is a key property that influences their toxicity and, consequently, their biological response, because it defines the fate of nanoparticles in the human body, as well as in the surrounding environment [75–78]. The solubility/dissolution of nanomaterials is frequently confused with their dispersion ability. Dissolution is defined as the dynamic process during which a particle dissolves in the matrix medium, in order to form a homogeneous solution or mixture [79]; this occurs when the constituent atoms or molecules have a specific solubility in the local environment. During this process, molecules from the surface of the dissolving nanomaterial are transferred to the solution forming a diffusion layer, which is the volume between the bulk solution and the solid nanomaterial surface that involves solvated molecules. The nanoparticle dissolution depends on the size [80,81] and the surface area [82,83], the surface morphology [77], the surface energy [84], the possible adsorbed species and the state of aggregation of the nanoparticles [85], as well as on the properties of the diffusion layer and the possible solute concentration in the suspending medium [79]. Furthermore, the dissolution kinetics depend on the size and, thus, the surface area as well, explaining why the dissolution of nanoparticles is faster and more extended in comparison with macroscopic particles of the same material [86,87].

Nanoparticle antibacterial properties [88], toxicity [89], biomedical characteristics, and environmental impact [90] are strongly associated with their dissolution, since highly-toxic ions such as Zn^{2+}, Cu^{2+}, Cd^{2+}, Ag^+, etc. may be delivered to the solution [91–94]. It is possible, however, that a complex suspension—involving partially dissolved nanoparticles, free ions dissolved from the nanoparticles, and adsorbed ions on the nanoparticle surface—may be produced through the dissolution process in the surrounding media [95,96]. Figure 5 schematically illustrates that the metal oxide nanomaterial toxicity may originate from [88] the nanoparticles themselves, the released ions, or the combination of both, while adsorption of metal ions on the nanoparticles also affects toxicity. Moreover, since the nanoparticle surface interacts directly with biological systems, surface area is a key parameter of their biological effect [97].

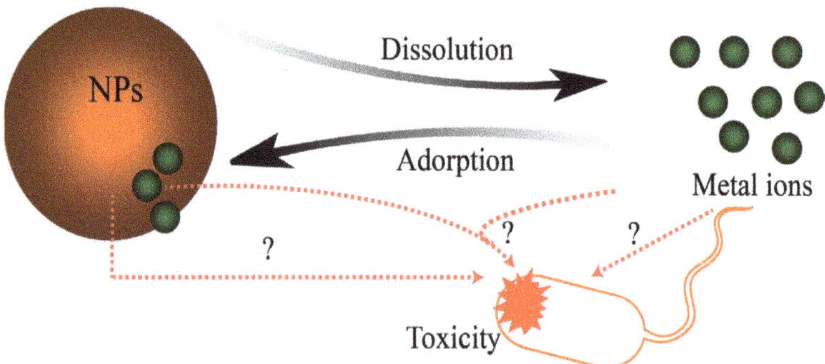

Figure 5. Nanoparticle toxicity can be attributed to the nanoparticles themselves, to released ions from the nanoparticles, or the combination of both. The procedures of dissolution and adsorption are both considered to contribute to the nanoparticle toxicity (Reprinted with permission from ref. [88]. Copyright 2016 Elsevier).

Generally, the dissolution of nanoparticles increases as the particle size decreases [98–101]. ZnO nanoparticles, however, do not exhibit major differences in their dissolution characteristics when compared to particles of micron size [102]; both nanoparticles and microparticles showed an 80% dissolution when added in Osterhout's medium. It has also been reported in the literature that decreasing the particle size can reduce the extent of, or even prohibit, dissolution; when the dissolution of hydroxyapatite nanoparticles was studied as a function of particle size, it was observed that it was the larger particles that were prone to dissolution [103]. The dissolution of silver (Ag) nanoparticles, which affects their antibacterial properties, depends on their size. The smaller the Ag nanoparticles, the higher the dissolution rate, provided that aggregation of the nanoparticles is avoided, since this may lead to sedimentation. The formation of a passivation layer (e.g., an oxide layer) can inhibit their dissolution and, thus, their antibacterial activity [104]. The effects of the concentration and size of nanomaterials on the release of silver ions from citrate-capped Ag nanoparticles in a common hydroponic nutrient medium (quarter-strength Hoagland medium) was investigated, and the kinetics of ion release was accounted for by a kinetic model within hard sphere collision theory using the Arrhenius equation; thus, providing insight into the mechanisms of the ion release kinetics from the Ag nanoparticles [105]. Moreover, when the dissolution in water of PVP-stabilized and citrate-stabilized Ag nanoparticles was investigated [106], it was observed that the concentration of released silver ions was limited, whereas the dissolution rate and degree depended on the functionalization of the particles and on storage temperature. The dissolution is not only affected by the nanoparticle size, but by their shape and surface morphology as well [107]; when different shapes of CuO nanoparticles (spherical and rod shaped) were investigated, it was found that spherical nanoparticles dissolved faster and to a greater extent compared to rod shaped particles. The kinetics of dissolution due to oxidative etching of Pt nanoparticles of cubic and icosahedral shapes in aqueous solutions was investigated using a mixture of $HAuCl_4$ and KCl as oxidative agent. Figure 6 shows the morphological changes of the icosahedral and the cubic Pt nanoparticles over a period of one hour. The shape of the nanoparticles was dramatically changed as dissolution proceeded. The corners became round and, after 1 h, the cube dissolved completely, while a small part of the icosahedron remained [108].

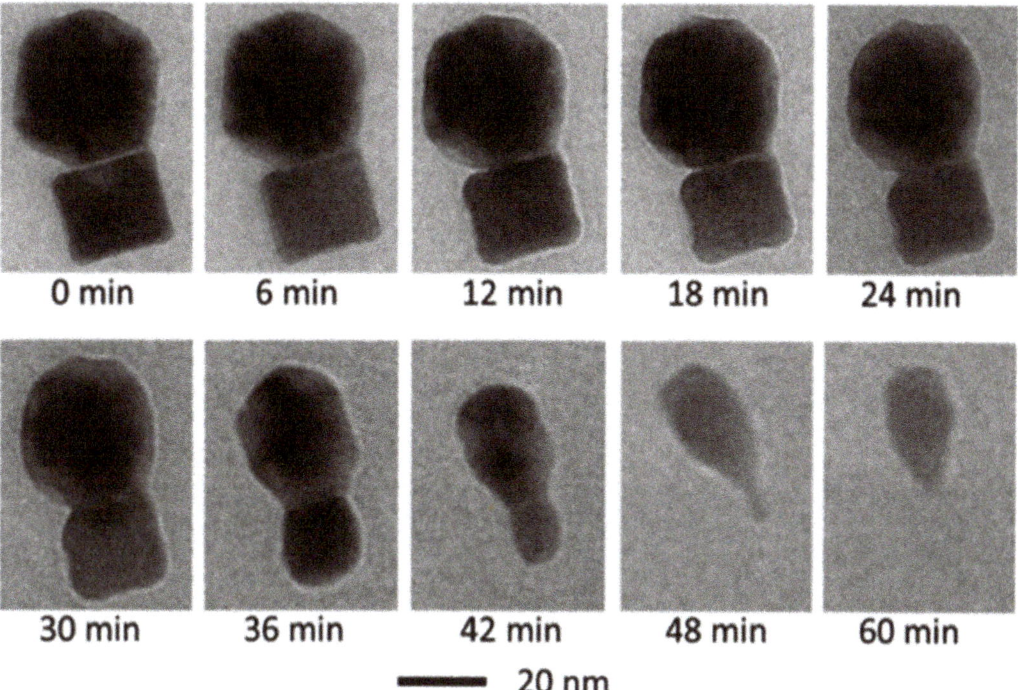

Figure 6. Morphological changes of icosahedral and cubic Pt Nanoparticles due to dissolution in the presence of aqueous solutions with a mixture of HAuCl$_4$ and KCl. (Reprinted with permission from ref. [108]. Copyright 2017 American Chemical Society).

Nanoparticle dissolution is also affected by the parameters of the surrounding media, including pH, water hardness, ionic strength, temperature, and the presence of detergents or organic compounds [7,109]. For example, complete dissolution of CuO nanoparticles was observed in the presence of media enriched in amino acids [110], whereas cysteine was found to increase the Ag nanoparticle dissolution [111]. The solubility of copper-based nanoparticles was enhanced at low pH [112], whereas it was observed that ZnS nanoparticles showed the highest solubility at lower pH (in the range 9–10) and for the smallest particle size [113]. Moreover, at pH 7 (in DMEM), ZnO nanoparticles dissolved significantly more after 48 and 72 h when compared to suspensions at pH 4 (in Milli-Q water). When the ZnO nanoparticle accumulation inside A-431 cells was investigated, the authors presented arguments that the toxicity could be attributed to the nanometric size until 24 h of exposure, whereas, after 24 h (up to the 72 h of exposure was studied), both released Zn^{2+} ions and nanoparticles played an important role in the toxicity [83].

The dissolution of nanoparticles is strongly related with their bioavailability, degree of uptake, and toxicity [114]. The toxicity of nanoparticles is related to their chemical characteristics and surface chemistry [115,116]; this is due to the possibility of releasing toxic ions and/or the production of reactive oxygen species (ROS) [117]. Toxic effects through the production of ROS are very likely to occur for nanoparticles of small size and, thus, of large reactive area. Nevertheless, when the dissolution of nanoparticles takes place during the cell culture, it is difficult to identify the origin of the toxic effects. The toxicity of a number of particles was tested in relation to their dissolution. The authors categorized the nanoparticles into soluble (Ca$_3$(PO$_4$)$_2$, Fe$_2$O$_3$, ZnO) and insoluble (CeO$_2$, TiO$_2$, ZrO$_2$), and studied the cytotoxicity on two different cells lines; it was found that, for high dissolution, the toxic effects were considerably higher compared to those for little or no dissolution [118].

The solubility of ZnO nanoparticles, with an emphasis on the toxicological effects of zinc ions, has been widely studied [119]. It has been reported that the higher the nanoparticle dose, the more the cell nuclei are condensed, leading to cell apoptosis [120]. ROS, such as hydrogen peroxide, superoxide anions, hydroxyl radicals, and organic hydroperoxides, can be produced in an aqueous suspension of ZnO nanoparticles; these ROS can cause injury to cells, whereas they also display a strong antibacterial activity [100]. Cytotoxicity studies of ZnO, CeO_2, and TiO_2 nanomaterials and their relation to dissolution suggested that the toxicity induced by ZnO nanoparticles is due to the dissolution of the ZnO nanoparticles in the aqueous environment and the release of Zn^+ in the culture medium, which is associated with high levels of ROS. On the other hand, CeO_2 showed a cytoprotective behavior by suppressing ROS production; this led to cellular resistance to the oxidative stress. Finally, TiO_2 was considered inert, since it did not result in toxic effects on mammalian cells [121]. To evaluate the toxicity in marine diatoms, ZnO nanoparticle dissolution has been examined in seawater; the toxicity was attributed to the ZnO dissolution that released zinc cations [122]. Even inert nanoparticles can induce ROS under living conditions; this is due to their ability to target mitochondria. A number of cellular events can be influenced by ROS, such as signal transduction, proliferation rate, gene expression, and protein redox regulation. At high ROS levels, cells may be damaged by altering proteins, deoxidizing lipids, or disrupting DNA, which can even lead to cancer due to gene transcription modulation [120,123]. The dissolution of ZnO nanoparticles, their uptake, and the routes they follow to enter LoVo cells has also been investigated. It was found that ZnO nanoparticles can enter LoVo cells by passive diffusion, endocytosis, or both, according to their agglomeration state. When ZnO nanoparticles contact the acidic pH of the lysosomes inside the cells, zinc ions are released. These ions together with the presence of ZnO nanoparticles produce ROS that cause DNA damages. Thus, the ZnO nanoparticle toxicity is attributed to a combination of the presence of the particles and of the zinc ions [124]. ZnO nanoparticle dissolution has been studied in various biologically relevant solutions, including HEPES, MOPS, and PIPES, where it was discovered that the buffers affect the dissolution kinetics and toxicity of the nanoparticles. Experiments on cell viability have shown that the use of buffers decreases the viability of Jurkat leukemic cells after the introduction of ZnO nanoparticles [125].

The dissolution of silver nanoparticles starts immediately upon exposure to the particular medium and continues for several hours. The oxidative dissolution of Ag is also responsible for the toxicity of the nanoparticles, which is ion- and particle-related [77]. The oxygen present induces the formation of Ag_2O on the surface of the silver nanoparticles and the release of silver cations in the aqueous solution. Moreover, low pH and smaller particle size enhance the Ag nanoparticle dissolution [126]. In general, different forms of silver may be contained within a suspension of Ag nanoparticles, such as free or complexed Ag^+ and Ag^+ adsorbed on the nanoparticles. The state of Ag nanoparticles in pure water or an aqueous nitric acid environment was investigated for a range of pHs, between 0.5 and 6.5 [127]; the findings suggest that the dissolution of silver nanoparticles depends on the particle size, since larger particles did not dissolve in nitric acid for concentrations up to 4 M, whereas faster reaction rates occurred with increasing temperature. The effect of chlorine anions on Ag nanoparticle dissolution, generation of ROS, and toxicity of Ag nanoparticles has also been investigated, since chlorine anions are the most common anions in aqueous systems. It was found that high concentrations of chlorine anions facilitate the dissolution and toxicity of the nanoparticles, because of the formation of Ag–Cl complexes [117]. Ag nanoparticle toxicity has also been examined for both positive and negative surface charges. The results revealed that the nanoparticles with positive charge were less toxic to tumor cell lines, even though they exhibited ion release rates similar to those of negatively charged nanoparticles. However, the cytotoxicity of Ag nanoparticles is a combination of events, which include, apart from the surface charge, the release of silver ions, the dissolution rate, and the activity of biological molecules [128].

Nanoparticles that dissolve in the medium before their uptake by organisms may have ion channels as a route for cellular entry [96]. The nanoparticles that resist complete dissolution follow other routes to influence the fate of cells, such as endocytosis, ion transportation, or both. Nanoparticle dissolution may also occur after cell uptake and inside the cells (intracellular dissolution); this is strongly dependent on the nanoparticle shape [129]. This dissolution mechanism shows how nanoparticles bypass the good protection of mammalian cells, as well as how heavy metal ions conduct themselves inside cells. Copper metal nanoparticles stabilized using a carbon layer were tested for the effects of nanoparticle dissolution on cytotoxicity and were compared to the behavior of copper oxide nanoparticles. The influence of pH on the solubility was studied using artificial buffer solutions of pH 5.5 and pH 7.4. At neutral pH, almost no free copper ions could be detected after 3 days in the cell culture medium, confirming the stability of the particles. However, at an acidic pH of 5.5, as found inside lysosomes, the copper oxide particles dissolved rapidly, whereas the fairly stable carbon-coated copper particles released copper to the surrounding medium. Thus, intracellular dissolution was attributed to pH effects [99].

Dissolution of nanoparticles is one of the main contributors to particle toxicity. The dissolution process may occur inside or outside cells. Nanoparticles dissolve mainly by releasing ions, which are possibly toxic for living organisms. Nanoparticle dissolution can be affected by the chemistry, size, shape, and surface coating of nanoparticles, as well as the type of media, the pH, and the solution characteristics of the surrounding environment.

2.3. Hydrophilicity–Hydrophobicity

The hydrophilic/hydrophobic behavior of nanomaterials is mainly associated with their chemical features, such as composition and surface charge, as well as their surface coating characteristics, stability, and surface reactivity. The wetting characteristics of nanoparticles are critical for their biological application [130,131] and are often strongly related to their biocompatibility and their dispersion and interaction with biomolecules [132]. The hydrophobic interaction is generally thought to be the strongest among all long-range non-covalent interactions in all aqueous systems, as well as in biological ones. It is advantageous for the adsorption of biomolecules, promotes the interaction and adhesion with cell membranes by increasing nanoparticle uptake for cellular delivery, and adjusts the release rate of drugs [133,134].

The modification of the wetting characteristics of a nanoparticle surface can be realized during either the nanoparticle synthesis or by the post-preparation of surface coatings on the nanoparticles using appropriate polymers or surfactants. Synthetic procedures in the presence of block or graft copolymers with hydrophilic segments can lead to hydrophilic surface coatings; polymeric surfactants used include poly(ethylene glycol) (PEG), poloxamers, poloxamines, polysaccharides, and nonionic surfactants, such as polysorbate 80 (Tween 80) [130]. Alternatively, post-preparation coating of the nanoparticle surface using hydrophilic polymers or surfactants is commonly achieved through chemisorption or covalent attachment of polymers or surfactants with a functional end-group to a reactive surface (grafting-to) or by in situ polymerization of monomers from immobilized initiators onto the nanoparticle surface (grafting-from) [135]. Hydrophilic homopolymers and copolymers and other coupling agents are also used to affect, both the nanoparticle morphology, and its surface modification, as well as to introduce specific functional groups on the nanoparticle surface; these agents can be silane coupling agents, titanate coupling agents, and organophosphonic acids [136,137].

Among all known nanomaterials, silver nanoparticles exhibit the highest biocompatibility and antimicrobial activity. One synthetic method utilizes the thermal reduction of $AgNO_3$ in the presence of oleylamine as a reducing and capping agent [131]; the adsorption of oleylamine on the surface of the nanoparticles makes them hydrophobic, as illustrated in Figure 7. In order to increase the dispersibility of such hydrophobic nanoparticles in water, a facile phase transfer mechanism has been developed using pluronic F-127, a biocompatible block copolymer [131]. Modifying the Ag nanoparticles surface utilizing PVP allows the formation of suspensions stable in polar solvents [60], whereas using an amphiphilic surfactant, such as oleic acid, allows suspensions stable in polar solvents, in non-polar solvents, and in polar/non-polar interface layers [61,62].

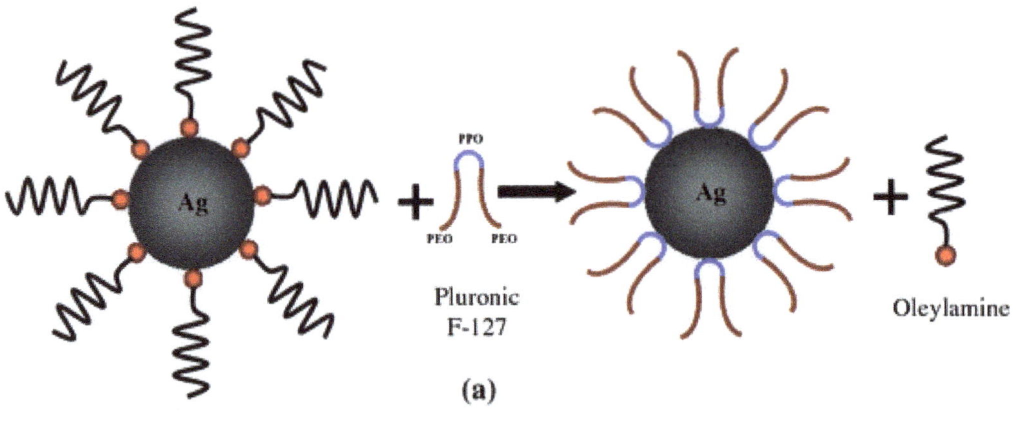

Figure 7. (**a**) The process of modifying the wetting characteristics of Ag nanoparticles from hydrophobic to hydrophilic using pluronic F-127 surfactant. (**b**) Ag nanoparticles before and after the phase transfer (Reprinted with permission from ref. [131]. Copyright 2010 Springer).

Single and multi-walled carbon nanotubes (CNTs), with diameters between 0.4 and 2 nm, and 2 and 100 nm, respectively, could potentially be utilized in a wide range of biological and biomedical applications. One of the main technical obstacles for the use of CNTs in these fields is their extremely low dispersibility in aqueous solutions. A number of methods have been used to alter the surface of CNTs, in order to modify their wettability and introduce a hydrophilic character, with the most common being functionalization with hydrophilic polymers [138]. Oxidative acid treatment can introduce nanotube-bound carboxyl acids, thus, enabling defect-targeted functionalization. Esterification, amidation, ionic interaction treatments, and sidewall-targeted functionalization of CNTs are most commonly realized by surface-attaching hydrophilic polymeric or oligomeric species onto nanotubes. PEG, poly(vinyl alcohol) (PVA) and poly(propionylethylenimine-co-ethylenimine) (PPEI-EI) have been utilized to functionalize SWCNTs. The hydrophobicity of CNTs can also be modified using non-covalent or covalent modification with carbohydrates (monosaccharides and polysaccharides), proteins, and nucleic acids [139]. Short double-stranded DNAs and certain RNAs have been used to directly disperse individual SWCNTs in water [140], where the interactions of nucleic acid with the SWCNTs in the aqueous media originate from the stacking of the bases of the nucleic acids on the nanotube surface with the hydrophilic sugar-phosphate backbone pointing towards the solution, to achieve solubility in water. The use of sodium dodecyl sulfate (SDS) as a dispersing agent allows the preparation of hydrophilic CNTs. The hydrophobic hydrocarbon segment of SDS interacts with the CNTs, where the hydrophilic sulfate group causes a high negative surface charge and steric repulsion that improves the stability of the CNT/SDS dispersion [64]. Hydrophilic MWCNTs decorated with magnetic nanoparticles have also been prepared by first synthesizing poly(acrylic acid)-functionalized MWCNTs (PAA-g-MWCNTs) and then decorating these with magnetic nanoparticles, utilizing chemical co-precipitation of Fe^{2+} and Fe^{3+} onto the outer surfaces of the PAA-g-MWCNTs; they exhibited an exceptional dispersion ability in water, as well as high magnetic susceptibility [141].

Silica nanoparticles are well known for their hydrophilicity and biocompatibility. However, often it is necessary to make them very hydrophilic. Generally, the presence of silanol groups on the surface of SiO_2 makes nanoparticles more hydrophilic and, consequently, more easily dispersible in aqueous media [142]. The addition of organosilane compounds containing PEG chains onto silica nanoparticles led to highly hydrophilic and more easily dispersible nanoparticles [143]. Alternatively, silica nanoparticles can be modified with other polymers soluble in water, such as poly(oxyethylene methacrylate) (POEM) and poly(styrene sulfonic acid) (PSSA) [135]. In this case, the process includes three steps: activation of the silanol surface groups of the SiO_2 nanoparticles, surface alteration to chlorine (-Cl) groups, and grafting-from polymerization of the polymer chains. The nanoparticles after modification exhibited better dispersibility compared to the unmodified ones [135]. Furthermore, polystyrene-functionalized silica nanoparticles have been prepared via radical polymerization of styrene monomer onto nanoparticles possessing vinyl groups, with benzoyl peroxide as the initiator, resulting in PS-g-SiO_2 particles. These PS-g-SiO_2 nanoparticles were easily dispersed in organic solvents such as methylbenzene, whereas when deposited onto a silicon wafer, they resulted in a superhydrophobic surface [144]. Hydrophilic silica nanoparticle surfaces have also been turned hydrophobic with the addition of alumina sol. The degree of hydrophilicity of the produced silica-alumina nanoparticles was controlled by changing the proportion of alumina. It was shown that the nanoparticles modified with 1, 2, and 5% alumina gained 5, 2, and 1% weight in water compared to the unmodified particles, where the gain was 8% [145].

Production of nanoparticles with hydrophilic composition and hydrophobic properties at the nanoscale level has been attempted by employing surface topology engineering (Figure 8). This takes advantage of the fact that surface roughness affects the wettability behavior. Thus, mesoporous hollow silica (MHS) nanospheres with controlled surface roughness (rough mesoporous hollow silica, RMHS) have been produced by introducing silica shell particles with smaller sizes of O(10nm) onto MHS with relatively larger sizes of O(100 nm). These rough MHS nanoparticles exhibited an unexpected hydrophobicity in comparison with the respective MHS with no roughness, and this led to higher adsorption of a range of hydrophobic molecules and the sustained release of hydrophilic drugs [134].

Hydrophobic barium sulfate ($BaSO_4$) nanoparticles were produced using a one step process that combined their synthesis and surface modification [137]. The nanoparticles were produced by a precipitation reaction of calcium chloride and ammonium sulfate in a aqueous solution using the modifying agent octadecyl dihydrogen phosphate (n-$C_{18}H_{37}OPO_3H_2$, ODP). The produced nanoparticles were hydrophobic because of the formation of a thin layer of barium alkyl phosphates on the nanoparticle surface during synthesis. It is noted that barium alkyl phosphates control the particle size and morphology of nanoparticles as well.

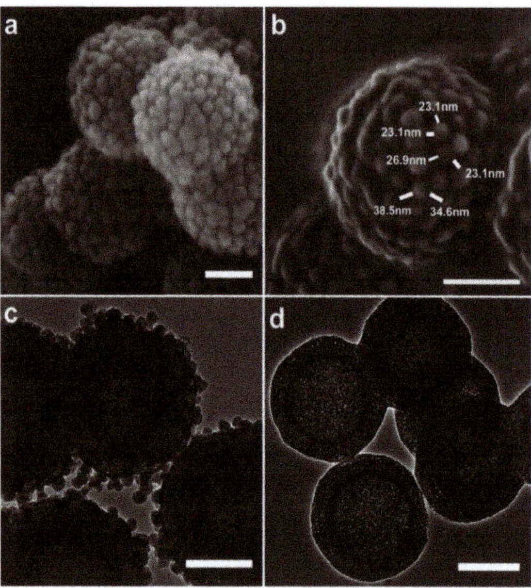

Figure 8. Morphology of the surface of RHMS and MHS nanoparticles. (**a**) SEM image of RMHS, (**b**) high-resolution SEM (HRSEM) image of RMHS, illustrating the distances between neighboring shell silica nanospheres, (**c**,**d**) HRTEM images of RMHS and MHS, respectively. Scale bar = 200 nm (Reprinted with permission from ref. [134]. Copyright 2015 American Chemical Society).

Iron oxide nanoparticles are of great importance in biomedical applications, such as bioimaging, drug delivery, cellular therapy, etc., due to the possibility of surface modification and their low toxicity [146,147]. With no surface coating, the surfaces of these nanoparticles are hydrophobic, and exhibit a large surface to volume ratio [148,149]. These particles tend to agglomerate because of hydrophobic interactions and form large clusters in aqueous media, which also significantly affects their magnetic properties. To overcome this, a variety of surface coatings have been employed to alter the nanoparticle surface, whereas, for effective stabilization, often a very high surface density for the coating is required. One approach, is to add some stabilizer, such as a surfactant or a polymer, at the time of preparation, to prevent aggregation of the nanoscale particulates. Alternatively, the particles can be modified after precipitation. Among the most common surface modifiers are synthetic (e.g., PEG, PVP, PAA, PVA, etc.) or natural polymers (e.g., dextran, chitosan and gelatin), fatty acids, polypeptides, and inorganic coatings [150].

When nanoparticles come into contact with biological fluids, they are coated with proteins within seconds; therefore, cells or tissues almost never interact with the bare particles [151,152]. The protein–nanoparticle interactions, which form the so-called nanoparticle–protein 'corona', have a key role in nanomedicine [153]. The proteins that are present in the plasma are adsorbed onto the nanoparticle surface, depending on the nanoparticle surface characteristics; this is crucial for their in vivo distribution [154]. The hydrophobicity of the nanoparticles affects both the quantity and the composition of the plasma protein adsorbed layer. Nanoparticles with decreasing surface hydrophobicity were studied with respect to their influence on plasma protein adsorption [155]. Latex particles with different hydrophobicities were used as model colloidal carriers; it was found that, when the surface hydrophobicity decreased, the quantity of adsorbed proteins decreased and the changes in the obtained protein adsorption patterns deteriorated. The hydrophobicity of copolymer nanoparticles (70–700 nm) was controlled via the co-monomer ratio of N-isopropyl-acrylamide and N-tert-butyl acrylamide (NIPAM/BAM) in the copolymer synthesis, where the NIPAM-rich particle was the most hydrophilic, and the adsorption of human serum albumin (HSA) onto these nanoparticles was investigated. The more hydrophobic nanoparticles were completely covered with a single layer of HAS, whereas particles with 25% BAM or less exhibited very little binding of HSA [150].

In the fields of nanomedicine and therapeutics, the successful cell uptake of nanoparticles and, consequently, the interaction of nanoparticles with the cell membrane is vital. The wetting characteristics of nanoparticles play a key role in cell uptake, since their interaction with the cell membrane depends not only on their shape, surface chemistry, and geometry but also on their hydrophobicity [156,157]. Small molecule nanoparticles (SMNPs), prepared by self-assembly of π-conjugated oligomers with varying degrees of hydrophobicity, were electroporated into live HeLa cells. It was observed that the more-hydrophilic SMNPs disassembled and dispersed upon cellular uptake cell, whereas the hydrophobic ones remained intact within the cells [158]. It has been shown that the bioactivity of synthetic nanoparticles can be improved with the introduction of hydrophilic co-monomers in the hydrophobic segment; the introduction of 2-hydroxyethyl methacrylate in the hydrophobic block of a poly(ethylene glycol)-block-poly(α-tocopheryl succinate) reduces the hydrophobicity of the corresponding nanoparticles and increases their bioactivity [159]. TiO_2 nanoparticles, which are used in oral applications, were tested for their wetting behavior in relation to their cell–nanoparticle interactions. The viability of epithelial cells, when in contact with either hydrophobic or hydrophilic nanoparticles, was not affected. However, the hydrophobic nanoparticles aligned to the cell membrane, wrapped up and were found in endosomes and lysosomes, while the hydrophilic nanoparticles directly entered the cells and were found in the cytoplasm [160].

3. How the Key Parameters Affect Functionalities with Respect to Applications

3.1. Cellular Uptake

3.1.1. Mechanisms of Cellular Uptake

Nanomaterials that circulate in a multicellular living organism interact with its components in a fundamentally different way compared to the soluble small molecules or micron-scale particles that are recognized by the immune system [161–164]. Materials at the nanoscale can interact with the endogenous cellular machinery through active energy-dependent processes that selectively move substances against their electrochemical gradient across cell membranes [165–172]. Endocytosis is the mechanism of actively transporting cargoes into the cell in transport vesicles derived from the plasma membrane [165]. The different mechanisms of endocytosis are generally classified as phagocytosis and pinocytosis. Phagocytosis is the predominant mechanism used mainly by macrophages and less frequently by nonprofessional phagocytes, including epithelial cells, fibroblasts, and endothelial cells [173]. Pinocytosis is present in all types of cells, in forms such as macropinocytosis, which enables the uptake of large NPs that seems impossible via other endocytosis pathways [174]; caveolae-dependent endocytosis; clathrin-dependent endocytosis; and clathrin- and caveolae-independent endocytosis, with the last three forms referred to as receptor-mediated endocytosis [175–177]. The phenomena taking place at this nano–bio interface result in the modulation of cell fate, the induction or prevention of mutations, the initiation of cell–cell communication, and the modulation of cell structure [178,179].

It has been extensively reported in the literature that the uptake of nanoparticles by the cells depends on the nanoparticle characteristics, including the size and/or shape, the surface charge, and surface hydrophobicity [178,180]; on the possible sedimentation of large and dense particles, on the properties of the protein corona of the individual nanoparticles [161,162,166,181–188]; and, finally, on the cycle phase of the living cell [189]. The nanoparticle properties mainly designate their endocytosis route, but, in many cases, the cell can internalize the nanoparticles by utilizing distinct mechanisms, which are also related to these parameters, as illustrated in Figure 9 [190].

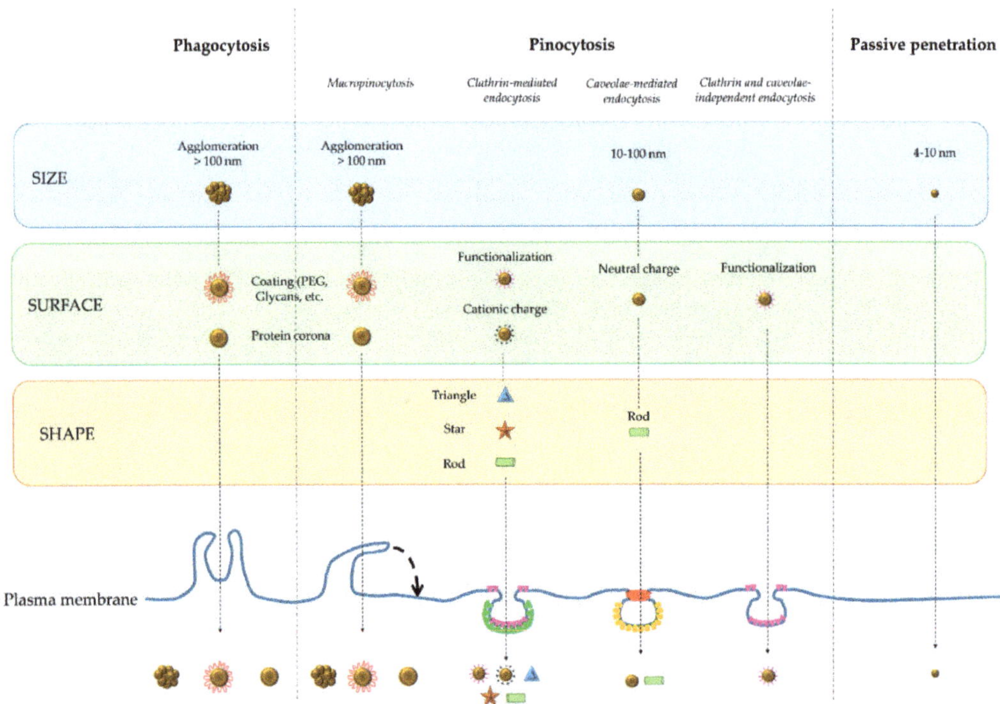

Figure 9. Different cellular internalization mechanisms in relation to the nanoparticle properties, such as size, surface functionality, and shape. The cell can internalize the nanoparticles by using different mechanisms, taking into account the same parameters [190].

The effect of size on the cellular uptake of nanomaterials is a central issue in the field of Nanobiology [191]. In this context, for the development of suitable cell-tracking and drug-carrier nanoparticle systems, nanoparticle size is considered an important parameter, since it determines the mechanism and rate of cellular uptake of the nanoparticle and its ability to permeate through tissues [192,193]. An equation has been formulated to calculate the minimum radius of a nanoparticle (R_{min}) required for full wrapping; this R_{min} is determined by the energy released from the ligand-receptor binding (adhesion strength) and the energy needed to bend the membrane (membrane rigidity). Thus, the dependence of cellular uptake on the nanoparticle size and shape has been extensively investigated [194].

3.1.2. Effects of Geometrical Characteristics on Cellular Uptake

Well-dispersed amorphous silica nanoparticles were utilized to investigate their uptake, localization, and cytotoxic effects in mouse keratinocytes (HEL-30) [195]. In that study, the cells were cultured for 24 h using different concentrations of SiO_2 nanoparticles with an 30–535 nm average particle size; the cells were assessed for particle uptake and biochemical changes. TEM analysis revealed that all silica particles were successfully taken up into the cells independently of size and were localized into the cytoplasm. Moreover, the interplay between silica nanoparticles of different sizes affecting the cellular uptake with Hela cells in serum-free medium has recently been reported [196]. When the cells were co-exposed to silica nanoparticles of different sizes, the bigger nanoparticles significantly promoted the cellular uptake of the smaller ones, while the smaller nanoparticles inhibited the cellular uptake of the larger ones. In fact, this was observed, even when the effects of size were very small or undetectable in the single-exposure experiments.

When surface-functionalized pomegranate-like ferrimagnetic nanoclusters (40–85 nm) were used in vitro, it was shown that the proliferation of spleenocytes, as well as the cytokine production, were consistent with the regulation of immune system cells based on size; it was inferred that small clusters mainly drive immune-stimulatory and inflammatory responses, while large ones could lead to immune-suppressive and anti-inflammatory actions [197].

The effects of the size and surface charge of polymeric nanoparticles on cellular uptake and biodistribution have been investigated [185]. Murine macrophages were found to more efficiently phagocytose nanoparticles with a large size and high surface charge. Even minor differences in the size and/or the surface charge of the nanoparticles had a significant impact on their cellular uptake activating different mechanisms in the endocytosis process. In vivo biodistribution indicated that 150-nm nanoparticles with small negative charge showed a tendency to accumulate more efficiently in tumors [185].

The cellular interactions of biologically-active gold nanoparticles as a function of size in the range of 15–55 nm with alveolar macrophages were evaluated. These cells, as professional phagocytes, are the first line of host defense in the lungs, and their potential role in initiating oxidative stress has also been studied. In vitro exposure resulted in morphologically unusual sizes and adherence characteristics, with significant uptake of nanoparticles at high doses after 24 h [198].

Significant differences were observed concerning the uptake of colloidal gold nanoparticles of different sizes and shapes [181]. More specifically, the intracellular concentrations of rod-shaped nanoparticles (74 × 14 nm) differed from those of either 74 or 14 nm spherical nanoparticles. These results were attributed to the difference in the curvature and the active surface area between rod-shaped and spherical nanoparticles: the rod-shaped nanoparticles actually have a larger contact area with the cell membrane receptors than the spherical ones when the longitudinal axis of the rods interacts with the receptors. An alternative explanation is related to differences in the distribution of the surfactant molecules adsorbed on surfaces with different curvatures during the synthesis of the nanoparticles, which may affect the homogeneity of the serum protein coating and, thus, the effective binding to receptors [181].

Generally, it is suggested that the receptor–ligand binding constants, the receptor recycling rates, and exocytosis can be mediated by the size and the shape of the nanoparticles. A significant number of studies have shown that geometry, in addition to the size of nanoparticles, determines the rate of uptake and, importantly, the uptake mechanism used by nanoparticles. More specifically, experimental studies using different cell types have shown that spherical nanoparticles undergo a higher cellular uptake than rod-shaped nanoparticles [181,182,199]. Moreover, some cylindrical nanoparticles of different materials, such as carbon nanotubes, iron oxide, and polymers, have demonstrated enhanced circulation and retention times compared to their spherical counterparts [200–203]. The in vitro responses of U87 glioblastoma cells to various types of gold nanomaterials (13-nm spheres, 50-nm spheres, and 40-nm stars) conjugated with siRNA were studied; a much higher uptake efficiency was observed for the 50-nm spheres and the 40-nm stars when compared to the 13-nm spheres, as illustrated in Figure 10 [204].

The geometry of nanoparticles appears to also affect the mechanism of their endocytosis. Cellular uptake inhibition experiments indicated that the endocytosis of spherical silica nanoparticles is mainly carried out by a clathrin-mediated mechanism, while most of their rod-like counterparts penetrate the cell membrane via macropinocytosis or phagocytosis [205]. However, functionalization of the nanoparticles seems to modify the manner of their internalization [206,207].

Saturation of the intracellular nanoparticle concentration within hours has been reported [181,208], whereas other reports indicated saturation after several days [209–211]. Moreover, the kinetics and the saturation concentrations were reported to strongly depend on the nanoparticle dimensions [181]; however, the saturation rate of their uptake seemed to depend on the number of available free proteins, which are not adsorbed on the nanoparticle surface in the medium, since these unbound proteins may compete for the receptor binding sites of the cell surface with those proteins adsorbed on the nanoparticle surface.

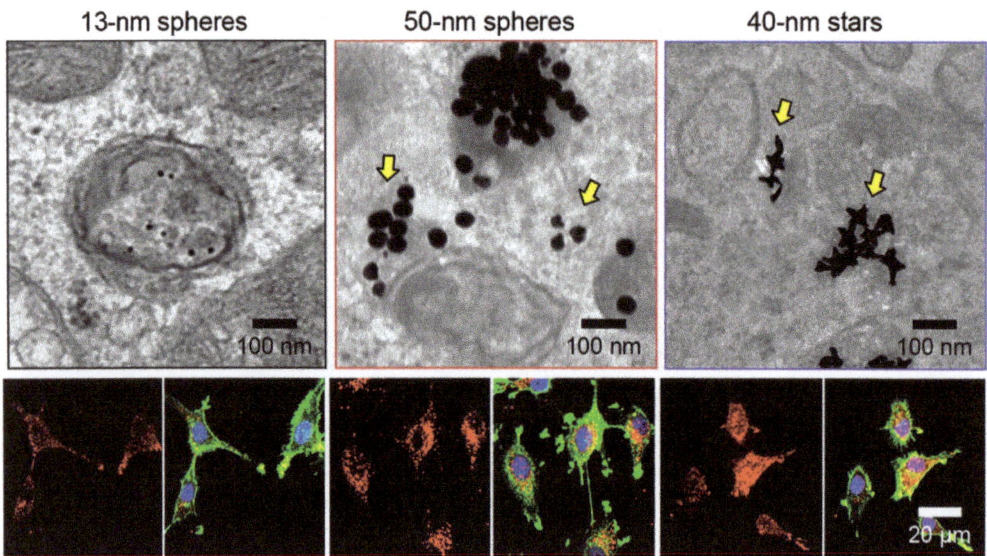

Figure 10. Dependence of the yield of cellular uptake and the intracellular distribution of gold nanoparticle–siRNA constructs on nanomaterial size and shape. In vitro response of U87 glioblastoma cells to various types of nanoconstructs. Transmission electron microscopy (TEM) images (**top row**) and confocal fluorescence microscopy images (**bottom row**) revealing the 13-nm spheres located within endocytic vesicles, with the 50-nm spheres and 40-nm stars being aggregated, and some being outside of the endocytic vesicles (yellow arrows in top row). In the fluorescence images, the actin cytoskeleton and the nucleus were stained with Alexa Fluor 594 Phalloidin (green) and DAPI (blue), respectively, whereas the nanoconstructs were labeled with Cy5 (red) (Reprinted with permission from Ref. [204]. Copyright 2017 American Chemical Society).

In order to avoid complications due to the sedimentation of nanoparticles in typical cell cultures, upright and inverted cell culture configurations were utilized. These kind of cell experiments illustrate that the cellular internalization of gold nanoparticles depends on their sedimentation and diffusion velocities and not on their size, shape, surface coating, density, and initial concentration. It was also found that more nanoparticles were endocytosed in the upright configuration than in the inverted one, whereas larger differences in uptake between the two configurations were observed for nanoparticles exhibiting faster sedimentation rates. It is, therefore, considered that for in vitro studies with large and/or heavy nanoparticles, sedimentation needs to be taken into serious consideration.

3.1.3. Effects of Surface Charge and Surface Coating on Cellular Uptake

Experimental and theoretical studies have investigated the effect of charge, hydrophobicity, and interfacial forces on the interaction between nanoparticles and lipid bilayer assemblies, in order to understand the interactions of the nanoparticles with the membrane and the mechanisms that affect their cellular influx, as well as the cytotoxicity and inflammatory effects [180,212–215].

Molecular dynamics simulations confirmed that electrostatic interactions dominate over the hydrophobic ones when considering nanoparticles, with the bilayer with charged nanoparticles interacting more favorably than their uncharged counterparts. More specifically, the adhesion of anionic nanoparticles more strongly influences the membrane structure when compared to cationic nanoparticles, which can promote local disorder in the area of adhesion, as well as membrane-wrapping phenomena [216,217]. In another study, computed results indicated that the initial orientation of non-spherical nanoparticles can be significantly affected by surface charge density; thus, enhancement of the translocation rate and maximizing the cell adhesion can be achieved by engineering the interplay of nanoparticle shape and surface charge density [218].

Additionally, a number of experimental studies have elucidated the impact of surface charge on the interaction between nanoparticles and cell membranes. In agreement with theoretical models, it has been shown experimentally that cationic nanoparticles strongly bind to the cell membrane, through electrostatic interactions with the lipid phosphate groups, increasing the surface tension of the membrane and resulting in the formation of pores [219]. It has also been reported that negatively or positively charged nanoparticles preferentially interacting with the choline-phosphate dipole (N^+/P^- terminus) of the lipid membranes, respectively, could cause the surface reconstruction of phospholipid membranes [220]. Charged nanoparticles tend to adsorb more proteins from the serum compared to neutral nanoparticles [180]. It was demonstrated that large amounts of plasma proteins were adsorbed on positively- or negatively-charged decorated gold nanoparticles, whereas relatively few proteins adsorbed onto neutral ones [221]. Mesoporous silica nanoparticles (MSNs), such MSNs modified with two different silanes, in order to produce mixed-charge amino-phosphonate pseudo-zwitterionic MSNs under physiological conditions (ZMSN-1.5) and of PEGylated MSNs were studied with respect to their internalization by flow cytometry and laser scanning confocal microscopy experiments. It was shown that cell uptake was drastically reduced for the functionalized nanoparticles, both for the pseudo-zwitterionic ZMSN-1.5 and for the PEGylated ones; this is illustrated in Figure 11 [222].

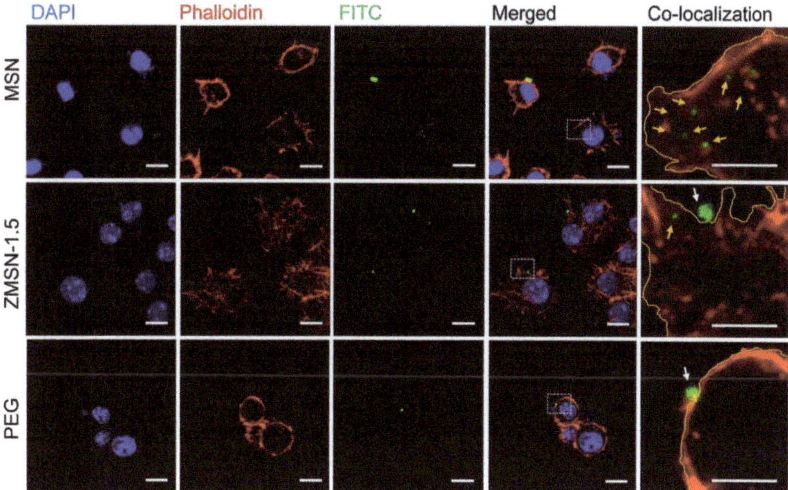

Figure 11. Dependence of the cellular uptake of bare mesoporous silica nanoparticles (MSNs), pseudo-zwitterionic ZMSN-1.5, and control PEGylated MSNs by RAW 264.7 macrophages. Laser scanning confocal microscopy images of the nuclei (DAPI), membrane (Phalloidin), and nanoparticle (FITC) emission channels are shown. Merged images and high magnification merged red-green channels overlain allow co-localizing the different systems studied. In the co-localization right row area, selection of region of interest was made with FiJi, marking in yellow the cell membrane border. Internalized nanoparticles are highlighted with yellow arrows, while those located in the outer area are marked with white ones (Scale bar: 10 μm, 5 μm for co-localization row) (Reprinted with permission from Ref. [222]. Copyright 2019 Elsevier).

Molecular dynamics computer simulation has suggested that the insertion of hydrophobic nanoparticles could lead to deformation and heterogeneity of the lipid bilayer, but that this would not cause membrane leakage, while semi-hydrophilic nanoparticles appear to be energetically absorbed on the surface of the bilayer, thus, inducing their endocytosis [223]. In other theoretical or experimental studies, different nanoparticles were used to investigate the influence of hydrophobicity on the elastic properties of cell membranes, on the stability of pre-existing pores in the lipid bilayer, on membrane penetration, and, therefore, on cell function [224–228].

Surface functionalization of nanoparticles by modifying their surface chemistry, charge, and hydrophobicity can obviously alter their targeting efficacy and cellular uptake rates. Indeed, increasing the number of amino groups ($-NH_2$), which enhances the positive surface charge, was shown to increase the internalization of nanoparticles into cells. However, the presence of –COOH functional groups, which increases the negative charge, enhances their further uptake into the endosomal compartments [229,230]. In different studies, it has also been reported that functionalized nanoparticles, such as polydopamine functionalized nanoparticle-aptamer bioconjugates, folic acid-functionalized nanoparticles, and poly(diallyldimethyl ammonium chloride)-coated gold nanorods, have better targeting efficacy and higher efficiency of internalization by cells [231–233].

As already mentioned, nanoparticles enter the cells through active processes because of their ability to interact with the cellular machinery. When the nanoparticles come into contact with biological fluids, such as the serum of a cell, a selective layer of proteins and other biomolecules adsorbs on their surface within a few seconds, forming the so-called corona [234], which mediates, in situ, the interactions with cells. As a consequence, one nanomaterial may cause a very different biological outcome when exposed to cells in the presence or absence of a preformed corona. More specifically, silica nanoparticles exhibited stronger adhesion to the cell membrane and higher internalization efficiency when they were exposed to cells in the absence of serum, as compared to those in a medium containing serum, where a corona was formed. The different conditions of exposure not only affected the levels of uptake but resulted in variation in the location of the intracellular nanoparticles and their impact on the cells. It is important to note that certain studies showed that, after just 1-h of exposure, a corona of very different nature can be formed on the nanoparticles exposed to cells in the absence of serum. This different outcome was attributed to the different adhesion and surface properties under the two conditions [234]. The protein adsorption capability is also affected by the nanoparticle properties. For example, both surface roughening and hydrophobic modification of the nanoparticles enhance the protein adsorption capacity and affect the cellular uptake performance; however, the relative importance of the two contributions depends on the cell type [235,236].

3.1.4. Role of Cell Type on Cellular Uptake

The role of cell cycle in the cellular uptake and dilution of nanoparticles in a cell population has also been investigated, as illustrated in Figure 12 [189,237]. It has been observed that the cellular uptake of nanoparticles is also influenced by the cell cycle phase Although more-or-less similar rates of nanoparticle internalization by the cells were observed for different phases of the cell cycle, after 24 h, the concentration of nanoparticles in the cells could be ranked according to the different phases, as follows: G2/M > S > G0/G1, where G0 is the resting phase, G1 is the phase during which the cell increases its size, S the phase when the cell synthesizes DNA, G2 the one it synthesizes proteins to prepare for cell division, and M the phase when the cell divides and the two daughter cells enter the G1 phase. During cell division, nanoparticles that are internalized by the cells are not exported but are split between daughter cells. Thus, it was indicated that, in a cell population, the dose of internalized nanoparticles in each cell varied as the cell advanced through the cell cycle.

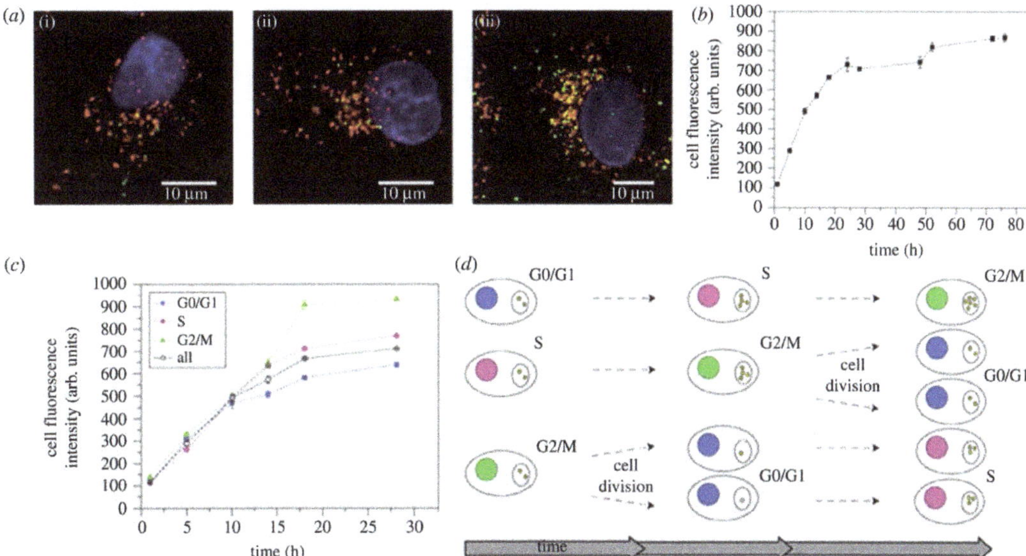

Figure 12. Dependence of the internalization of ~40 nm carboxylated polystyrene nanoparticles (25 µg/mL in cMEM) in A549 human lung carcinoma cells on the cell cycle phase for exposures up to 72 h. (**a**): Confocal microscopy images after cell exposure to nanoparticles for (**i**) 5, (**ii**) 24, and (**iii**) 72 h show the nanoparticle accumulation in the lysosomes. Blue: cell nuclei (DAPI); red: lysosomes (LAMP1 antibody); green: nanoparticles. (**b**): Mean cell fluorescence intensity as acquired by flow cytometry as a function of time. (**c**): Mean fluorescence intensities as a function of time of A549 cells in the G0/G1, S and G2/M phases, respectively. (**d**) Schematic of populations of the G0/G1, S, and G2/M phases by cells and consequences for cellular NP content as a function of time (Adapted with permission from ref. [189] and ref. [237]. Copyright 2012 Nature Publishing Group and 2013 Royal Society Publishing.

In general, nanoparticles, due to their ability to be endocytosed, cause completely different cell responses from bulk surfaces of the same material. In spite of what has been achieved so far in the materials and nanotechnology fields, a complete understanding from a biological point of view is still missing. In this context, emerging technologies such as omics, high-throughput screening systems, and organ-on-a-chip technologies, in synergy with computational approaches, should enable, not only the analysis and documentation of large amounts of data, but also the decoding of nano–cell interactions [178,238].

3.2. Optical and Electronic Properties and Catalytic Activity

The nanometer size of manufactured nanomaterials results in very interesting and very important size effects that affect their chemical, structural, thermal, spectroscopic, electronic, magnetic, and mechanical properties; these effects are on top of any possible influence of the chemistry of their bulk crystals. This is schematically illustrated in Figure 13 [239].

Moreover, a single manufactured nanomaterial (MNM) may function differently in various systems; thus, it is important to carefully design MNMs to develop devices with enhanced performance, safety, and stability for both humans and the environment. While material chemistry and nanomaterial size and shape play a significant role in the core properties of an inorganic nanoparticle, the selection of ligand molecules, which functionalize the surface of the MNMs, is of great significance for their colloidal function and stability [240]. In this part of the work, a series of studies on the key properties of MNMs affecting the functionalities relative to applications are discussed; the emphasis is on the electronic and optical properties and the catalytic activity of materials and devices. It is noted that most functionalities of this type are correlated with the MNM's key properties.

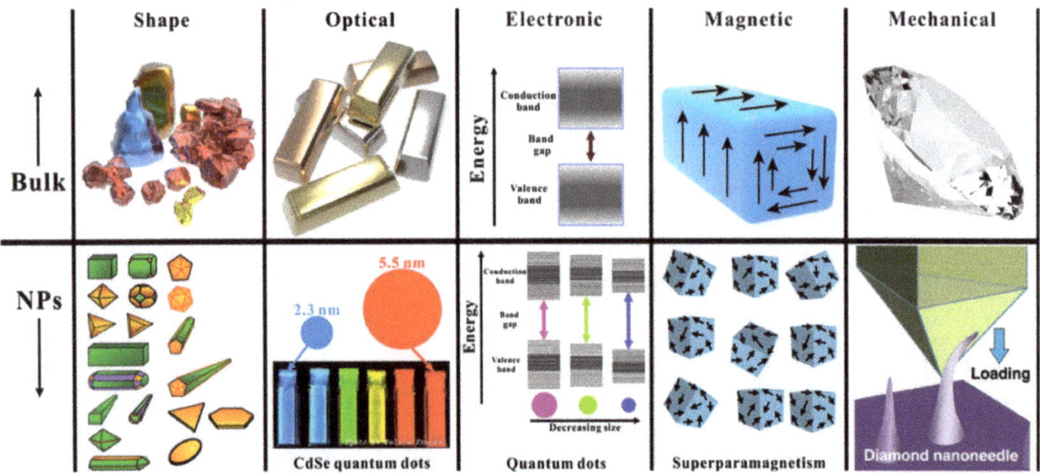

Figure 13. Schematic comparison between bulk materials and nanomaterials: nanoparticles with varying mechanical, electronic, optical, and magnetic properties, due to their different size and shape [239].

3.2.1. Catalytic Properties

Generally, the catalytic properties of nanomaterials are far superior compared to bulk materials. ZnO nanomaterials are characterized as possible candidates for transistors, solar cells, light-emitting diodes, sensors, nano-lasers, photocatalysts, and antimicrobial agents because of their good stability, low cost, high excitation binding energy (60 meV), wide band gap (3.37 eV), and widespread availability [241]. Moreover, ZnO properties could be enhanced by doping with elements such as Mg [242], Al [243], and Cu [244,245]. In particular, Cu-doping of ZnO nanomaterials improved the optical properties by creating impurity levels localized in the optical energy band gap [246]. Furthermore, the optical energy band gap is reduced when the average size of the crystallites decreases, because Cu ions are incorporated into the ZnO structure [247]. The catalytic activity of ZnO nanomaterials in the presence of light has been widely investigated for environmental applications (e.g., purification), and this was found to depend on the oxygen vacancies and the morphology of ZnO. Specifically, the photocatalytic performance of ZnO nanodisks for the decomposition of methylene blue dye was enhanced because of the higher population of (0001) crystal plane structures [248]. Furthermore, ZnO nanorods with a cone of small aspect ratio are more effective in the photocatalytic degradation of organic pollutants than ZnO nanorods with a cone of large aspect ratio and ZnO microrods that are short-and-fat [249]. Moreover, ZnO nanosheets and nanoflowers demonstrated a much higher photocatalytic activity for the degradation of methyl orange than ZnO nanospheres [250]. The decomposition of volatile organic compounds, such as butane, was investigated, taking advantage of the photocatalytic activity of ZnO nanomaterials over multi-channel porous alumina ceramic membranes coated with ZnO nanoparticles, nanorods, and nanowires; the activity depended strongly on the shape of the nanomaterial used [251]. It was reported that ZnO nanowires showed a higher catalytic activity than ZnO nanoparticles or nanorods and, most importantly, the process did not result in unwanted byproducts such as propane, acetaldehyde, and acetylene. Moreover, better carbon balance and selectivity towards carbon oxides were obtained with the ZnO nanowires and nanorods than with nanoparticles. ZnO structure, shape, and crystallite size are also important parameters for their antimicrobial performance [252]. ZnO nanoflowers showed enhanced photocatalytic activity in Escherichia coli and Staphylococcus aureus inactivation compared to ZnO nanorods or nanospheres.

The optimization of catalytic performance requires the adjustment of both catalytic activity and mass transfer. Various bioinspired inner-mobile multifunctional ZnO/CdS heterostructures have been synthesized, with their artificial cilia mimicking natural ciliary motion (assisted by external magnetic fields and internal magnetism). Such a synthesis resulted in a three-times better photocatalytic performance of mobile arrays compared to static arrays [253].

3.2.2. Sensing Behavior

A bioelectrochemical sensing interface can be engineered with functional nanomaterials, so as to develop novel electro-chemical biosensors with enhanced performance in terms of simplicity, sensitivity, selectivity, and stability [254]. It should be noted that the use of functional nanomaterials for the development of novel biosensors takes advantage of nanomaterial properties such as conductivity, high surface area, and improved catalytic activity; and such properties depend on the size and shape of the nanomaterials, which control, e.g., the optical properties of metal nanoparticles [255], the electrical conductivity of the carbon nanomaterials [256], as well as the electrocatalytic properties of nano-carbons and metal nanoparticles [257], etc.

Carbon nanomaterials (CNMs) exhibit unique electrical, optical, thermal, mechanical, and chemical properties and are, thus, extensively applied in photovoltaic, electronic, optoelectronic, and sensing devices. A more recent application of CNMs in the biosensing field is their use in the area of electrochemical aptasensors (ECASs) [258]. ECASs use aptamers (short single-stranded oligonucleotides of DNA or RNA), selected through a systematic evolution of ligands using an exponential enrichment technique (from a random oligonucleotide library), as recognition elements and exhibit the advantages of low cost, simple operation, fast response, and high sensitivity. A concentration- or activity-related electrochemical signal is produced by the transducers as a result of the recognition reaction. Clinical diagnosis via DNA analysis, immunoassay, or enzymatic sensing, as well as for environmental monitoring, including ocean and atmospheric pollutants, are the main detection strategies [258].

The use of carbon nanomaterials significantly increases the detection efficiency of sensors, in terms of sensitivity, selectivity, and stability, and has become one of the current development strategies for ECASs-based sensing platforms. The excellent electrical conductivity and high specific surface area of the CNMs allow them to function as electronic conductive matrices and immobilization platforms for the aptamers [258,259]. These properties depend on the atomic structures of the different CNMs, such as graphene, graphene oxide, carbon nanotubes, etc., as well as on their interactions with other nanomaterials, such as chitosan, silica, or gold nanoparticles. In particular, carbon nanotubes (CNTs) are commonly used as catalyst carriers or backing layers. CNTs demonstrate an enhanced electro-catalytic activity and a very large surface area to volume ratio, with multi-walled carbon nanotubes (MWCNTs) being used more often in ECASs applications than single-walled carbon nanotubes (SWCNTs). Moreover, combining CNTs with other nanomaterials (e.g., gold nanoparticles, reduced graphene oxide, dendrimers, chitosan, etc.) can further enhance the carrier content and stability of enzymes and proteins. Graphene, graphene oxide, and reduced graphene oxide have also been utilized in ECASs [260,261], with the main differences in this application originating from their significantly different electrical conductivities; the effectiveness of these three types of CNMs follows their ranking of conductivities, with graphene being preferable for ECASs development, followed by reduced graphene oxide and, then, graphene oxide.

Improved device performance and notably enhanced electrical properties were reported when SWCNTs were assembled into aligned arrays with full surface coverage (via the Langmuir–Schaefer method). The intrinsic mobility of the CNTs was preserved for a semiconducting nanotube purity of 99% and full surface coverage and, thus, for high packing density [262].

The use of carbon nanomaterials to construct functional composites was reviewed [263], and effective methods were presented to achieve light harvesting and conversion, effective phonon transport along a particular direction, and rapid ion and electron motion in structural electrodes through the chemical grafting of functional groups to improve their reactivity and thermal stability [263]. Moreover, novel optical-triggered graphene-based actuators were fabricated with a bilayer structure including chitosan and polyethylene (PE) over a large area [264]. The graphene nanosheets played the role of a connecting bridge between light and the conversion of light energy at the nanoscale.

The hybridization of different types of carbon nanomaterials has been utilized to enable many different properties and performances beyond that of the individual nanomaterials, for example in electrochemical or analytical devices. Hybrid nanomaterial systems are, in principle, designed to develop more efficient sensors. Each nanomaterial exhibits its own advantages for various applications; thus, it is important to involve synergies due to the presence of the different nanomaterials, so as to complement each other in the hybrid system [265,266]. For example, graphene–inorganics composites that take advantage of the properties of both graphene and the inorganic elements (e.g., gold nanoparticles) enable even higher active surface areas and enhanced rates of electron transfer. Thus, functional hybrids are developed based on graphene nanosheets, in order to take advantage of the electrical, optical, and catalytic properties of graphene and enhance its performance in analytical chemistry and electrochemistry [256].

MWCNT-modified electrodes have been used to investigate the electrochemical oxidation of nicotinamide adenine dinucleotide (NADH) and to elucidate their respective mechanisms of oxidation [257]; the study compared the behavior with cases when boron-doped diamond and glassy carbon electrodes were used, as well as with cases when edge plane and basal pyrolytic graphite electrodes were utilized, which allowed the reactive sites of carbon nanotubes to be deduced. It was concluded that electron transfer was more facile with samples containing a higher proportion of edge plane defects, compared to basal plane graphite electrodes. It was, thus, indicated that electroanalytical sensors with carbon-based electrodes should optimally possess a large proportion of edge plane sites, for achieving the best detection limits, whereas edge plane pyrolytic graphite electrodes can conveniently replace CNT-modified electrodes for routine sensing of NADH, due to their simple preparation process, low detection limit, low susceptibility to fouling of the electrode, and insensitivity to interference from ascorbic acid. It was demonstrated that an electrode produced fully of edge plane graphite (disc of pyrolytic graphite with the disc surface facing parallel with the edge plane) displayed high levels of electro-catalytic activity for different electroanalytical tasks, including gas sensing [267] and thiol oxidation [268].

Carbon nanotubes exhibit a quantum electron confinement normal to the nanotube axis, thus, being able to transport electrons over long lengths [269]. They have great potential as biomolecule immobilization platforms. According to some studies, CNTs/polymer nanostructured composites developed on electrodes can improve the analytical performance of amperometric biosensors [270,271]. Such composites display percolation behavior, by remarkably enhancing the electrode conductivity. Moreover, the CNTs thermal and electrical conductivity and their electrocatalytic activity can be modified by doping of the CNTs with elements such as K, B, Ce, N, Si, P, etc. [272,273].

Furthermore, multifunctional CNTs offer routes towards the production of smart and high-performance sensors, logic gates, and similar optoelectronic devices [274]. By combining CNTs with photochromic molecules, and in particular by decorating them, reversible changes in the geometrical structure, the electronic properties, and the nanoscale mechanics triggered by light can be achieved [274]. As a result, there is control of the local variation in the optical, electrostatic, and mechanical environment with light illumination. For example, azobenzenes blended with CNTs and polymers are used to form nanocomposites possessing light-induced conductance switching properties; such nanocomposites are good candidates for electro-optical memories, smart packaging, and smart window applications [275]. A graphene/azobenzene/Au heterostructure switch was found to further induce the reversible modification of the electrical and quantum properties of the Dirac fermions of graphene [276]. Furthermore, a hybrid system of chemically grafted spiropyrans to CNTs was utilized to regulate horseradish peroxidase (HRP) activity via light illumination. This resulted in enhancement of the catalytic activity of HRP and was used as a label-free colorimetric lysozyme assay with a detection limit of 30 nM. This high selectivity approach can be applied to regulate the activity of other natural proteins using light [277].

3.2.3. Optoelectronic Properties

Certain nanomaterials are used as biomolecular labels because they exhibit unique optical properties. They amplify biorecognition signals and enhance the biosensor sensitivity [269]. Various nanoparticles, including metal, oxide, or semiconductor nanoparticles and their composites, have been widely used in the fields of biosensors and electrochemical sensors [278]. The majority of the nanoparticles possess a high isoelectric point (IEP), favoring electrostatic protein adsorption with low IEP. Thus, they are promising supports for protein immobilization. A cholesterol biosensor consists of an interfacial layer of gold nanoparticles, which is used for immobilizing cholesterol oxidase on gold electrode surfaces. Here, gold nanoparticles provided an environment for the enhanced electrocatalytic activity of cholesterol oxidase and, thus, improved the stability of the biosensor [279]. The gold nanoparticles were found to favor the analytical performance of the cholesterol biosensors; this was attributed to the biocompatibility of the gold nanoparticle-based immobilization matrices, to assist proteins in retaining their biological activity for long periods and, thus, improve the stability of the biosensor [269]. The enhancement of the sensitivity and selectivity of the biosensor was mainly due to the electrocatalytic activity of the gold nanoparticles; gold nanoparticles improved the conductivity of the electrodes and facilitated the electron transfer between the electrode and the enzyme redox center. Gold nanoparticles on flat electrode surfaces may also partially penetrate the enzyme matrix and, thus, come closer to the enzyme redox center, which further aids the electron transfer pathway.

Interesting nanomaterials include the helical carbon nanofibers (CNFs), with excellent optical, electromagnetic, and mechanical properties, due to their unique spiral structure; aiming at applications such as microwave absorbing materials and electrode materials [280]. To improve the optical, physical, mechanical, and chemical properties of CNFs, more functional building blocks were incorporated, to form CNF-based composites. An example is the in situ synthesized mesoporous N-CNFs containing graphitic-C_3N_4 (g-C_3N_4), in which the strong coupling between the components of the CNFs enabled the final material to have an efficient optical storage performance, improved charge separation, and multi-dimensional electron transport path; thus, improving the performance of hydrogenation production, as well as the performance in photocatalytic and optoelectronic applications [281].

Another application of gold nanoparticles in the medical field is in cardiac tissue engineering, due to their controlled geometrical, surface, chemical, and optical properties [282]. Additionally, gold nanoparticles enhance the electrical conductivity of nanocomposites. High electrical conductivity, acceptable biocompatibility, the capability for surface modification, nanotopography, and innate optical properties make this nanoparticle type a desirable nanostructure for cardiac scaffolds [283].

Metal oxide nanoparticles are able to achieve low detection limits in analysis, due to their electron transfer [284,285]. Moreover, the capability for enhanced adsorption of the biomolecules leads to high biosensor stability. Cerium oxide (CeO_2), iron oxide (Fe_3O_4), zinc oxide (ZnO), and titanium oxide (TiO_2) nanoparticles have been exploited for improving sensor performance [286–288]. Gold nanoparticles exhibit outstanding optical properties as well; this is due to the surface plasmon resonance (SPR) phenomenon, when the light interacts with the collective oscillations of electrons on the gold nanoparticle surface at a certain light wavelength [269]. This depends on the shape, size, and state of aggregation of the gold nanoparticles. An important application is in the field of detection assays, where an alteration of the light extinction that results from the aggregation of gold nanoparticles upon analyte addition is used as the optical signal [289].

The incorporation of nanoparticles into various building blocks within the solar cell architecture, in order to enhance photovoltaic performance and stability, has also been reported [290]. It was observed that the conversion efficiency of solar cells with silicon nanocrystals was 5.3-times higher than one with only titania (TiO_2) particles, contributing to further light absorption and, thus, to an improvement of the conversion efficiency. Further incorporation of nanoparticles such as Ag and Au, produced via laser ablation in liquids, into the active/hole transport layer interface of P3HT:PCBM bulk heterojunction solar cells was reported to lead to an enhanced conversion efficiency [291]. The role of ligand coatings on nanoparticles in the photovoltaic performance has also been discussed, as illustrated in Figure 14 [292].

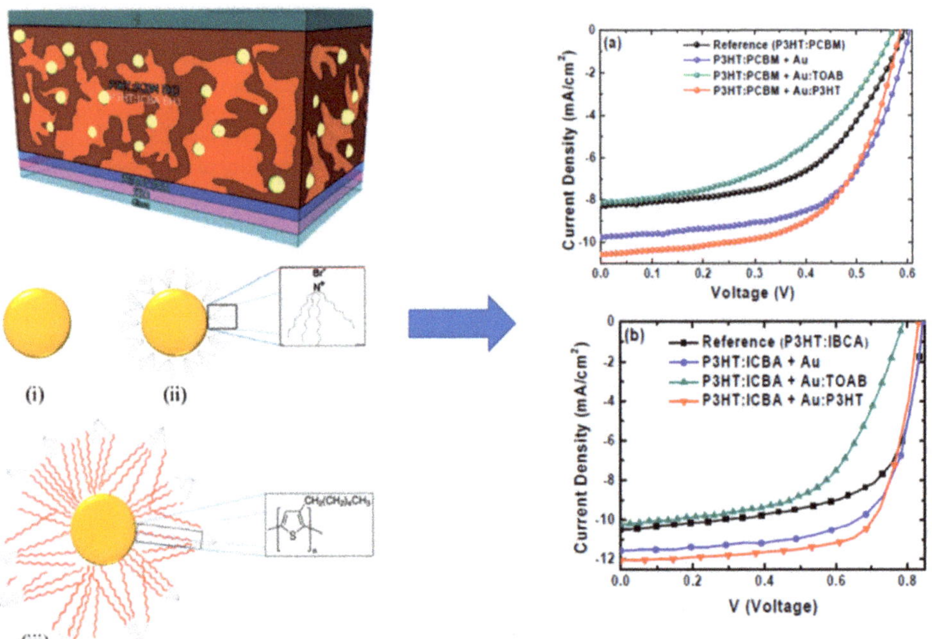

Figure 14. Schematic representation of a bulk heterojunction organic photovoltaic cell with three kinds of nanoparticles within the active layer: (**i**) bare, (**ii**) TOAB-functionalized, and (**iii**) P3HT-functionalized. J−V curves of the devices with configurations (**a**) ITO/PEDOT:PSS/P3HT:PCBM/Al and (**b**) ITO/PEDOT:PSS/P3HT:ICBA/Ca/Al, respectively (Reprinted with permission from ref. [240], Copyright 2019 American Chemical Society) with the original data from Ref. [292], Copyright 2015 American Chemical Society). Nomenclature: ITO: indium tin oxide; PEDOT: poly(3,4-ethylenedioxythiophene); PSS: poly(styrene sulfonate); P3HT: poly(3-hexylthiophene-2,5-diyl); PCBM: [6,6]-phenyl-C_{61}-butyric acid methyl ester; ICBA: indene-C60 bisadduct.

The chemical, optical, electrical, thermal, and magnetic properties of magnetic nanoparticles can also be exploited in various steps of analytical processes, including sample treatment, chromatographic techniques, and detection [293]. Iron oxides (Fe_2O_3 and Fe_3O_4) and their corresponding ferrites (e.g., $MnFe_2O_4$ or $CoFe_2O_4$) are commonly utilized because of their biological compatibility, the simple preparation processes, and high magnetic moment relative to other nanoparticles based on metals and alloys (e.g., Mn_3O_4, Co, Ni, FePt), which exhibit rapid oxidation in air and/or potential cytotoxicity. Magnetic nanoparticles can be modified with inorganic, organic, or biochemical compounds to improve their physicochemical behavior. For example, hybrid magnetic nanoparticles are developed by the combination of Fe_3O_4 nanoparticles and carbon, metallic, polymeric, or silica nanoparticles for the manufacturing of electrodes, thus improving their electrocatalytic properties, among others [294]. Such electrodes are advantageous, due to their large surface area, low resistance to electronic transmission, and ability to adsorb (bio)chemical analytes, which make them useful in electrochemical systems. The main advantages of magnetic nanoparticles in this area are the increase of electrocatalytic activity, the minimization of deterioration of the electrode surfaces, and the simplification of the immobilization process [293].

Last, we would like to point out that in this mini review we have mostly discussed the behavior of single nanoparticles, and not nanoparticle assemblies [295]. The formation of the latter is mostly induced by the very high surface energy of the nanoparticles, because of their high specific surface area; this provides the driving force for the spontaneous aggregation of the nanoparticles, which would decrease the Gibbs free energy of the system and would lead to large assemblies. In these cases, the performance of the nanoparticles for various applications will be based on the coupling of, and cooperation among, individual nanoparticles, rather than on their individual properties; this collective behavior would, of course, depend on the interparticle interactions that would determine their structural arrangement in space [239]. Such nanoparticle assemblies may lead to a plethora of practical applications, such as sensing, energy storage, strong materials, catalysis, therapies, etc. Moreover, introducing different nanoparticles into a superlattice can lead to substitutional doping when the size of the two types of nanoparticles are similar, in an analogy to the classical doping process where atomic impurities are intentionally added to a host material to significantly modify its properties; the electronic properties of such doped superlattices are significantly influenced by the presence and density of the nanoparticle dopants, leading to highly tunable nanomaterials [296].

4. Concluding Remarks

Nanotechnology, which deals with the understanding and control of matter in dimensions between about 1 and 100 nanometers and where unique phenomena allow new applications, has enabled the development of a variety of nanomaterials with unique properties, aimed at various applications. Thus, it becomes apparent that the interaction of nanomaterials with their environment is governed by different mechanisms and leads to new responses.

To summarize the main points of this literature review, the key parameters of manufactured nanomaterials that play an important role for each of the functionalities are outlined below:

The **dispersion ability** of the nanomaterials is a key issue affecting their behavior. Nanoparticles form, in general, aggregates and/or agglomerates in water or other aqueous media; SiO_2 nanoparticles are the only exception, where the primary particle size is detected in certain cases. The dispersion ability is affected by the particle chemical composition, the existence of an appropriate surface coating, the surface charge, as well as by the dispersion media, whereas it depends only weakly on their shape and crystallinity. The particle size is not that crucial in determining dispersibility, except when nanoparticles and particles with radii larger than 300–400 nm are compared, because of the influence of gravity. Apart from the nanoparticles themselves, the presence of organic moieties in the solution (e.g., proteins), the solution pH and its ionic strength affect dispersibility.

The **hydrophobicity/hydrophilicity** of nanoparticles and other manufactured nanomaterials depends on their chemical characteristics (chemistry, surface charge) and their surface coating (characteristics, surface reactivity and stability). Besides the effects of hydrophobicity/hydrophilicity on the dispersibility, with hydrophilic nanoparticles being more easily dispersed in aqueous media than hydrophobic ones, nanoparticle hydrophobicity/hydrophilicity is also very important for their biocompatibility. Hydrophobic nanoparticles can be rendered hydrophilic by appropriate modification of their surface using surfactants or various hydrophilic polymers.

Solubility/dissolution of the nanoparticles implicates the release of ions from the nanomaterials into the solution. It is a function of the nanoparticle characteristics, such as chemistry, composition, size and surface area, surface coating, and crystallinity. It is also affected by the pH and the temperature of the solution. The dissolution of nanoparticles affects their antimicrobial activity and biocompatibility.

The physicochemical properties of nanoparticles, such as size, shape, and surface properties, control the internalization pathways, thus, playing a pivotal role in **cellular uptake**. In biomedical applications of nanoparticles, their coating modification has been shown to affect the modulation of their cellular internalization. It is important to take into consideration the possible sedimentation of large and/or dense particles and their diffusion velocities when in vitro studies are performed utilizing large and/or heavy nanomaterials. Moreover, the formation of a protein corona on the nanomaterial surface and its composition play an important role in the possible cellular uptake.

Individual nanomaterials can play various roles in devices in the field of biosensing. Depending on the desired application, their main key parameters should be designed and tuned carefully, whereas composite systems are frequently used to enhance the performance with regards to detection, stability, and duration. **The optical and electronic properties and the catalytic activity of the nanomaterials** are functionalities that depend on their size and shape, whereas the organization of the individual nanomaterials in a hybrid affects the general performance of the various devices.

All of the above findings are illustrated in the two following Tables. Table 1 demonstrates how the three functionalities that have been discussed are affected by the main key parameters, whereas Table 2 shows how the key parameters influence the final properties (optical, electronic, and catalytic properties and the cell uptake). The key parameters discussed have been grouped into six categories, i.e., as geometrical, chemical, crystallinity, morphological, coating related, and test medium related parameters. In the tables, we have introduced the notation of two stars (**) to illustrate that a parameter is a 'priority'; i.e., it significantly determines a particular functionality/property, and the notation of one star (*) to illustrate that a parameter is 'of importance'; i.e., it is important but it does not determine the behavior by itself. According to Table 1, it is clear that key parameters like the chemical composition, the existence of a surface coating, and the test medium are of utmost importance related to all functionalities, whereas the significance of the others should be deduced case by case. As far as Table 2 is concerned, it is the size, the shape, the chemical composition, and the surface charge of nanoparticles that influence, in general, all properties.

One should also point out that an inter-relation exists between the parameters and the functionalities, and this significantly affects the final properties and, thus, the applications in which the nanomaterials are used. Moreover, it is noted that the Nanotechnology Characterization Laboratory (NCL) at the National Cancer Institute USA, which has assessed more than one hundred and thirty different types of nanomaterials, including metal oxides, fullerenes, liposomes, dendrimers, polymers, quantum dots, and gold colloids, came to the conclusion that hydrophobicity (which is a 'functionality'), and size and surface charge (which are 'key parameters') are the main factors that influence nanomaterial biocompatibility [297].

Table 1. How the key parameters of nanomaterials affect performance.

| | KEY PARAMETERS | PERFORMANCE | | |
		Dispersion	Solubility/ Dissolution	Hydrophobicity/ Hydrophilicity
Geometrical	Particle Size (e.g., hydrodynamic radius and polydispersity index)	*	**	*
	Shape	*	**	*
	Aspect Ratio	*	*	

Table 1. *Cont.*

| | KEY PARAMETERS | PERFORMANCE | | |
		Dispersion	Solubility/ Dissolution	Hydrophobicity/ Hydrophilicity
Chemical	Composition	**	**	**
	Surface charge/ζ potential	**	*	*
Crystallinity	Crystal structure/Crystallinity	*		
Morphological	Topology (e.g., core shell, etc.)			
	Porosity		*	
	Surface area	*	*	*
	Roughness		*	*
Coating	Chemistry, Thickness, Topology	**		*
	Surface Coating Stability		**	**
	Surface reactivity		**	**
Test Medium	Kind	**	**	**
	pH	**	**	**
	Ionic Strength	**		*

**: a key parameter designated as 'a priority' (see text); *: a key parameter designated as 'of importance' (see text).

Table 2. How the key parameters of nanomaterials affect their applications.

| | KEY PARAMETERS | APPLICATIONS | | | |
		Cellular Uptake	Optical Properties	Electronic Properties	Catalytic Activity/ Biorecognition
Geometrical	Particle Size (e.g., hydrodynamic radius and polydispersity index)	**	**	**	**
	Shape	**	**	**	**
	Aspect Ratio	**	*	*	*
Chemical	Composition	**	*	**	**
	Surface charge/ζ potential	**	**	**	
Crystallinity	Crystal structure/Crystallinity		*	*	*
Morphological	Topology (e.g., core shell, etc.)				
	Porosity				
	Surface area	*	*	*	*
	Roughness	*			
Coating	Chemistry, Thickness, Topology	**	*	*	*
	Surface Coating Stability		*	*	
	Surface reactivity	*	*	*	*
Test Medium	Kind	**	*		
	pH	*			
	Ionic Strength	*			

**: a key parameter designated as 'a priority' (see text); *: a key parameter designated as 'of importance' (see text).

5. Challenges and Prospects

In order to advance knowledge in the area of the physicochemical properties/ functionalities of nanoparticles, on how these are determined by their key parameters, and, more importantly, on how these influence their behavior and their potential to induce, or not induce, toxicity to both humans and the environment, as well as their ultimate fate more focused research is still needed in this area. Despite the plethora of related works, there are still many open challenges with regards to the interrelationships between the physicochemical main key parameters of nanoparticles and their functionalities, which are considered as very important aspects for enhancing their safety early on in the design process.

Such challenges include:

- Understanding the interdependence between the bulk properties of the materials (i.e., in their pristine state) versus the respective properties when the materials exist in nanodimensions within a particular medium, i.e., dispersed in a biological fluid
- Developing different production/manufacturing routes and different residues
- Understanding and, possibly, modifying different experimental conditions, e.g., instruments, protocols, in vitro versus in vivo methodologies
- Improving the measuring tools for site-specific or local assessment of nanomaterials, e.g., high resolution imaging, 3D reconstruction, data acquisition processes

To improve the design of a nanomaterial, one needs to consider the use of innovative tools to probe the dynamic biophysicochemical interactions. The adoption and optimization of both theoretical and experimental characterization methods, which are traditionally utilized for characterizing the properties of bulk materials, for studies of the environment surrounding nanomaterials and the resulting interfaces is mandatory. This will also be helped by simple and widely accessible laboratory equipment.

Research is, therefore, needed at the interface of different disciplines, such as engineering, physics, chemistry, biology, and medicine. This research should aim at the advanced chemical synthesis of new nanostructures with precisely defined biophysicochemical characteristics and properties, at the development of nanostructures that will replace biological structures, and at addressing the knowledge gaps concerning the possible health and safety effects of exposure to manufactured nanomaterials. Such research will be able to give prominence to nanomedicine as a promising stakeholder in the field of diagnosis, imaging, treatment, therapeutics, and regenerative medicine.

Author Contributions: Conceptualization, all authors; methodology, all authors; writing—original draft preparation, K.C., P.K., G.K. and A.R.; writing—review and editing, S.H.A., K.C. and E.S.; supervision, S.H.A. and E.S.; funding acquisition, S.H.A. and E.S. All authors have read and agreed to the published version of the manuscript.

Funding: The work was partially supported by the European Union within the NANoREG (Grant Agreement Number 310584) and NanoReg2 (Grant Agreement Number 646221) projects. Part of this article was included in Deliverable D.6.6 of project NANoREG and in Deliverable D3.3 of project NanoReg2.

Institutional Review Board Statement: Not applicable.

Informed Consent Statement: Not applicable.

Data Availability Statement: Data presented in this review article are available from the authors of the cited publications.

Acknowledgments: We would like to thank Adrienne Sips, Cornelle Noorlander, and Lya Hernandez of the Institute for Public Health and the Environment (Rijksinstituut voor Volksgezondheid en Milieu, RIVM), The Netherlands, and Thies Oosterwijk of TNO, The Netherlands, for valuable discussions. We would also like to thank Tom van Teunenbroek of the Ministry of Infrastructure and the Environment, The Netherlands, for introducing us to the subject of the NANoREG and NanoReg2 projects.

Conflicts of Interest: The authors declare no conflict of interest.

References

1. Potocnik, J. Commission recommendation of 18 October 2011 on the definition of nanomaterial. *Off. J. Eur. Communities Legis.* **2011**, *275*, 38–40. [CrossRef]
2. Hochella, M.F., Jr.; Mogk, D.W.; Ranville, J.; Allen, I.C.; Luther, G.W.; Marr, L.C.; McGrail, P.B.; Murayama, M.; Qafoku, N.P.; Rosso, K.M.; et al. Natural, incidental, and engineered nanomaterials and their impacts on the Earth system. *Science* **2019**, *363*, 1414. [CrossRef]
3. Stone, V.; Nowack, B.; Baun, A.; Van den Brink, N.; Von der Kammer, F.; Dusinska, M.; Handy, R.; Hankin, S.; Hassellöv, M.; Joner, E.; et al. Nanomaterials for environmental studies: Classification, reference material issues, and strategies for physico-chemical characterization. *Sci. Total Environ.* **2010**, *408*, 1745–1754. [CrossRef]
4. Salieri, B.; Turner, D.A.; Nowack, B.; Hischier, R. Life cycle assessment of manufactured nanomaterials: Where are we? *NanoImpact* **2018**, *10*, 108–120. [CrossRef]
5. Giustia, A.; Atlurib, R.; Tsekovskad, R.; Gajewicze, A.; Apostolova, M.D.; Battistelli, C.L.; Bleeker, E.A.J.; Bossa, C.; Bouillard, J.; Dusinska, M.; et al. Nanomaterial grouping: Existing approaches and future recommendations. *NanoImpact* **2019**, *16*, 100182. [CrossRef]
6. Otterstedt, J.-E.; Brandreth, D.A. *Small Particles Technology*; Springer: Boston, MA, USA, 1998. [CrossRef]
7. Nel, A.E.; Mädler, L.; Velegol, D.; Xia, T.; Hoek, E.M.V.; Somasundaran, P.; Klaessig, F.; Castranova, V.; Thompson, M. Understanding biophysicochemical interactions at the nano–bio interface. *Nat. Mater.* **2009**, *8*, 543–557. [CrossRef]
8. Spahr, M.E.; Rothon, R. Carbon Black as a Polymer Filler. In *Fillers for Polymer Applications*; Rothon, R., Ed.; Springer: Cham, Switzerland, 2017; pp. 261–291. [CrossRef]
9. Powers, K.W.; Brown, S.C.; Krishna, V.B.; Wasdo, S.C.; Moudgil, B.M.; Roberts, S.M. Research Strategies for Safety Evaluation of Nanomaterials. Part VI. Characterization of Nanoscale Particles for Toxicological Evaluation. *Toxicol. Sci.* **2006**, *90*, 296–303. [CrossRef]
10. Kharisov, B.I.; Rasika Dias, H.V.; Kharissova, O.V.; Vázquez, A.; Pena, Y.; Gómez, I. Solubilization, dispersion and stabilization of magnetic nanoparticles in water and non-aqueous solvents: Recent trends. *RSC Adv.* **2014**, *4*, 45354. [CrossRef]
11. Iijima, M. Surface modification techniques toward controlling the dispersion stability and particle-assembled structure of slurries. *J. Cer. Soc. Jpn.* **2017**, *125*, 603–607. [CrossRef]
12. Peijnenburg, W.J.G.M.; Baalousha, M.; Chen, J.; Chaudry, Q.; Von der Kammer, F.; Kuhlbusch, T.A.J.; Lead, J.; Nickel, C.; Quik, J.T.K.; Renker, M.; et al. A Review of the Properties and Processes Determining the Fate of Engineered Nanomaterials in the Aquatic Environment. *Crit. Rev. Environ. Sci. Technol.* **2015**, *45*, 2084–2134. [CrossRef]
13. Derjaguin, B.; Landau, L.D. Theory of the stability of strongly charged lyophobic sols and of the adhesion of strongly charged particles in solutions of electrolytes. *Acta Physicochim. URSS* **1941**, *14*, 633–662. [CrossRef]
14. Verwey, E.J.W.; Overbeek, J.T.G. *Theory of the Stability of Lyophobic Colloids*; Elsevier: Amsterdam, The Netherlands, 1948.
15. Halamoda-Kenzaoui, B.; Ceridono, M.; Colpo, P.; Valsesia, A.; Urban, P.; Ojea-Jimenez, I.; Gioria, S.; Gilliland, D.; Rossi, F.; Kinsner-Ovaskainen, A. Dispersion behaviour of silica nanoparticles in biological media and its influence on cellular uptake. *PLoS ONE* **2015**, *10*, e0141593. [CrossRef]
16. Glomstad, B.; Zindler, F.; Jenssen, B.M.; Booth, A.M. Dispersibility and dispersion stability of carbon nanotubes in synthetic aquatic growth media and natural freshwater. *Chemosphere* **2018**, *201*, 269–277. [CrossRef]
17. Gutierrez, E.R.; Kamens, R.M.; Tolocka, M.; Sexton, K.; Jaspers, I. A comparison of three dispersion media on the physicochemical and toxicological behavior of TiO$_2$ and NiO nanoparticles. *Chem. Biol. Interact.* **2015**, *236*, 74–81. [CrossRef]
18. Wang, Z.; Luo, Z.; Yan, Y. Dispersion and sedimentation of titanium dioxide nanoparticles in freshwater algae and daphnia aquatic culture media in the presence of arsenate. *J. Exp. Nanosci.* **2018**, *13*, 119–129. [CrossRef]
19. Ji, Z.; Jin, X.; George, S.; Xia, T.; Meng, H.; Wang, X.; Suarez, E.; Zhang, H.; Hoek, E.M.V.; Godwin, H.; et al. Dispersion and Stability Optimization of TiO$_2$ Nanoparticles in Cell Culture Media. *Environ. Sci. Technol.* **2010**, *44*, 7309–7314. [CrossRef]
20. Sager, T.M.; Porter, D.W.; Robinson, V.A.; Lindsley, W.G.; Schwegler-Berry, D.E.; Castranova, V. Improved method to disperse nanoparticles for in vitro and in vivo investigation of toxicity. *Nanotoxicology* **2007**, *1*, 118–129. [CrossRef]
21. Vranic, S.; Gosens, I.; Jacobsen, N.R.; Jensen, K.A.; Bokkers, B.; Kermanizadeh, A.; Stone, V.; Baeza-Squiban, A.; Cassee, F.R.; Tran, L.; et al. Impact of serum as a dispersion agent for in vitro and in vivo toxicological assessments of TiO$_2$ nanoparticles. *Arch. Toxicol.* **2017**, *91*, 353–363. [CrossRef]
22. Strojan, K.; Leonardi, A.; Bregar, V.B.; Krizaj, I.; Svete, J.; Pavlin, M. Dispersion of nanoparticles in different media importantly determines the composition of their protein corona. *PLoS ONE* **2017**, *12*, e0169552. [CrossRef]
23. Wiogo, H.T.R.; Lim, M.; Bulmus, V.; Yun, J.; Amal, R. Stabilization of magnetic iron oxide nanoparticles in biological media by fetal bovine serum (FBS). *Langmuir* **2011**, *27*, 843–850. [CrossRef]
24. Anders, C.B.; Chess, J.J.; Wingett, D.G.; Punnoose, A. Serum proteins enhance dispersion stability and influence the cytotoxicity and dosimetry of ZnO nanoparticles in suspension and adherent cancer cell models. *Nanoscale Res. Lett.* **2015**, *10*, 448. [CrossRef]
25. Sun, Y.; Devore, D.; Ma, X.; Yuan, Y.; Kohn, J.; Qian, J. Promotion of dispersion and anticancer efficacy of hydroxyapatite nanoparticles by the adsorption of fetal bovine serum. *J. Nanopart. Res.* **2019**, *21*, 267. [CrossRef]
26. Sabuncu, A.C.; Grubbs, J.; Qian, S.; Abdel-Fattah, T.M.; Stacey, M.W.; Beskok, A. Probing nanoparticle interactions in cell culture media. *Colloids Surf. B Biointerfaces* **2012**, *95*, 96–102. [CrossRef] [PubMed]

27. Lee, J.; Han, K.; Koo, J. A novel method to evaluate dispersion stability of nanofluids. *Int. J. Heat Mass Transf.* **2014**, *70*, 421–429. [CrossRef]
28. Tantra, R.; Jing, S.; Pichaimuthu, S.K.; Walker, N.; Noble, J.; Hackley, V.A. Dispersion stability of nanoparticles in ecotoxicological investigations: The need for adequate measurement tools. *J. Nanopart. Res.* **2011**, *13*, 3765–3780. [CrossRef]
29. Prathna, T.C.; Chandrasekaran, N.; Mukherjee, A. Studies on aggregation behaviour of silver nanoparticles in aqueous matrices: Effect of surface functionalization and matrix composition. *Colloids Surf. A Physicochem. Eng. Asp.* **2011**, *390*, 216–224. [CrossRef]
30. Sharma, V.; Verma, D.; Okram, G.S. Influence of surfactants, particle size and dispersion medium on surface plasmon resonance of silver nanoparticles. *J. Phys. Condens. Matter* **2020**, *32*, 145302. [CrossRef] [PubMed]
31. Murdock, R.C.; Braydich-Stolle, L.; Schrand, A.M.; Schlager, J.J.; Hussain, S.M. Characterization of Nanomaterial Dispersion in Solution Prior to In Vitro Exposure Using Dynamic Light Scattering Technique. *Toxicol. Sci.* **2008**, *101*, 239–253. [CrossRef]
32. Bondarenko, O.; Juganson, K.; Ivask, A.; Kasemets, K.; Mortimer, M.; Kahru, A. Toxicity of Ag, CuO and ZnO nanoparticles to selected environmentally relevant test organisms and mammalian cells in vitro: A critical review. *Arch. Toxicol.* **2013**, *87*, 1181–1200. [CrossRef]
33. De Jong, W.H.; Hagens, W.I.; Krystek, P.; Burger, M.C.; Sips, A.J.A.M.; Geertsma, R.E. Particle size-dependent organ distribution of gold nanoparticles after intravenous administration. *Biomaterials* **2008**, *29*, 1912–1919. [CrossRef]
34. Goršak, T.; Makovec, D.; Javornik, U.; Belec, B.; Kralj, S.; Lisjak, D. A functionalization strategy for the dispersion of permanently magnetic barium-hexaferrite nanoplatelets in complex biological media. *Coll. Surf. A* **2019**, *573*, 119–127. [CrossRef]
35. Al-Hamadani, Y.A.J.; Chu, K.H.; Son, A.; Heo, J.; Her, N.; Jang, M.; Park, C.M.; Yoon, Y. Stabilization and dispersion of carbon nanomaterials in aqueous solutions: A review. *Sep. Purif. Technol.* **2015**, *156*, 861–874. [CrossRef]
36. Buford, M.C.; Hamilton, R.F., Jr.; Holian, A. A comparison of dispersing media for various engineered carbon nanoparticles. *Part. Fibre Toxicol.* **2007**, *4*, 6. [CrossRef]
37. Gao, J.; Youn, S.; Hovsepyan, A.; Llaneza, V.L.; Wang, Y.; Bitton, G.; Bonzongo, J.-C.J. Dispersion and Toxicity of Selected Manufactured Nanomaterials in Natural River Water Samples: Effects of Water Chemical Composition. *Environ. Sci. Technol.* **2009**, *43*, 3322–3328. [CrossRef]
38. Deguchi, S.; Alargova, R.G.; Tsujii, K. Stable Dispersions of Fullerenes, C60 and C70, in Water Preparation and Characterization. *Langmuir* **2001**, *17*, 6013–6017. [CrossRef]
39. Kamiya, H.; Iijima, M. Surface modification and characterization for dispersion stability of inorganic nanometer-scaled particles in liquid media. *Sci. Technol. Adv. Mater.* **2010**, *11*, 044304. [CrossRef] [PubMed]
40. Laurent, S.; Forge, D.; Port, M.; Roch, A.; Robic, C.; Van der Elst, L.; Muller, R.N. Magnetic Iron Oxide Nanoparticles: Synthesis, Stabilization, Vectorization, Physicochemical Characterizations, and Biological Applications. *Chem. Rev.* **2008**, *108*, 2064–2110. [CrossRef] [PubMed]
41. Heinz, H.; Pramanik, C.; Heinz, O.; Ding, Y.; Mishra, R.K.; Marchon, D.; Flatt, R.J.; Estrela-Lopis, I.; Llop, J.; Moya, S.; et al. Nanoparticle decoration with surfactants: Molecular interactions, assembly and applications. *Surf. Sci. Rep.* **2017**, *72*, 1–58. [CrossRef]
42. Chen, Y.; Renner, P.; Liang, H. Dispersion of nanoparticles in lubricating oil: A critical review. *Lubricants* **2019**, *7*, 7. [CrossRef]
43. Kievit, F.M.; Zhang, M. Surface Engineering of Iron Oxide Nanoparticles for Targeted Cancer Therapy. *Acc. Chem. Res.* **2011**, *44*, 853–862. [CrossRef]
44. Lisuzzo, L.; Cavallaro, G.; Lazzara, G.; Milioto, S.; Parisi, F.; Stetsyshyn, Y. Stability of halloysite, imogolite and boron nitride nanotubes in solvent media. *Appl. Sci.* **2018**, *8*, 1068. [CrossRef]
45. Ortelli, S.; Costa, A.L.; Blosi, M.; Brunelli, A.; Badetti, E.; Bonetto, A.; Hristozov, D.; Marcomini, A. Colloidal characterization of CuO nanoparticles in biological and environmental media. *Environ. Sci. Nano* **2017**, *4*, 1264–1272. [CrossRef]
46. Khairul, M.A.; Shah, K.; Doroodchi, E.; Azizian, R.; Moghtaderi, B. Effects of surfactant on stability and thermo-physical properties of metal oxide nanofluids. *Int. J. Heat Mass Transf.* **2016**, *98*, 778–787. [CrossRef]
47. Cacua, K.; Ordonez, F.; Zapata, C.; Herrera, B.; Pabon, E.; Buitrago-Sierra, R. Surfactant concentration and pH effects on the zeta potential values of alumina nanofluids to inspect stability. *Colloids Surf. A* **2019**, *583*, 123960. [CrossRef]
48. Das, P.K.; Islam, N.; Santra, A.K.; Ganguly, R. Experimental investigation of thermophysical properties of Al2O3-water nanofluid: Role of surfactants. *J. Mol. Liq.* **2017**, *237*, 304–312. [CrossRef]
49. Arasu, V.A.; Kumar, D.D.; Khan, I.A. Experimental investigation of thermal conductivity and stability of TiO2-Ag/water nanocompositefluid with SDBS and SDS surfactants. *Thermochim. Acta* **2019**, *678*, 178308. [CrossRef]
50. Zhai, Y.; Li, L.; Wang, J.; Li, Z. Evaluation of surfactant on stability and thermal performance of Al2O3-ethylene glycol (EG) nanofluids. *Powder Technol.* **2019**, *343*, 215–224. [CrossRef]
51. Chakraborty, S.; Sarkar, I.; Behera, D.K.; Pal, S.K.; Chakraborty, S. Experimental investigation on the effect of dispersant addition on thermal and rheological characteristics of TiO2 nanofluid. *Powder Technol.* **2017**, *307*, 10–24. [CrossRef]
52. Loosli, F.; Stoll, S. Effect of surfactants, pH and water hardness on the surface properties and agglomeration behavior of engineered TiO2 nanoparticles. *Environ. Sci. Nano* **2017**, *4*, 203–211. [CrossRef]
53. Krishnamoorty, A.; Varghese, S. Role of surfactants on the stability on nano Zinc Oxide despersions. *Part. Sci. Technol.* **2017**, *35*, 67–70. [CrossRef]
54. Ordóñez, F.; Chejne, F.; Pabón, E.; Cacua, K. Synthesis of ZrO2 nanoparticles and effect of surfactant on dispersion and stability. *Ceram. Int.* **2020**, *46*, 11970–11977. [CrossRef]

55. Chen, Y.; Gao, Q.; Chen, W.; Wu, F.; Yang, Y.; Werner, D.; Tao, S.; Wang, X. A mechanistic study of stable dispersion of titanium oxide nanoparticles by humic acid. *Water Res.* **2018**, *135*, 85–94. [CrossRef] [PubMed]

56. Baccile, N.; Noiville, R.; Stievano, L.; Van Bogaert, I. Sophorolipids-functionalized iron oxide nanoparticles. *Phys. Chem. Chem. Phys.* **2013**, *15*, 1606–1620. [CrossRef] [PubMed]

57. Brunelli, A.; Badetti, E.; Basei, G.; Izzo, F.C.; Hristozov, D.; Marcomini, A. Effects of organic modifiers on the colloidal stability of TiO$_2$ nanoparticles. A methodological approach for NPs categorization by multivariate statistical analysis. *NanoImpact* **2018**, *9*, 114–123. [CrossRef]

58. Fabrega, J.; Luoma, S.N.; Tyler, C.R.; Galloway, T.S.; Lead, J.R. Silver nanoparticles: Behaviour and effects in the aquatic environment. *Environ. Int.* **2011**, *37*, 517–531. [CrossRef]

59. Shoults-Wilson, W.A.; Reinsch, B.C.; Tsyusko, O.V.; Bertsch, P.M.; Lowry, G.V.; Unrine, J.M. Effect of silver nanoparticle surface coating on bioaccumulation and reproductive toxicity in earthworms (*Eisenia fetida*). *Nanotoxicology* **2011**, *5*, 432–444. [CrossRef] [PubMed]

60. Yan, N.; Zhang, J.G.; Tong, Y.Y.; Yao, S.Y.; Xiao, C.X.; Li, Z.C.; Kou, Y. Solubility adjustable nanoparticles stabilized by a novel PVP based family: Synthesis, characterization and catalytic properties. *Chem. Commun.* **2009**, *45*, 4423–4425. [CrossRef]

61. Li, D.G.; Chen, S.H.; Zhao, S.Y.; Hou, X.M.; Ma, H.Y.; Yang, X.G. A study of phase transfer processes of Ag nanoparticles. *Appl. Surf. Sci.* **2002**, *200*, 62–67. [CrossRef]

62. Seo, D.; Yoon, W.; Park, S.; Kim, R.; Kim, J. The preparation of hydrophobic silver nanoparticles via solvent exchange method. *Colloids Surf. A Physicochem. Eng. Asp.* **2008**, *313*, 158–161. [CrossRef]

63. Vaisman, L.; Wagner, H.D.; Marom, G. The role of surfactants in dispersion of carbon nanotubes. *Adv. Colloid Interface Sci.* **2006**, *128–130*, 37–46. [CrossRef]

64. Jiang, L.; Gao, L.; Sun, J. Production of aqueous colloidal dispersions of carbon nanotubes. *J. Colloid Inter. Sci.* **2003**, *260*, 89–94. [CrossRef]

65. Sabuncu, A.C.; Kalluri, B.S.; Qian, S.; Stacey, M.W.; Beskok, A. Dispersion state and toxicity of MWCNTs in cell culture medium with different T80 concentrations. *Colloids Surf. B Biointerfaces* **2010**, *78*, 36–43. [CrossRef]

66. Jiang, J.; Oberdörster, G.; Biswas, P. Characterization of size, surface charge, and agglomeration state of nanoparticle dispersions for toxicological studies. *J. Nanopart. Res.* **2009**, *11*, 77–89. [CrossRef]

67. Domingos, R.F.; Tufenkji, N.; Wilkinson, K.J. Aggregation of Titanium Dioxide Nanoparticles: Role of a Fulvic Acid. *Environ. Sci. Technol.* **2009**, *43*, 1282–1286. [CrossRef]

68. French, R.A.; Jacobson, A.R.; Kim, B.; Isley, S.L.; Lee Penn, R.; Baveye, P.C. Influence of Ionic Strength, pH, and Cation Valence on Aggregation Kinetics of Titanium Dioxide Nanoparticles. *Environ. Sci. Technol.* **2009**, *43*, 1354–1359. [CrossRef]

69. Metin, C.O.; Lake, L.W.; Miranda, C.R.; Nguyen, Q.P. Stability of aqueous silica nanoparticle dispersions. *J. Nanopart. Res.* **2011**, *13*, 839–850. [CrossRef]

70. Gallego-Urrea, J.A.; Perez-Holmberg, J.; Hassellöv, M. Influence of different types of natural organic matter on titania nanoparticle stability: Effects of counter ion concentration and pH. *Environ. Sci. Nano* **2014**, *1*, 181–189. [CrossRef]

71. Wang, X.-J.; Li, X.; Yang, S. Influence of pH and SDBS on the Stability and Thermal Conductivity of Nanofluids. *Energy Fuels* **2009**, *23*, 2684–2689. [CrossRef]

72. Li, X.; Zhu, D.; Wang, X. Evaluation on dispersion behavior of the aqueous copper nano-suspensions. *J. Colloid Inter. Sci.* **2007**, *310*, 456–463. [CrossRef]

73. Qi, W.; Yu, J.; Zhang, Z.; Xu, H.-N. Effect of pH on the aggregation behavior of cellulose nanocrystals in aqueous medium. *Mater. Res. Express* **2019**, *6*, 125078. [CrossRef]

74. Hoshino, A.; Fujioka, K.; Oku, T.; Suga, M.; Sasaki, Y.F.; Ohta, T.; Yasuhara, M.; Suzuki, K.; Yamamoto, K. Physicochemical Properties and Cellular Toxicity of Nanocrystal Quantum Dots Depend on Their Surface Modification. *Nano Lett.* **2004**, *4*, 2163–2169. [CrossRef]

75. Hoseini, S.M.; Hedayati, A.; Mirghaed, A.T.; Ghelichpour, M. Toxic effects of copper sulfate and copper nanoparticles on minerals, enzymes, thyroid hormones and protein fractions of plasma and histopathology in common carp *Cyprinus carpio*. *Exp. Toxicol. Pathol.* **2016**, *68*, 493–503. [CrossRef]

76. Chakraborty, S.; Nair, A.; Paliwal, M.; Dybowska, A.; Misra, S.K. Exposure media a critical factor for controlling dissolution of CuO nanoparticles. *J. Nanopart. Res.* **2018**, *20*, 331. [CrossRef]

77. Zhang, W.; Xiao, B.; Fang, T. Chemical transformation of silver nanoparticles in aquatic environments: Mechanism, morphology and toxicity. *Chemosphere* **2018**, *191*, 324–334. [CrossRef]

78. Yu, S.; Liu, J.; Yin, Y.; Shen, M. Interactions between engineered nanoparticles and dissolved organic matter: A review on mechanisms and environmental effects. *J. Environ. Sci.* **2018**, *63*, 198–217. [CrossRef]

79. Born, P.; Klaessg, F.C.; Landry, T.D.; Moudgil, B.; Pauluhn, J. Research strategies for safety evaluation of nanomaterials, Part V: Role of dissolution in biological fate and effects of nanoscale particles. *Toxicol. Sci.* **2006**, *90*, 23–32. [CrossRef]

80. Silva, T.; Pokhrel, L.R.; Dubey, B.; Tolaymat, T.M.; Maier, K.J.; Liu, X. Particle size, surface charge and concentration dependent ecotoxicity of three organo-coated silver nanoparticles: Comparison between general linear model-predicted and observed toxicity. *Sci. Total Environ.* **2014**, *468–469*, 968–976. [CrossRef]

81. Wang, Z.Q.; Xue, Y.Q.; Cui, Z.X.; Duan, H.J.; Xia, X.Y. The Size Dependence of Dissolution Thermodynamics of Nanoparticles. *Nano* **2016**, *11*, 16501001–165010010. [CrossRef]

82. Braun, K.; Pochert, A.; Beck, M.; Fiedler, R.; Gruber, J.; Linden, M. Dissolution kinetics of mesoporous silica nanoparticles in different simulated body fluids. *J. Sol Gel Sci. Technol.* **2016**, *79*, 319–327. [CrossRef]

83. Senapati, V.A.; Kumar, A. ZnO nanoparticles dissolution, penetration and toxicity in human epidermal cells. Influence of pH. *Environ. Chem. Lett.* **2018**, *16*, 1129–1135. [CrossRef]

84. Finnegan, M.; Zhang, H.; Banfield, J. Phase Stability and Transformation in Titania Nanoparticles in Aqueous Solutions Dominated by Surface Energy. *J. Phys. Chem. C* **2007**, *111*, 1962–1968. [CrossRef]

85. Li, X.; Lenhart, J.J.; Walker, H.W. Aggregation Kinetics and Dissolution of Coated Silver Nanoparticles. *Langmuir* **2012**, *28*, 1095–1104. [CrossRef]

86. David, C.A.; Calceran, J.; Rey-Castro, C.; Puy, J.; Companys, E.; Salvador, J.; Monne, J.; Wallace, R.; Vakourov, A. Dissolution Kinetics and Solubility of ZnO Nanoparticles Followed by AGNES. *J. Phys. Chem. C* **2012**, *116*, 11758–11767. [CrossRef]

87. Ely, D.R.; Garcia, R.E.; Thommes, M. Ostwald–Freundlich diffusion-limited dissolution kinetics of nanoparticles. *Powder Technol.* **2014**, *257*, 120–123. [CrossRef]

88. Wang, D.; Lin, Z.; Wang, T.; Yao, Z.; Qin, M.; Zheng, S.; Lu, W. Where does the toxicity of metal oxide nanoparticles come from: The nanoparticles, the ions, or a combination of both? *J. Hazard. Mater.* **2016**, *308*, 328–334. [CrossRef]

89. Lee, I.C.; Ko, J.W.; Park, S.H.; Lim, J.O.; Shin, I.S.; Moon, C.; Kim, S.H.; Heo, J.D.; Kim, J.C. Comparative toxicity and biodistribution of copper nanoparticles and cupric ions in rats. *Int. J. Nanomed.* **2016**, *11*, 2883–2900. [CrossRef]

90. McGillicuddy, E.; Murray, I.; Kavanagh, S.; Morrison, L.; Fogarty, A.; Cormican, M.; Dockery, P.; Prendergast, M.; Rowan, N.; Morris, D. Silver nanoparticles in the environment: Sources, detection and ecotoxicology. *Sci. Total Environ.* **2017**, *575*, 231–246. [CrossRef]

91. Aruoja, V.; Dubourguier, H.C.; Kasemets, K.; Kahru, A. Toxicity of nanoparticles of CuO, ZnO and TiO$_2$ to microalgae *Pseudokirchneriella subcapitata*. *Sci. Total Environ.* **2009**, *407*, 1461–1468. [CrossRef]

92. Zoroddu, M.A.; Medici, S.; Ledda, A.; Nurch, V.M.; Lachowicz, J.I.; Peana, M. Toxicity of Nanoparticles. *Curr. Med. Chem.* **2014**, *21*, 3837–3853. [CrossRef]

93. Moreno-Garrido, I.; Perez, S.; Blasco, J. Toxicity of silver and gold nanoparticles on marine microalgae. *Mar. Environ. Res.* **2005**, *111*, 60–73. [CrossRef]

94. Hou, J.; Wu, Y.; Li, X.; Wei, B.; Li, S.; Wang, X. Toxic effects of different types of zinc oxide nanoparticles on algae, plants, invertebrates, vertebrates and microorganisms. *Chemosphere* **2018**, *193*, 852–860. [CrossRef]

95. Lubick, N. Nanosilver toxicity: Ions, nanoparticles or both? *Environ. Sci. Technol.* **2008**, *42*, 8617. [CrossRef]

96. Misra, S.K.; Dybowska, A.; Berhanu, D.; Luoma, S.N.; Valsami-Jones, E. The complexity of nanoparticle dissolution and its importance in nanotoxicological studies. *Sci. Total Environ.* **2012**, *438*, 225–232. [CrossRef]

97. Fard, K.J.; Jafari, S.; Eghbal, M.A. A Review of Molecular Mechanisms Involved in Toxicity of Nanoparticles. *Adv. Pharm. Bull.* **2015**, *5*, 447–454. [CrossRef]

98. Sue, K.; Murata, K.; Kimura, K.; Arai, K. Continuous synthesis of zinc oxide nanoparticles in supercritical water. *Green Chem.* **2003**, *5*, 659–662. [CrossRef]

99. Studer, A.M.; Limbach, L.K.; Duc, L.V.; Krumeich, F.; Athanassiou, E.K.; Gerber, L.C.; Moch, H.; Stark, W.J. Nanoparticle cytotoxicity depends on intracellular solubility: Comparison of stabilized copper metal and degradable copper oxide nanoparticles. *Toxicol. Lett.* **2010**, *197*, 169–174. [CrossRef]

100. Raghupathi, K.R.; Koodali, R.T.; Manna, A.C. Size-dependent bacterial growth inhibition and mechanism of antibacterial activity of zinc oxide nanoparticles. *Langmuir* **2011**, *27*, 4020–4028. [CrossRef]

101. Yang, X.; Gondikas, A.P.; Marinakos, S.M.; Auffan, M.; Liu, K.; Hsu-Kim, H.; Meyer, J.N. Mechanism of silver nanoparticle toxicity is dependent on dissolved silver and surface coating in *Caenorhabditis elegans*. *Environ. Sci. Technol.* **2012**, *46*, 1119–1127. [CrossRef]

102. Mortimer, M.; Kasemets, K.; Kahru, A. Toxicity of ZnO and CuO nanoparticles to ciliated protozoa *Tatrahymena thermophile*. *Toxicology* **2010**, *269*, 182–189. [CrossRef]

103. Tang, R.; Wang, L.; Nancollas, G.H. Size-effects in the dissolution of hydroxyapatite: An understanding of biological demineralization. *J. Mater. Chem.* **2004**, *14*, 2341–2346. [CrossRef]

104. Le Ouay, B.; Stellacci, F. Antibacterial activity of silver nanoparticles: A surface science insight. *Nano Today* **2015**, *10*, 339–354. [CrossRef]

105. Zhang, W.; Yao, Y.; Sullivan, N.; Chen, Y. Modeling the primary size effects on citrate-coated silver nanoparticles on their ion release kinetics. *Environ. Sci. Technol.* **2011**, *45*, 4422–4428. [CrossRef]

106. Kittler, S.; Greulich, C.; Diendorf, J.; Koller, M.; Epple, M. Toxicity of silver nanoparticles during storage because of slow dissolution under release of silver ions. *Chem. Mater.* **2010**, *22*, 4548–4554. [CrossRef]

107. Misra, S.K.; Dybowska, A.; Berhanu, D.; Croteau, M.N.; Luoma, S.N.; Boccaccini, A.R.; Valsami-Jones, E. Isotopically modified nanoparticles for enhanced detection in bioaccumulation studies. *Environ. Sci. Technol.* **2012**, *46*, 1216–1222. [CrossRef]

108. Wu, J.; Gao, W.; Yang, H.; Zuo, J. Dissolution Kinetics of Oxidative Etching of Cubic and Icosahedral Platinum Nanoparticles Revealed by in Situ Liquid Transmission Electron Microscopy. *ACS Nano* **2017**, *11*, 1696–1703. [CrossRef]

109. Li, X.; Lenhart, J.J.; Walker, H.W. Dissolution-Accompanied Aggregation Kinetics of Silver Nanoparticles. *Langmuir* **2010**, *26*, 16690–16698. [CrossRef]

110. Gunawan, C.; Teoh, W.Y.; Marquis, C.P.; Amal, R. Cytotoxic origin of copper (II) oxide nanoparticles: Comparative studies with micron-sized particles, leachate and metal salts. *ACS Nano* **2011**, *5*, 7214–7225. [CrossRef]

111. Gondikas, A.P.; Morris, A.; Reinsch, B.C.; Marinakos, S.M.; Lowry, G.V. Cysteine-induced modifications of zero-valent silver nanoparticles implications for particle surface chemistry, aggregation, dissolution and silver speciation. *Environ. Sci. Technol.* **2012**, *46*, 7037–7045. [CrossRef]

112. Conway, J.R.; Adeleye, A.S.; Gardea-Torresdey, J.; Keller, A.A. Aggregation, Dissolution, and Transformation of Copper Nanoparticles in Natural Waters. *Environ. Sci. Technol.* **2015**, *49*, 2749–2756. [CrossRef]

113. Zhang, H.; Chen, B.; Banfield, J.F. Particle Size and pH Effects on Nanoparticle Dissolution. *J. Phys. Chem. C* **2010**, *114*, 14876–14884. [CrossRef]

114. Gaiser, B.K.; Fernandes, T.F.; Jepson, M.; Lead, J.R.; Tyler, C.R. Assessing exposure, uptake and toxicity of silver and cerium dioxide nanoparticles form contaminated environments. *Environ. Health* **2009**, *8*, S2. [CrossRef]

115. Buzea, C.; Pacheco, I.I.; Robbie, K. Nanomaterials and nanoparticles: Sources and toxicity. *Biointerphases* **2007**, *2*, MR17–MR71. [CrossRef]

116. Savoly, Z.; Hracs, K.; Pemmer, B.; Streli, C.; Zaray, G.; Nagy, P.I. Uptake and toxicity of nano-ZnO in the plant-feeding nematode, *Xiphinema vuittenezi*: The role of dissolved zinc and nanoparticle-specific effects. *Environ. Sci. Pollut. Res.* **2016**, *23*, 9669–9678. [CrossRef]

117. Li, Y.; Zhao, J.; Shanh, E.; Xia, X.; Niu, J.; Crittenden, J. Effects of Chloride Ions on Dissolution, ROS Generation, and Toxicity of Silver Nanoparticles under UV Irradiation. *Environ. Sci. Technol.* **2018**, *52*, 4842–4849. [CrossRef] [PubMed]

118. Brunner, T.J.; Wick, P.; Manser, P.; Spohn, P.; Grass, R.N.; Limbach, L.K.; Bruinink, A.; Stark, W.J. In vitro cytotoxicity of oxide nanoparticles: Comparison to asbestos, silica and the effect of particle solubility. *Environ. Sci. Technol.* **2006**, *40*, 4374–4381. [CrossRef]

119. Liu, J.; Feng, X.; Wei, L.; Chen, L.; Song, B.; Shao, L. The toxicology of ion-shedding zinc oxide nanoparticles. *Crit. Rev. Toxicol.* **2016**, *46*, 348–384. [CrossRef]

120. Sruthi, S.; Mohanan, P.V. Investigation on cellular interactions of astrocytes with zinc oxide nanoparticles using rat C6 cell lines. *Colloid Surf. B* **2015**, *133*, 1–11. [CrossRef]

121. Xia, T.; Kovochich, M.; Liong, M.; Madler, L.; Gilbert, B.; Shi, H.; Yeh, J.I.; Zink, J.I.; Nel, A.E. Comparison of the mechanism of toxicity of zinc oxide and cerium oxide nanoparticles based on dissolution and oxidative stress properties. *ACS Nano* **2008**, *2*, 2121–2123. [CrossRef]

122. Wong, S.W.Y.; Leung, P.T.Y.; Djurusie, A.B.; Leung, K.M.U. Toxicities of nano zinc oxide to five marine organisms: Influences to aggregate size and ion solubility. *Anal. Bioanal. Chem.* **2010**, *396*, 609–618. [CrossRef]

123. Sharifi, S.; Behzadi, S.; Laurent, S.; Forrest, M.L.; Stroeve, P.; Mahmoudi, M. Toxicity of nanomaterials. *Chem. Soc. Rev.* **2012**, *41*, 2323–2343. [CrossRef]

124. Condello, M.; De Beratdis, B.; Ammendolia, M.G.; Barone, F.; Condello, G.; Degan, P.; Meschini, S. ZnO nanoparticle tracking from uptake to genotoxic damage in human colon carcinoma cells. *Toxicol. In Vitro* **2016**, *35*, 169–179. [CrossRef] [PubMed]

125. Eixenberger, J.; Anders, C.; Hermann, R.; Brown, R.; Reddy, K.M. Rapid dissolution of ZnO nanoparticles induced by biological buffers significantly impacts cytotoxicity. *Chem. Res. Toxicol.* **2017**, *30*, 1641–1651. [CrossRef] [PubMed]

126. Levard, C.; Hotze, E.M.; Lowry, G.V.; Brown, G.E. Environmental transformation of silver nanoparticles: Impact on stability and toxicity. *Environ. Sci. Technol.* **2012**, *46*, 6900–6914. [CrossRef] [PubMed]

127. Elzey, S.; Grassian, V.H. Agglomeration, isolation and dissolution of commercially manufactured silver nanoparticles in aqueous environment. *J. Nanopart. Res.* **2010**, *12*, 1945–1958. [CrossRef]

128. Barbasz, A.; Ocwieja, M.; Roman, M. Toxicity of silver nanoparticles towards tumoral human cell lines U-937 and HL-60. *Colloids Surf. B Biointerfaces* **2017**, *156*, 397–404. [CrossRef]

129. Graf, C.; Nordmeyer, D.; Sengstock, C.; Ahlberg, S.; Diendorf, J.; Raabe, J.; Epple, M.; Koller, M.; Lademann, J.; Vogt, A.; et al. Shape-Dependent Dissolution and Cellular Uptake of Silver Nanoparticles. *Langmuir* **2018**, *34*, 1506–1519. [CrossRef]

130. Mohanaj, V.J.; Chen, Y. Nanoparticles—A Review. *Trop. J. Pharm. Res.* **2006**, *5*, 561–573. [CrossRef]

131. Chudasama, B.; Vala, A.K.; Andhariya, N.; Mehra, R.V.; Upadhaya, R.V. Highly bacterial resistant silver nanoparticles: Synthesis, and antibacterial activities. *J. Nanopart. Res.* **2010**, *12*, 1677–1685. [CrossRef]

132. Calvo, P.; Remunan-Lopez, C.; Vila-Jato, J.L.; Alonso, M.J. Novel hydrophilic chitosan-polyethylene oxide nanoparticles as protein carriers. *J. Appl. Polym. Sci.* **1997**, *63*, 125–132. [CrossRef]

133. Moghaddam, M.S.; Heiny, M.; Shastri, V.P. Enhanced cellular uptake of nanoparticles by increasing the hydrophobicity of poly (lactic acid through copolymerization with cell-membrane-lipid components. *Chem. Commun.* **2015**, *51*, 14605–14608. [CrossRef]

134. Nor, Y.A.; Nui, Y.; Karmakar, S.; Zhou, L.; Hu, C.; Zhang, J.; Zhang, H.; Yu, M.; Mahony, D.; Mitter, N.; et al. Shaping nanoparticles with hydrophilic compositions and hydrophobic properties as nanocarriers for antibiotic delivery. *ACS Cent. Sci.* **2015**, *1*, 328–334. [CrossRef]

135. Park, J.T.; Seo, J.A.; Ahn, S.H.; Kim, J.H.; Kang, S.W. Surface modification of silica nanoparticles with hydrophilic polymers. *J. Ind. Eng. Chem.* **2010**, *16*, 517–522. [CrossRef]

136. Qi, L.; Colfen, H.; Antonietti, M. Control of barite morphology by double hydrophilic block copolymers. *Chem. Mater.* **2000**, *12*, 2392–2403. [CrossRef]

137. Bala, H.; Fu, W.; Guo, Y.; Zhao, J.; Jiang, Y.; Ding, X.; Yu, K.; Li, M.; Wang, Z. In situ preparation and surface modification of barium sulfate nanoparticles. *Colloid Surf. A* **2006**, *274*, 71–76. [CrossRef]

138. Mallakpour, S.; Soltanian, S. Surface functionalization of carbon nanotubes: Fabrication and applications. *RSC Adv.* **2016**, *6*, 109916–109935. [CrossRef]

139. Lin, Y.; Taylor, S.; Li, K.; Shiral Fernando, K.A.; Qu, L. Advances toward bioapplications of carbon nanotubes. *J. Mater. Chem.* **2004**, *14*, 527–541. [CrossRef]

140. Zheng, M.; Jagota, A.; Semke, E.D.; Diner, B.A.; McLean, R.S.; Lustig, S.R.; Richardson, R.E.; Tassi, N.G. DNA-assisted dispersion and separation of carbon nanotubes. *Nat. Mater.* **2003**, *2*, 338–342. [CrossRef] [PubMed]

141. Yang, D.; Yang, F.; Hu, J.; Long, J.; Wang, C. Hydrophilic multi-walled carbon nanotubes decorated with magnetite nanoparticles as lymphatic targeted drug delivery vehicles. *Chem. Commun.* **2009**, *45*, 4447–4449. [CrossRef] [PubMed]

142. Sandra, S.; Zhang, P.; Wang, K.; Tapec, R.; Tan, W. Conjugation of biomolecules with luminophore-doped silica nanoparticles for photostable biomarkers. *Anal. Chem.* **2001**, *73*, 4988–4993. [CrossRef] [PubMed]

143. Wang, L.; Wang, K.; Sandra, S.; Zhao, X.; Hilliard, L.R.; Smith, J.E.; Wu, J.R.; Tan, W.H. Watching silica nanoparticles glow in the biological world. *Anal. Chem.* **2006**, *78*, 647–654. [CrossRef]

144. Sun, X.L.; Fan, Z.P.; Zhang, L.D.; Wnag, L.; Wei, Z.J. Superhydrophobicity of silica nanoparticles modified with polystyrene. *Appl. Surf. Sci.* **2011**, *257*, 2308–2312. [CrossRef]

145. Abd Ellateif, T.M.A.; Maitra, S. Some studies on the surface modification of sol-gel derived hydrophilic Silica nanoparticles. *Int. J. Nano Dimens.* **2017**, *8*, 97–106. [CrossRef]

146. Ling, D.; Lee, N.; Hyeon, T. Chemical Synthesis and Assembly of Uniformly Sized Iron Oxide Nanoparticles for Medical Applications. *Acc. Chem. Res.* **2015**, *48*, 1276–1285. [CrossRef] [PubMed]

147. Zanganeh, S.; Hutter, G.; Spitler, R.; Lenkov, O.; Mahmoudi, M.; Shaw, A.; Pajarinen, J.S.; Nejadnik, H.; Goodman, S.; Moseley, M.; et al. Iron oxide nanoparticles inhibit tumour growth by inducing pro-inflammatory macrophage polarization in tumour tissues. *Nat. Nanotechol.* **2016**, *11*, 986–994. [CrossRef]

148. Woo, K.; Hong, J. Surface Modification of Hydrophobic Iron Oxide Nanoparticles for Clinical Applications. *IEEE Trans. Magn.* **2005**, *41*, 4137–4139. [CrossRef]

149. Park, I.Y.; Piao, Y.; Lee, N.; Yoo, B.; Kim, B.H.; Choi, S.H.; Hyeon, T. Transformation of hydrophobic iron oxide nanoparticles to hydrophilic and biocompatible maghemite nanocrystals for use as highly efficient MRI contrast agent. *J. Mater. Chem.* **2011**, *21*, 11472–11477. [CrossRef]

150. Gupta, A.K.; Gupta, M. Synthesis and surface engineering of iron oxide nanoparticles for biomedical applications. *Biomaterials* **2005**, *26*, 3995–4021. [CrossRef]

151. Lindan, S.; Lynch, I.; Thulin, E.; Nilson, H.; Dawson, K.A.; Linse, S. Systematic investigation of the thermodynamics of HSA adsorption to N-iso-Propylacrylamide/N-tert-Butylacrylamide copolymer nanoparticles. Effects of particle size and hydrophobicity. *Nano Lett.* **2007**, *7*, 914–920. [CrossRef]

152. Mayano, D.F.; Goldsmith, M.; Solfeill, D.J.; Landesman-Kilo, D.; Miranda, O.R. Nanoparticle hydrophobicity dictates immune response. *J. Am. Chem. Soc.* **2012**, *134*, 3965–3967. [CrossRef]

153. Lynch, I.; Dawson, K.A. Protein-nanoparticle interactions. *Nano Today* **2008**, *3*, 40–47. [CrossRef]

154. Verma, A.; Stellacci, F. Effect of surface properties of nanoparticle-cell interactions. *Small* **2010**, *6*, 12–21. [CrossRef] [PubMed]

155. Gessner, A.; Waicz, R.; Lieske, A.; Raulke, B.R.; Mader, K. Nanoparticles with decreasing surface hydrophobicities: Influence on plasma protein adsorption. *Int. J. Pharm.* **2000**, *196*, 245–249. [CrossRef]

156. Zhang, X. Gold Nanoparticles: Recent Advances in the Biomedical Applications. *Cell Biochem. Biophys.* **2015**, *72*, 771–775. [CrossRef] [PubMed]

157. Su, C.F.; Merlitz, H.; Rabbel, H.; Sommer, J.U. Nanoparticles of various degrees of hydrophobicity interacting with lipid membranes. *J. Phys. Chem. Lett.* **2017**, *8*, 4069–4076. [CrossRef] [PubMed]

158. Van Onzen, A.H.A.M.; Albertazzi, L.; Schenning, A.P.H.J.; Milroy, L.G.; Brunsveld, L. Hydrophobicity determines the fate of self-assembled fluorescent nanoparticles in cells. *Chem. Commun.* **2017**, *53*, 1626–1629. [CrossRef] [PubMed]

159. Palao-Suay, R.; Aquilar, M.R.; Parra-Ruiz, F.J.; Maji, S.; Hoogenboom, R.; Rohner, N.A.; Thomas, S.N.; Roman, J.S. Enhanced Bioactivity of α-Tocopheryl Succinate Based Block Copolymer Nanoparticles by Reduced Hydrophobicity. *Macromol. Biosci.* **2016**, *16*, 1824–1837. [CrossRef]

160. Teubl, B.J.; Schimpel, C.; Leitinger, G.; Bauer, B.; Frohlich, E.; Zimmer, A.; Roblegg, E. Interactions between nano-TiO$_2$ and the oral cavity: Impact of nanomaterial surface hydrophilicity/hydrophobicity. *J. Hazard. Mater.* **2015**, *286*, 298–305. [CrossRef]

161. Jiang, W.; Kim, B.Y.S.; Rutka, J.T.; Chan, W.C.W. Nanoparticle-Mediated Cellular Response is Size-Dependent. *Nat. Nanotechnol.* **2008**, *3*, 145–150. [CrossRef]

162. Gratton, S.E.A.; Ropp, P.A.; Pohlhaus, P.D.; Luft, J.C.; Madden, V.J.; Napier, M.E.; DeSimone, J.M. The Effect of Particle Design on Cellular Internalization Pathways. *Proc. Natl. Acad. Sci. USA* **2008**, *105*, 11613–11618. [CrossRef]

163. Oberdörster, G. Safety Assessment for Nanotechnology and Nanomedicine: Concepts of Nanotoxicology. *J. Intern. Med.* **2010**, *267*, 89–105. [CrossRef]

164. Salvati, A.; Åberg, C.; Dos Santos, T.; Varela, J.; Pinto, P.; Lynch, I.; Dawson, K.A. Experimental and theoretical comparison of intracellular import of polymeric nanoparticles and small molecules: Toward models of uptake kinetics. *Nanomed. Nanotechnol. Biol. Med.* **2011**, *7*, 818–826. [CrossRef]

165. Conner, S.D.; Schmid, S.L. Regulated Portals of Entry into the Cell. *Nature* **2003**, *422*, 37–44. [CrossRef] [PubMed]

166. Rejman, J.; Oberle, V.; Zuhorn, I.S.; Hoekstra, D. Size-Dependent Internalization of Particles via the Pathways of Clathrin- and Caveolae-Mediated Endocytosis. *Biochem. J.* **2004**, *377*, 159–169. [CrossRef]

167. Xing, X.; He, X.; Peng, J.; Wang, K.; Tan, W. Uptake of Silica-Coated Nanoparticles by HeLa Cells. *J. Nanosci. Nanotechnol.* **2005**, *5*, 1688–1693. [CrossRef] [PubMed]

168. Kim, J.-S.; Yoon, T.-J.; Yu, K.-N.; Noh, M.S.; Woo, M.; Kim, B.-G.; Lee, K.-H.; Sohn, B.-H.; Park, S.-B.; Lee, J.-K.; et al. Cellular Uptake of Magnetic Nanoparticle Is Mediated through Energy-Dependent Endocytosis in A549 Cells. *J. Vet. Sci.* **2006**, *7*, 321–326. [CrossRef] [PubMed]

169. Lu, J.; Liong, M.; Sherman, S.; Xia, T.; Kovochich, M.; Nel, A.; Zink, J.; Tamanoi, F. Mesoporous Silica Nanoparticles for Cancer Therapy: Energy-Dependent Cellular Uptake and Delivery of Paclitaxel to Cancer Cells. *NanoBiotechnology* **2007**, *3*, 89–95. [CrossRef]

170. Mayor, S.; Pagano, R.E. Pathways of Clathrin-Independent Endocytosis. *Nat. Rev. Mol. Cell Biol.* **2007**, *8*, 603–612. [CrossRef]

171. Doherty, G.J.; McMahon, H.T. Mechanisms of Endocytosis. *Annu. Rev. Biochem.* **2009**, *78*, 857–902. [CrossRef]

172. Shapero, K.; Fenaroli, F.; Lynch, I.; Cottell, D.C.; Salvati, A.; Dawson, K.A. Time and Space Resolved Uptake Study of Silica Nanoparticles by Human Cells. *Mol. BioSyst.* **2011**, *7*, 371–378. [CrossRef]

173. Hillaireau, H.; Couvreur, P. Nanocarriers' entry into the cell: Relevance to drug delivery. *Cell. Mol. Life Sci.* **2009**, *66*, 2873–2896. [CrossRef]

174. Kuhn, A.D.; Vanhecke, D.; Michen, B.; Blank, F.; Gehr, P.; Petri-Fink, A.; Rothen-Rutishauser, B. Different endocytotic uptake mechanisms for nanoparticles in epithelial cells and macrophages. *Beilstein J. Nanotechnol.* **2014**, *5*, 1625–1636. [CrossRef]

175. Rappoport, J. Focusing on clathrin-mediated endocytosis. *Biochem. J.* **2008**, *412*, 415–423. [CrossRef] [PubMed]

176. Wang, J.; Byrne, D.J.; Napier, E.M.; DeSimone, M.J. More effective nanomedicines through particle design. *Small* **2011**, *7*, 1919–1931. [CrossRef]

177. Manzanares, D.; Ceña, V. Endocytosis: The Nanoparticle and Submicron Nanocompounds Gateway into the Cell. *Pharmaceutics* **2020**, *12*, 371. [CrossRef]

178. Behzadi, S.; Serpooshan, V.; Tao, W.; Hamaly, A.M.; Alkawareek, Y.M.; Dreaden, C.E.; Brown, D.; Alaaldin, M.A.; Farokhzad, C.O.; Mahmoudi, M. Cellular uptake of nanoparticles: Journey inside the cell. *Chem. Soc. Rev.* **2017**, *46*, 4218–4244. [CrossRef]

179. Zhang, R.; Qin, X.; Kong, F.; Chen, P.; Panc, G. Improving cellular uptake of therapeutic entities through interaction with components of cell membrane. *Drug Deliv.* **2019**, *26*, 328–342. [CrossRef]

180. Sun, H.; Jiang, C.; Wu, L.; Bai, B.; Shumei Zhai, S. Cytotoxicity-Related Bioeffects Induced by Nanoparticles: The Role of Surface Chemistry. Front. *Bioeng. Biotechnol.* **2019**, *7*, 414. [CrossRef] [PubMed]

181. Chithrani, B.D.; Ghazani, A.A.; Chan, W.C.W. Determining the size and shape dependence of gold nanoparticle uptake into mammalian cells. *Nano Lett.* **2006**, *6*, 662–668. [CrossRef] [PubMed]

182. Chithrani, B.D.; Chan, W.C.W. Elucidating the mechanism of cellular uptake and removal of protein-coated gold nanoparticles of different sizes and shapes. *Nano Lett.* **2007**, *7*, 1542–1550. [CrossRef]

183. Aggarwal, P.; Hall, J.B.; McLeland, C.B.; Dobrovolskaia, M.A.; McNeil, S.E. Nanoparticle interaction with plasma proteins as it relates to particle biodistribution, biocompatibility and therapeutic efficacy. *Adv. Drug Deliv. Rev.* **2009**, *61*, 428–437. [CrossRef]

184. Cho, E.C.; Au, L.; Zhang, Q.; Xia, Y. The effects of size, shape, and surface functional group of gold nanostructures on their adsorption and internalization by cells. *Small* **2010**, *6*, 517–522. [CrossRef]

185. He, C.; Hu, Y.; Yin, L.; Tang, C.; Tin, C. Effects of particle size and surface charge on cellular uptake and biodistribution of polymeric nanoparticles. *Biomaterials* **2010**, *31*, 3657–3666. [CrossRef] [PubMed]

186. Lesniak, A.; Campbell, A.; Monopoli, M.P.; Lynch, I.; Salvati, A.; Dawson, K.A. Serum heat inactivation affects protein corona composition and nanoparticle uptake. *Biomaterials* **2010**, *31*, 9511–9518. [CrossRef] [PubMed]

187. Xia, X.-R.; Monteiro-Riviere, N.A.; Riviere, J.E. An index for characterization of nanomaterials in biological systems. *Nat. Nanotechol.* **2010**, *5*, 671–675. [CrossRef] [PubMed]

188. Cho, E.C.; Zhang, Q.; Xia, Y. The effect of sedimentation and diffusion on cellular uptake of gold nanoparticles. *Nat. Nanotechol.* **2011**, *6*, 385–391. [CrossRef] [PubMed]

189. Kim, J.; Åberg, C.; Salvati, A.; Dawson, K.A. Role of cell cycle on the cellular uptake and dilution of nanoparticles in a cell population. *Nat. Nanotechol.* **2012**, *7*, 62–68. [CrossRef]

190. Panzarini, E.; Mariano, S.; Carata, E.; Mura, F.; Rossi, M.; Dini, L. Intracellular Transport of Silver and Gold Nanoparticles and Biological Responses: An Update. *Int. J. Mol. Sci.* **2018**, *19*, 1305. [CrossRef]

191. Lu, F.; Wu, S.-H.; Hung, Y.; Mou, C.-Y. Size Effect on Cell Uptake in Well-Suspended, Uniform Mesoporous Silica Nanoparticles. *Small* **2009**, *5*, 1408–1413. [CrossRef] [PubMed]

192. Tallury, P.; Payton, K.; Santra, S. Silica-based multimodal/multifunctional nanoparticles for bioimaging and biosensing applications. *Nanomedicine* **2008**, *3*, 579–592. [CrossRef]

193. Smith, A.M.; Duan, H.W.; Mohs, A.M.; Nie, S.M. Bioconjugated quantum dots for in vivo molecular and cellular imaging. *Adv. Drug. Deliv. Rev.* **2008**, *60*, 1226–1240. [CrossRef]

194. Johannes, L.; Mayor, S. Induced domain formation in endocytic invagination, lipid sorting, and scission. *Cell* **2010**, *142*, 507–510. [CrossRef] [PubMed]

195. Yu, K.O.; Grabinski, C.M.; Schrand, A.M.; Murdock, R.C.; Wang, W.; Gu, B.; Schlager, J.J.; Hussain, S.M. Toxicity of amorphous silica nanoparticles in mouse keratinocytes. *J. Nanopart. Res.* **2009**, *11*, 15–24. [CrossRef]

196. Li, L.; Xi, W.-S.; Su, Q.; Li, Y.; Yan, G.-H.; Liu, Y.; Wang, H.; Cao, A. Unexpected Size Effect: The Interplay between Different-Sized Nanoparticles in Their Cellular Uptake. *Small* **2019**, *15*, 1901687. [CrossRef]

197. Kostopoulou, A.; Brintakis, K.; Fragogeorgi, E.; Anthousi, A.; Manna, L.; Begin-Colin, S.; Billotey, C.; Ranella, A.; Loudos, G.; Athanassakis, I.; et al. Iron oxide colloidal nanoclusters as theranostic vehicles and their interactions at the cellular level. *Nanomaterials* **2018**, *8*, 315. [CrossRef] [PubMed]

198. Carlson, C.; Hussain, S.M.; Schrand, A.M.; Braydich-Stolle, L.K.; Hess, K.L.; Jones, R.L.; Schlager, J.J. Unique cellular interaction of silver nanoparticles: Size-dependent generation of reactive oxygen species. *J. Phys. Chem. B* **2008**, *112*, 13608–13619. [CrossRef]

199. Champion, J.A.; Mitragotri, S. Shape Induced Inhibition of Phagocytosis of Polymer Particles. *Phar. Res.* **2009**, *26*, 244–249. [CrossRef]

200. Geng, Y.; Dalhaimer, P.; Cai, S.; Tsai, R.; Tewari, M.; Minko, T.; Discher, D.E. Shape effects of filaments versus spherical particles in flow and drug delivery. *Nat. Nanotechol.* **2007**, *2*, 249–255. [CrossRef]

201. Liu, Z.; Cai, W.; He, L.; Nakayama, N.; Chen, K.; Sun, X.; Chen, X.; Dai, H. In vivo biodistribution and highly efficient tumour targeting of carbon nanotubes in mice. *Nat. Nanotechol.* **2007**, *2*, 47–52. [CrossRef]

202. Park, J.-H.; Von Maltzahn, G.; Zhang, L.; Schwartz, M.P.; Ruoslahti, E.; Bhatia, S.N.; Sailor, M.J. Magnetic Iron Oxide Nanoworms for Tumor Targeting and Imaging. *Adv. Mater.* **2008**, *20*, 1630–1635. [CrossRef]

203. Park, J.-H.; Von Maltzahn, G.; Zhang, L.; Derfus, A.M.; Simberg, D.; Harris, T.J.; Ruoslahti, E.; Bhatia, S.N.; Sailor, M.J. Systematic surface engineering of magnetic nanoworms for in vivo tumor targeting. *Small* **2009**, *5*, 694–700. [CrossRef]

204. Yue, J.; Feliciano, T.J.; Li, W.; Lee, A.; Odom, T.W. Gold Nanoparticle Size and Shape Effects on Cellular Uptake and Intracellular Distribution of siRNA Nanoconstructs. *Bioconjug. Chem.* **2017**, *28*, 1791–1800. [CrossRef] [PubMed]

205. Herd, H.; Daum, N.; Jones, A.T.; Huwer, H.; Ghandehari, H.; Lehr, C.M. Nanoparticle geometry and surface orientation influence mode of cellular uptake. *ACS Nano* **2013**, *7*, 1961–1973. [CrossRef]

206. Dag, A.; Zhao, J.C.; Stenzel, M.H. Origami with ABC Triblock Terpolymers Based on Glycopolymers: Creation of Virus-Like Morphologies. *ACS Macro Lett.* **2015**, *4*, 579–583. [CrossRef]

207. Li, Z.; Sun, L.; Zhang, Y.; Dove, A.P.; O'Reilly, R.K.; Chen, G. Shape Effect of Glyco-Nanoparticles on Macrophage Cellular Uptake and Immune Response. *ACS Macro Lett.* **2016**, *5*, 1059–1064. [CrossRef] [PubMed]

208. Wilhelm, C.; Gazeau, F.J.; Roger, J.; Pons, J.N.; Bacri, J.-C. Interaction of anionic superparamagnetic nanoparticles with cells: Kinetic analyses of membrane adsorption and subsequent internalization. *Langmuir* **2002**, *18*, 8148–8155. [CrossRef]

209. Cho, E.C.; Xie, J.; Wurm, P.A.; Xia, Y. Understanding the role of surface charges in cellular adsorption versus internalization by selectively removing gold nanoparticles on the cell surface with a I_2/KI etchant. *Nano Lett.* **2009**, *9*, 1080–1084. [CrossRef] [PubMed]

210. Lin, J.; Zhang, H.; Chen, Z.; Zheng, Y. Penetration of Lipid Membranes by Gold Nanoparticles: Insights into Cellular Uptake, Cytotoxicity, and Their Relationship. *ACS Nano* **2010**, *4*, 5421–5429. [CrossRef] [PubMed]

211. Trono, J.D.; Mizuno, K.; Yusa, N.; Matsukawa, T.; Yokoyama, K.; Uesaka, M. Size, concentration and incubatio dependence of gold nanoparticle uptake into pancreas cancer cells and its future application to X-ray drug delivery system. *J. Radiat. Res.* **2011**, *52*, 103–109. [CrossRef] [PubMed]

212. Rozenberg, B.A.; Tenne, R. Polymer-Assisted Fabrication of Nanoparticles and Nanocomposites. *Prog. Polym. Sci.* **2008**, *33*, 40–112. [CrossRef]

213. Abulateefeh, S.R.; Spain, S.G.; Thurecht, K.J.; Aylott, J.W.; Chan, W.C.; Garnett, M.C.; Alexander, C. Enhanced uptake of nanoparticle drug carriers via a thermoresponsive shell enhances cytotoxicity in a cancer cell line. *Biomater. Sci.* **2013**, *1*, 434–442. [CrossRef] [PubMed]

214. Zhu, M.; Nie, G.; Meng, H.; Xia, T.; Nel, A.; Zhao, Y. Physicochemical properties determine nanomaterial cellular uptake, transport, and fate. *Acc. Chem. Res.* **2013**, *46*, 622–631. [CrossRef] [PubMed]

215. Usman, M.; Zaheer, Y.; Younis, M.R.; Demirdogen, R.E.; Hussain, S.Z.; Sarwar, Y.; Rehman, M.; Khan, W.S.; Ihsan, A. The effect of surface charge on cellular uptake and inflammatory behavior of carbon dots. *Colloids Interface Sci. Commun.* **2020**, *35*, 100243. [CrossRef]

216. Li, Y.; Gu, N. Thermodynamics of Charged Nanoparticle Adsorption on Charge-Neutral Membranes: A Simulation Study. *J. Phys. Chem. B* **2010**, *114*, 2749–2754. [CrossRef]

217. Lin, H.-C.; Lin, H.-H.; Kao, C.-Y.; Yu, A.L.; Peng, W.-P.; Chen, C.-H. Quantitative measurement of nano-/microparticle endocytosis by cell mass spectrometry. *Angew. Chem. Int. Ed.* **2010**, *49*, 3460–3464. [CrossRef]

218. Nangia, S.; Sureshkumar, R. Effects of Nanoparticle Charge and Shape Anisotropy on Translocation through Cell Membranes. *Langmuir* **2012**, *28*, 17666–17671. [CrossRef]

219. Leroueil, P.R.; Berry, S.A.; Duthie, K.; Han, G.; Rotello, V.M.; McNerny, D.Q.; Baker, J.R.; Orr, B.G.; Banaszak Holl, M.M. Wide varieties of cationic nanoparticles induce defects in supported lipid bilayers. *Nano Lett.* **2008**, *8*, 420–424. [CrossRef]

220. Wang, B.; Zhang, L.; Bae, S.C.; Granick, S. Nanoparticle-induced surface reconstruction of phospholipid membranes. *Proc. Natl. Acad. Sci. USA* **2008**, *105*, 18171–18175. [CrossRef] [PubMed]

221. Deng, Z.J.; Liang, M.; Toth, I.; Monteiro, M.; Minchin, R.F. Plasma protein binding of positively and negatively charged polymer-coated gold nanoparticles elicits different biological responses. *Nanotoxicology* **2013**, *7*, 314–322. [CrossRef] [PubMed]

222. Encinas, N.; Angulo, M.; Carlos Astorga, C.; Montserrat Colilla, M.; Izquierdo-Barba, I.; Vallet-Regi, M. Mixed-charge pseudo-zwitterionic mesoporous silica nanoparticles with low-fouling and reduced cell uptake properties. *Acta Biomater.* **2019**, *84*, 319–329. [CrossRef]

223. Li, Y.; Chen, X.; Gu, N. Computational Investigation of Interaction between Nanoparticles and Membranes: Hydrophobic/Hydrophilic Effect. *J. Phys. Chem. B* **2008**, *112*, 16647–16653. [CrossRef]

224. Wong-Ekkabut, J.; Baoukina, S.; Triampo, W.; Tang, I.-M.; Tieleman, D.P.; Monticelli, L. Computer simulation study of fullerene translocation through lipid membranes. *Nat. Nanotechnol.* **2008**, *3*, 363–368. [CrossRef]

225. Alexeev, A.; Uspal, W.E.; Balazs, A.C. Harnessing janus nanoparticles to create controllable pores in membranes. *ACS Nano* **2008**, *2*, 1117–1122. [CrossRef] [PubMed]

226. Jing, B.; Zhu, Y.J. Disruption of Supported Lipid Bilayers by Semihydrophobic Nanoparticles. *J. Am. Chem. Soc.* **2011**, *133*, 10983–10989. [CrossRef]

227. Olubummo, A.; Schulz, M.; Lechner, B.D.; Scholtysek, P.; Bacia, K.; Blume, A.; Kressler, J.; Binder, W.H. Controlling the localization of polymer-functionalized nanoparticles in mixed lipid/polymer membranes. *ACS Nano* **2012**, *6*, 8713–8727. [CrossRef] [PubMed]

228. Lee, H.-Y.; Shin, S.H.R.; Abezgauz, L.L.; Lewis, S.A.; Chirsan, A.M.; Danino, D.D.; Bishop, K.J.M. Integration of Gold Nanoparticles into Bilayer Structures via Adaptive Surface Chemistry. *J. Am. Chem. Soc.* **2013**, *135*, 5950–5953. [CrossRef] [PubMed]

229. Holzapfel, V.; Lorenz, M.; Weiss, C.K.; Schrezenmeier, H.; Landfester, K.; Mailänder, V. Synthesis and biomedical applications of functionalized fluorescent and magnetic dual reporter nanoparticles as obtained in the miniemulsion process. *J. Phys. Condens. Matt.* **2006**, *18*, S2581–S2594. [CrossRef]

230. Alexis, F.; Pridgen, E.; Molnar, L.K.; Farokhzad, O.C. Factors affecting the clearance and biodistribution of polymeric nanoparticles. *Mol. Pharm.* **2008**, *5*, 505–515. [CrossRef]

231. Qiu, Y.; Liu, Y.; Wan, L.; Xu, L.; Bai, R.; Ji, Y.; Wu, X.; Zhao, Y.; Li, Y.; Chen, C. Surface chemistry and aspect ratio mediated cellular uptake of Au nanorods. *Biomaterials* **2010**, *31*, 7606–7619. [CrossRef]

232. Tao, W.; Zhang, J.; Zeng, X.; Liu, D.; Liu, G.; Zhu, X.; Liu, Y.; Yu, Q.; Huanget, L.; Mei, L. Blended nanoparticle system based on miscible structurally similar polymers: A safe, simple, targeted, and surprisingly high efficiency vehicle for cancer therapy. *Adv. Healthc. Mater.* **2015**, *4*, 1203–1214. [CrossRef]

233. Tao, W.; Zeng, X.; Wu, J.; Zhu, X.; Yu, X.; Zhang, X.; Zhang, J.; Liu, G.; Mei, L. Polydopamine-based surface modification of novel nanoparticle-aptamer bioconjugates for in vivo breast cancer targeting and enhanced therapeutic effects. *Theranostics* **2016**, *6*, 470–484. [CrossRef] [PubMed]

234. Lesniak, A.; Fenaroli, F.; Monopoli, M.P.; Åberg, C.; Dawson, K.A.; Salvati, A. Effects of the Presence or Absence of a Protein Corona on Silica Nanoparticle Uptake and Impact on Cells. *ACS Nano* **2012**, *6*, 5845–5857. [CrossRef] [PubMed]

235. Niu, Y.; Yu, M.; Meka, A.; Liu, Y.; Zhang, J.; Yang, Y.; Yu, C. Understanding the contribution of surface roughness and hydrophobic modification of silica nanoparticles to enhanced therapeutic protein delivery. *J. Mater. Chem. B* **2016**, *4*, 212–219. [CrossRef] [PubMed]

236. Piloni, A.; Wong, C.K.; Fan, C.; Lord, M.; Walther, A.; Stenzel, M.H. Surface roughness influences the protein corona formation of glycosylated nanoparticles and alter their cellular uptake. *Nanoscale* **2019**, *11*, 23259–23267. [CrossRef] [PubMed]

237. Treuel, L.; Jiang, X.; Nienhaus, G.U. New views on cellular uptake and trafficking of manufactured nanoparticles. *J. R. Soc. Interface* **2013**, *10*, 20120939. [CrossRef]

238. Fadeel, B.; Farcal, L.; Hardy, B.; Vázquez-Campos, S.; Hristozov, D.; Marcomini, A.; Lynch, I.; Valsami-Jones, E.; Alenius, H.; Savolainen, K. Advanced tools for the safety assessment of nanomaterials. *Nat. Nanotechol.* **2018**, *13*, 537–543. [CrossRef]

239. Huang, C.; Chen, X.; Xue, Z.; Tie Wang, T. Effect of structure: A new insight into nanoparticle assemblies from inanimate to animate. *Sci. Adv.* **2020**, *6*, eaba1321. [CrossRef]

240. Heuer-Jungemann, A.; Feliu, N.; Bakaimi, I.; Hamaly, M.; Alkilany, A.; Chakraborty, I.; Masood, A.; Casula, M.F.; Kostopoulou, A.; Oh, E.; et al. The Role of Ligands in the Chemical Synthesis and Applications of Inorganic Nanoparticles. *Chem. Rev.* **2019**, *119*, 4819–4880. [CrossRef]

241. Moezzi, A.; McDonagh, A.M.; Cortie, M.B. Zinc oxide particles: Synthesis, properties and applications. *Chem. Eng. J.* **2012**, *185–186*, 1–22. [CrossRef]

242. Pradeev Raj, K.; Sadaiyandi, K.; Kennedy, A.; Sagadevan, S.; Chowdhury, Z.Z.; Johan, M.R.B.; Aziz, F.A.; Rafique, R.F.; Selvi, R.T.; Bala, R.R. Influence of Mg doping on ZnO nanoparticles for enhanced photocatalytic evaluation and antibacterial analysis. *Nanoscale Res. Lett.* **2018**, *13*, 229. [CrossRef]

243. Anopchenko, A.; Gurung, S.; Tao, L.; Arndt, C.; Lee, H.W.H. Atomic layer deposition of ultra-thin and smooth Al-doped ZnO for zero-index photonics. *Mater. Res. Express* **2018**, *5*, 014012. [CrossRef]

244. Huang, J.-H.; Chen, J.-X.; Tu, Y.-F.; Tian, Y.; Zhou, D.; Zheng, G.; Sang, J.-P.; Fu, Q.-M. Preparation and photocatalytic activity of CuO/ZnO composite nanostructured films. *Mater. Res. Express* **2018**, *6*, 015035. [CrossRef]

245. Lavín, A.; Sivasamy, R.; Mosquera, E.; Morel, M.J. High proportion ZnO/CuO nanocomposites: Synthesis, structural and optical properties, and their photocatalytic behavior. *Surf. Inter.* **2019**, *17*, 100367. [CrossRef]

246. Rojas-Michea, C.; Morel, M.; Gracia, F.; Morelle, G.; Mosquera, E. Influence of copper doping on structural, morphological, optical, and vibrational properties of ZnO nanoparticles synthesized by sol gel method. *Surf. Interfaces* **2020**, *21*, 100700. [CrossRef]

247. Othman, A.A.; Ali, M.A.; Ibrahim, E.M.M.; Osman, M.A. Influence of Cu doping on structural, morphological, photoluminescence, and electrical properties of ZnO nanostructures by ice-bath assisted sonochemical method. *J. Alloys Compd.* **2016**, *683*, 399–411. [CrossRef]

248. Farbod, M.; Jafarpoor, E. Hydrothermal synthesis of different colors and morphologies of ZnO nanostructures and comparison of their photocatalytic properties. *Ceram. Int.* **2014**, *40*, 6605–6610. [CrossRef]

249. Zhang, L.; Yang, H.; Ma, J.; Li, L.; Wang, X.; Zhang, L.; Tian, S.; Wang, X. Controllable synthesis and shape-dependent photocatalytic activity of ZnO nanorods with a cone and different aspect ratios and of short-and-fat ZnO microrods by varying the reaction temperature and time. *Appl. Phys. A* **2010**, *100*, 1061–1067. [CrossRef]

250. Xie, J.; Wang, H.; Duana, M.; Zhang, L. Synthesis and photocatalysis properties of ZnO structures with different morphologies via hydrothermal method. *Appl. Surf. Sci.* **2011**, *257*, 6358–6363. [CrossRef]

251. Gandhi, M.S.; Mok, Y.S. Shape-dependent plasma-catalytic activity of ZnO nanomaterials coated on porous ceramic membrane for oxidation of butane. *Chemosphere* **2014**, *117*, 440–446. [CrossRef]

252. Talebian, N.; Amininezhad, S.M.; Doudi, M. Controllable synthesis of ZnO nanoparticles and their morphology-dependent antibacterial and optical properties. *J. Photochem. Photobiol. B* **2013**, *120*, 66–73. [CrossRef]

253. Peng, F.; Zhou, Q.; Zhang, D.; Lu, C.; Ni, Y.; Kou, J.; Wang, J.; Xu, Z. Bio-inspired design: Inner-motile multifunctional ZnO/CdS heterostructures magnetically actuated artificial cilia film for photocatalytic hydrogen evolution. *Appl. Catal. B Environ.* **2015**, *165*, 419–427. [CrossRef]

254. Jia, X.; Dong, S.; Wang, E. Engineering the bioelectrochemical interface using functional nanomaterials and microchip technique toward sensitive and portable electrochemical biosensors. *Biosens. Bioelectron.* **2016**, *76*, 80–90. [CrossRef]

255. Mayer, K.M.; Hafner, J.H. Localized surface plasmon resonance sensors. *Chem. Rev.* **2011**, *111*, 3828–3857. [CrossRef]

256. Guo, S.; Dong, S. Graphene nanosheet: Synthesis, molecular engineering, thin film, hybrids, and energy and analytical applications. *Chem. Soc. Rev.* **2011**, *40*, 2644–2677. [CrossRef] [PubMed]

257. Banks, C.E.; Compton, R.G. Exploring the electrocatalytic sites of carbon nanotubes for NADH detection: An edge plane pyrolytic graphite electrode study. *Analyst* **2005**, *130*, 1232–1239. [CrossRef]

258. Wang, Z.; Yu, J.; Gui, R.; Jin, H.; Xia, Y. Carbon nanomaterials-based electrochemical aptasensors. *Biosens. Bioelectron.* **2016**, *79*, 136–149. [CrossRef] [PubMed]

259. Ma, X.; Li, X.; Zhang, W.; Meng, F.; Wang, X.; Yanan Qin, Y.; Zhang, M. Carbon-Based Nanocomposite Smart Sensors for the Rapid Detection of Mycotoxins. *Nanomaterials* **2021**, *11*, 2851. [CrossRef] [PubMed]

260. Pumera, M. Electrochemistry of graphene: New horizons for sensing and energy storage. *Chem. Rec.* **2009**, *9*, 211–223. [CrossRef]

261. Mao, H.Y.; Laurent, S.; Chen, W.; Akhavan, O.; Imani, M.; Ashkarran, A.A.; Mahmoudi, M. Graphene: Promises, Facts, Opportunities, and Challenges in Nanomedicine. *Chem. Rev.* **2013**, *113*, 3407–3424. [CrossRef]

262. Cao, Q.; Han, S.-J.; Tulevski, G.S.; Zhu, Y.; Lu, D.D.; Haensch, W. Arrays of single-walled carbon nanotubes with full surface coverage for high-performance electronics. *Nat. Nanotechnol.* **2013**, *8*, 180–186. [CrossRef]

263. Li, Z.; Wang, L.; Li, Y.; Feng, Y.; Feng, W. Carbon-based functional nanomaterials: Preparation, properties and applications. *Compos. Sci. Technol.* **2019**, *179*, 10–40. [CrossRef]

264. Wu, C.Z.; Feng, J.; Peng, L.L.; Ni, Y.; Liang, H.Y.; He, L.H.; Xie, Y. Large-area graphene realizing ultrasensitive photothermal actuator with high transparency: New prototype robotic motions under infrared-light stimuli. *J. Mater. Chem.* **2011**, *21*, 18584–18591. [CrossRef]

265. Wang, X.; Yang, T.; Feng, Y.; Jiao, K.; Li, G. A Novel Hydrogen Peroxide Biosensor Based on the Synergistic Effect of Gold-Platinum Alloy Nanoparticles/Polyaniline Nanotube/Chitosan Nanocomposite Membrane. *Electroanalysis* **2009**, *21*, 819–825. [CrossRef]

266. Ahmad, M.; Gan, L.; Pan, C.; Zhu, J. Controlled synthesis and methanol sensing capabilities of Pt-incorporated ZnO nanospheres. *Electrochim. Acta* **2010**, *55*, 6885–6891. [CrossRef]

267. Banks, C.E.; Goodwin, A.; Heald, C.G.R.; Compton, R.G. Exploration of gas sensing possibilities with edge plane pyrolytic graphite electrodes: Nitrogen dioxide detection. *Analyst* **2005**, *130*, 280–282. [CrossRef]

268. Moore, R.R.; Banks, C.E.; Compton, R.G. Electrocatalytic detection of thiols using an edge plane pyrolytic graphite electrode. *Analyst* **2004**, *129*, 755–758. [CrossRef] [PubMed]

269. Saxena, U.; Das, A.B. Nanomaterials towards fabrication of cholesterol biosensors: Key roles and design approaches. *Biosens. Bioelectron.* **2016**, *75*, 196–205. [CrossRef] [PubMed]

270. Rodriguez, M.C.; Rubianes, M.D.; Rivas, G.A. Highly Selective Determination of Dopamine in the Presence of Ascorbic Acid and Serotonin at Glassy Carbon Electrodes Modified with Carbon Nanotubes Dispersed in Polyethylenimine. *J. Nanosci. Nanotechnol.* **2008**, *8*, 6003–6009. [CrossRef] [PubMed]

271. Canbay, E.; Sahin, B.; Kiran, M.; Akyilmaz, E. MWCNT–cysteamine–Nafion modified gold electrode based onmyoglobin for determination of hydrogen peroxide and nitrite. *Bioelectrochemistry* **2014**, *101*, 126–131. [CrossRef]

272. Cruz-Silva, E.; Lopez-Urias, F.; Munoz-Sandoval, E.; Sumpter, B.G.; Terrones, H.; Charlier, J.C.; Meunier, V.; Terrones, M. Electronic Transport and Mechanical Properties of Phosphorus- and Phosphorus−Nitrogen-Doped Carbon Nanotubes. *ACS Nano* **2009**, *3*, 1913–1921. [CrossRef]

273. Xu, X.; Jiang, S.; Hu, Z.; Liu, S. Nitrogen-Doped Carbon Nanotubes: High Electrocatalytic Activity toward the Oxidation of Hydrogen Peroxide and Its Application for Biosensing. *ACS Nano* **2010**, *4*, 4292–4298. [CrossRef]

274. Zhang, X.; Hou, L.; Samorì, P. Coupling carbon nanomaterials with photochromic molecules for the generation of optically responsive materials. *Nat. Commun.* **2016**, *7*, 11118. [CrossRef]
275. Schneidera, V.; Strunskusa, T.; Elbahrib, M.; Faupel, F. Light-induced conductance switching in azobenzene based near-percolated single wall carbon nanotube/polymer composites. *Carbon* **2015**, *90*, 94–101. [CrossRef]
276. Margapoti, E.; Strobel, P.; Asmar, M.M.; Seifert, M.; Juan Li, J.; Sachsenhauser, M.; Ceylan, Ö.; Palma, C.-A.; Barth, J.V.; Garrido, J.A.; et al. Emergence of photoswitchable states in a graphene-azobenzene-Au platform. *Nano Lett.* **2014**, *14*, 6823–6827. [CrossRef] [PubMed]
277. Song, Y.; Xu, C.; Wei, W.; Ren, J.; Qu, X. Light regulation of peroxidase activity by spiropyran functionalized carbon nanotubes used for label-free colorimetric detection of lysozyme. *Chem. Commun.* **2011**, *47*, 9083–9085. [CrossRef]
278. Katz, E.; Willner, I. Integrated nanoparticle-biomolecule hybrid systems: Synthesis, properties, and applications. *Angew. Chem. Int. Ed.* **2004**, *43*, 6042–6108. [CrossRef]
279. Saxena, U.; Chakraborty, M.; Goswami, P. Covalent immobilization of cholesterol oxidase on self-assembled gold nanoparticles for highly sensitive amperometric detection of cholesterol in real samples. *Biosens. Bioelectron.* **2011**, *26*, 3037–3043. [CrossRef]
280. Zhou, X.; Wang, Y.; Gong, C.; Liu, B.; Wei, G. Production, structural design, functional control, and broad applications of carbon nanofiber-based nanomaterials: A comprehensive review. *Chem. Eng. J.* **2020**, *402*, 126189. [CrossRef]
281. Han, Q.; Wang, B.; Gao, J.; Qu, L.T. Graphitic carbon nitride/nitrogen-rich carbon nanofibers: Highly efficient photocatalytic hydrogen evolution without cocatalysts. *Angew. Chem. Int. Ed.* **2016**, *55*, 10849–10853. [CrossRef]
282. Dykman, L.; Khlebtsov, N. Gold nanoparticles in biomedical applications: Recent advances and perspectives. *Chem. Soc. Rev.* **2012**, *41*, 2256–2282. [CrossRef] [PubMed]
283. Ashtari, K.; Nazari, H.; Ko, H.; Tebon, P.; Akhshik, M.; Akbari, M.; Alhosseini, S.N.; Mozafari, M.; Mehravi, B.; Soleimani, M.; et al. Electrically conductive nanomaterials for cardiac tissue engineering. *Adv. Drug Deliv. Rev.* **2019**, *144*, 162–179. [CrossRef] [PubMed]
284. Zhou, H.; Gan, X.; Wang, J.; Zhu, X.; Li, G. Hemoglobin-based hydrogen peroxide biosensor tuned by the photovoltaic effect of nano titanium dioxide. *Anal. Chem.* **2005**, *77*, 6102–6104. [CrossRef]
285. Sharifi, E.; Salimi, A.; Shams, E.; Noorbakhsh, A.; Amini, M.K. Shape-dependent electron transfer kinetics and catalytic activity of NiO nanoparticles immobilized onto DNA modified electrode: Fabrication of highly sensitive enzymeless glucose sensor. *Biosens. Bioelectron.* **2014**, *56*, 313–319. [CrossRef]
286. Feng, K.J.; Yang, Y.H.; Wang, Z.J.; Jiang, J.H.; Shen, G.L.; Yu, R.Q. A nano-porous CeO$_2$/Chitosan composite film as immobilization matrix for colorectal cancer DNA sequence-selective electrochemical biosensor. *Talanta* **2006**, *70*, 561–565. [CrossRef]
287. Chauhan, N.; Pundir, C.S. Amperometric determination of acetylcholine—A neurotransmitter, by chitosan/gold-coated ferric oxide nanoparticles modified gold electrode. *Biosens. Bioelectron.* **2014**, *61*, 1–8. [CrossRef]
288. Wang, W.; Hao, Q.; Bao, L.; Lei, J.; Wang, Q.; Ju, H. Quantum dot-functionalized porous ZnO nanosheets as a visible light induced photoelectrochemical platform for DNA detection. *Nanoscale* **2014**, *6*, 2710–2717. [CrossRef]
289. Aslan, K.; Lakowicz, J.R.; Geddes, C.D. Nanogold-plasmon-resonance-based glucose sensing. *Anal. Biochem.* **2004**, *330*, 145–155. [CrossRef]
290. Stratakis, E.; Kymakis, E. Nanoparticle-based plasmonic organic photovoltaic devices. *Mater. Today* **2013**, *16*, 133–146. [CrossRef]
291. Spyropoulos, G.D.; Stylianakis, M.M.; Stratakis, E.; Kymakis, E. Organic bulk heterojunction photovoltaic devices with surfactant-free Au nanoparticles embedded in the active layer. *Appl. Phys. Lett.* **2012**, *100*, 213904. [CrossRef]
292. Kymakis, E.; Spyropoulos, G.D.; Fernandes, R.; Kakavelakis, G.; Kanaras, A.G.; Stratakis, E. Plasmonic Bulk Heterojunction Solar Cells: The Role of Nanoparticle Ligand Coating. *ACS Photonics* **2015**, *2*, 714–723. [CrossRef]
293. Ríos, Á.; Zougagh, M. Recent advances in magnetic nanomaterials for improving analytical processes. *Trends Anal. Chem.* **2016**, *84*, 72–83. [CrossRef]
294. Bagherzadeh, M.; Pirmoradian, M.; Riahi, F. Electrochemical detection of Pb and Cu by using DTPA Functionalized magnetic nanoparticles. *Electrochim. Acta* **2014**, *115*, 573–580. [CrossRef]
295. Hooshmand, N.; El-Sayed, M.A. Collective multipole oscillations direct the plasmonic coupling at the nanojunction interfaces. *Proc. Natl. Acad. Sci. USA* **2019**, *116*, 19299–19304. [CrossRef]
296. Cargnello, M.; Johnston-Peck, A.C.; Diroll, B.T.; Wong, E.; Datta, B.; Damodhar, D.; Doan-Nguyen, V.V.; Herzing, A.A.; Kagan, C.R.; Murray, C.B. Substitutional doping in nanocrystal superlattices. *Nature* **2015**, *524*, 450–453. [CrossRef] [PubMed]
297. McNeil, S.E. Nanoparticle therapeutics: A personal perspective. *WIREs Nanomed. Nanobiotechnol.* **2009**, *1*, 264–271. [CrossRef]

nanomaterials

Article

Promoting the Selectivity of Pt/m-ZrO$_2$ Ethanol Steam Reforming Catalysts with K and Rb Dopants

Michela Martinelli [1], Richard Garcia [2], Caleb D. Watson [2], Donald C. Cronauer [3], A. Jeremy Kropf [3] and Gary Jacobs [2,4,*]

[1] Center for Applied Energy Research, University of Kentucky, 2540 Research Park Drive, Lexington, KY 40511, USA; michela.martinelli@uky.edu

[2] Department of Biomedical Engineering and Chemical Engineering, The University of Texas at San Antonio, 1 UTSA Circle, San Antonio, TX 78249, USA; g123richard@gmail.com (R.G.); caleb.watson378@gmail.com (C.D.W.)

[3] Argonne National Laboratory, Lemont, IL 60439, USA; dccronauer@anl.gov (D.C.C.); kropf@anl.gov (A.J.K.)

[4] Department of Mechanical Engineering, The University of Texas at San Antonio, 1 UTSA Circle, San Antonio, TX 78249, USA

* Correspondence: gary.jacobs@utsa.edu; Tel.: +1-210-458-7080

Abstract: The ethanol steam reforming reaction (ESR) was investigated on unpromoted and potassium- and rubidium-promoted monoclinic zirconia-supported platinum (Pt/m-ZrO$_2$) catalysts. Evidence from in situ diffuse reflectance infrared Fourier transform spectroscopy (DRIFTS) characterization indicates that ethanol dissociates to ethoxy species, which undergo oxidative dehydrogenation to acetate followed by acetate decomposition. The acetate decomposition pathway depends on catalyst composition. The decarboxylation pathway tends to produce higher overall hydrogen selectivity and is the most favored route at high alkali loading (2.55 wt.% K and higher or 4.25 wt.% Rb and higher). On the other hand, decarbonylation is a significant route for the undoped catalyst or when a low alkali loading (e.g., 0.85% K or 0.93% Rb) is used, thus lowering the overall H$_2$ selectivity of the process. Results of in situ DRIFTS and the temperature-programmed reaction of ESR show that alkali doping promotes forward acetate decomposition while exposed metallic sites tend to facilitate decarbonylation. In previous work, 1.8 wt.% Na was found to hinder decarbonylation completely. Due to the fact that 1.8 wt.% Na is atomically equivalent to 3.1 wt.% K and 6.7 wt.% Rb, the results show that less K (2.55% K) or Rb (4.25% Rb) is needed to suppress decarbonylation; that is, more basic cations are more efficient promoters for improving the overall hydrogen selectivity of the ESR process.

Keywords: ethanol steam reforming; potassium; rubidium; basicity; zirconia; XANES; DRIFTS

Citation: Martinelli, M.; Garcia, R.; Watson, C.D.; Cronauer, D.C.; Kropf, A.J.; Jacobs, G. Promoting the Selectivity of Pt/m-ZrO$_2$ Ethanol Steam Reforming Catalysts with K and Rb Dopants. *Nanomaterials* **2021**, *11*, 2233. https://doi.org/10.3390/nano11092233

Academic Editor: Simon Freakley

Received: 30 July 2021
Accepted: 25 August 2021
Published: 29 August 2021

1. Introduction

In recent decades, the catalytic steam reforming of hydrocarbons such as natural gas has been the most economically competitive method to produce hydrogen in the chemical industry. However, this method is not sustainable as the feedstock is a fossil source, and significant amounts of CO$_2$ are produced in the process. Thus, the development of new sustainable reforming technologies from renewable feedstocks (e.g., biomass-derived oxygenates) is necessary for reducing net greenhouse gas emissions [1]. In this scenario, researchers are focusing on several renewable feedstocks such as ethanol, polyols, and dimethyl ether [2–5]. Among these renewable feedstocks, bio-ethanol is very attractive because of its favorable hydrogen content, wide abundance, low toxicity, and ability to be easily stored for transportation and portable power [3]. Moreover, while ethanol production (e.g., from sugar cane or corn) currently competes with food production, cellulosic ethanol is currently under development. The overall ethanol steam reforming (ESR) reaction, which occurs at 350–650 °C, can be summarized by Equation (1):

$$C_2H_5OH + 3H_2O \rightarrow 2CO_2 + 6H_2 \tag{1}$$

Acetaldehyde, ethylene, and methane are formed during ESR. However, the concentration of these species must be minimized to achieve higher efficiency of hydrogen production and avoid carbon formation and consequently catalyst deactivation [6]. These intermediates and byproducts can be converted via different pathways, depending on catalyst structure as well as the conditions used in the reactor [3].

Ethanol steam reforming catalysts are often transition metals such as copper, cobalt, and nickel, or noble metals such as platinum, palladium, rhodium, gold, or ruthenium; combinations of metals have also been used [7,8]. Among the transition metals, nickel and cobalt are the most common. However, catalyst deactivation by coke deposition is a major issue for these catalysts [6,8–14]. Noble metals such as Pt and Rh have better resistance to coke deposition as well as high activity, but they are much higher in cost [8,15–18]. These metals are typically supported on basic, acidic, or inert supports [7]. Oxides that form surface defects through partial reduction (e.g., ceria, zirconia, ceria–zirconia mixtures, and metal-promoted ceria catalysts) have also been investigated because of their high oxygen mobility, their ability to dissociate water or ROH (i.e., where R is an alkyl group - methanol, ethanol, etc.), and their capability to shuttle O-bound intermediates on the catalyst surface [19,20]. Ciambelli et al. [21] observed that Pt/CeO_2 exhibited higher activity as compared to Pt/Al_2O_3 for ESR in the temperature range of 300–450 °C. He et al. [22] obtained similar findings. The high activity of Pt/CeO_2 was due in part to the ability of Pt to facilitate the formation of defect sites on the partially reducible oxide support. ROH molecules can then dissociate on these defect sites [23], which is comparable to the adsorption of H_2O at defect sites that result in the formation of bridging OH groups [24–26].

Recently, the alkali promotion of noble or transition metal catalysts has been investigated. Alkali doping can impact many catalyst properties, including activity, stability, resistance to coke formation, selectivity, and surface acidity/basicity [6,27–42]. The effect of alkali on nickel-based catalysts is contradictory. Frusteri et al. [33,35] found that Li and Na adversely affect the nickel dispersion, whereas they improve the extent of reduction of nickel. In contrast, no effect on either Ni dispersion or catalyst morphology was detected for K-promoted catalysts. Moreover, the authors examined the influence of alkali loading on catalyst performance. Lithium and potassium enhanced catalyst stability mainly by depressing nickel sintering, whereas carbon laydown did not seem to be influenced by adding K. However, a different effect of potassium was observed by Slowik et al. [6]. In that case, potassium promotion of Ni/CeO_2 was not found to protect the catalyst against the formation of carbon deposits and did not improve stability during ESR.

Improvements in stability/activity were reported for cobalt-based catalysts, and this is mainly related to the inhibition of carbon deposition [27–31,38]. Recently, Grzybek et al. [27] studied the alkali surface state (location, dynamics) by Species Resolved Thermal Alkali Desorption (SR-TAD). Movement of potassium from cobalt to alumina was observed during both activation and ESR. This phenomenon stabilizes small cobalt crystallites by hindering their detachment from the catalyst surface, which would otherwise result in encapsulation by the growing carbonaceous deposit. Furthermore, Grzybek et al. [28] found that potassium loadings from 0.1 to 4 wt.% improve the activity of their catalyst by enhancing C–C bond scission, but 0.3 wt.% is the optimal loading to maximize the selectivity to H_2 and CO_2, the most desirable products. The beneficial effect of K can also be related to improvement in reduction of Co^{2+} to Co^0, the stabilization of acetate species, and the suppression of methane formation [31].

To our knowledge, not many studies are available on the effect of alkali for noble metal catalysts, especially at low temperatures [34,40–42]. Low potassium loading (0.2%) was found to decrease the initial conversion but improves the stability for Rh/CeO_2-ZrO_2. At higher loading (5%), the catalyst activity is negligible [42]. Dömök et al. explored doping potassium to Pt/Al_2O_3 catalyst, and they found that increasing potassium content progressively decreases the ethanol conversion and changes the product distribution toward a higher selectivity to CH_4 and CO_2 as compared to 1% Pt/Al_2O_3 [34]. Furthermore, potassium destabilizes adsorbed acetate, promoting its decomposition to CO_2 and CH_4 at a lower temperature; on the other hand, acetate species were more stable on the undoped catalyst, decomposing at ~420 °C [34].

In our prior study of Na-doping to Pt/ZrO_2 [40,41], a similar trend was obtained, as acetate decomposed at a lower temperature (100–150 °C) when 1.8–2.5 wt.% Na was added to the formulation as compared to the undoped catalyst. Our focus is on the low-temperature conversion of ethanol with steam to H_2, CO_2, and CH_4, with the latter being reformed in a conventional methane reforming (e.g., autothermal reforming). With this decarboxylation route Equation (1), which involves the steps below Equations (2) and (3), higher hydrogen selectivity is expected as compared to the decarbonylation route Equation (4):

Decarboxylation pathway:

$$C_2H_5OH + H_2O \rightarrow 2H_2 + CH_4 + CO_2 \tag{2}$$

$$CH_4 + 2H_2O \rightarrow 4H_2 + CO_2 \tag{3}$$

Decarbonylation pathway

$$C_2H_5OH + H_2O \rightarrow 4H_2 + 2CO \tag{4}$$

In that work, DRIFTS experiments revealed that the acetate decomposition pathway depends on the Na loading. Forward direct acetate decomposition to CH_4 and carbonate Equation (2) is the most favorable pathway at high sodium loading (1.8 or 2.5 wt.%), whereas the unselective decarbonylation route occurs for the unpromoted catalyst or at low sodium loading (0.5 wt.%) and is promoted by metallic sites.

The question remains as to whether more basic alkali metals might further improve the decarboxylation selectivity over that of decarbonylation during ESR. To that end, the effect of potassium and rubidium loading on the relative rates of decarboxylation/decarbonylation was investigated. $Pt/m-ZrO_2$ catalyst was promoted by the following potassium loadings: 0% (reference), 0.85%, 1.70%, 2.55%, 3.40%, 4.25%, and 8.50 wt.%; whereas rubidium was added with the following loadings: 0.93%, 1.86%, 2.79%, 3.72%, 4.65%, 5.59%, and 9.29%. Atomically equivalent loadings allowed for some comparisons between the K- and Rb-promoted catalysts, as well as with catalysts prepared in our earlier study using Na as the dopant [40,41]. These systems were characterized by N_2 physisorption, transmission electron microscopy (TEM), X-ray absorption near edge spectroscopy (XANES), extended X-ray absorption fine structure spectroscopy (EXAFS), hydrogen temperature programmed reduction (H_2-TPR), temperature-programmed ESR, and DRIFTS. Catalyst activity and selectivity were measured at steady-state using a fixed bed tubular reactor.

2. Materials and Methods

2.1. Catalyst Preparation

Various potassium and rubidium loadings on 2% Pt/ZrO_2 were prepared via incipient wetness impregnation (IWI). Firstly, monoclinic ZrO_2 (Alfa Aesar, Haverhill, MA, USA) was impregnated by 2% Pt with an aqueous solution of tetraamine platinum (II) nitrate (Alfa Aesar, Haverhill, MA, USA) followed by drying and calcination at 350 °C (four hours, muffle furnace). Then, the appropriate amount of KNO_3 (Alfa Aesar, Haverhill, MA, USA) or $RbNO_3$ (Alfa Aesar, Haverhill, MA, USA) was added via IWI. Promoted catalysts were dried and recalcined using the same conditions.

2.2. Catalyst Characterization

Surface area and pore size were determined using a Micromeritics 3-Flex instrument (Micromeritics, Norcross, GA, USA). The BJH method was used to calculate the average pore diameter and specific volume. Samples were pre-treated at 160 °C at 50 mTorr for no less than 12 h.

H_2 temperature-programmed reduction (TPR) plots of the catalysts were obtained using an Altamira AMI-300R (Altamira, Pittsburgh, PA, USA) instrument employing a thermal conductivity detector (TCD). During experiments, approximately 200 mg of catalyst was loaded into the U-tube reactor, and then a mixture of 10% H_2 in Ar (30 cm^3/min) (Airgas, San Antonio, TX, USA) was flowed while the temperature was ramped from 50 to 1000 °C at a heating rate of 10 °C/min.

Transmission electron microscopy (TEM) and scanning transmission microscopy (STEM) were conducted with an FEI Talos F200X scope (Thermo Fisher Scientific, Waltham, MA, USA) equipped with bright field (BF), dark-field (DF) 2, DF 4, and high-angle annular dark-field (HAADF) detectors. Imaging was performed using a field emission gun with an accelerating voltage of 200 kV and a high-speed Ceta 16M camera. The elemental distributions were determined via FEI super energy-dispersive X-ray spectroscopy (EDX) (Thermo Fisher Scientific, Waltham, MA, USA). Velox software was utilized to process the data. Before analysis, samples were first treated in H_2 at 350 °C for 1 hour and then cooled to room temperature, followed by passivation with 1% O_2 (balance N_2). Then, the reduced/passivated samples were sonicated for 30 min in ethanol. A sample of this suspension was dropped onto a lacey carbon-covered Cu grid (300 mesh) and dried in air for 12 h.

Temperature-programmed reaction/desorption analyses were carried out with the Altamira AMI-300R unit. For each experiment, the catalyst was reduced at 300 °C using 10 cm^3/min H_2 (Airgas, San Antonio, TX, USA) and 20 cm^3/min argon (Airgas, San Antonio, TX, USA) for 1 h. After cooling the catalyst to 50 °C in flowing Ar, ethanol was pumped at a rate of 100 mL/min across the catalyst for 10 min, and then Ar was flowed at 30 cm^3/min for 20 min. Following this purge, water was bubbled in He at a rate of 30 cm^3/min for 10 min, and then another purge in Ar was performed. Finally, Ar was flowed at 30 cm^3/min while the temperature was increased to 700 °C. This allowed analysis of ESR under transient conditions. CH_4 evolution was measured to examine the effect of K or Rb on C–C cleaving.

A Nicolet (Thermo Fisher, Waltham, MA, USA) iS-10 Fourier Transform infrared spectrometer, coupled with a Harrick Scientific (Pleasantville, New York, NY, USA) Praying Mantis accessory, was used for the temperature desorption/reaction experiments. The catalyst was reduced at 300 °C with a 1:1 mixture of H_2:He at a flow rate of 200 cm^3/min for 1 h and then cooled to 50 °C in hydrogen. Next, helium was used to bubble ethanol at a flow rate of 75 cm^3/min for 15 min. Subsequently, He was bubbled through water (31 °C water bath), giving an H_2O concentration of 4.4% with a flow rate of 75 cm^3/min. Temperature was stepped in 50 °C increments from 50 °C to 500 °C. At each step, 512 scans were taken at a resolution of 4.

Temperature programmed reduction with X-ray absorption near edge spectroscopy (TPR-XANES) was performed at the Materials Research Collaborative Access Team (MR-CAT) beamline located at the Advanced Photon Source, Argonne National Laboratory. Sample amount was optimized for the platinum L_3 edge. The quartz reactor was held in a clamshell furnace located on a positioning table, and the beam passed through six samples in a sequential manner with 20 μm accuracy for repeat scans. He was flowed through the catalysts for more than 5 min at a flow rate of 30 mL/min. Pure hydrogen was then passed through the sample at a flow rate of 30 mL/min, and a heating rate of 0.83 °C/min was started for the furnace to achieve 300 °C. After soaking at this temperature for 1 h, samples were cooled to room temperature and scanned for both the XANES and EXAFS regions at both the platinum L_3 (11.564 keV) and L_2 edges (13.273 keV), so that the L_3-L_2 edge difference procedure could be applied [43]. Spectra were recorded in transmission mode, and the respective metallic foil was measured in a concurrent manner for the purpose of energy calibration. The Pt L_3 edge data were in the range of 11.400–12.700 keV, and Pt L_2 edge data were in the range of 13.100–13.850 keV. Standard data reduction was conducted with WinXAS (Version 2.0, Thorsten Ressler, Berlin, Germany) [44], while fittings were carried out for EXAFS with Atoms (Copyright 2001, Department of Physics, University of Washington, Seattle, WA, USA) [45], FEFF8 (Version 8.20, Department of Physics, University of Washington, Seattle, WA, USA) [46], and FEFFIT (Copyright 2001, Department of Physics, University of Washington, Seattle, WA, USA) [46] software over Δk = 3–10 Å^{-1} and ΔR = 1.85–3.25 Å. A typical analysis included post-edge background subtraction (Victoreen function), pre-edge and post-edge background subtraction (degree 1 polynomials), normalization based on the edge jump, conversion to k-space with background subtraction using a cubic spline, and applying a Fourier transform of the χ(k) function to R-space.

2.3. Catalytic Activity

The activity of the catalysts was tested in a fixed bed reactor (stainless steel tubular reactor, I.D. 0.444 in.); more information on the experimental set-up is reported in our previous study [47]. Briefly, 80 mg of catalyst (63–106 μm) was diluted with 300 mg of SiO$_2$ beads and activated using 100 cm^3/min H$_2$ at 350 °C for 1 h. Next, the temperature was cooled to 300 °C, and the gas was changed to a mixture containing 26.1% H$_2$O, 2.9% C$_2$H$_5$OH (balance N$_2$) at P = 1 atm, gas hourly space velocity (GHSV) = 190,560 Ncm^3/min/g$_{cat}$. The unpromoted catalyst was also tested at different GHSV in order to compare the selectivities among the catalysts at similar conversion. The products were then passed through a cold trap (held at 5 °C) to collect condensable compounds. The condensable products were analyzed by SRI (SRI Instruments, Torrance, CA, USA) 8610 GC equipped with HayeSep Q-column, whereas the gas products were analyzed by Inficon micro-GC Fusion equipped by molecular sieve, alumina, plot-u, and OV-1. Ethanol conversion Equation (5), carbon selectivity Equation (6), and H$_2$ yield Equation (7) were calculated using the following formulas:

$$X_{C_2H_5OH} = 1 - \frac{F^{out}_{C_2H_5OH}}{F^{in}_{C_2H_5OH}} \tag{5}$$

$$S_i = \frac{n_i \cdot F_i^{out,\ prod}}{\sum_i n_i \cdot F_i^{out,prod}} \tag{6}$$

$$H_2\ yield = \frac{F^{out,\ prod}_{H2}}{6 \cdot F^{in}_{C2H5OH}} \tag{7}$$

where $F^{in}_{C_2H_5OH}$ is the molar feed rate of ethanol, $F^{out}_{C_2H_5OH}$ is the effluent molar flow rate, n_i is carbon number, $F^{out,\ prod}_{H2}$ is the effluent molar flow rate of H$_2$, and $F_i^{out,\ prod}$ is effluent molar flow rate of the C-containing species (i = CO, CO$_2$, C$_2$H$_6$, C$_2$H$_4$, C$_3$H$_6$, C$_2$H$_4$O).

3. Results and Discussion

Surface area and porosity results for un-promoted and K and Rb-promoted catalysts are provided in Table 1. Both K and Rb series showed that low alkali loadings caused a slight decrease in the surface area, whereas high loadings dramatically decreased it (i.e., below what is expected from the decrease due to the added mass of alkali). For example, the surface area dropped from 89.7 m^2/g$_{cat}$ (unpromoted) to 34.7 and 58.2 m^2/g$_{cat}$ for 8.5% K and 9.29% Rb, respectively. This decrease in the surface area suggests that pore blocking is more significant at high alkali loading. Increasing alkali doping progressively diminished the pore volume, but little impact was observed for the average pore diameter of the Rb-promoted catalyst. In contrast, above 2.55% K, the average diameter increased systematically, suggesting preferential blocking of narrower pores by K at higher loadings.

Table 1. Surface area, porosity, and Pt particle size for K and Rb-promoted catalysts (adapted from [48,49] with permission from Elsevier (copyright, 2020) and MDPI (copyright, 2021)).

Sample ID	A_s (BET) (m^2/g)	V_p (BJH Des) (cm^3/g)	D_p (BJH Des) (Å)	Est. Pt Diam (nm)	Est. % Pt Disp. (%)
Pt/ZrO$_2$ (K-series)	82.8	0.276	103	1.0 */0.92 **	88
0.85% K-Pt/ZrO$_2$	78.3	0.260	101	1.2 */1.1 **	82
1.70% K-Pt/ZrO$_2$	72.2	0.249	100	1.3 */1.2 **	79
2.55% K-Pt/ZrO$_2$	68.4	0.245	103	2.7 */2.6 **	47
3.40% K-Pt/ZrO$_2$	59.6	0.219	105	3.0 */3.0 **	42
4.25% K-Pt/ZrO$_2$	53.8	0.200	109	3.6 */3.6 **	35
8.50% K-Pt/ZrO$_2$	34.7	0.139	123	-	-
Pt/ZrO$_2$ (Rb-series)	89.7	0.260	95	0.8 */0.72 **	94
0.55% Rb-Pt/ZrO$_2$	87.9	0.268	96	-	-
0.93% Rb-Pt/ZrO$_2$	91.6	0.275	93	0.86 */0.78 **	92
1.86% Rb-Pt/ZrO$_2$	88.7	0.262	94	1.0 */0.93 **	87
2.79% Rb-Pt/ZrO$_2$	86.7	0.260	93	1.1 */0.99 **	85
4.65% Rb-Pt/ZrO$_2$	72.3	0.235	95	1.3 */1.2 **	77
9.29% Rb-Pt/ZrO$_2$	58.2	0.202	102	2.0 */1.9 **	56

* Estimated from Jentys assuming spherical cluster morphology [50]. ** Estimated from Marinkovic et al. [51].

Estimated Pt diameter and dispersion obtained by EXAFS fittings are also reported in Table 1. Interestingly, the diameter of the Pt cluster increases with the alkali loading, and the effect is more pronounced for the K-promoted catalyst. This results in a decreasing trend in Pt dispersion, which drops from 88% and 94% for the unpromoted catalysts to 35% and 56% for 4.25% K-Pt/ZrO$_2$ and 9.29% Rb-Pt/ZrO$_2$, respectively.

STEM-EDX images for Rb-promoted catalysts are shown in Figure 1. Both platinum and rubidium were well dispersed for 0.93% Rb-2% Pt/ZrO$_2$, and all clusters observed were below 3 nm. By increasing the rubidium loading to 9.29%, STEM-EDX images (Figure 1, bottom) show rubidium was well dispersed, whereas platinum particles tended to form agglomerates of several Pt domains. Some agglomerates were on the order of 10 nm, while the domains were typically below 3 nm. The spatial distribution of Pt and Rb also suggests, especially in the case of 9.29% Rb, that there is a strong possibility of contact between the two elements.

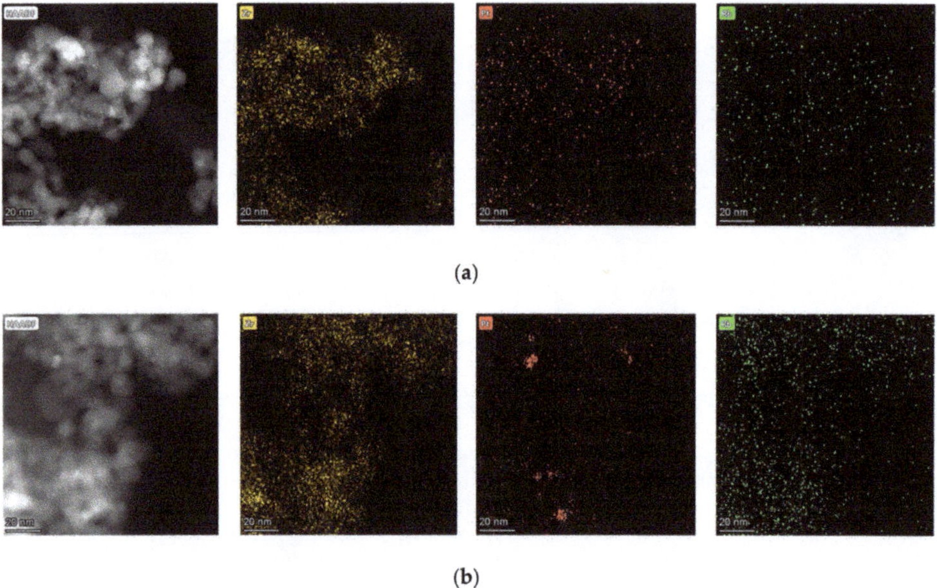

Figure 1. Transmission electron microscopy (TEM) and scanning transmission microscopy with energy-dispersive X-ray spectroscopy (STEM-EDX) images for the (**a**) 0.93% Rb-2% Pt/ZrO$_2$ catalyst and (**b**) the 9.29% Rb-2% Pt/ZrO$_2$ catalyst.

Catalyst reducibility was investigated by TPR (Figure 2). The unpromoted Pt/zirconia catalyst had a small hydrogen uptake in the range of 150–200 °C, indicative of reduction of platinum oxide to metal, which occurs around 200 °C, and Pt-catalyzed defect formation [46]. Pt accelerates the decomposition of surface carbonates and facilitates the formation of oxygen vacancies and bridging OH groups [25,52,53]. The hydrogen uptake during TPR further increased by adding potassium or rubidium.

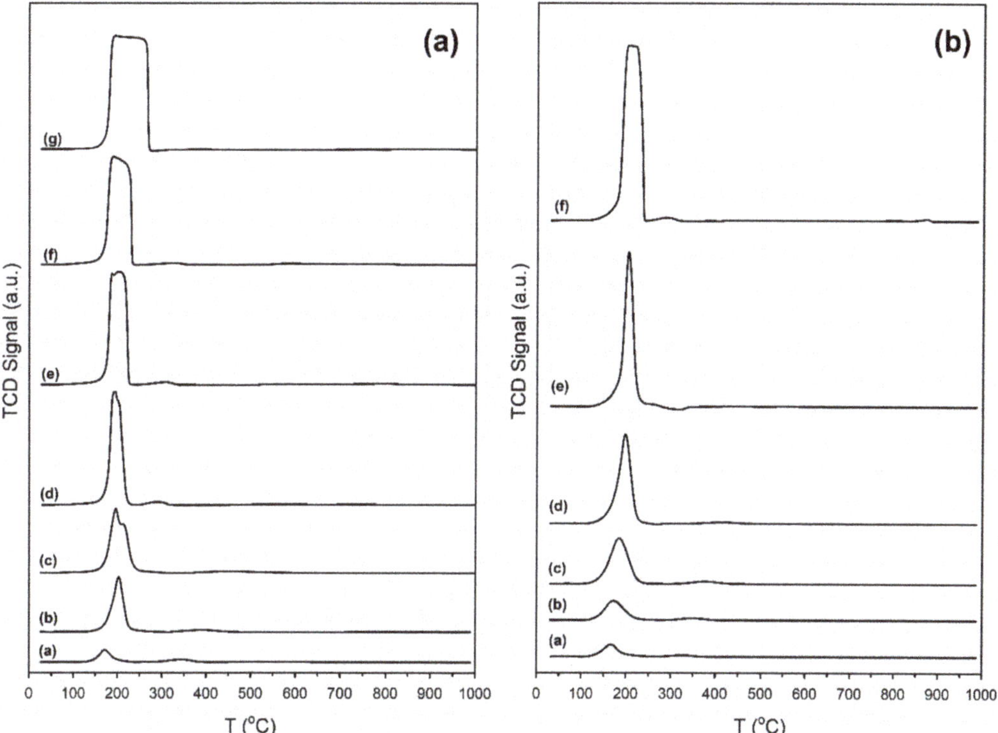

Figure 2. (**a**): TPR profiles for K-promoted catalysts: (a) 2% Pt/ZrO$_2$, (b) 0.85% K-2% Pt/ZrO$_2$, (c) 1.70% K-2% Pt/ZrO$_2$, (d) 2.55% K-2% Pt/ZrO$_2$, (e) 3.40% K-2% Pt/ZrO$_2$, (f) 4.25% K-2% Pt/ZrO$_2$, and (g) 8.50% K-2% Pt/ZrO$_2$; (**b**): TPR profile for Rb-promoted catalysts: (h) 0.55% Rb-2% Pt/ZrO$_2$, (i) 1.86% Rb-2% Pt/ZrO$_2$, (j) 2.79% Rb-2% Pt/ZrO$_2$, (k) 4.65% Rb-2% Pt/ZrO$_2$, and (l) 9.29% Rb-2% Pt/ZrO$_2$. Figure 2a reprinted from [48] with permission from Elsevier, copyright 2020. Figure 2b reprinted from [48] with permission from MDPI, copyright 2021.

Temperature programmed reduction with mass spectrometry (TPR-MS) spectra for K and Rb-promoted catalysts, reported in our previous works [48,49], showed that the evolution of carbon monoxide occurs concurrently with hydrogen uptake, especially for high alkali loading. This indicates the presence of surface carbonates on the catalyst before H$_2$ activation, as previously noted [52,53]. Doping with K or Rb increases surface basicity, such that more carbonate (i.e., adsorbed carbon dioxide—which is acidic) decomposes from the catalyst as the alkali loading is increased. Decomposition of these surface carbonates through Pt-assisted decarbonylation is easily monitored by in situ DRIFTS (Figure 3) with the generation of CO via Pt carbonyl species.

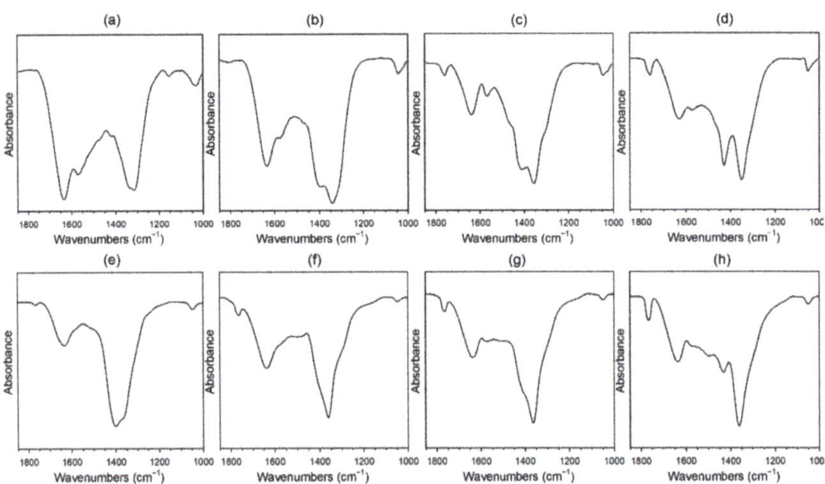

Figure 3. DRIFTS spectra of carbonate decomposition during activation for (**a**) 2% Pt/ZrO$_2$, and the same doped with (**b**) 0.93% Rb, (**c**) 4.65% Rb, (**d**) 9.29% Rb, (**e**) 0.85% K, (**f**) 2.55% K, (**g**) 4.25% K, (**h**) 8.50% K. Figure 3a–d reprinted from [49] with permission from MDPI, copyright 2021. Figure 3e,f reprinted from [48] with permission from Elsevier, copyright 2020.

Table 2 provides assignments from the open literature regarding adsorbed species during ESR. Ethanol adsorption during DRIFTS was found to produce two bands in the range 1000–1200 cm^{-1} (Figures 4–7 for K series and Figures 8–11 for Rb series), which are assigned to ethoxy species. These species are formed by the dissociative adsorption of ethanol on the catalyst surface [54–56]. Type II ethoxy species located at surface defects on zirconia exhibit a low wavenumber υ(CO) band at ~1050 cm^{-1}, while Type I ethoxy species are associated with unreduced sites on zirconia and positioned at higher wavenumbers [23]. Acetate produced several observable bands: symmetric υ(OCO) stretching (1300 cm^{-1}), asymmetric υ(OCO) stretching (1510 cm^{-1}), and υ(C–H) stretching bands (2700–3100 cm^{-1}). These assignments were confirmed by comparing our results with the literature [2]. The observed formation of acetate likely suggests that ethoxy species underwent oxidative dehydrogenation.

Table 2. Main bands in cm^{-1} observed for unpromoted, K-promoted, and Rb-promoted 2% Pt/ZrO$_2$ catalysts.

Bands	0% K	0.85% K	2.55% K	4.25% K	0% Rb	0.93% Rb	4.65% Rb	9.29% Rb
				50 °C				
ν(CO) ethoxy	1100, 1070, 1056	1092, (1065), 1051	1103, 1058	1107, (1067) 1056	1101, 1072, 1057	1099, (1067), 1055	1098, 1056	1105, 1057
ν(CH) ethox/acet	2970, 2928, 2897, 2868	2970, 2927, (2894), 2872	2969, 2926, (2897), 2876	(2977), 2965, (2934, 2881), 2868	2973, 2929, (2896), 2873, (2854)	2973, 2929, (2898), 2873, (2856)	2973, 2931, 2898, 2879, 2858	(2989), 2970, (2955), 2932, 2902, 2880, 2868
ρ(CH$_3$) ethoxy	1156, 1116	1148, 1124	(1147, 1125)	-	1154, (1133), 1118	(1163, 1150, 1128–1113)	-	-
ν_a(OCO) acetate	1562	1564, (1507)	1577, (1519)	1567–1513	1564	1560	1572, (1549–1512)	1578
ν_s(OCO) acetate	(1467), 1433	1467, (1431), 1417	(1490, 1464) 1412	1460	(1470), 1433	(1487) 1474, 1443, 1428	1472, (1445), 1408	-
δ_s(CH$_3$) acetate	1381, (1357), 1344, (1274)	1376, (1351) 1335, (1314–1269)	1377, 1358, (1340, 1280)	1358, 1310, 1271	1381, (1358, 1343, 1276)	1381, 1357, 1343, 1327–1295, 1275	(1372), 1360, (1339)	1355
ν(CO) Pt-CO	2055, (2037), 2012, 1990–1810	2051, 2030, (2018–1870)	1930 (2080–1800)	-	2051, (2032), 2015, 1978–1870	(2053, 2046–1985, 1985–1840), 1951	(2063–1985, 1963), 1927, (1950–1830)	-

Table 2. *Cont.*

Bands	0% K	0.85% K	2.55% K	4.25% K	0% Rb	0.93% Rb	4.65% Rb	9.29% Rb
200 °C								
ν_a(OCO) acetate	**1556**, (1470)	(1569), **1557**, (1543–1508), 1467	**1580**, (1549, 1523, 1508), 1471	**1580**, (1550–1444)	(1566), **1554**, (1470)	**1556**, (1525–1470)	1633, **1579**, (1549, 1533–1465)	**1580**
ν_s(OCO) acetate	**1439**	**1437**, 1380	**1408**	1422, **1408**	**1439**	**1437**, (1383)	**1404**	1437, **1404**
δ_s(CH₃) acetate	1346	1332, (1303), 1273	(1351), 1334, (1297)	(1350), (1330)	1347	1339, (1330-1312, 1297, 1274)	(1328), 1300	1355, 1339, (1335–1267)
ν(CH) acetate	2965, **2937**, 2876, 2862	3000, 2984, 2966, **2931**, 2872	2971, **2928**, (2897, 2878, 2858)	(2965), 2937, 2869	2965, **2936**, (2917–2892), 2877	2965, **2937**, (2920–2905, 2881	2986, 2967, **2927**	(2997), 2965, (2935), 2925, (2904, 2886–2860)
500 °C								
ν_a(OCO) carbonate	1556	1622, 1576–1490	1620, (1568–1492)	(1604), 1551, (1530, 1519, 1503)	1550–1500	1580–1495, 1556	1627, (1581), (1550–1430)	1608, (1592, 1566, 1550)
ν_s(OCO) carbonate	(1470), 1439	1454	(1468)	(1463, 1445, 1425)	1473, 1442	1490-1410, 1444	(1550–1430)	1439
ν_s(OCO) carbonate	(1405–1367)	1370, (1342) 1297, (1276–1225)	(1386, 1353) 1307	1400-1355, 1329, (1300–1200)	1395, 1363	(1373), 1340, (1300), 1271, (1256)	(1365, 1333, 1291)	(1355), 1317

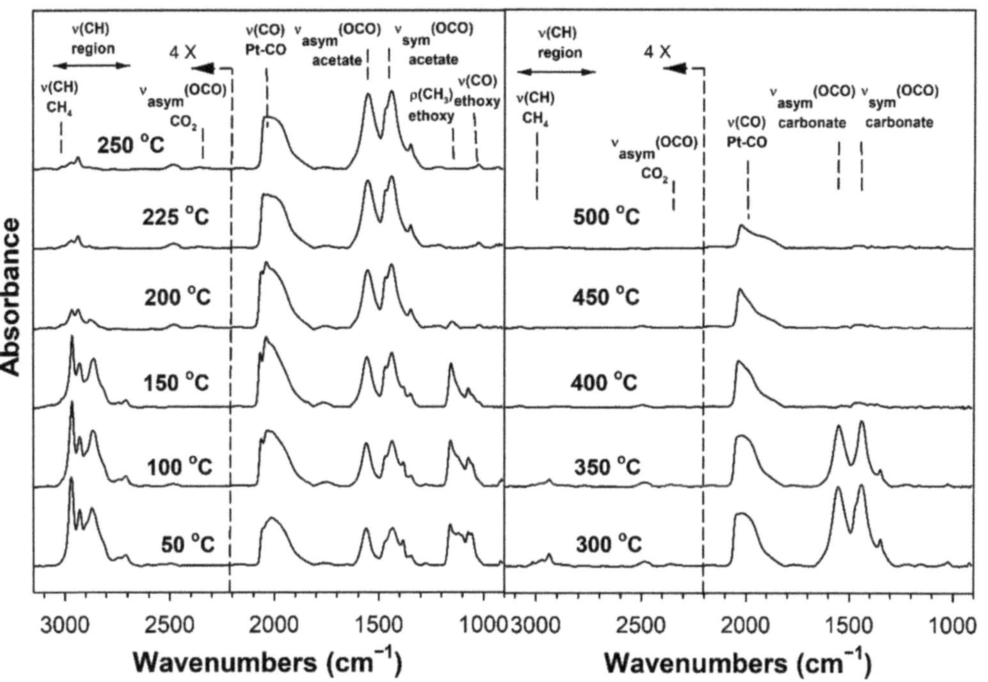

Figure 4. DRIFTS spectra of transient ESR over 2% Pt/m-ZrO₂ (K-series).

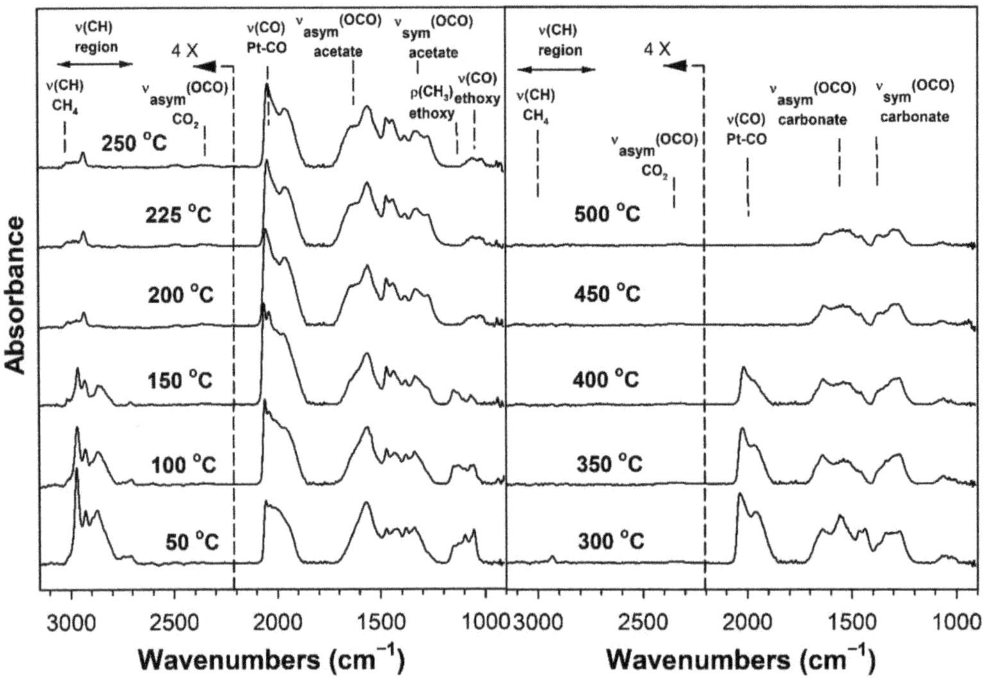

Figure 5. DRIFTS spectra of transient ESR over 0.85% K-2% Pt/m-ZrO$_2$.

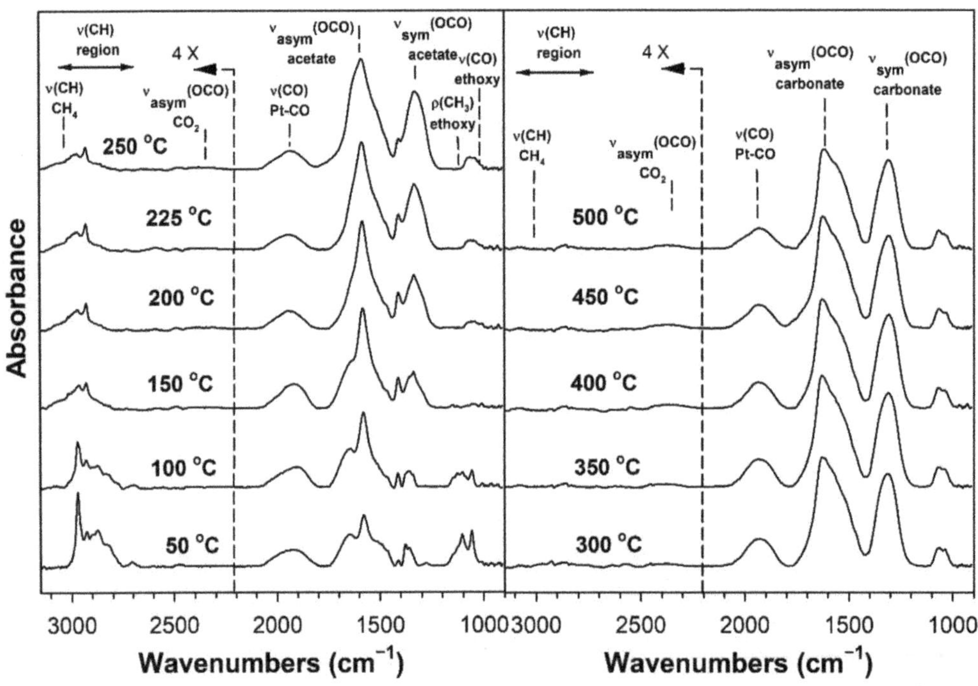

Figure 6. DRIFTS spectra of transient ESR over 2.55% K-2% Pt/m-ZrO$_2$.

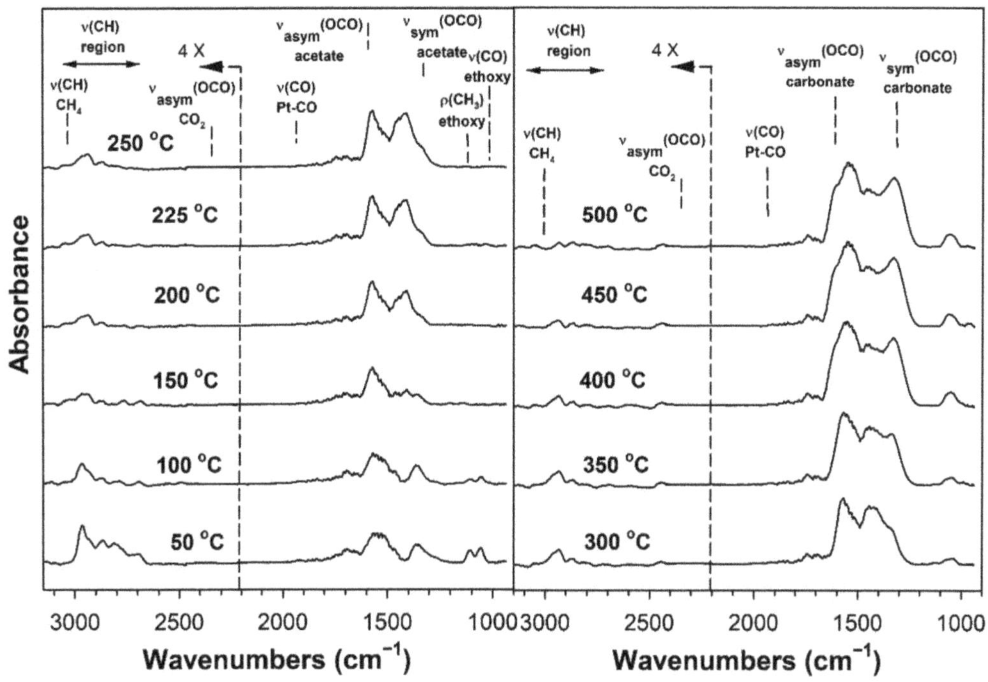

Figure 7. DRIFTS spectra of transient ESR over 4.25% K-2% Pt/m-ZrO$_2$.

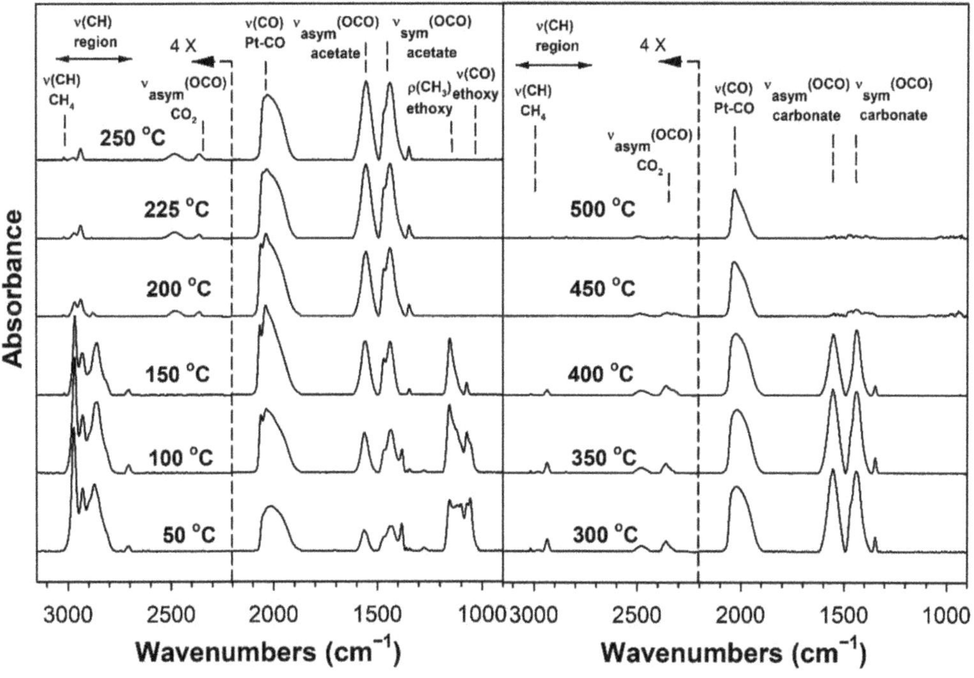

Figure 8. DRIFTS spectra of transient ESR over 2% Pt/m-ZrO$_2$ (Rb-series).

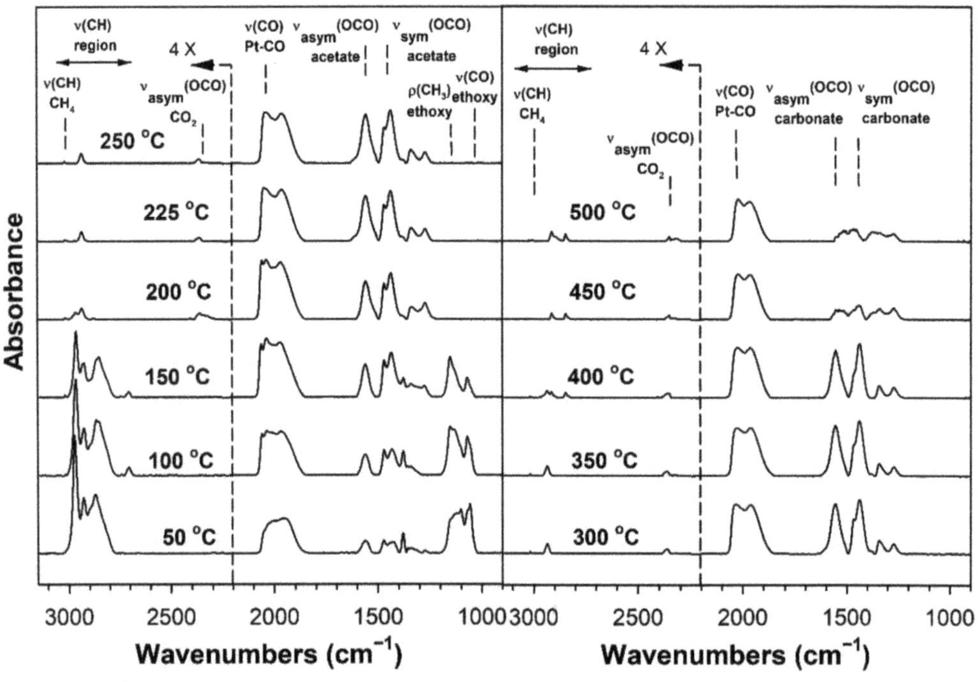

Figure 9. DRIFTS spectra of transient ESR over 0.93% Rb-2% Pt/m-ZrO$_2$.

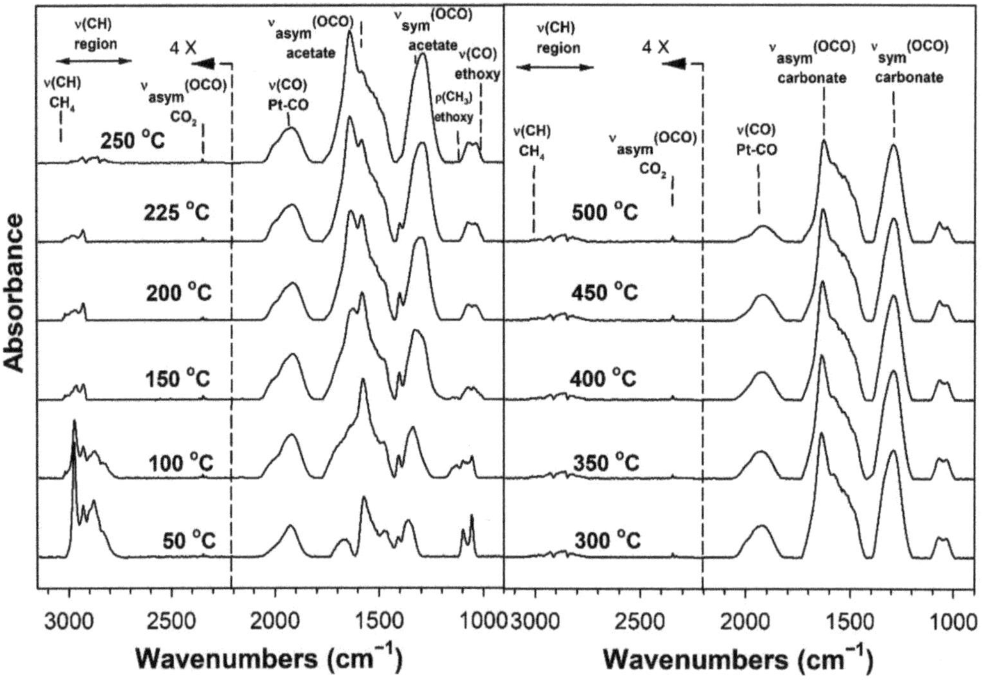

Figure 10. DRIFTS spectra of transient ESR over 4.65% Rb-2% Pt/m-ZrO$_2$.

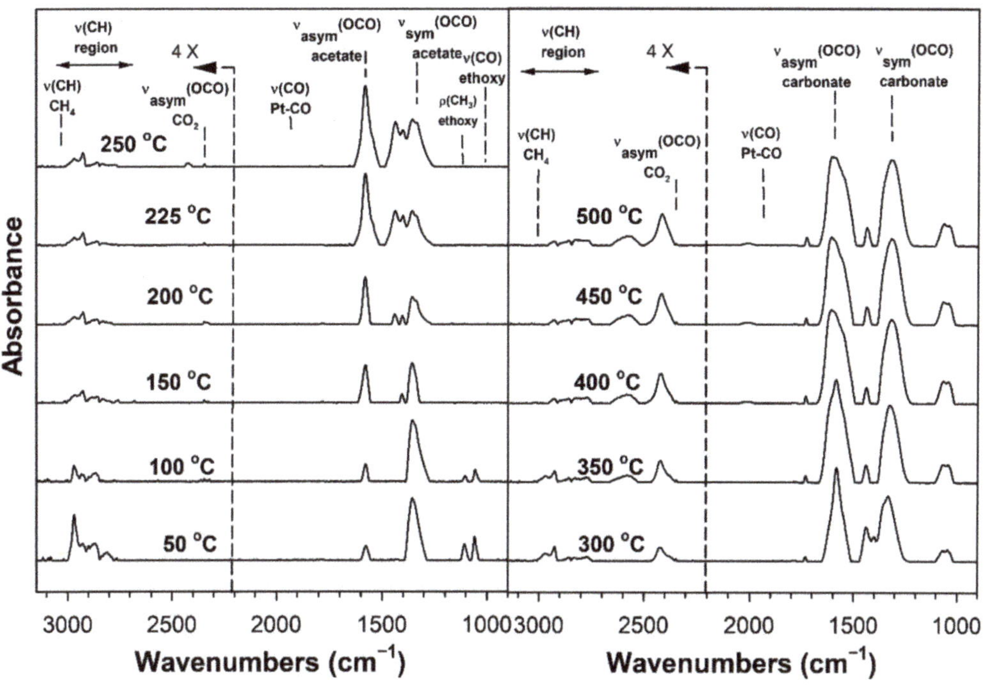

Figure 11. DRIFTS spectra of transient ESR over 9.29% Rb-2% Pt/m-ZrO$_2$.

DRIFTS of transient ESR was conducted on unpromoted, K-promoted, and Rb-promoted catalysts to shed further light on the possible mechanism (Figures 4–11). Bands of ethoxy species, formed from dissociative adsorption of ethanol, were observed in the range of 50–150 °C for the unpromoted catalyst (Figures 4 and 8 for K-doped and Rb-doped series, respectively). Increasing the temperature further, the surface concentration of ethoxy species decreased until they were virtually completely decomposed, while, simultaneously, there was a concomitant increase in the intensity of bands assigned to acetate; this suggests that ethoxy species underwent oxidative dehydrogenation to acetate—during this stage, CO$_2$ gas was not produced. At the same time, the magnitude of the Pt-CO band increased, suggesting that decarbonylation occurred to a certain extent. After the amount of acetate attained a maximum of approximately 250–300 °C, further steam reforming afforded CH$_4$ and CO$_2$ in addition to CO. The detection of CH$_4$ and CO$_2$ is consistent with the forward decomposition of acetate. The acetate species nearly completely decomposed by 400 °C.

The effect of alkali promotion was also explored by conducting DRIFTS and varying the K loading (Figures 5–7) and Rb loading (Figures 9–11). DRIFTS revealed that the ethoxy intermediate reacted most rapidly on the 2.55% K- and 4.65% Rb-doped catalyst, evidenced by the fact that it was entirely converted to acetate by 150 °C through oxidative dehydrogenation. In contrast, at low-potassium (0.85%) or -rubidium (0.93%) loading, the ethoxy species exhibited greater stability, as they had significantly decomposed by only 200 °C, very similar to what was observed in the case of the undoped 2% Pt/m-ZrO$_2$ catalyst. The acetate species is observed for all the potassium or rubidium loadings, suggesting it is a likely key intermediate during ESR. Once acetate is formed, however, DRIFTS results showed some differences among catalysts in terms of the selectivity of acetate decomposition, which depended on potassium or rubidium loading. The Pt–carbonyl bands for undoped, 0.85% K, and 0.93% Rb catalysts were at higher wavenumbers and at a significantly increased intensity, reaching maxima at 150–200 °C for the undoped, 150 °C for the 0.85% K doped, and 150–200 °C for the 0.93% Rb catalysts, respectively. This is reasonable to expect since acetate decomposes through different pathways depending on the level of alkali promotion. DRIFTS results suggest that acetate decarboxylation is preferred at high K (2.55% K and 4.25% K) or Rb (4.65% and 9.29% Rb) loading, while the decarbonylation is more favored on the unpromoted catalyst and lower loading alkali-doped catalysts (0.85% K and 0.93% Rb). The dependence of the two different decomposition pathways on alkali loading was also observed for the analogous Na-doped system for both ESR [40,41] and methanol steam reforming [57]. In the latter case, the analogous formate intermediate species were observed (formed from oxidative dehydrogenation of methoxy species), which decarboxylated at 2.5% Na loading, whereas it decarbonylated at low Na loading (i.e., 0.25%, 0.5%, 1%). This decarboxylation pathway enhanced CO_2 selectivity and CO conversion, making 1.8% Na-2.5% Na the optimum loading range for H_2 production.

DRIFTS results showed that acetate species decomposed at a lower temperature for the 2.55% K-doped catalyst and 4.65% Rb-doped catalyst as compared to the unpromoted catalysts, as a significant band for CH_4 was detected at 150 °C (with a slight signal even at 100 °C), while a less intense band for CH_4 was observed at 200 °C for the unpromoted catalysts (with a slight signal even detected at 150 °C). Thus, the acetate C–C bond breaks at lower temperatures for the K- and Rb-promoted catalysts.

This bond is analogous to the formate C–H band that is seen during water–gas shift or methanol steam reforming. In our prior WGS investigations, increasing Na, K, Rb, or Cs loading shifted the ν(CH) band to lower wavenumbers. At the same time, the difference between the ν(OCO) bands for asymmetric and symmetric stretching increased with alkali loading. This suggests that changes in surface basicity may be responsible for the CH bond weakening of formate. In the case of ESR, C–C bond weakening is not easily measured. However, we indeed see an increase in the wavenumber difference between the ν(OCO) bands for asymmetric and symmetric stretching. From Table 2, they are: 0% K, Δ = 117 cm^{-1}; 0.85% K, Δ = 120 cm^{-1}; 2.55% K, Δ = 172 cm^{-1}; 4.25% K, Δ = 172 cm^{-1}; 0% Rb, Δ = 115 cm^{-1}; 0.93% Rb, Δ = 119 cm^{-1}; 4.65% Rb, Δ = 175 cm^{-1}; and 9.29% Rb, Δ = 176 cm^{-1}. In all cases, the results suggest that the alkali promotes the weakening of the respective bond, emphasizing once again the analogous nature of the two alkali-doped catalyst systems. At 4.25% K or 9.29% Rb doping level, the catalyst surface was significantly blocked with the alkali, thereby creating a bottleneck for the formation of, and subsequent decomposition of, the intermediates. In addition to the CO_2 that evolved, residual carbonates are clearly observed on the catalysts with K or Rb, as bands such as at 1620, 1574–1439 cm^{-1}, and ~1315 cm^{-1} O–C–O stretching modes of carbonates [58].

Interestingly, the optimum K loading (2.55%) corresponds to a very similar weight percent as the optimal Na loading (2.5%) from prior work [41], meaning that the optimal K loading occurred at 60% of the optimal Na loading atomically. Some works have shown that at excessive alkali loadings on Pt/ZrO_2, the surface of the Pt nanoparticles is blocked, inhibiting the role of Pt in hydrogen transfer reactions during LT-WGS [59,60]. This effect provides a reasonable explanation for the results obtained, given that K$^+$ is considerably larger than Na$^+$; the surface of the Pt nanoparticles becomes covered by the alkali metal at considerably lower atomic loadings for potassium-promoted catalysts. Additionally, it has been shown that at high alkali loadings, catalyst basicity promotes the formation of the carbonate intermediate, which is the precursor to CO_2 formation; however, it was also found to impede CO_2 liberation during LT-WGS [60,61]. This may be due to higher basicity since CO_2 is an acidic molecule that is therefore adsorbed more strongly and/or it may be due to the fact that the alkali obstructs the metallic function, which is known to facilitate carbonate decomposition. Due to the fact that potassium is a more basic promoter than sodium, one would expect this effect to cause CO_2 liberation to be significantly hindered for potassium relative to sodium. However, potassium is able to achieve similar promotion to sodium, and it does so at a lower atomic loading; this may be due to its lower electronegativity.

DRIFTS experiments showed that the optimal loading for the Rb-promoted catalyst is 4.65%, which is 80% and 50% of the optimal K and Na atomically loading, respectively.

This trend suggests that a lower alkali atomically loading is required for the C–C bond scission of intermediate acetate when the alkali electronegativity decreases. Indeed, the electronegativity of Rb is 0.706 compared to 0.869 for Na and 0.734 for K [62]. A similar trend also was observed in our previous work [48]. Decreasing the electronegativity, the ν(C–H) formate band progressively shifted to lower wavenumbers, which was associated with a faster formate decomposition in the presence of steam [59,63,64].

Figures 12 and 13 show the TPD-MS profiles for methane over the K and Rb series of catalysts, respectively. The main peak for the 2% Pt/ZrO$_2$ catalyst is at 391 °C, with a very minor peak at 200 °C. The position of the main peak did not change to a significant degree for the 0.85% K- and 1.70% K-doped catalysts, with peaks at 383 °C and 360 °C, respectively; minor peaks at low temperature were also observed at 100–190 °C and 115–170 °C. However, a significant shift to lower temperature occurred once the dopant loading reached 2.55% K, where the temperature was 270 °C for the main peak, with a low-temperature peak at 130 °C. At 3.40% K, 4.25% K, and 8.50% K doping levels, the main peak decreased in intensity and shifted to higher temperatures of 314 °C, 327 °C, and 345 °C, respectively, while the low-temperature peak remained at a similar temperature. The results indicate that cleaving of the C–C bond of acetate species during ESR is not only promoted by alkali addition but also that the K-doping loading is close to the optimum at 2.55% K. For the Rb series, increasing Rb loading to 1.86% Rb decreases the main CH$_4$ evolution signal from 391 °C to 384 °C. In the range of 2.79% Rb–4.65% Rb, the peak at 370 °C is attenuated, and a new low-temperature peak at 160–180 °C emerges. At 5.58% and 9.29% Rb loadings, the CH$_4$ signal is attenuated overall, and by 9.3% Rb, the signal of the higher temperature peak has increased to 430 °C due to blocking of the catalyst surface by excessive alkali. DRIFTS and TPD-MS show that alkali promotion had a beneficial effect on the C–C bond scission of the intermediate acetate. However, the electronic effect of alkali on the structure of Pt metal is not well understood. Different hypotheses can be formulated: (a) charge transfer, which may result in a change in the white line intensity in the presence of K [65]; (b) an electrostatic effect, which could cause bond weakening in adsorbed species on the catalyst surface [66]; (c) a Fermi level electronic perturbation [43]; or (d) an alteration in bond strengths due to changes in the acidity/basicity of the catalyst. Confounding the analysis of (a) is particle size effects [67,68], which tend to cause the binding energy and white line to both increase with decreasing particle size. To gain insight into whether K or Rb promotion leads to charge transfer from the alkali to Pt, the Pt L-3 XANES spectra can be analyzed. As shown in Figure 14 for the K-series and Figure 15 for the Rb-series, during reduction in hydrogen at 350 °C and after cooling in hydrogen to ambient conditions, it is evident that there are differences in the XANES spectra of catalysts having no or low alkali content versus those with high alkali content. However, the white line intensity is also affected by the size of Pt particles, so the difference between the L-3 XANES and L-2 XANES can be studied to remove this effect (Figures 14 and 15). If K or Rb promotion causes electron charge transfer from the alkali to Pt, then the Pt L$_3$-L$_2$ XANES difference should decrease in magnitude as a function of alkali loading. However, this trend is not observed, indicating that neither potassium nor rubidium is likely transferring electron charge density to platinum. Despite this, it is still suggested that the alkali and Pt are in direct contact, which can be seen through the TPR-XANES and TPR-EXAFS reported in our previous studies [48,49]. The TPR-XANES and TPR-EXAFS revealed that as the loading of K or Rb increases, the reduction of PtO is hindered, which is likely the result of the covering of platinum by the alkali. It is also expected that if the alkali donated electron density to Pt particles, a relaxation in the edge energy should occur (related to the binding energy) [69,70]. Figures 16 and 17 reveal that in comparing catalysts with the Pt^0 foil, no such shift was detected.

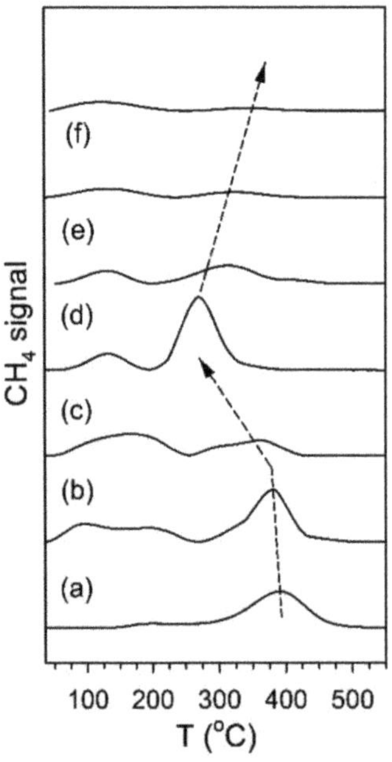

Figure 12. TPD-MS of ethanol steam reforming over (a) 2% Pt/ZrO$_2$, (b) 0.85% K-2% Pt/ZrO$_2$, (c) 1.70% K-2% Pt/ZrO$_2$, (d) 2.55% K-2% Pt/ZrO$_2$, (e) 3.40% K-2% Pt/ZrO$_2$, (f) 4.25% K-2% Pt/ZrO$_2$, and (g) 8.50% K-2% Pt/ZrO$_2$.

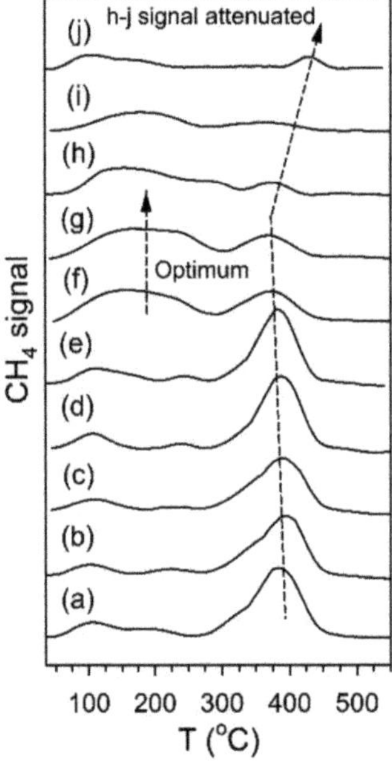

Figure 13. TPD-MS of ethanol steam reforming over (a) 2% Pt/ZrO$_2$, (b) 0.37% Rb-2% Pt/ZrO$_2$, (c) 0.74% Rb-2% Pt/ZrO$_2$, (d) 0.93% Rb-2% Pt/ZrO$_2$, (e) 1.86% Rb-2% Pt/ZrO$_2$, (f) 2.79% Rb-2% Pt/ZrO$_2$, (g) 3.72% Rb-2% Pt/ZrO$_2$, (h) 4.65% Rb-2% Pt/ZrO$_2$, (i) 5.58% Rb-2% Pt/ZrO$_2$, (j) 9.29% Rb-2% Pt-ZrO$_2$.

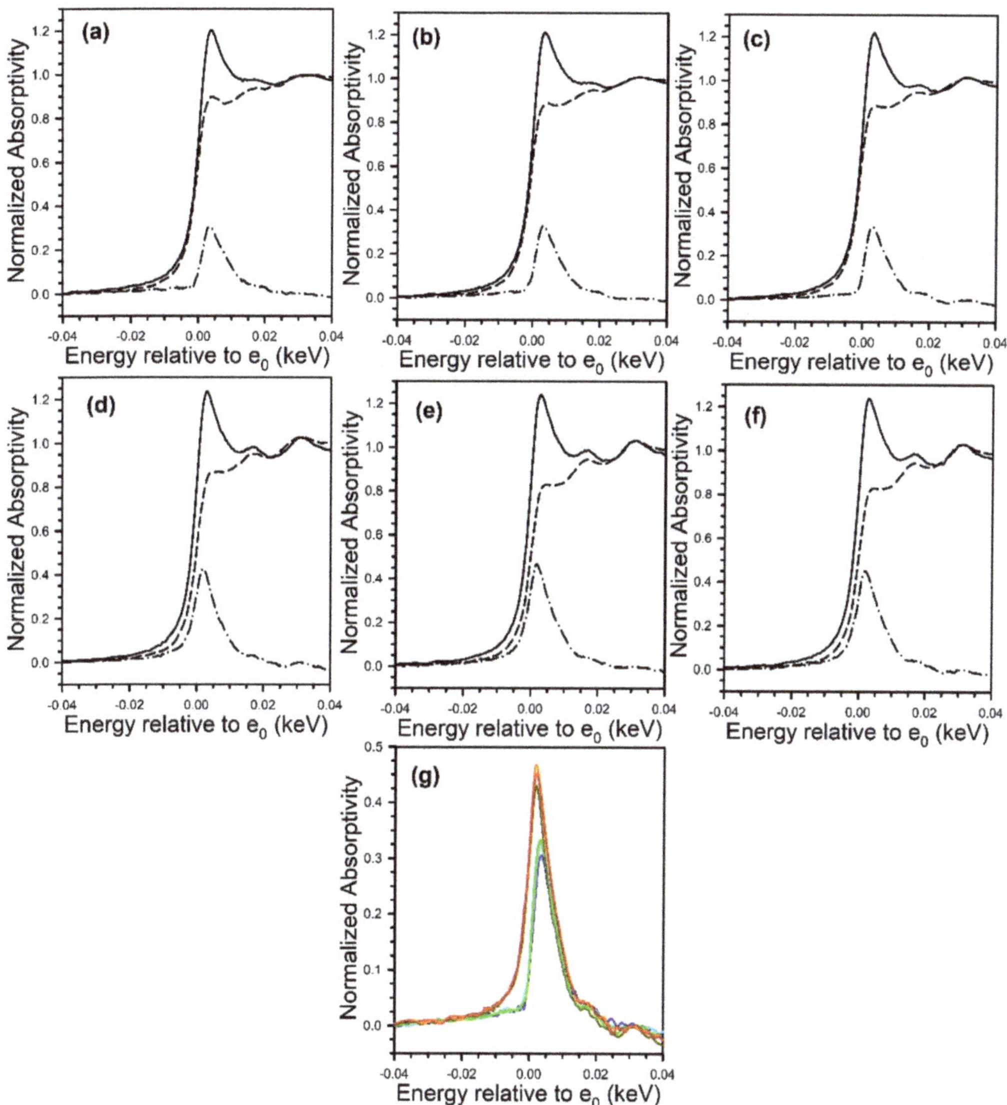

Figure 14. XANES and L_3-L_2 XANES difference spectra at the (dashed line) Pt L_2 edge and (solid line) Pt L_3 edge following reduction in pure hydrogen and cooling to ambient temperature, including 2% Pt/ZrO_2 with: (**a**, blue) 0% K; (**b**, cyan) 0.85% K; (**c**, green) 1.70% K; (**d**, dark yellow) 2.55% K; (**e**, orange) 3.40% K; and (**f**, red) 4.25% K. (**g**) Overlays of L_3-L_2 difference spectra, showing an increase in intensity with K loading. No evidence for e⁻ transfer to Pt from K⁺ was found.

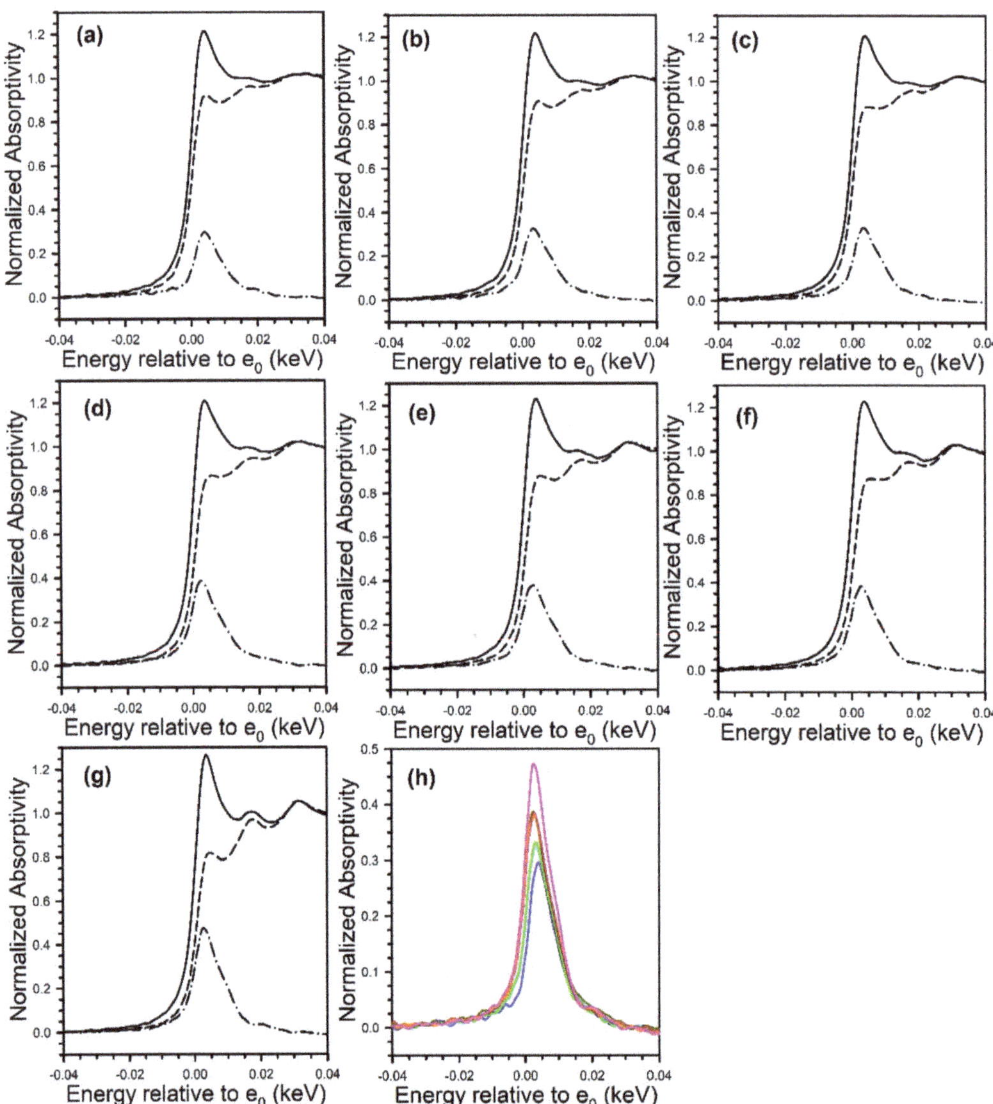

Figure 15. XANES spectra at the Pt (solid line) L_3 edge and (dashed) line L_2 edge, as well as (dash-dotted line) the L_3-L_2 difference spectra of (**a**, blue) 2% Pt/ZrO_2, (**b**, cyan) 0.93% Rb-2% Pt/ZrO_2, (**c**, green) 1.86% Rb-2% Pt/ZrO_2, (**d**, dark yellow) 2.79% Rb-2% Pt/ZrO_2, (**e**, orange) 4.65% Rb-2% Pt/ZrO_2, (**f**, red) 5.58% Rb-2% Pt/ZrO_2, and (**g**, pink) 9.3% Rb-2% Pt/ZrO_2. (**h**) Overlays of L_3 2212 L_2 difference spectra, showing an increase in intensity with Rb loading. No evidence for e⁻ transfer to Pt from Rb⁺ was found, which should result in an opposite trend. Figure 15 reprinted from [49] with permission from MDPI, copyright 2021.

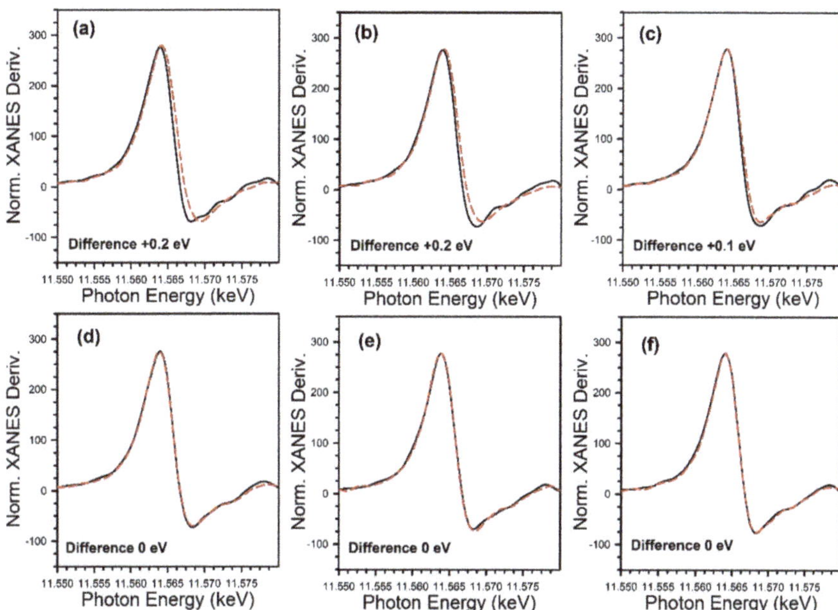

Figure 16. XANES derivative spectra at the Pt L$_3$ edge of (solid line) the Pt^0 foil and (red line) the catalysts following reduction in pure hydrogen, including 2% Pt/ZrO$_2$ with: (**a**) 0% K; (**b**) 0.85% K; (**c**) 1.70% K; (**d**) 2.55% K, (**e**) 3.40% K; and (**f**) 4.25% K. No evidence for e$^-$ transfer to Pt from K$^+$ was found.

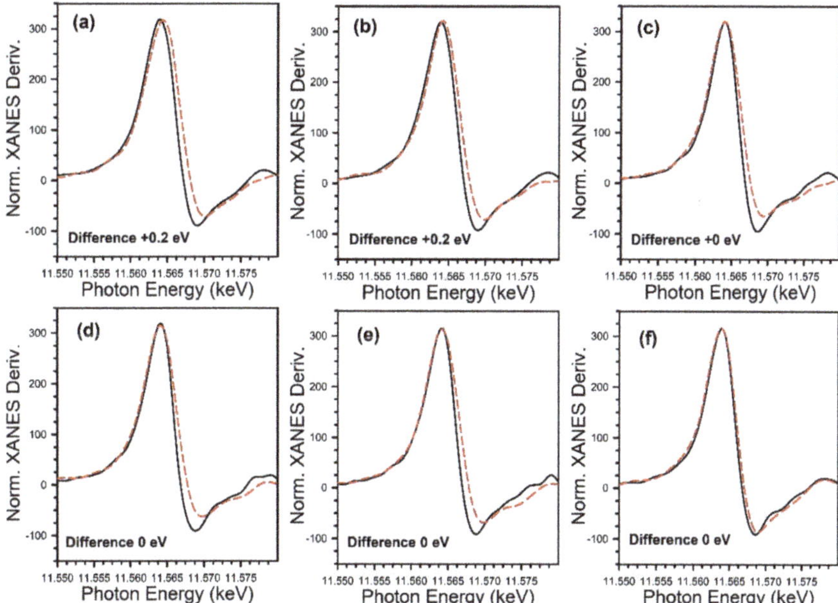

Figure 17. XANES derivative spectra at the Pt L$_3$ edge of (solid line) the Pt^0 foil and (red line) the catalysts following reduction in pure hydrogen, including 2% Pt/ZrO$_2$ with: (**a**) 0% Rb, (**b**) 0.93% Rb, (**c**) 1.86% Rb, (**d**) 2.79% Rb, (**e**) 4.65% Rb, and (**f**) 9.3% Rb-2% Pt/ZrO$_2$. No evidence for e$^-$ transfer to Pt from Rb$^+$ was found.

The catalytic activity data for K and Rb-series are reported in Table 3, whereas the H_2 yield and product selectivity trends of the unpromoted catalyst at different conversions are reported in Table 4. The addition of potassium or rubidium progressively decreased the ethanol conversion. A similar trend was observed for sodium in our previous work [41]. Catalysts with very high alkali loading (i.e., 4.25% K or 9.29% Rb) exhibited negligible catalytic activity. TEM-EDX and EXAFS fittings showed that Pt clusters aggregated at higher loading. Moreover, the alkali likely partially covers the platinum particles, as evidenced by a decreasing $\nu(CO)$ band at higher K or Rb alkali loading (virtually disappearing at the highest loadings). However, the most interesting effect of alkali promotion on Pt/ZrO_2 is related to product selectivity. Indeed, CO is only detected among the products for unpromoted and low alkali doping (0.85% K and 0.93% Rb), whereas no CO is produced at higher alkali loading. Alkali promotion decreased the acetaldehyde selectivity, which is 3.49% for the unpromoted, whereas it is lower than 1.5% for the K- or Rb-promoted catalyst. The activity results confirm that different pathways occur depending on the potassium or rubidium loading, as already pointed out by DRIFTS. The decarbonylation route is present for the unpromoted catalyst and the catalysts with low alkali loading, whereas the decarboxylation route (where acetate decomposes to CH_4 and a carbonate species, which further decomposes to CO_2) completely dominates when the alkali loadings reach 2.55% K and 4.93% Rb; a similar effect occurred at 1.80% Na in our prior work [41]. Moreover, 2.55% K corresponds to 5.57% Rb and 1.50% Na on an atomic loading basis. Therefore, the results suggest that decarboxylation is improved by increasing the basicity of the alkali (moving down the Group 1 column) because the atomic loading of potassium to stave off decarbonylation is lower than that of sodium and higher than that of rubidium.

Table 3. ESR catalytic activity for K-series and Rb-series (300 °C, 1 atm, 190,560 Ncc/h/g_{cat}, feed: C_2H_5OH 2.98% H_2O 26.14% N_2 70.88%).

Catalyst	Conv. C_2H_5OH (%)	H_2 Yield (%)	C Selectivity (%)						
			CH_4	CO_2	CO	C_2H_6	C_2H_4	C_3H_6	CH_3CHO
2% Pt/ZrO$_2$	86.91	14.26	45.20	28.5	21.16	0.92	0.39	0.34	3.49
0.85% K-2% Pt/ZrO$_2$	60.01	13.84	46.86	40.59	11.14	0.26	-	-	1.14
2.55% K-2% Pt/ZrO$_2$	47.02	12.66	48.75	50.72	-	-	-	-	0.76
4.55% K-2% Pt/ZrO$_2$	27.30	3.17	49.48	49.08	-	-	-	-	1.42
0.93% Rb-2% Pt/ ZrO$_2$	59.60	13.65	46.81	35.78	15.55	0.31	0.11	-	1.43
4.25% Rb-2% Pt/ZrO$_2$	38.97	8.70	53.56	46.27	-	-	-	-	0.16
9.29% Rb-2% Pt/ZrO$_2$	11.19	0.18	12.18	87.81	-	-	-	-	-

Table 4. ESR catalytic activity for 2% Pt/ZrO$_2$ catalyst at different C_2H_5OH conversion (300 °C, 1 atm, feed: C_2H_5OH 2.98% H_2O 26.14% N_2 70.88%).

Catalyst	Conv. C_2H_5OH (%)	H_2 Yield (%)	C Selectivity (%)						
			CH_4	CO_2	CO	C_2H_6	C_2H_4	C_3H_6	CH_3CHO
2% Pt/ZrO$_2$	86.91	14.26	45.20	28.5	21.16	0.92	0.39	0.34	3.49
	58.55	8.53	47.46	14.86	28.37	0.37	0.57	0.36	7.98
	48.41	4.81	55.87	11.56	22.07	0.24	0.49	-	9.73
	30.89	3.05	60.79	7.13	19.70	-	0.48	-	11.81

However, Table 4 shows that the product selectivity of Pt/ZrO_2 changes to a degree as a function of conversion. Therefore, in order to place alkali effects on a firmer footing, it is necessary to compare the product selectivity and H_2 yield at the same conversion level. As shown in Table 5, the alkali-promoted catalysts have approximately 60–62% higher H_2 yield as compared to the unpromoted catalyst when compared at similar ethanol conversion. This is due in part to enhanced decarboxylation over decarbonylation, as the CO_2 selectivity is increased by 240–273%, while the CO selectivity is diminished by 45–61%.

While the 0.85% K and 0.93% Rb catalysts provide similar H_2 yields, 0.85% K corresponds to 1.86% Rb. The fact that a similar improvement in H_2 yield occurred over that of the unpromoted 2% Pt/ZrO_2 catalyst at an atomic loading of Rb that is 50% of the atomic loading of K indicates that higher basicity alkali promoters are more effective at facilitating the more selective decarboxylation pathway.

Table 5. ESR catalytic activity for select catalysts at the same C_2H_5OH conversion (300 °C, 1 atm, feed: C_2H_5OH 2.98% H_2O 26.14% N_2 70.88%) for selectivity comparison.

Catalyst	Conv. C_2H_5OH (%)	H_2 Yield (%)	C Selectivity (%)						
			CH_4	CO_2	CO	C_2H_6	C_2H_4	C_3H_6	CH_3CHO
2% Pt/ZrO_2	58.55	8.53	47.46	14.86	28.37	0.37	0.57	0.36	7.98
0.85% K-2% Pt/ZrO_2	60.01	13.84	46.86	40.59	11.14	0.26	-	-	1.14
0.93% Rb-2% Pt/ ZrO_2	59.60	13.65	46.81	35.78	15.55	0.31	0.11	-	1.43

4. Conclusions

The addition of potassium or rubidium to Pt/ZrO_2 progressively decreased the surface area and the pore volume because of some pore blocking. Platinum particle size was ~1 nm for the unpromoted and lower alkali loading (0.85% K and 0.93% Rb), while aggregation occurred at higher alkali loading. TEM-EDX of 9.29% Rb-Pt/ZrO_2 showed Pt aggregates of 10 nm. The difference between Pt L_3–L_2 XANES spectra indicates that neither potassium nor rubidium is likely transferring electron charge density to platinum. Moreover, no relaxation effect on the edge jump energy was observed with the addition of K or Rb.

DRIFTS experiments were carried out to investigate the mechanism. The results suggest that the catalysts have similar steps during ESR: dissociation of ethanol to produce an ethoxy species, oxidative dehydrogenation of ethoxy species to acetate, and acetate decomposition. The forward decomposition of acetate to CH_4 and carbonate (the precursor to CO_2) is facilitated by the presence of K or Rb, and there is an optimum alkali loading for facilitating the C–C scission of acetate. This is inferred from the temperature at which CH_4 evolution occurs as well as a systematic increase in the difference in band position for ν (OCO) asymmetric and symmetric stretching for acetate, which occurs with increasing the alkali loading. Ethoxy species are more stable on the unpromoted catalyst and acetate decomposition, which is associated with the formation of methane, occurs at a higher temperature. Methane formation was detected at 391 °C for 2% Pt/ZrO_2 in TPD-MS, whereas it occurred at 270 °C for 2.55% K and in multiple peaks (160–180 °C and 370 °C) for 2.79–4.65% Rb. Moreover, DRIFTS experiments and catalytic activity testing point out the existence of different pathways for acetate decomposition depending on the alkali loading. Decarboxylation is the most favored route at high alkali loading (2.55 and 4.25 wt.%). In this pathway, acetate decomposes in the forward direction, yielding CH_4 and a carbonate species, which further decomposes to CO_2. In contrast, the unselective decarbonylation pathway occurs to a significant extent for the unpromoted catalyst and the catalyst having low alkali loading (e.g., 0.85% K or 0.93% Rb). By increasing the alkali basicity by switching from K to Rb, lower loadings of alkali enabled virtually complete blocking of the non-selective decarbonylation pathway. Results of unpromoted, K-promoted, and Rb-promoted catalysts compared at similar conversion further confirmed the promoting effect of the alkali, as well as the basicity trend.

Author Contributions: Conceptualization, catalyst preparation, catalyst characterization, formal analysis, supervision, and writing, G.J. Catalyst preparation, catalyst characterization, formal analysis, and writing, R.G. and C.D.W. Reaction testing, characterization, formal analysis, conceptualization, and writing, M.M. Catalyst preparation, supervision, and resources, D.C.C. Catalyst characterization, data curation, resources, and supervision, A.J.K. All authors have read and agreed to the published version of the manuscript.

Funding: This work is supported by the USDA National Institute of Food and Agriculture, Interdisciplinary Hands-on Research Traineeship and Extension Experiential Learning in Bioenergy/Natural Resources/Economics/Rural project, U-GREAT (Under Graduate Research, Education And Training) program (2016-67032-24984). This research was also supported by the National Science Foundation through Grant Award 1832388.

Institutional Review Board Statement: Not applicable.

Informed Consent Statement: Not applicable.

Data Availability Statement: Not applicable.

Acknowledgments: Argonne's research was supported in part by the US Department of Energy (DOE), Office of Fossil Energy, National Energy Technology Laboratory (NETL). Advanced photon source was supported by the US Department of Energy, Office of Science, Office of Basic Energy Sciences, under contract number DE-AC02-06CH11357. MRCAT operations are supported by the Department of Energy and the MRCAT member institutions. CAER research was supported by the Commonwealth of Kentucky. Research carried out at UTSA was supported by UTSA, the State of Texas, and the STARs program. Caleb D. Watson would like to acknowledge funding from the UTSA College of Engineering Scholarship. Gary Jacobs would like to thank UTSA and the State of Texas for financial support through startup funds.

Conflicts of Interest: The authors declare no conflict of interest.

References

1. Li, D.; Li, X.; Gong, J. Catalytic Reforming of Oxygenates: State of the Art and Future Prospects. *Chem. Rev.* **2016**, *116*, 11529–11653. [CrossRef]
2. Mattos, L.V.; Jacobs, G.; Davis, B.H.; Noronha, F.B. Production of Hydrogen from Ethanol: Review of Reaction Mechanism and Catalyst Deactivation. *Chem. Rev.* **2012**, *112*, 4094–4123. [CrossRef] [PubMed]
3. Ni, M.; Leung, D.Y.C.; Leung, M.K.H. A review on reforming bio-ethanol for hydrogen production. *Int. J. Hydrogen Energy* **2007**, *32*, 3238–3247. [CrossRef]
4. Vaidya, P.D.; Rodrigues, A.E. Glycerol Reforming for Hydrogen Production: A Review. *Chem. Eng. Tech.* **2009**, *32*, 1463–1469. [CrossRef]
5. Tran, N.H.; Kannangara, G.S.K. Conversion of glycerol to hydrogen rich gas. *Chem. Soc. Rev.* **2013**, *42*, 9454–9479. [CrossRef]
6. Słowik, G.; Greluk, M.; Rotko, M.; Machocki, A. Evolution of the structure of unpromoted and potassium-promoted ceria-supported nickel catalysts in the steam reforming of ethanol. *Appl. Catal. B Environ.* **2018**, *221*, 490–509. [CrossRef]
7. Contreras, J.L.; Salmones, J.; Colín-Luna, J.A.; Nuño, L.; Quintana, B.; Córdova, I.; Zeifert, B.; Tapia, C.; Fuentes, G.A. Catalysts for H_2 production using the ethanol steam reforming (a review). *Int. J. Hydrogen Energy* **2014**, *39*, 18835–18853. [CrossRef]
8. Ogo, S.; Sekine, Y. Recent progress in ethanol steam reforming using non-noble transition metal catalysts: A review. *Fuel Proc. Tech.* **2020**, *199*, 106238. [CrossRef]
9. Frusteri, F.; Freni, S.; Spadaro, L.; Chiodo, V.; Bonura, G.; Donato, S.; Cavallaro, S. H_2 production for MC fuel cell by steam reforming of ethanol over MgO supported Pd, Rh, Ni and Co catalysts. *Catal. Commun.* **2004**, *5*, 611–615. [CrossRef]
10. Song, H.; Ozkan, U.S. Ethanol steam reforming over Co-based catalysts: Role of oxygen mobility. *J. Catal.* **2009**, *261*, 66–74. [CrossRef]
11. Ferencz, Z.; Varga, E.; Puskás, R.; Kónya, Z.; Baán, K.; Oszkó, A.; Erdőhelyi, A. Reforming of ethanol on Co/Al$_2$O$_3$ catalysts reduced at different temperatures. *J. Catal.* **2018**, *358*, 118–130. [CrossRef]
12. Gaudillere, C.; González, J.J.; Chica, A.; Serra, J.M. YSZ monoliths promoted with Co as catalysts for the production of H_2 by steam reforming of ethanol. *Appl. Catal. A Gen.* **2017**, *538*, 165–173. [CrossRef]
13. Ángel-Soto, J.; Martínez-Rosales, M.; Ángel-Soto, P.; Zamorategui-Molina, A. Synthesis, characterization and catalytic application of Ni catalysts supported on alumina–zirconia mixed oxides. *Bullet. Mat. Sci.* **2017**, *40*, 1309–1318. [CrossRef]
14. Campos, C.H.; Pecchi, G.; Fierro, J.L.G.; Osorio-Vargas, P. Enhanced bimetallic Rh-Ni supported catalysts on alumina doped with mixed lanthanum-cerium oxides for ethanol steam reforming. *Molec. Catal.* **2019**, *469*, 87–97. [CrossRef]
15. Liguras, D.K.; Kondarides, D.I.; Verykios, X.E. Production of hydrogen for fuel cells by steam reforming of ethanol over supported noble metal catalysts. *Appl. Catal. B Environ.* **2003**, *43*, 345–354. [CrossRef]
16. Yamazaki, T.; Kikuchi, N.; Katoh, M.; Hirose, T.; Saito, H.; Yoshikawa, T.; Wada, M. Behavior of steam reforming reaction for bio-ethanol over Pt/ZrO$_2$ catalysts. *Appl. Catal. B Environ.* **2010**, *99*, 81–88. [CrossRef]
17. De Lima, S.M.; Silva, A.M.; da Cruz, I.O.; Jacobs, G.; Davis, B.H.; Mattos, L.V.; Noronha, F.B. H_2 production through steam reforming of ethanol over Pt/ZrO$_2$, Pt/CeO$_2$ and Pt/CeZrO$_2$ catalysts. *Catal. Today* **2008**, *138*, 162–168. [CrossRef]
18. Bilal, M.; Jackson, S.D. Ethanol steam reforming over Rh and Pt catalysts: Effect of temperature and catalyst deactivation. *Catal. Sci. Technol.* **2013**, *3*, 754–766. [CrossRef]

19. De Lima, S.M.; Silva, A.M.; Graham, U.M.; Jacobs, G.; Davis, B.H.; Mattos, L.V.; Noronha, F.B. Ethanol decomposition and steam reforming of ethanol over CeZrO$_2$ and Pt/CeZrO$_2$ catalyst: Reaction mechanism and deactivation. *Appl. Catal. A Gen.* **2009**, *352*, 95–113. [CrossRef]

20. Jacobs, G.; Davis, B.H. In situ DRIFTS investigation of the steam reforming of methanol over Pt/ceria. *Appl. Catal. A Gen.* **2005**, *285*, 43–49. [CrossRef]

21. Ciambelli, P.; Palma, V.; Ruggiero, A. Low temperature catalytic steam reforming of ethanol. 1. The effect of the support on the activity and stability of Pt catalysts. *Appl. Catal. B Environ.* **2010**, *96*, 18–27. [CrossRef]

22. He, Z.; Yang, M.; Wang, X.; Zhao, Z.; Duan, A. Effect of the transition metal oxide supports on hydrogen production from bio-ethanol reforming. *Catal. Today* **2012**, *194*, 2–8. [CrossRef]

23. Jacobs, G.; Keogh, R.A.; Davis, B.H. Steam reforming of ethanol over Pt/ceria with co-fed hydrogen. *J. Catal.* **2007**, *245*, 326–337. [CrossRef]

24. Shido, T.; Iwasawa, Y. Reactant-Promoted Reaction Mechanism for Water-Gas Shift Reaction on Rh-Doped CeO$_2$. *J. Catal.* **1993**, *141*, 71–81. [CrossRef]

25. Jacobs, G.; Graham, U.M.; Chenu, E.; Patterson, P.M.; Dozier, A.; Davis, B.H. Low-temperature water–gas shift: Impact of Pt promoter loading on the partial reduction of ceria and consequences for catalyst design. *J. Catal.* **2005**, *229*, 499–512. [CrossRef]

26. Laachir, A.; Perrichon, V.; Badri, A.; Lamotte, J.; Catherine, E.; Lavalley, J.C.; El Fallah, J.; Hilaire, L.; Le Normand, F.; Quéméré, E.; et al. Reduction of CeO$_2$ by hydrogen. Magnetic susceptibility and Fourier-transform infrared, ultraviolet and X-ray photoelectron spectroscopy measurements. *J. Chem. Soc. Faraday Trans.* **1991**, *87*, 1601–1609. [CrossRef]

27. Grzybek, G.; Greluk, M.; Indyka, P.; Góra-Marek, K.; Legutko, P.; Słowik, G.; Turczyniak-Surdacka, S.; Rotko, M.; Sojka, Z.; Kotarba, A. Cobalt catalyst for steam reforming of ethanol–Insights into the promotional role of potassium. *Int. J. Hydrogen Energy* **2020**, *45*, 22658–22673. [CrossRef]

28. Grzybek, G.; Góra-Marek, K.; Patulski, P.; Greluk, M.; Rotko, M.; Słowik, G.; Kotarba, A. Optimization of the potassium promotion of the Co | α-Al$_2$O$_3$ catalyst for the effective hydrogen production via ethanol steam reforming. *Appl. Catal. A Gen.* **2021**, *614*, 118051. [CrossRef]

29. Llorca, J.; Homs, N.S.; Sales, J.; Fierro, J.-L.G.; Ramírez de la Piscina, P. Effect of sodium addition on the performance of Co–ZnO-based catalysts for hydrogen production from bioethanol. *J. Catal.* **2004**, *222*, 470–480. [CrossRef]

30. Espinal, R.; Taboada, E.; Molins, E.; Chimentao, R.J.; Medina, F.; Llorca, J. Cobalt hydrotalcites as catalysts for bioethanol steam reforming. The promoting effect of potassium on catalyst activity and long-term stability. *Appl. Catal. B Environ.* **2012**, *127*, 59–67. [CrossRef]

31. Ogo, S.; Shimizu, T.; Nakazawa, Y.; Mukawa, K.; Mukai, D.; Sekine, Y. Steam reforming of ethanol over K promoted Co catalyst. *Appl. Catal. A Gen.* **2015**, *495*, 30–38. [CrossRef]

32. Yoo, S.; Park, S.; Song, J.H.; Kim, D.H. Hydrogen production by the steam reforming of ethanol over K-promoted Co/Al$_2$O$_3$–CaO xerogel catalysts. *Molec. Catal.* **2020**, *491*, 110980. [CrossRef]

33. Frusteri, F.; Freni, S.; Chiodo, V.; Spadaro, L.; Di Blasi, O.; Bonura, G.; Cavallaro, S. Steam reforming of bio-ethanol on alkali-doped Ni/MgO catalysts: Hydrogen production for MC fuel cell. *Appl. Catal. A Gen.* **2004**, *270*, 1–7. [CrossRef]

34. Dömök, M.; Baán, K.; Kecskés, T.; Erdőhelyi, A. Promoting Mechanism of Potassium in the Reforming of Ethanol on Pt/Al$_2$O$_3$ Catalyst. *Catal. Lett.* **2008**, *126*, 49–57. [CrossRef]

35. Frusteri, F.; Freni, S.; Chiodo, V.; Spadaro, L.; Bonura, G.; Cavallaro, S. Potassium improved stability of Ni/MgO in the steam reforming of ethanol for the production of hydrogen for MCFC. *J. Power Sources* **2004**, *132*, 139–144. [CrossRef]

36. Sharma, Y.C.; Kumar, A.; Prasad, R.; Upadhyay, S.N. Ethanol steam reforming for hydrogen production: Latest and effective catalyst modification strategies to minimize carbonaceous deactivation. *Renew. Sustain. Energy Rev.* **2017**, *74*, 89–103. [CrossRef]

37. Greluk, M.; Rybak, P.; Słowik, G.; Rotko, M.; Machocki, A. Comparative study on steam and oxidative steam reforming of ethanol over 2KCo/ZrO$_2$ catalyst. *Catal. Today* **2015**, *242*, 50–59. [CrossRef]

38. Banach, B.; Machocki, A. Effect of potassium addition on a long term performance of Co–ZnO–Al$_2$O$_3$ catalysts in the low-temperature steam reforming of ethanol: Co-precipitation vs citrate method of catalysts synthesis. *Appl. Catal. A Gen.* **2015**, *505*, 173–182. [CrossRef]

39. Słowik, G.; Gawryszuk-Rżysko, A.; Greluk, M.; Machocki, A. Estimation of Average Crystallites Size of Active Phase in Ceria-Supported Cobalt-Based Catalysts by Hydrogen Chemisorption vs TEM and XRD Methods. *Catal. Lett.* **2016**, *146*, 2173–2184. [CrossRef]

40. Martinelli, M.; Watson, C.D.; Jacobs, G. Sodium doping of Pt/m-ZrO$_2$ promotes C–C scission and decarboxylation during ethanol steam reforming. *Int. J. Hydrogen Energy* **2020**, *45*, 18490–18501. [CrossRef]

41. Martinelli, M.; Castro, J.D.; Alhraki, N.; Matamoros, M.E.; Kropf, A.J.; Cronauer, D.C.; Jacobs, G. Effect of sodium loading on Pt/ZrO$_2$ during ethanol steam reforming. *Appl. Catal. A Gen.* **2021**, *610*, 117947. [CrossRef]

42. Roh, H.-S.; Platon, A.; Wang, Y.; King, D.L. Catalyst deactivation and regeneration in low temperature ethanol steam reforming with Rh/CeO$_2$–ZrO$_2$ catalysts. *Catal. Lett.* **2006**, *110*, 1–6. [CrossRef]

43. Ramaker, D.E.; Mojet, B.L.; Garriga Oostenbrink, M.T.; Miller, J.T.; Koningsberger, D.C. Contribution of shape resonance and Pt-H EXAFS in the Pt L$_{2,3}$ X-ray absorption edges of supported Pt particles: Application and consequences for catalyst characterization. *Phys. Chem. Chem. Phys.* **1999**, *1*, 2293–2302. [CrossRef]

44. Ressler, T. WinXAS: A Program for X-ray Absorption Spectroscopy Data Analysis under MS-Windows. *J. Synchrotron Rad.* **1998**, *5*, 118–122. [CrossRef] [PubMed]

45. Ravel, B. ATOMS: Crystallography for the X-ray absorption spectroscopist. *J. Synchrotron Radiat.* **2001**, *8*, 314–316. [CrossRef] [PubMed]

46. Newville, M.; Ravel, B.; Haskel, D.; Rehr, J.J.; Stern, E.A.; Yacoby, Y. Analysis of multiple-scattering XAFS data using theoretical standards. *Phys. B Cond. Matt.* **1995**, *208–209*, 154–156. [CrossRef]

47. Martinelli, M.; Jacobs, G.; Graham, U.M.; Shafer, W.D.; Cronauer, D.C.; Kropf, A.J.; Marshall, C.L.; Khalid, S.; Visconti, C.G.; Lietti, L.; et al. Water-gas shift: Characterization and testing of nanoscale YSZ supported Pt catalysts. *Appl. Catal. A Gen.* **2015**, *497*, 184–197. [CrossRef]

48. Watson, C.D.; Martinelli, M.; Cronauer, D.C.; Kropf, A.J.; Marshall, C.L.; Jacobs, G. Low temperature water-gas shift: Optimization of K loading on Pt/m-ZrO$_2$ for enhancing CO conversion. *Appl. Catal. A Gen.* **2020**, *598*, 117572. [CrossRef]

49. Watson, C.D.; Martinelli, M.; Cronauer, D.C.; Kropf, A.J.; Jacobs, G. Low Temperature Water-Gas Shift: Enhancing Stability through Optimizing Rb Loading on Pt/ZrO$_2$. *Catalysts* **2021**, *11*, 210. [CrossRef]

50. Jentys, A. Estimation of mean size and shape of small metal particles by EXAFS. *Phys. Chem. Chem. Phys.* **1999**, *1*, 4059–4063. [CrossRef]

51. Marinkovic, N.S.; Sasaki, K.; Azic, R.R. Nanoparticle size evaluation of catalysts by EXAFS: Advantages and limitations. *Zast. Mater.* **2016**, *57*, 101–109. [CrossRef]

52. Chenu, E.; Jacobs, G.; Crawford, A.C.; Keogh, R.A.; Patterson, P.M.; Sparks, D.E.; Davis, B.H. Water-gas shift: An examination of Pt promoted MgO and tetragonal and monoclinic ZrO$_2$ by in situ DRIFTS. *Appl. Catal. B Environ.* **2005**, *59*, 45–56. [CrossRef]

53. Pigos, J.M.; Brooks, C.J.; Jacobs, G.; Davis, B.H. Low temperature water-gas shift: Characterization of Pt-based ZrO$_2$ catalyst promoted with Na discovered by combinatorial methods. *Appl. Catal. A Gen.* **2007**, *319*, 47–57. [CrossRef]

54. Yee, A.; Morrison, S.J.; Idriss, H. A Study of Ethanol Reactions over Pt/CeO$_2$ by Temperature-Programmed Desorption and in Situ FT-IR Spectroscopy: Evidence of Benzene Formation. *J. Catal.* **2000**, *191*, 30–45. [CrossRef]

55. Yee, A.; Morrison, S.J.; Idriss, H. A Study of the Reactions of Ethanol on CeO$_2$ and Pd/CeO$_2$ by Steady State Reactions, Temperature Programmed Desorption, and In Situ FT-IR. *J. Catal.* **1999**, *186*, 279–295. [CrossRef]

56. Mattos, L.V.; Noronha, F.B. Hydrogen production for fuel cell applications by ethanol partial oxidation on Pt/CeO$_2$ catalysts: The effect of the reaction conditions and reaction mechanism. *J. Catal.* **2005**, *233*, 453–463. [CrossRef]

57. Martinelli, M.; Jacobs, G.; Graham, U.M.; Davis, B.H. Methanol Steam Reforming: Na Doping of Pt/YSZ Provides Fine Tuning of Selectivity. *Catalysts* **2017**, *7*, 148. [CrossRef]

58. Binet, C.; Daturi, M.; Lavalley, J.-C. IR study of polycrystalline ceria properties in oxidised and reduced states. *Catal. Today* **1999**, *50*, 207–225. [CrossRef]

59. Martinelli, M.; Alhraki, N.; Castro, J.D.; Matamoros, M.E.; Jacobs, G. *New Dimensions in Production and Utilization of Hydrogen*, 1st ed.; Nanda, S., Vo, D.V., Nguyen-Tri, P., Eds.; Elsevier International Publishing: Amsterdam, The Netherlands, 2020; pp. 143–160.

60. Evin, H.N.; Jacobs, G.; Ruiz-Martinez, J.; Graham, U.M.; Dozier, A.; Thomas, G.; Davis, B.H. Low Temperature Water–Gas Shift/Methanol Steam Reforming: Alkali Doping to Facilitate the Scission of Formate and Methoxy C–H Bonds over Pt/ceria Catalyst. *Catal. Lett.* **2008**, *122*, 9–19. [CrossRef]

61. Evin, H.N.; Jacobs, G.; Ruiz-Martinez, J.; Thomas, G.A.; Davis, B.H. Low Temperature Water–Gas Shift: Alkali Doping to Facilitate Formate C–H Bond Cleaving over Pt/Ceria Catalysts—An Optimization Problem. *Catal. Lett.* **2008**, *120*, 166–178. [CrossRef]

62. Allen, L.C. Electronegativity is the average one-electron energy of the valence-shell electrons in ground-state free atoms. *J. Amer. Chem. Soc.* **1989**, *111*, 9003–9014. [CrossRef]

63. Pigos, J.M.; Brooks, C.J.; Jacobs, G.; Davis, B.H. Low temperature water–gas shift: The effect of alkali doping on the CH bond of formate over Pt/ZrO$_2$ catalysts. *Appl. Catal. A Gen.* **2007**, *328*, 14–26. [CrossRef]

64. Martinelli, M.; Jacobs, G.; Shafer, W.D.; Davis, B.H. Effect of alkali on CH bond scission over Pt/YSZ catalyst during water-gas-shift, steam-assisted formic acid decomposition and methanol steam reforming. *Catal. Today* **2017**, *291*, 29–35. [CrossRef]

65. Menacherry, P.V.; Haller, G.L. Electronic effects and effects of particle morphology in n-hexane conversion over zeolite-supported platinum catalysts. *J. Catal.* **1998**, *177*, 175–188. [CrossRef]

66. Fukunaga, T.; Ponec, V. On the role of additives to platinum catalysts for reforming reactions. *Appl. Catal. A Gen.* **1997**, *154*, 207–219. [CrossRef]

67. Bazin, D.; Sayers, D.; Rehr, J.J.; Mottet, C. Numerical Simulation of the Platinum L$_{III}$ Edge White Line Relative to Nanometer Scale Clusters. *J. Phys. Chem. B* **1997**, *101*, 5332–5336. [CrossRef]

68. Dai, Y.; Gorey, T.J.; Anderson, S.L.; Lee, S.; Lee, S.; Seifert, S.; Winans, R.E. Inherent Size Effects on XANES of Nanometer Metal Clusters: Size-Selected Platinum Clusters on Silica. *J. Phys. Chem. C* **2017**, *121*, 361–374. [CrossRef]

69. Haller, G.L. New catalytic concepts from new materials: Understanding catalysis from a fundamental perspective, past, present, and future. *J. Catal.* **2003**, *216*, 12–22. [CrossRef]

70. Mojet, B.L.; Miller, J.T.; Ramaker, D.E.; Koningsberger, D.C. A new model describing the metal-support interaction in noble metal catalysts. *J. Catal.* **1999**, *186*, 373–386. [CrossRef]

nanomaterials

Article

Improved Water–Gas Shift Performance of Au/NiAl LDHs Nanostructured Catalysts via CeO$_2$ Addition

Margarita Gabrovska [1,*], Ivan Ivanov [1], Dimitrinka Nikolova [1], Jugoslav Krstić [2], Anna Maria Venezia [3], Dorel Crişan [4], Maria Crişan [4], Krassimir Tenchev [1], Vasko Idakiev [1] and Tatyana Tabakova [1,*]

[1] Institute of Catalysis, Bulgarian Academy of Sciences, 1113 Sofia, Bulgaria; bogoev@ic.bas.bg (I.I.); dimi_nik@abv.bg (D.N.); tenchev@ic.bas.bg (K.T.); idakiev@ic.bas.bg (V.I.)

[2] Department of Catalysis and Chemical Engineering, Institute of Chemistry, Technology and Metallurgy, University of Belgrade, 11000 Belgrade, Serbia; jkrstic@nanosys.ihtm.bg.ac.rs

[3] Istituto per lo Studio dei Materiali Nanostrutturati, CNR, 90146 Palermo, Italy; annamaria.venezia@cnr.it

[4] Ilie Murgulescu Institute of Physical Chemistry, Romanian Academy, 060021 Bucharest, Romania; dcrisan@icf.ro (D.C.); mcrisan@icf.ro (M.C.)

* Correspondence: margarita.gabrovska@abv.bg (M.G.); tabakova@ic.bas.bg (T.T.); Tel.: +359-2-979-3578 (M.G.); +359-2-979-2528 (T.T.)

Abstract: Supported gold on co-precipitated nanosized NiAl layered double hydroxides (LDHs) was studied as an effective catalyst for medium-temperature water–gas shift (WGS) reaction, an industrial catalytic process traditionally applied for the reduction in the amount of CO in the synthesis gas and production of pure hydrogen. The motivation of the present study was to improve the performance of the Au/NiAl catalyst via modification by CeO$_2$. An innovative approach for the direct deposition of ceria (1, 3 or 5 wt.%) on NiAl-LDH, based on the precipitation of Ce^{3+} ions with 1M NaOH, was developed. The proposed method allows us to obtain the CeO$_2$ phase and to preserve the NiAl layered structure by avoiding the calcination treatment. The synthesis of Au-containing samples was performed through the deposition–precipitation method. The as-prepared and WGS-tested samples were characterized by X-ray powder diffraction, N$_2$-physisorption and X-ray photoelectron spectroscopy in order to clarify the effects of Au and CeO$_2$ loading on the structure, phase composition, textural and electronic properties and activity of the catalysts. The reduction behavior of the studied samples was evaluated by temperature-programmed reduction. The WGS performance of Au/NiAl catalysts was significantly affected by the addition of CeO$_2$. A favorable role of ceria was revealed by comparison of CO conversion degree at 220 °C reached by 3 wt.% CeO$_2$-modified and ceria-free Au/NiAl samples (98.8 and 83.4%, respectively). It can be stated that tuning the properties of Au/NiAl LDH via CeO$_2$ addition offers catalysts with possibilities for practical application owing to innovative synthesis and improved WGS performance.

Keywords: Ni–Al layered double hydroxides; gold catalyst; CeO$_2$ addition; water–gas shift reaction

Citation: Gabrovska, M.; Ivanov, I.; Nikolova, D.; Krstić, J.; Venezia, A.M.; Crişan, D.; Crişan, M.; Tenchev, K.; Idakiev, V.; Tabakova, T. Improved Water–Gas Shift Performance of Au/NiAl LDHs Nanostructured Catalysts via CeO$_2$ Addition. *Nanomaterials* **2021**, *11*, 366. https://doi.org/10.3390/nano11020366

Academic Editor: John Vakros

Received: 8 January 2021

Accepted: 25 January 2021

Published: 2 February 2021

1. Introduction

Among the key drivers responsible for the renewed interest in the water–gas shift (WGS) reaction is associated with the growing hydrogen production. Hydrogen is considered an efficient sustainable energy carrier and an alternative carbon-free fuel that could substitute fossil fuels in the near future. The utilization of pure hydrogen in energy conversion technologies as fuel cells is anticipated to ensure an environmentally friendly way to satisfy global energy needs [1,2].

The manufacture of pure hydrogen for fuel cell applications requires the consideration that the presence of CO in the produced H$_2$-rich synthesis gas could irreversibly destroy the metal anode in the fuel cells. Therefore, the purification of synthesis gas by CO subtraction is of particular importance, and is commonly attained via conversion of CO by water vapor, referred to as WGS reaction (CO + H$_2$O $\leftrightarrow$ CO$_2$ + H$_2$, $\Delta H = -41.2$ kJ mol^{-1}).

The last is conventionally applied to reduce the CO level and to enrich synthesis gas with hydrogen as well. Since WGS reaction is reversible and moderately exothermic, it is thermodynamically favored at low temperatures and kinetically at high ones; thus, the equilibrium CO conversion decreases with increasing the reaction temperature. In regard to industrial application, aiming to achieve high CO conversion, the up-to-date WGS reaction is performed through a two-stage WGS converter system using both different temperatures and catalysts. There is a high temperature shift stage (350–450 °C, Fe_2O_3-Cr_2O_3, residual CO of about 2–5%), followed by a low temperature shift stage (180–250 °C, Cu-ZnO-Al_2O_3, residual CO content < 1%) with interstage cooling [3–7].

A variety of catalysts combining noble or transition metals (Pt, Rh, Au, Pd, Fe, Cu, Ni, Co) with different oxides or mixed oxides can catalyze the WGS reaction, as described in many scientific reviews and numerous papers. The process parameters and recent advances in WGS catalysis, the design of novel and effective catalyst compositions, the preparation approaches, the nature of the support, the catalyst precursors and promoter additives (if there) as well as their impact on the performance in the WGS reaction have been debated and clarified [5–9].

Amid WGS catalysts, the supported gold ones have played a key role for more than three decades [10–15]. Because of their unique features, such as stability in oxidizing surroundings, non-pyrophority, no need of additional activation pre-treatment, appreciable WGS activity and specific electronic and structural peculiarity, these catalytic formulations have proven to be among the most auspicious options to replace the conventional WGS catalysts for small-scale applications in the low-temperature range (180–260 °C). It is also important to note that the choice of support used is of crucial importance for the WGS behavior of supported gold nanoparticles. By way of illustration, supported gold on non-reducible oxides (SiO_2, Al_2O_3 or MgO) showed lower activity as opposed to those on reducible oxides (Fe_2O_3, CeO_2, doped-CeO_2, ZrO_2, TiO_2, etc.). These oxides significantly improve gold catalyst activity in the low-temperature WGS region [3,16–18].

Ceria (CeO_2) is among the most extensively studied reducible supports. The considerable performance of ceria and ceria-based mixed oxides in WGS reaction is related to its high capacity to store and release oxygen by Ce^{3+}/Ce^{4+} surface sites, surface and bulk oxygen vacancies and redox properties. The exceptional exchange between Ce^{3+} and Ce^{4+} oxidation states, the easy change from Ce^{4+} (CeO_2) under oxidizing conditions to Ce^{3+} (Ce_2O_3) under net reducing conditions and vice versa are typical for ceria. Nonstoichiometric CeO_{2-y} can be formed by oxygen release and reduction of Ce^{4+} to Ce^{3+}, with the concomitant formation of oxygen vacancies within the crystal structure. The specific features and application of CeO_2 in catalysis have been revealed in numerous reviews and papers [19–24]. Moreover, CeO_2 maintains high dispersion of supported metals and assists the noble metal reduction, oxidation and WGS reaction as well [19]. On the other hand, nanoscaled ceria containing high concentrations of Ce^{3+} and oxygen vacancies is considered as a preferred carrier for the Au catalysts for low-temperature WGS reaction. The first reports about the unique WGS activity of Au/CeO_2 at low temperatures were accomplished by the Flytzani-Stephanopoulos [25] and Andreeva [26] groups. After that, many efforts have been focused on the elucidation of the effect of ceria and ceria-doped oxides on WGS performance of gold catalysts depending on the ceria particle size, surface area, surface structure, presence of dopants, etc. [10,15,27–36].

Among the various carriers of gold catalysts, layered double hydroxides (LDHs) have been stated as appropriate supports of efficient WGS catalysts [37–40]. In recent years, considering the metal availability and economic considerations, an object of research investigations has been the design of nickel-based LDHs as precursors of WGS catalysts [41–47]. NiAl LDHs, also called takovite-like compounds $Ni_6Al_2(OH)_{16}CO_3\cdot4H_2O$ [48] are layered nanometric materials, which are members of a large family of natural or synthetic inorganic anionic clay-like layered compounds. They are built from a two-dimensional structure consisting of successively alternating positively charged NiAl hydroxide layers $[Ni^{2+}_{1-x}Al^{3+}_x(OH)_2]^{x+}$ and an interlayer space $[A^{n-}_{x/n}\cdot mH_2O]$ containing charge compensating exchangeable anions (CO_3^{2-}, SO_4^{2-}, NO_3^{-}, Cl^{-}, etc.) and m water molecules. The layered structure assumes uniform distribution of the octahedrally coordinated Ni^{2+} and Al^{3+} ions within the hydroxide sheets. One key feature of LDHs is the ability to exchange various metals in the metal hydroxide sheets, which permits tailoring of the catalyst to a specific function. Upon heating, the layered structure loses the interlayered anions and water molecules and finally forms NiAl mixed metal oxides, $Ni^{2+}(Al^{3+})O$, characterized by nanoscaled crystal size, high surface area and well-distributed both cations in situ [49,50].

The incorporation of rare earth elements into the structure of the LDHs is very attractive because they can affect the catalytic, electrical and magnetic properties. However, the incorporation of cerium into NiAl hydroxide-like layers is disputed due to the large ionic radius of Ce^{3+} (1.15 Å) in comparison with Ni^{2+} (0.69 Å) and Al^{3+} (0.51 Å) [51]. This is the reason that there are few publications about the synthesis of cerium containing NiAl LDHs. Sanati and Rezvani [52] reported the synthesis of Ni/AlCe LDH by precipitation with triethylamine in an autoclave without nitrogen. The authors supposed that the Ce^{3+} ions do not fill any octahedral site in the hydroxide-like layers, beingdeposited on the LDH surface, making a separate CeO_2 phase. Another procedure for synthesis of the Ni/AlCe LDH compound by the co-precipitation method also followed by hydrothermal treatment in the autoclave for 48 h at 120 °C was presented [53]. A successful introduction of Ce^{3+} cations into the brucite-like layers of the layered structure was testified.

An important aspect to consider is the approach of CeO_2 synthesis. The main chemical methods for the preparation of nanoscaled ceria including homogeneous precipitation, sol-gel process, sonochemical synthesis, thermal decomposition, hydrothermal synthesis, microwave synthesis, microemulsion and solvothermal method were studied by several researchers and recently described in detail by Reni and Nesaraj [54].

Among these techniques, the precipitation method has been extensively studied because it is an unpretentious process, easy scale-up and low cost. For example, Matijević and Hsu [55] prepared sub-micron crystalline oxidicarbonate $Ce_2O\text{-}(CO_3)_2 \cdot H_2O$ particles using cerium nitrate and urea. The sample calcined at 400 °C for 2 h converted to a cubic CeO_2. Chen and Chen [56] synthesized CeO_2 particles from cerium nitrate with hexamethylenetetramine (HMT), which can hydrolyze slowly to yield ammonia as an alternative to urea. The difference between both methods can be explained by the different ligands. In the Ce-HMT method, ligands are OH^{-} ions, while in the Ce-urea method, the major ligands are OH^{-} ions and CO_3^{2-} ions. The formation of $Ce_2O\text{-}(CO_3)_2$ rather than CeO_2 in the Ce-urea method is the result of the competition between OH and CO_3^{2-}. It seems that the urea-based method is unsuitable for the precipitation of pure Ce^{3+}-containing compounds. Zhou et al. [57] obtained CeO_2 particles of about 4 nm from cerium nitrate and ammonia precipitant at pH value $\approx$ 9 under oxygen bubbling into the reactor for oxidation of the Ce^{3+} to Ce^{4+} ions. The dried at room temperature precipitates directly yielded CeO_2 particles. The synthesis of nanocrystalline CeO_2 powders via a carbonate precipitation method, using ammonium carbonate (AC) as the precipitant and cerium nitrate hexahydrate as a cerium source was reported [58]. It was found that the precursors synthesized at AC/Ce^{3+} molar ratio (R) of $2 < R \leq 3$ are basic carbonates ($Ce(OH)CO_3 \cdot 2.5H_2O$), whereas those produced at $R > 3$ are ammonium cerous double carbonates (($NH_4)_x Ce(CO_3)_{1.5+x/2} \cdot yH_2O$, $x \leq 1.0$). The CeO_2 powder calcined at 700 °C can be densified to $\approx$99% of the theoretical value using isothermal sintering at 1000 °C for 2 h. Another method is based on the preparation of CeO_2 particles from cerium chloride and sodium hydroxide with the presence of hydrogen peroxide under various pH conditions from 6 to 12 [59]. Uekawa et al. [60] obtained 7–9 nm CeO_2 particles starting from cerium nitrate in the polyethylene glycol solution.

The majority of the abovementioned methods are focused on the type of CeO_2 precursors, presence of ligands and additives. In most cases, the formation of CeO_2 is achieved after temperature treatment.

We propose a simple and innovative approach for the direct deposition of ceria on NiAl-LDH, based on the precipitation of Ce^{3+} ions with NaOH. The suggested method allows us to obtain CeO_2 phase and to preserve the NiAl layered structure by avoiding the need for capping ligand, hydrothermal and calcination treatment. Briefly, the interaction of $Ce(NO_3)_3 \cdot 6H_2O$ with alkaline solution immediately leads to the formation of $Ce(OH)_3$ hydroxide precipitate, due to the extremely low solubility constant (7×10^{-21}) [61]. In alkaline environment, Ce^{3+} ions are oxidized to hydrated Ce^{4+} ions [62], and then further hydrolyzed to form a $[Ce(OH)_x(H_2O)_y]^{(4-x)+}$ complex [63,64]. The complex is deprotonated by water molecules to CeO_2 [62]. It is obvious that the presence of OH^- ions strongly affects the oxidation of Ce^{3+} to Ce^{4+} ions. With regard to this finding, NaOH was chosen for precipitation of Ce^{3+} ions on NiAl-LDH.

Regarding our recent studies on the suitability of NiAl ($Ni^{2+}/Al^{3+} = 2.5/1$) LDH as a carrier of gold-containing WGS catalysts [44] and the unique properties of ceria, the goal of the current work was to obtain novel catalytic systems of improved WGS performance via CeO_2 addition. A modification of NiAl with CeO_2 was performed, aiming to affect the dispersion of gold species. The role of the CeO_2 dopant (1, 3 or 5 wt.%) and gold presence (3 wt.%) in the studied NiAl LDHs were estimated by a comparative analysis of the catalytic and reduction behavior, phase compositions, structural and surface properties of the CeO_2-modified and gold-containing NiAl samples, with their unmodified analogues, before and after the activity tests.

2. Materials and Methods

2.1. Reagents

All the reagents, nickel nitrate hexahydrate $(Ni(NO_3)_2 \cdot 6H_2O)$, aluminum nitrate nonahydrate $(Al(NO_3)_3 \cdot 9H_2O)$, cerium nitrate hexahydrate $(Ce(NO_3)_3 \cdot 6H_2O)$, gold(III) chloride trihydrate $(HAuCl_4 \cdot 3H_2O)$, sodium carbonate anhydrous (Na_2CO_3) and sodium hydroxide (NaOH) were of "pro analyze" purity grade, provided by Sigma-Aldrich (Steinheim, Germany), and were used for the synthesis as received. All the solutions were prepared with fresh distilled water.

2.2. Sample Preparation

2.2.1. Adjusting the Procedure for NiAl-LDH Modification with CeO_2

Synthesis of CeO_2

The CeO_2 sample was prepared by a wet chemical precipitation method using $Ce(NO_3)_3 \cdot 6H_2O$ and NaOH at room temperature. An appropriate amount of nitrate salt was placed in a reaction vessel and dissolved in distilled water under vigorous stirring. The pH of nitrate solution was adjusted to a value of 12 by dropwise addition of 1M NaOH. The immediately formed suspension was treated for 2 h under constant stirring conditions (pH = 12 and room temperature). The resultant precipitate was filtered and washed with distilled water to complete the removal of NO_3^- ions (testing with a solution of diphenylamine in H_2SO_4), then was dried at 105 °C for 20 h and designated CeO_2-105. Subsequently, a part of the dried solid was calcined at 250 °C for 2 h and marked CeO_2-250. The choice of calcination temperature is associated with our goal to preserve the NiAl layered structure, which was further modified with CeO_2.

Synthesis of NiAl-LDH

The carbonate form of NiAl-LDH with a Ni^{2+}/Al^{3+} molar ratio of 2.5/1 was obtained through the co-precipitation method at 80 °C and constant pH = 8 by the usage of two aqueous solutions: mixed 0.5M Ni-Al nitrate and a 0.9M Na_2CO_3 as a precipitant.

A certain volume of distilled water was loaded into a five-necked glass reactor supplied with a stirrer, pH electrode, thermocouple and reflux condenser. The water was heated to 80 °C and pH value of 8.0 was tuned with Na_2CO_3 solution. The proper amounts of both mixed Ni-Al and Na_2CO_3 solutions were simultaneously entered into the reactor by means of two peristaltic pumps with the reactants feed flow rate of 1 L h^{-1} under stirring at 260 rpm. The gained slurry was aged for 60 min under the controlled conditions, filtered off and carefully washed with hot distilled water until the pH of the filtrate decreased to

≈6–7 and the absence of NO_3^- ions. The precipitate was dried at 105 °C for 20 h, named the as-synthesized sample, and denoted as NiAl. Part of this sample was heated at 250 °C for 2 h, and coded as NiAl-250.

2.2.2. Synthesis of CeO_2-Modified NiAl-LDH

The synthesized NiAl-LDH would be further modified with CeO_2 by precipitation of $Ce(NO_3)_3 \cdot 6H_2O$ with alkaline solution. Because of that, initially, the stability of NiAl-LDH in an alkaline medium was inspected. The solid was treated with 1M NaOH solution under vigorous stirring at room temperature for 2 h. The washed sample (pH of filtrate ≈ 6–7) was dried at 105 °C and named NiAl-1MNaOH. Part of this sample was heated at 250 °C for 2 h (NiAl-1MNaOH-250).

The influence of the precipitating agent concentration was examined by precipitation of 5 wt.% CeO_2 on NiAl-LDH using 0.1M and 1M NaOH solution. The obtained modified samples were denoted as 5CeNiAl-0.1MNaOH and 5CeNiAl-1MNaOH, respectively. After calcination at 250 °C for 2 h, they were marked as 5CeNiAl-0.1MNaOH-250 and 5CeNiAl-1MNaOH-250, respectively.

Three CeO_2-modified NiAl-LDH samples, containing 1, 3 and 5 wt.% CeO_2, were prepared by the direct deposition of ceria over the NiAl-LDH suspended in distilled water, based on the precipitation of Ce^{3+} ions with 1M NaOH, following the procedure depicted above. The synthesized samples were designated as xCeNiAl, where x represents wt.% CeO_2, for example, 1CeNiAl.

2.2.3. Deposition of Gold on the Surface of NiAl and xCeNiAl LDHs

Gold-comprising samples were obtained by deposition–precipitation of gold over the NiAl and xCeNiAl LDHs. The solids were suspended in distilled water through ultrasound. The deposition of gold (3 wt.%) was performed by simultaneous addition of aqueous solutions of 0.06 M $HAuCl_4 \cdot 3H_2O$ and 0.2 M Na_2CO_3 into the reactor at 60 °C and pH = 7.0 under stirring at 250 rpm and reactant feed flow rate of 0.15 L h^{-1}. After aging under the same conditions for 60 min, the sample was filtered and carefully washed with distilled water until the absence of Cl$^-$ ions. The gained Au-containing materials were dried under vacuum at 80 °C and designated as Au/NiAl and Au/xCeNiAl.

Since in all studied samples, the Ni^{2+}/Al^{3+} molar is the same (2.5/1) and the content of gold is also identical, for convenience, they will not be indicated in the text.

2.3. Sample Characterization Methods

The chemical composition of the as-synthesized materials was defined by inductively coupled plasma atomic emission spectroscopy (ICP-AES) by a JY 38 spectrometer (Horiba Jobin—Yvon, Longjumeau, France) after appropriate acid treatment.

The phase composition of the as-synthesized and post-WGS reaction samples (spent catalysts) was determined by the powder X-ray diffraction (PXRD) technique. The PXRD patterns were obtained using a Bruker D8 Advance powder diffractometer (Bruker-AXS, Karlsruhe, Germany) employing CuKα radiation (U = 40 kV and I = 40 mA) and LynxEye detector (Bruker-AXS, Karlsruhe, Germany). Scans were performed for 2θ values from 5° to 90° with a step of 0.04° 2θ. The crystalline phases were identified by means of International Centre for Diffraction Data (ICDD) powder diffraction files. The unit cell parameters and mean size of the coherently scattering domains (LVol-FWHM) were obtained through the analysis of line positions and profile broadening by using the fundamental parameters peak shape description, including appropriate corrections for the instrumental broadening and diffractometer geometry with the program TOPAS V4.2 (Bruker-AXS, Karlsruhe, Germany).

The texture characteristics of the as-synthesized and spent catalysts were determined through N_2 adsorption—desorption measurements conducted at a low temperature (−196 °C) with a Sorptomatic 1990 (Thermo Finnigan, Milan, Italy) apparatus. Prior to the measurements, the samples were degassed for 2 h at room temperature followed for 36 h at 80 °C under a vacuum. Specific surface area (*SSA*) values were calculated according to the Brunauer–Emmett–Teller (BET) method from the linear part of the N_2 adsorption isotherms [65]. Micropore volume (V_{mic}) values were calculated using the Dubinin–Radushkevich method [66]. The mesopore volume (V_{meso}) and the mesopore size distribution were estimated by the Barrett, Joyner and Halenda (BJH) method [67] from the desorption branch using the Lecloux standard isotherm [68]. Total pore volume (V_{tot}) was estimated from the N_2 volume physisorbed for relative pressure (p/p_0) of 0.99.

The temperature-programmed reduction (TPR) in the as-synthesized LDHs was accomplished in the measurement cell of a SETARAM model DSC-111 differential scanning calorimeter (SETARAM, Caluire, France). The temperature was linearly raised from 25 to 700 °C at a heating rate of 10 °C min^{-1}. The TPR experiments were performed by a gas mixture of 10% H_2 in Ar at a flow rate of 20 cm^3 min^{-1}. The selected experimental conditions are in agreement with the criteria recommended by Monti and Baiker [69] to avoid mass transfer and temperature control limitations.

The X-ray photoelectron spectra (XPS) were recorded with a VG Microtech ESCA 3000 Multilab (VG Scientific, Sussex, UK), equipped with a dual Mg/Al anode. The spectra were excited by the unmonochromatized Al Kα source (1486.6 eV) run at 14 kV and 15 mA. The experimental procedure has been previously described [70]. The constant charging of the samples was removed by referencing all the energies to the C 1s set at 285.1 eV, arising from the adventitious carbon. Analyses of the peaks were carried out with the Casa XPS software. Atomic concentrations were calculated from peak intensity using the sensitivity factors provided by the software. The binding energy values were quoted with a precision of $\pm$0.15 eV and the atomic percentage with a precision of $\pm$10%.

2.4. WGS Activity Evaluation

The catalytic tests in WGS reaction were carried out in a flow reactor at atmospheric pressure and temperature interval 120–300 °C by stepwise increase in the reaction temperature using 0.5 cm^3 samples in the 0.63–0.80 mm particle size range and gas feed composition of 3.37 vol.% CO, 25.01 vol.% H_2O and 71.62 vol.% Ar with space velocity of 4000 h^{-1} (gas flow rate 2 L h^{-1}). The gas flow rate was controlled by a mass flow controller AALBORG Model GFC17 (AALBORG, Orangeburg, NY, USA). The temperature controller "COMECO" Model RT1800 (COMECO, Plovdiv, Bulgaria) was used for temperature control in the reactor. The control in the thermostatic box was carried out by "COMECO" (Model RT38), aiming to prevent water vapor condensation. A syringe pump RAZEL model R-99 (Razel Scientific Instruments, a part of Mansfield Research and Development, Saint Albans, VT, USA) was used for control of the water vapor concentration. The measurements were performed within four consecutive days (about 40 h). The test of each sample includes: (i) first day—activation of the catalysts by stepwise increase in the reaction temperature in the reaction gas mixture; (ii) second day—temperature dependence of CO conversion; (iii) third day—tests at different space velocities and CO/H_2O ratio; (iv) fourth day— repeated temperature dependence of CO conversion. The WGS activity was expressed by the degree of CO conversion after the reaching of a stationary CO conversion (at every 20 °C step) The outlet CO concentration was analyzed by a Uras 3G (Hartmann & Braun AG, Frankfurt am Main, Germany) gas analyzer. The CO conversion degree was calculated based on the inlet and outlet CO concentration.

3. Results and Discussions

3.1. Tuning the Procedure for NiAl-LDH Modification with CeO_2

We started with the synthesis of CeO_2 aiming to verify the literature reports that ceria can be obtained directly by precipitation of Ce^{3+} ions with alkaline solution. Part of the CeO_2 powder prepared by the above-described synthesis procedure was additionally calcined at 250 °C, for the reason of comparison.

The samples' structural features were estimated by means of PXRD analysis (Figure 1). The patterns of dried (CeO_2-105) and calcined CeO_2 (CeO_2-250) powders comprise reflections at 2θ = 28.5°, 33.1°, 47.5°, 56.3°, 58.9°, 69.4°, 76.6° and 78.9°, which are attributed to (111), (200), (220), (311), (222), (400), (331) and (420) planes, characteristic of the ceria phase in the cubic crystal structure of the fluorite type (ICDD-PDF file 00-034-0394). The diffractograms of both samples seem very similar in the peak angle location and intensity. The calculated unit cell parameter (a) and the total volume (V) of CeO_2 crystal lattice (Table 1) confirm this observation, showing comparable values to those of the standard CeO_2 oxide (ICDD-PDF file 00-034-0394). The results evidenced that we obtained a well-crystallized monophasic ceria structure in the dried as well as in the calcined samples.

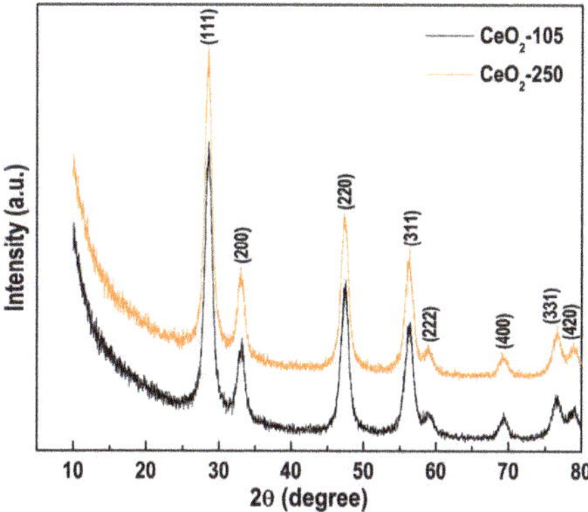

Figure 1. Powder X-ray diffraction (PXRD) patterns of unsupported CeO_2, thermally treated at 105 and 250 °C.

Table 1. Structural characteristics of the synthesized CeO_2 samples.

Sample	a (Å)	V (Å³)	L (nm)
CeO_2-105	5.4211(10)	159.32(9)	7.27(6)
CeO_2-250	5.4185(9)	159.09(8)	7.55(6)
CeO_2-standard	5.4113	158.46	-

Moreover, the PXRD analysis reveals comparable mean crystallite sizes (L) of ceria (Table 1), a little larger for the CeO_2-250, indicating slightly improved crystallinity of the sample after the calcination at 250 °C.

The obtained data provide experimental evidence for direct CeO_2 synthesis by precipitation of Ce^{3+} ions using 1M NaOH without further thermal treatment.

The recorded PXRD spectrum of the as-synthesized NiAl sample (Figure 2a) highlights the reflections at 11.38°, 23.09°, 34.90°, 39.41°, 46.84°, 60.92°, 62.29°, 66.18° and 72.60°. These can be related to (003), (006), (012), (015), (018), (110), (113), (116) and (202) planes, corresponding to the crystal lattice of the mineral takovite (ICDD-PDF file 00-015-0087), a nickel aluminum LDH with Ni/Al = 3.0/1 molar ratio.

Considering the peak positions of the takovite structure, PXRD analysis confirms a successful synthesis of the NiAl LDH structure. Additionally, the high crystallinity of NiAl sample is pointed out by the well-resolved doublet of the reflections at 2θ = 60.92° and 62.29° ((110) and (113) planes) that is connected to the high degree of crystallinity [49] and good ordering of cations in the hydroxide layers [71].

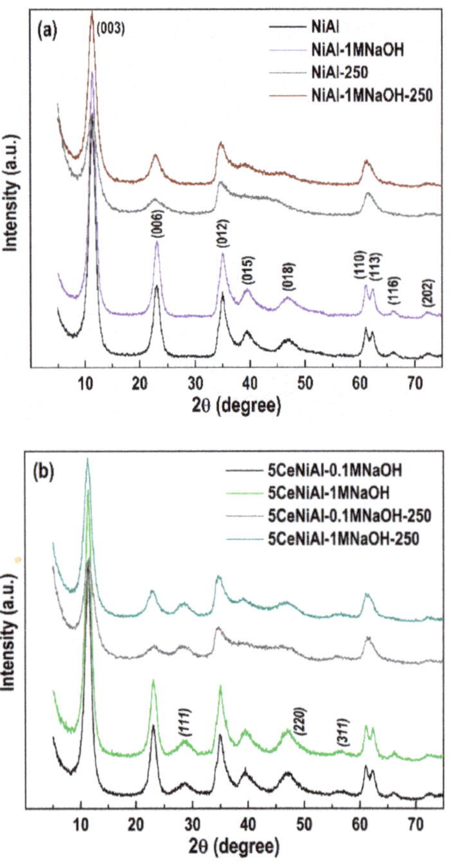

Figure 2. PXRD patterns of: (**a**) as-prepared NiAl, alkali treated NiAl-1MNaOH, calcined NiAl-250 and calcined alkali treated NiAl-1MNaOH-250 and (**b**) as-prepared 5CeNiAl-0.1MNaOH, and 5CeNiAl-1MNaOH and calcined 5CeNiAl-0.1MNaOH-250 and 5CeNiAl-1MNaOH-250. The diffraction lines of CeO_2 phase are marked in italics.

The stability of NiAl-LDH in a strongly alkaline medium was examined through the treatment with 1M NaOH solution under vigorous stirring at room temperature. The PXRD patterns of NiAl-1MNaOH sample display diffraction peaks, which wholly match the positions and intensities of those of the as-prepared NiAl. PXRD characterization designates that NiAl-LDH is stable in a highly alkaline environment, since the layered structure is completely preserved and the appearance of the additional new phases is not registered as well. As can be seen, the calcination at 250 °C of NiAl and NiAl-1MNaOH (Figure 2a) induces broadening of the takovite characteristic reflections, which appear almost at the same positions as those of the as-synthesized NiAl, disclosing reduced crystalline nature of the LDH. Actually, an intermediate metastable dehydrated takovite-like (TKl) structure with reduced peak number and intensities is created. This phenomenon is attributed to the reversible removal of physisorbed water on the TKl external surface and the interlayer water, weakly bonded with charge compensating CO_3^{2-} anions from the interlayer space [72]. Noticeably, the thermal treatment more appreciably affects the structure of the non-alkali employed sample (NiAl-250). The diffractogram of alkali operated calcined solid (NiAl-1MNaOH-250) suggests preservation of the layered structure to a greater extent than in non-alkali treated NiAl-250, confirming the stability of the NiAl sample in 1M NaOH.

The deposition of 5 wt.% CeO_2 on NiAl-LDH applying $Ce(NO_3)_3 \cdot 6H_2O$ and two NaOH solutions (0.1M NaOH and 1M NaOH) was performed in order to optimize the conditions for NiAl modification with CeO_2. PXRD analysis of 5CeNiAl-0.1MNaOH and 5CeNiAl-1MNaOH indicates that the usage of both NaOH solutions prompts the appearance of new diffraction lines, in addition to the TKl phase (Figure 2b). The reflections at $2\theta = 28.3°$, $47.4°$ and $56.2°$, which are ascribed to (111), (220) and (311) lines of the cubic ceria phase, respectively, are well organized in the case of 1M NaOH application. Similarly, the PXRD study of Ce-doped NiAl samples calcined at 250 °C (5CeNiAl-0.1MNaOH-250 and 5CeNiAl-1MNaOH-250) demonstrates the co-existence of NiAl-LDH and CeO_2 phases, also well defined when 1M NaOH is used. The comparison of PXRD patterns of dried and calcined Ce-doped NiAl samples disclosed a well-preserved layered structure when the precipitation of Ce^{3+} ions was performed with 1M NaOH. It is obvious that 1M NaOH as precipitation agent is more appropriate for ceria deposition over the NiAl-LDH because it contributes to obtaining better defined structures in comparison with the solution of lower NaOH concentration.

It may be summarized that the registration of CeO_2 in the uncalcined samples confirms the statement that oxidation of Ce^{3+} to Ce^{4+} ions still occurs in the wet precipitate. These findings allowed us to modify the NiAl-LDH with 1, 3 and 5 wt.% CeO_2 via Ce^{3+} ion precipitation with 1M NaOH at room temperature. Further, these samples were used as supports of deposited gold particles.

3.2. Chemical Analysis

The chemical composition of the obtained samples is presented in Table 2. The ICP analysis discloses that the Ni^{2+}/Al^{3+} molar ratio in all the as-synthesized compositions is identical to that in the mixed NiAl nitrate solution used for the NiAl preparation. The oxide form of the components is included for a clear understanding. The gold loading in the Au-containing samples is also listed in the Table.

Table 2. Sample notation and chemical composition of the as-synthesized samples.

Sample	Chemical Composition (wt.%)				Ni^{2+}/Al^{3+} Molar Ratio
	NiO	CeO_2	Al_2O_3	Au	
NiAl	78.55	-	21.45	-	2.5
1CeNiAl	77.76	1.0	21.24	-	2.5
3CeNiAl	76.19	3.0	20.81	-	2.5
5CeNiAl	74.62	5.0	20.38	-	2.5
Au/NiAl	76.30	-	20.70	3.0	2.5
Au/1CeNiAl	75.43	0.97	20.60	3.0	2.5
Au/3CeNiAl	73.90	2.91	20.19	3.0	2.5
Au/5CeNiAl	72.38	4.85	19.77	3.0	2.5

3.3. Water–Gas Shift Activity

Unlike generally accepted preliminary calcination of the catalysts, we conducted WGS activity tests directly over the as-synthesized samples aiming to preserve the layered Ni-Al structure. This concept is based on the catalytic measurements that demonstrated lower WGS activity of NiAl and 5CeNiAl after calcination even at a low temperature of 250 °C within 2 h (Figure 3a).

When the WGS reaction starts over the uncalcined samples, the layered structure begins to decompose with the reaction temperature increase, causing formation of the NiO phase. Considering the associative reaction mechanism [44], the NiO structure represents a "precursor" for creation of the active nickel species in the redox WGS reaction. Under the reaction conditions, partial hydroxylation of the NiO surface takes place, prompting the creation of active centers on the catalyst surface, namely, $Ni(OH)_2$ and $NiOOH$

structures containing Ni^{2+} and Ni^{3+} cations, respectively. It is understandable that the creation of these species on the surface of the uncalcined sample is accomplished prior to those of the calcined NiAl sample because of structural changes that occur after calcination (see Figure 2a). This, in turn, contributes to lower catalytic activity of calcined NiAl.

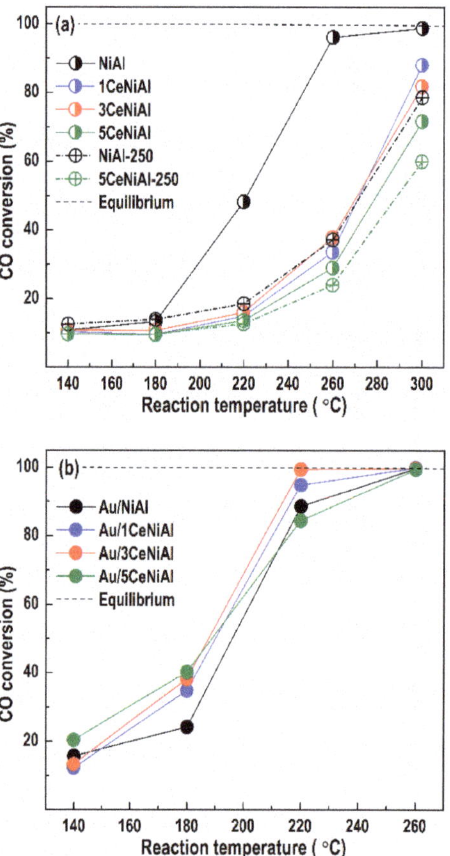

Figure 3. The temperature dependence of CO conversion during WGS reaction over: (**a**) NiAl and *x*CeNiAl samples and (**b**) Au/NiAl and Au/*x*CeNiAl samples. Calcined NiAl-250 and 5CeNiAl-250 are included for comparison in section (**a**).

In Figure 3, the catalytic activity data collected after catalyst activation are reported. As can be seen in Figure 3a, the calcined NiAl sample (NiAl-250) gains only 37.2% CO conversion at temperature of 260 °C versus the as-prepared NiAl analogue, which achieves 96.2% conversion at the same temperature. Similarly, the calcination of the CeO_2-doped sample (5CeNiAl-250) does not contribute to the increase in WGS activity. Therefore, in order to achieve high CO conversion at lower temperatures, the usage of uncalcined samples is more favorable than calcined ones.

Another observation is that the CeO_2 addition in the selected three concentrations does not enhance the catalytic activity of NiAl-LDH. On the other hand, the deposition of gold over the NiAl and *x*CeNiAl solids significantly improves their CO conversion, thus verifying the promoting effect of gold to activate the CO molecule. Comparison of the WGS activity of all gold-containing catalysts reveals that Au/3CeNiAl shows 99.7% CO conversion at 220 °C, i.e., CO almost reaches the equilibrium conversion degree. The rest of the catalysts demonstrate similar, but lower, activity at the same reaction temperature, following the order: Au/1CeNiAl > Au/NiAl > Au/5CeNiAl.

The stability of the catalysts was examined under different space velocities (Figure 4a) and different water vapor partial pressures (Figure 4b). Studying the impact of the space velocities on the degree of CO conversion at 220 °C (the temperature at which maximum WGS activity is achieved over Au/3CeNiAl) in the studied catalysts represents the effect of the CO conversions as a function of the contact time. The steady-state CO conversion as a function of space velocity is demonstrated in Figure 4a. The NiAl catalyst exhibits diminution in CO conversion with the contact time decrease, while this effect is practically negligible for the Ce-modified NiAl catalysts, especially at the highest space velocity of 8000 h^{-1}.

The impact of the contact time on WGS activity of Au-containing catalysts is more substantial (Figure 4a). They all display the same WGS activity at a space velocity of 2000 h^{-1}. The lowering of the contact time markedly decreases the activity of the Au/NiAl catalyst, showing 36% CO conversion at 8000 h^{-1}, but higher than that of the Au-free analogue (NiAl) at all studied space velocities. The increase in space velocity to 4000 h^{-1} leads to an insignificant change in CO conversion of the Au/1CeNiAl and Au/3CeNiAl catalysts and slightly decreases the CO conversion of Au/5CeNiAl. A further increase to 8000 h^{-1} induces diminution of the activity of all Au/xCeNiAl catalysts with 14 to 25%, namely, Au/3CeNiAl (14%) > Au/1CeNiAl (22%) > Au/5CeNiAl (25%).

The catalytic behavior in the WGS reaction also depends on the water partial pressure. The dependence of WGS activity on the water amount at 220 °C is shown in Figure 4b. The increase in water partial pressure (from 20 to 47.3 kPa) provokes an increase in CO conversion of both NiAl and Au/NiAl, more significantly in the case of the Au-containing sample (with 30%). The Ce-modified catalysts demonstrate independence from the value of the water partial pressure. Contrariwise, the CO conversion of Au-containing catalysts slightly increases as a function of the H_2O/CO ratio. It should be stressed that the presence of the highest content of water (47.3 kPa) in the reaction mixture slightly enhances the activity of Au/xCeNiAl catalysts, following again in the same order: Au/3CeNiAl > Au/1CeNiAl > Au/5CeNiAl. They show good tolerance toward a high concentration of water. It is known that high H_2O/CO ratios are typical under fuel processor conditions for low-temperature shift reactors, and the resistance to the presence of a high amount of steam could be considered as an important feature.

Figure 5 illustrates the temperature dependence of the CO conversion over all catalysts studied after the stability tests (influence of space velocity and water vapor partial pressure). It should be noted that after different treatments, the NiAl catalyst demonstrates WGS activity (Figure 5a) very similar to the initial (Figure 3a), thus showing stable performance. Ce-doped NiAl catalysts also repeated the initial CO conversion (Figure 3a) regardless of the CeO$_2$ content at all studied temperatures. The Au/3CeNiAl catalyst completely repeats the initial activity (Figure 5b) after the space velocity and the H_2O/CO ratio tests. It is interesting that at 220 °C, Au/1CeNiAl and Au/5CeNiAl catalysts increase their CO conversion by 3 and 10%, respectively (Figure 5b), in comparison with the initial ones (Figure 3b). Au/NiAl exhibits the lowest activity among the gold-containing catalysts.

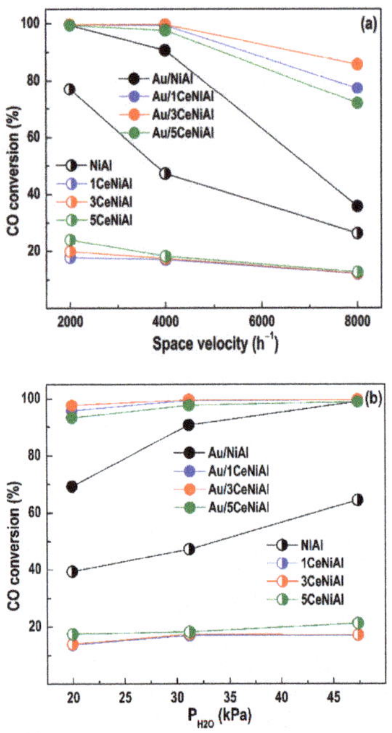

Figure 4. Effect of the (**a**) space velocity and (**b**) water vapor partial pressures on the degree of CO conversion at 220 °C over the studied catalysts.

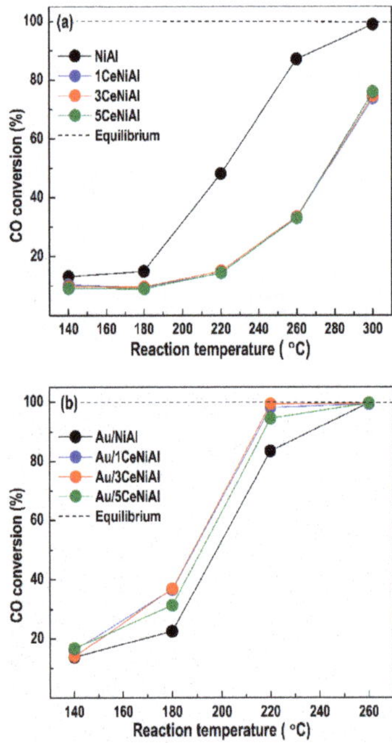

Figure 5. Temperature dependence of the CO conversion over the (**a**) NiAl and Ce-modified NiAl and (**b**) Au-containing NiAl and Ce-modified NiAl catalysts, studied after the stability tests.

The results obtained disclose that gold/ceria-NiAl catalytic compositions established high CO conversion in the low-temperature range (140–260 °C) after treatment at different space velocities and H_2O/CO ratios. Among them, the Au/3CeNiAl catalyst exhibits the highest and most stable CO conversion. As will be discussed later, this could be related to the higher gold dispersion (Table 4).

3.4. Structure and Phase Composition of As-Synthesized and Spent Catalysts

PXRD patterns of the as-synthesized NiAl and CeO_2-doped NiAl samples (Figure 6a) exhibit reflections at 2θ = 11.38, 23.09, 34.90, 39.41, 46.84, 60.92, 62.29, 66.18 and 72.60°, which are related to (003), (006), (012), (015), (018), (110), (113), (116) and (202) planes, respectively, characteristic of the stoichiometric takovite-type structure (ICDD-PDF file 00-015-0087).

As mentioned, additional diffraction lines at 2θ = 28.57, 47.44 and 56.31°, which are attributed to (111), (220) and (311) crystalline planes of cubic ceria phase (ICDD-PDF file 00-034-0394), respectively, were recorded (Figure 6a) after the precipitation of 5 wt.% CeO_2 over the NiAl LDH (sample 5CeNiAl). The formation of the separate ceria phase over NiAl is supported by the calculated unit cell parameter a_{CeO2} = 5.410(6) Å and total volume V_{CeO2} = 158.4(5) Å^3 of the CeO_2 crystal lattice in 5CeNiAl (Table 3), which are identical to those of the standard CeO_2 oxide (Table 1). The intensity and number of ceria diffraction lines decrease with the diminution of its amount to 3 wt.%, as seen in Figure 6a. The absence of a separate ceria phase in 1CeNiAl could be explained by the low CeO_2 content of 1 wt.% in the sample. Generally, the presence of ceria lowers the crystallinity of the parent NiAl solid, which is more pronounced in the sample with the highest CeO_2 loading, 5CeNiAl.

The deposition of gold over the NiAl and CeO_2-doped NiAl samples (Figure 6b) is proven by the presence of the reflections at 2θ = 38.2, 44.4, 64.6, 77.5 and 81.7°, which are ascribed to (111), (200), (220) (311) and (222) crystalline planes, respectively, character-istic of the face centered-cubic metal phase (ICDD-PDF file 00-004-0784), in addition to the LDH phase.

PXRD patterns of NiAl and CeO_2-doped NiAl LDHs were indexed in rhombohedral crystal symmetry with hexagonal cell setting of takovite-containing carbonate ions in the interlayer space. The analysis of the calculated unit cell parameters (a_{TK}, c_{TK}), total volume (V_{TK}) and mean crystallite size (L_{TK}) of the TKl phase discloses small differences as a function of the sample composition (Table 3).

It is obvious that precipitation of 1 wt.% CeO_2 over the NiAl LDH practically does not influence the parameters and crystallite size values of the sample. A very small increase in these is observed with an increase in the CeO_2 content to 5 wt.% CeO_2. The comparison of PXRD patterns in Figure 6b discloses that the deposition of gold over NiAl, 1CeNiAl and 3CeNiAl samples slightly decreases the crystallite size of the TKl phase in the corresponding Au-containing analogues. Additionally, all gold-supported NiAl samples possess Au^0 particles with average sizes in the range of $14-17$ nm. On the other hand, smaller Au^0 particles of 11 nm are registered for the Au/1CeNiAl solid, outlining high dispersion of the supported gold particles (Table 2).

The recorded specific PXRD patterns of TKl compounds in all studied solids as well as the similarity of the lattice parameters and the mean crystallite size values of the TKl phase in all the studied samples (Table 3) signify that the deposition of ceria and gold does not destroy the layered structure.

PXRD analyses performed before and after WGS tests show that under the influence of the reaction conditions, all the as-prepared samples undergo structure alterations, related to decomposition of the NiAl layered structure.

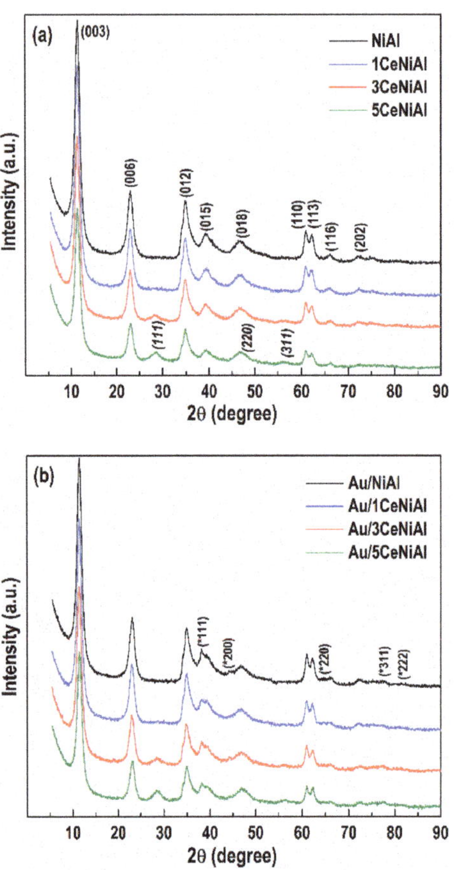

Figure 6. PXRD patterns of (**a**) Ce-modified NiAl samples and (**b**) Au-containing Ce-modified NiAl samples. The diffraction lines of CeO_2 phase are marked in italics, and those of gold phase are marked by asterisks.

Table 3. Structural characteristics of the as-synthesized samples.

Sample	a_{TK} (Å)	c_{TK} (Å)	V_{TK} (Å^3)	L_{TK} (nm)	L_{CeO2} (nm)	L_{Au} (nm)
NiAl	3.0333(6)	22.964(13)	182.98(13)	9.90	-	-
1CeNiAl	3.0334(6)	22.958(13)	182.98(13)	9.80	-	-
3CeNiAl	3.0336(6)	22.974(12)	183.07(12)	10.10	-	-
5CeNiAl	3.0338(6)	22.972(13)	183.11(13)	10.20	2.16	-
Au/NiAl	3.0343(7)	22.994(12)	183.34(12)	8.05	-	14.0
Au/1CeNiAl	3.0355(6)	23.026(11)	183.75(12)	9.10	1.90	11.0
Au/3CeNiAl	3.0339(6)	23.020(12)	183.49(13)	9.30	1.82	17.0
Au/5CeNiAl	3.0353(6)	23.017(13)	183.65(13)	10.10	2.18	16.0
* Takovite	3.0250	22.5950	179.06	-	-	-

* ICDD 00-015-0087.

PXRD study of the spent catalysts (Figure 7) reveals reflections at $2\theta = 37.2$, 43.3, 62.9, 75.4 and 79.4, indexed as (111), (200), (220), (311) and (222) crystalline planes of the cubic NiO phase (JCPDS file 00-047-1049), respectively. The solids decompose to NiO, better organized in NiAl-s with a crystallite size of 3.09 nm (Table 4). The observed broadening of the NiO diffraction lines is ascribed to the incorporation of Al^{3+} ions into the cubic framework of NiO, leading to lattice distortion [47].

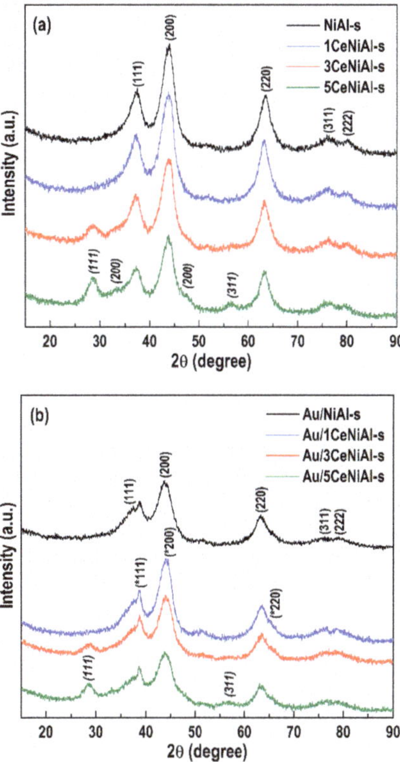

Figure 7. PXRD patterns of (**a**) spent Ce-doped NiAl samples and (**b**) Au-containing Ce-doped NiAl catalysts. The diffraction lines of CeO_2 phase are marked in italics, and asterisks mark those of gold phase.

A decrease in the crystallite size of NiO with an increase in CeO_2 from 1 to 5 wt.% was detected, outlining the role of ceria. This finding is more significant after gold deposition. The dimension of NiO phase for the gold-containing spent catalysts is in the range 2.2–2.6 nm, specifying a higher dispersion of the NiO phase in the presence of gold particles.

Table 4. Mean crystallite size of NiO, CeO_2 and Au in the spent samples.

Sample	L_{NiO} (nm)	L_{CeO2} (nm)	L_{Au} (nm)
NiAl-s	3.09	-	-
1CeNiAl-s	3.08	n.d.*	-
3CeNiAl-s	2.98	3.10	-
5CeNiAl-s	2.78	3.50	-
Au/NiAl-s	2.65	-	8.5
Au/1CeNiAl-s	2.63	n.d.	11.8
Au/3CeNiAl-s	2.41	4.95	5.8
Au/5CeNiAl-s	2.24	3.65	9.2

* n.d.—not detected

The comparison of gold crystallite sizes reveals that modification of Au/NiAl catalyst with 3 wt.% CeO_2 provides the highest dispersion of gold particles in spent Au/3CeNiAl-s catalyst (Table 4).

3.5. Catalyst Reduction Properties

TPR measurements were performed to evaluate the samples' reducibility and the effect of ceria and gold addition on oxygen mobility. Given that the experiments were carried out in as-prepared samples, the reduction process comprised thermal decomposition of takovite in the H_2 atmosphere followed by reduction of the obtained NiO (Figure 8). In accordance with the models proposed for decomposition and reduction of NiAl LDH compounds, [73], a well-resolved peak at 320 °C and a shoulder at 385 °C from the TPR profile of NiAl sample (Figure 8a) can be attributed to the reduction of easily reducible Ni^{2+} species from a NiO phase, which contains small amounts of Al^{3+} ions. The broad peak centered at 477 °C is associated with the reduction of the Ni^{2+} species, hardly bonded to Al^{3+} ions. This represents the reduction of a quasi-amorphous non-stoichiometric spinel-type phase, which decorates the surface of the NiO particles and/or acts as their support. Addition of ceria onto NiAl caused the appearance of new low-temperature features instead of the peak at 320 °C. The intensity and position of temperature maximum (T_{max}) of the peaks below 300 °C are closely related to the content of ceria. A broad peak at 250 °C was observed in the profile of the sample with 5 wt.% CeO_2, i.e., 5CeNiAl. A weaker peak at 275 °C was registered in the pattern of 3CeNiAl, while only a shoulder around 300 °C was visible for 1CeNiAl. All these peaks could be ascribed to the reduction of easily reducible Ni^{2+} species affected by ceria in close vicinity. Observed the profile of pure ceria (lab-prepared following the same procedure as in the case of ceria-modified NiAl), a large broad peak between 350 and 550 °C with T_{max} at 380 °C can be seen and attributed to the surface ceria oxygen reduction. Numerous papers have reported experimental evidence for the role of ceria in facilitating transition metal oxide reduction, but also the effect of metal oxides to enhance surface reduction of ceria. We suggest that these low-temperature components resulted from the combined contribution of Ni species and CeO_2 to the reducibility of the samples. Additionally, well-discernible maxima at about 350 and 380 °C were produced in the complex TPR profile with a maximum temperature in the interval from 477 (NiAl) to 487 °C (5CeNiAl). These components correspond to various types of Ni^{2+} species with different reducibility due to a complex interaction between the components of LDHs. Despite the appearance of weak low-temperature peaks, the shift to a higher temperature of reduction implies that ceria hampers reducibility, as reported recently by Swirk et al. [74]. This effect of ceria on reduction behavior could explain lower WGS activity of ceria-modified LDHs (Figure 3a).

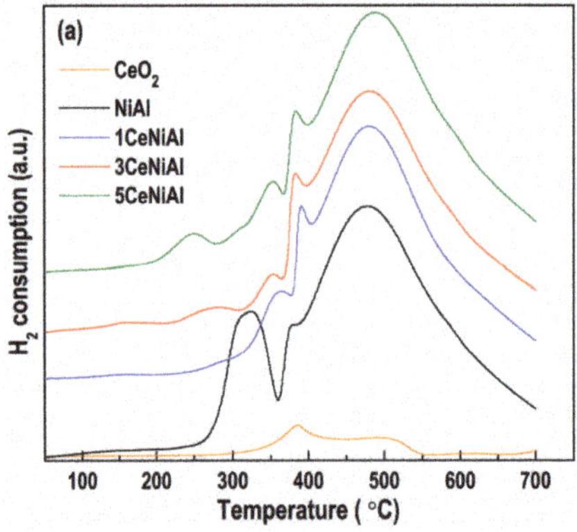

Figure 8. *Cont.*

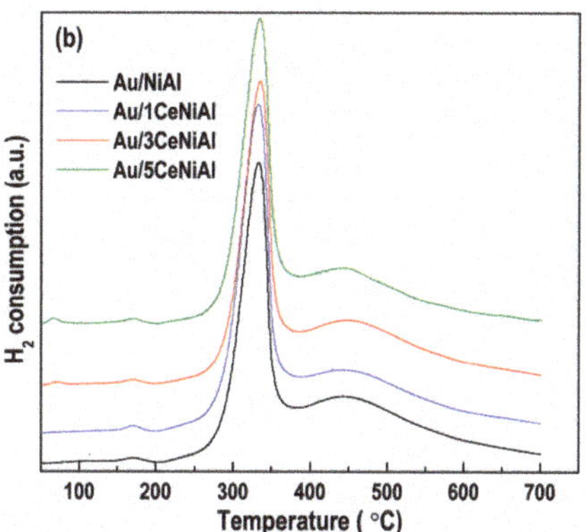

Figure 8. Temperature-programmed reduction (TPR) profiles of (**a**) Ce-doped NiAl samples and (**b**) Au-containing Ce-doped NiAl samples.

The presence of gold remarkably boosts the redox properties of all LDHs (Figure 8b). The broad high-temperature peaks were shifted significantly towards lower temperatures. The reduction profiles were dominated by narrowed peaks with very similar intensity and identical T_{max} = 335 °C, assigned to Ni^{2+} reduction, in agreement with the well-known ability of gold to improve reducibility of metal oxides by weakening the M–O bond [25,26]. Additionally, hydrogen dissociation occurs on small metallic gold particles, and the produced active hydrogen atoms can spill over onto the support and enhance the reduction. Very weak peaks at 65 (Au/5CeNiAl) and 71 °C (Au/3CeNiAl) correspond unequivocally to ceria surface layer reduction due to the dependence of their intensity and T_{max} position on the amount of ceria. The similarity in reduction behavior, respectively in oxygen mobility, correlates well with catalytic performance.

3.6. N_2 Physisorption Analysis

Knowledge of the textural characteristics of the as-prepared NiAl LDHs used as a carrier of the active species (Au) is necessary to understand the contributions of modifiers (Ce-species) and the active phase. However, this is insufficient to give an estimate of the magnitude of the impact of possible texture change on the properties and activity of the catalysts used in the WGS reaction. In addition, the influence of reaction conditions on the texture of the support (NiAl), ceria-modified support (CeNiAl) and catalyst (Au/CeNiAl) would remain hidden. Therefore, by measuring the physisorption of N_2 at −196 °C, a textural characterization was performed not only of freshly prepared samples of NiAl, 3CeNiAl, Au/NiAl and Au/3CeNiAl but also of their pairs used in catalytic tests. From all ceria-modified NiAl supports, 3CeNiAl was chosen for characterization because the corresponding gold-containing catalyst (Au/3CeNiAl) showed the best catalytic properties.

The N_2 isotherms of the freshly prepared and spent samples are shown in Figure 9, while the integral and differential distributions of mesopores are given in Figure 10, and all calculated textural parameters are given in Table 5.

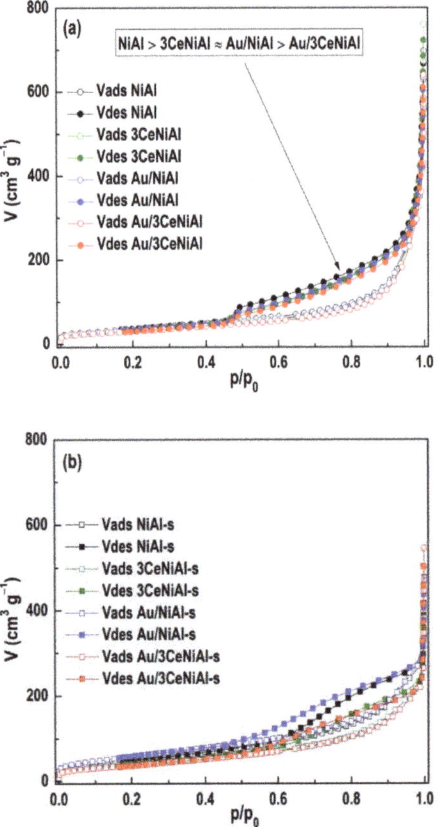

Figure 9. N_2 adsorption–desorption isotherms of (**a**) as-synthesized samples and (**b**) spent catalysts.

The shape of the isotherms of all four fresh synthesized materials (Figure 9a) is remarkably similar. The adsorption part of all isotherms has the same features: (i) a limited increase in the region of low relative pressures; (ii) an increase slightly concave toward the x-axis with p/p_0 increase, followed by an almost linear increase up to $p/p_0 \approx 0.8$; (iii) upraised convex towards the p/p_0 axis; and (iv) a significant increase in the highest values of relative pressure. The desorption part, from the higher relative pressure side, is almost parallel to the adsorption branch, while the hysteresis loop, obviously present at $p/p_0 \leq 0.95$, and existing from highest relative pressure in all isotherms, ends at $p/p_0 \approx 0.42$. All the above characteristics classify the isotherms of all four materials as type II with hysteresis loop H3, according to the IUPAC nomenclature [65], or as type IIb, as proposed by Rouquerol et al. [75]. This type of isotherm with a H3 hysteresis loop is typical of many materials, e.g., clays, pigments and cements, with aggregates of plate-like particles, which possess non-rigid slit-shaped pores.

The fact is that individually adding 3 wt.% CeO_2 or Au causes a slight reduction in the value of all textural parameters relative to NiAl (Table 5). This change seems to be independent of the species being added. For example, desorption branches of isotherms of Au/NiAl and 3CeNiAl are literally identical, while the decrease in the specific surface area of 20 m^2 g^{-1} for the Au/3CeNiAl catalyst compared to unmodified NiAl is almost the ideal sum of *SSA* losses with a single addition of CeO_2 or Au (12 m^2 g^{-1} + 9 m^2 g^{-1}). However, the similarity between the shapes of all four isotherms in Figure 9 indicates that the addition of only 3wt.% CeO_2 or gold, and even their combinations, in a total amount of 6 wt.% does not substantially alter the pore system present in unmodified NiAl LDH.

For the mesoporous region, shown in Figure 10, this is undoubtedly true for the 50–4 nm segment. The overlap of all four curves of the mesopore size distribution (PSD) is almost complete. Positions of maxima on these curves, at 3.8 and 5.2 nm, are also identical.

Table 5. Textural parameters.

Sample	SSA (m^2 g^{-1})	V_{mic} (m^3 g^{-1})	V_{meso} (m^3 g^{-1})	V_{tot} (m^3 g^{-1})
NiAl	138	0.048	0.496	0.575
NiAl-s	191	0.058	0.409	0.417
3CeNiAl	126	0.042	0.474	0.569
3CeNiAl-s	147	0.053	0.327	0.367
Au/NiAl	129	0.043	0.456	0.596
Au/NiAl-s	221	0.072	0.418	0.466
Au/3CeNiAl	118	0.041	0.446	0.534
Au/3CeNiAl-s	143	0.047	0.333	0.350

The difference in peak intensity is usually a consequence of the different pore volume of that diameter. However, the peak of the smaller diameter on the PSD curve does not have to originate from the pores that actually exist, but from the so-called "tensile strength effect" [76], which corresponds to the hysteresis curve closing position at about 0.42 p/p_0. Thus, the difference in the peak intensities at 3.8 nm may also be due to the existence of different pore contents in the segment below that diameter value.

Exposure of freshly prepared samples to reaction conditions leads to recognizable alteration in the shape of the isotherms of all four materials (Figure 9b), as well as to a significant change in their texture parameters (Table 5), but not all in the same way. First, although the deviation of the desorption branch from the almost parallel tracking of the adsorption branch is evident in all isotherms of the used samples, this deviation is unevenly expressed. A kind of a hump on the desorption branch in the region of relative pressure 0.9 to 0.6 is evident and most noticeable for spent Au/NiAl-s and NiAl-s samples, and barely visible for ceria modified support (3CeNiAl-s). Further, for all spent samples compared to freshly prepared ones, two of the textural parameters, SSA and V_{mic}, increase, while the other two, V_{meso} and V_{tot}, decrease. At the same time, the intensities of change are not nearly the same for samples that have the same tendency. For example, an increase in SSA reaches over 70% for the Au/ and only 17% for the 3CeNiAl-s support. Additionally, a similar trend of non-equal increase in micropore volume can be observed.

Comparing only the results of the final mesoporous volume of all four materials, it is obvious that they are grouped in two pairs: the first made of Au/NiAl and its support NiAl (0.418 vs. 0.409 cm^3 g^{-1}), and the second made of Au/3CeNiAl and its ceria-modified support (0.333 vs. 0.327 cm^3 g^{-1}). This type of grouping is also recognized when BET surface area values are taken into account. Although the addition of gold to NiAl and exposure to WGS reaction conditions increased the SSA value by 30 m^2 g^{-1} compared to NiAl-s, while its addition to 3CeNiAl caused almost no changes in the SSA compared to 3CeNiAl-s (reduction by 4 m^2 g^{-1}), the differences between these pairs are obvious.

The size distribution of the mesopores of materials used in the WGS reaction (Figure 10b) shows the existence of two peaks for all materials, just as for freshly prepared samples, although with a significantly changed ratio of peak intensity, namely, the peak corresponding to the smaller diameter decreased significantly and had a smaller contribution to the overall PSD profile in all samples. It is interesting to note that the positions of the maximum values corresponding to the pores of larger diameter for the used unmodified and cerium-modified support were moved to higher values by 0.8 and 2 nm, respectively.

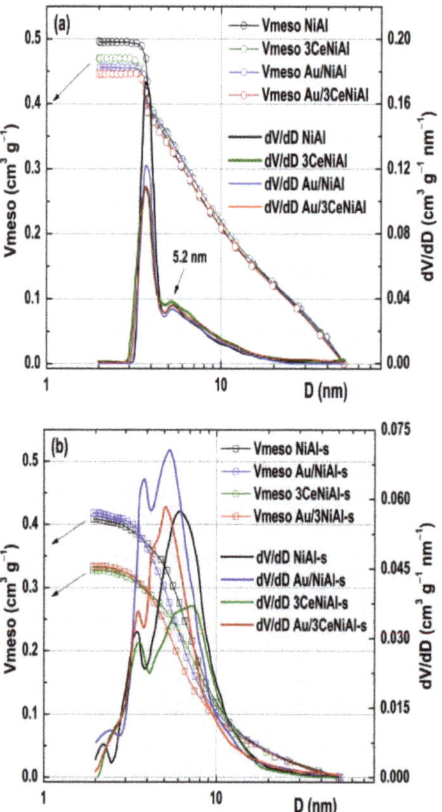

Figure 10. Mesopore size distribution (PSD) curves of (**a**) as-synthesized samples and (**b**) of spent catalysts.

The applied reaction conditions, mainly the temperatures, but also gaseous reactants and products lead to the decomposition of materials based on NiAl LDH, contributing to the change of their textural characteristics, regardless of whether they are supports or catalysts. The values of all textural parameters (SSA, V_{mic} and V_{meso}) are smaller in Au/3CeNiAl compared to unmodified Au/NiAl, regardless of whether pairs of freshly prepared samples or a sample used in catalytic tests are analyzed.

Therefore, based on the determined values of texture characteristics, it can be said with certainty that the improved activity of Au/3CeNiAl compared to Au/NiAl cannot be caused by a change in texture properties due to modification of NiAl support by ceria.

3.7. X-ray Photoelectron Spectroscopy (XPS)

The oxidation-reduction characteristic of the WGS reaction requires information to be obtained for the components' oxidation states on the catalysts surface of the most active Au/3CeNiAl catalysts compared to NiAl, 3CeNiAl and Au/NiAl samples in their as-prepared and postreaction state (spent state). The correlation between the surface species state and WGS activity was searched for. Attention was also given to another activity-determining factor, such as relative dispersion of Au and Ce over NiAl-LDH, and the Ni/Al ratio before and after WGS reaction.

The binding energy (BE) values of the as-prepared samples are summarized in Table 6. The main Ni2p$_{3/2}$ peak of the bare NiAl is characterized by the binding energy (BE) of 855.4 eV and the shake-up peak, 6 eV apart from the main Ni 2p$_{3/2}$ peak, corresponding to the Ni^{2+} state. Bearing in mind the nature of NiAl LDH, namely, the layered NiAl hydroxide sheet structure, the BE value is attributed to $Ni(OH)_2$ (855.3–856.6 eV) [77–80].

Table 6. Binding energies (eV) of the main elements' peaks and the oxygen species contribution in as-synthesized samples.

Sample	Ni 2p$_{3/2}$	Al 2p	O 1s Position (eV) and Contribution (%) *		Au 4f$_{7/2}$	Ce 3d$_{5/2}$
			O$_I$	O$_{II}$		
NiAl	855.4	74.2	529.3(11)	531.1(89)		
Au/NiAl	855.7	74.5	529.7(10)	531.7(90)	84.7	
3CeNiAl	855.3	74.2	529.3(16)	531.3(84)		882.2
Au/3CeNiAl	855.7	74.5	529.4(12)	531.6(88)	84.5	881.8

* The values in parentheses refer to the atomic percentage of the oxygen species.

The modification of NiAl material with 3 wt.% CeO_2 does not cause changes in the layered structure of 3CeNiAl, as is documented by XRD data. In contrast, the gold deposition in both Au/NiAl and Au/3CeNiAl shifts Ni 2p$_{3/2}$ peaks toward higher BEs by 0.3 eV, namely, 855.7 eV (Table 6 and Figure 11) more than the ± 0.15 eV experimental accuracy. A similar shift in the gold-containing samples is observed in the Al 2p binding energy (Table 6), assigned to Al^{3+} species. The shifts of Ni 2p and Al 2p binding energies reveal a stronger bond in the layered hydroxide structure. This statement is in agreement with the small lowering of the TKl phase mean crystallite size (L_{TK}) disclosed by XRD (Table 3). Through the curve fitting of the Ni 2p region in the spent catalysts, a lowering of BE for Ni^{2+} oxidation state compared to as-synthesized samples (Figure 11 and Table 7) is detected. Moreover, a second component at a higher binding energy of 856.7 ± 0.3 eV in all catalysts is identified, which is attributed to Ni^{3+}, likely as Ni−OOH species [44,78]. The lower Ni^{2+} BEs reveal changes in the NiAl structure due to WGS reaction, namely, destroying of the NiAl layered hydroxide structure and interaction between Al and Ni in mixed NiAlO oxide structures [81]. This was also suggested by PXRD analysis of spent catalysts. The smaller shift (0.2 eV) observed in the Ni2p$_{3/2}$ of the bare NiAl catalyst as compared to the larger shifts observed for the CeNiAl, the AuNiAl and the AuCeNiAl (0.5–0.6 eV) implies the significant impact of CeO_2 and gold promotion on Ni^{2+} state after the reaction is run.

In addition, a reciprocal effect of gold and ceria on the corresponding binding energy values is observed. Indeed, the main Au 4f$_{7/2}$ peak at 84.5 eV of the as-prepared Au/3CeNiAl sample discloses a slight shift as compared to 84.7 eV of the Au/NiAl. Additionally, the main Ce 3d$_{5/2}$ peak of the Au/3CeNiAl is detected at 881.8 eV, whereas it is detected at 882.1 eV in 3CeNiAl. It is known that the binding energy at 84.5 eV ± 0.5 eV of Au 4f$_{7/2}$ is typical of metallic gold, and the Ce 3d$_{5/2}$ BE value corresponds to the Ce^{4+} oxidation state [36]. Obviously, the simultaneous Au and Ce presence makes their interactions with NiAl weaker, thus suggesting their greater reactivity in the Au/3CeNiAl catalyst during redox WGS reaction. The BEs of the Au 4f$_{7/2}$ peak in the spent Au/NiAl and Au/3CeNiAl catalysts are the same. The variation of the Au 4f$_{7/2}$ binding energy of the spent catalysts with respect to the as-synthesized samples (Tables 6 and 7) is within the experimental error, confirming that the gold oxidation state is preserved after WGS reaction. The surface concentration of gold on all of the as-prepared samples is the same (0.4 at.%), and it diminishes in spent catalysts (0.2 at.%).

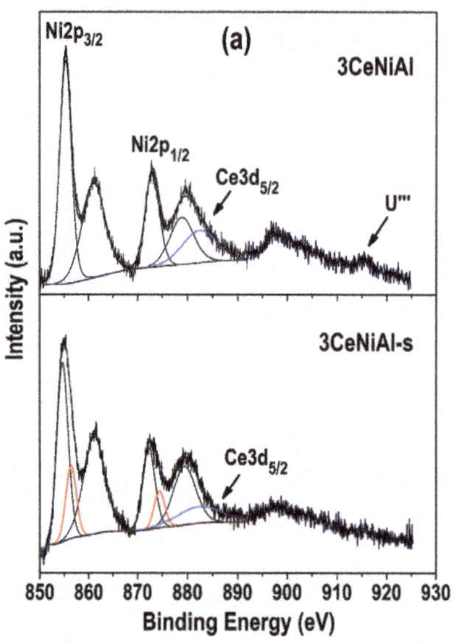

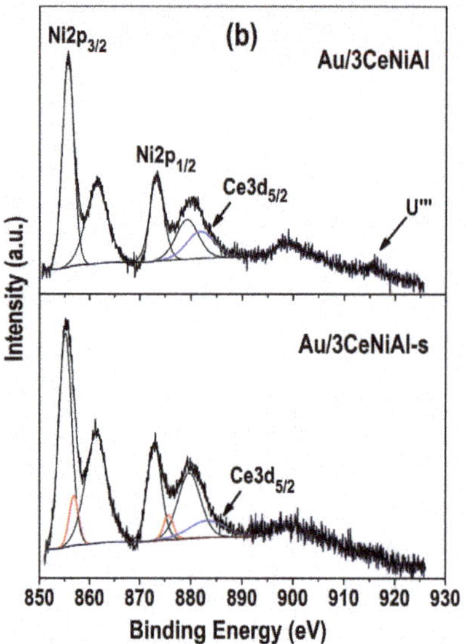

Figure 11. Ni 2p- and Ce 3d-photoelectron regions of the studied samples: (**a**) 3CeNiAl as-prepared and spent; (**b**) Au/3CeNiAl as-prepared and spent. The Ni^{3+} oxidation state is colored in red. The blue contour line outlines the Ce 3d signal.

Table 7. X-ray photoelectron spectra (XPS) analysis data of the spent catalysts.

Catalyst	Ni 2p$_{3/2}$ Position (eV) and Contribution (%)		Ni^{3+}/Ni^{2+} Ratio	O 1s Position (eV) and Contribution (%)		O$_{II}$/(O$_{II}$+O$_I$)	Au 4f$_{7/2}$	Ce3d$_{5/2}$
	Ni^{2+}	Ni^{3+}		O$_I$	O$_{II}$			
NiAl-s	855.2(90)	856.8(10)	0.11	530.5(58)	532.1(42)	0.42		
Au/NiAl-s	855.2(93)	856.9(7)	0.08	530.3(51)	531.8(49)	0.49	84.4	
3CeNiAl-s	854.7(84)	856.4(16)	0.19	530.1(68)	531.6(32)	0.32		881.9
Au/3CeNiAl-s	855.1(92)	856.9(8)	0.09	529.1(58)	531.7(42)	0.42	84.4	881.9

The careful inspection of the Ce 3d photoelectron regions indicates (Figure 11) that in both 3CeNiAl and Au/3CeNiAl samples in their as-prepared state, the detected highest binding energy component is situated at 915.3 and 915.7 eV, respectively. This component is signed as U′′′, typical of the Ce^{4+} oxidation state. The poor quality of the Ce 3d spectra due to the overlap with the Ni 2p and to the low CeO$_2$ content (3 wt.%) did not allow a reliable fitting. However, the position of the main Ce 3d$_{5/2}$ component and a rough estimate of the Ce at.% = 0.5 from the entire Ce 3d region partially overlapping with the Ni 2p$_{3/2}$ were obtained. It is well known that the intensity of U′′′ satellite gives information about Ce^{3+} contribution, by relating it to the total Ce 3d area [82,83]. Unfortunately, in the present spectra, variations in the intensity of the U′′′ peak are hardly detectable; therefore, no conclusive statement about the partial reduction of Ce^{4+} to Ce^{3+} induced by gold can be made [28]. However, concerning the as-prepared samples, the Ce 3d$_{5/2}$ BE shifting of 0.4 eV observed in Au/3CeNiAl as compared to the Au-free sample could be indicative of an Au-Ce^{3+} bond formed by the strain effect as claimed by a theoretical study based on density functional theory (DFT) calculations for an Au/CeO$_2$ system [84]. The data in Table 7 show that the Ce 3d$_{5/2}$ position is the same in both CeO$_2$-modified NiAl-LDHs after catalytic tests (3CeNiAl-s and Au/3CeNiAl-s). The visible change is the disappearance of the U′′′ peaks, confirming that CeO$_2$ undergoes reduction to Ce^{3+} ions during the WGS reaction.

The associative reaction mechanism of WGS over Au/NiAl catalysts was proven in our previous papers, involving redox Ni^{2+} ↔ Ni^{3+} transition on the catalyst surface as well as adsorption and activation of the CO molecule on Au particles [44,47]. The availability of nickel in both Ni^{2+} and Ni^{3+} oxidation states on the catalyst surface contributes to the high activity of the Au/NiAl catalyst in the studied temperature range. The role of reversible redox Ni^{2+} ↔ Ni^{3+} transition implies evaluation of the Ni^{3+}/Ni^{2+} ratio, which is presented in Table 7. The values clearly disclose that ceria addition increases the Ni^{3+}/Ni^{2+} ratio 1.72 times, while gold presence decreases it. Evidently, the gold keeps the catalyst surface more reduced, which is in correlation with the higher activities of Au/NiAl and Au/3CeNiAl compared to NiAl. The oxidized surface of the 3CeNiAl samples can be related to the lowest WGS activity of this catalyst.

The review of the O 1s photoelectron regions after the curve fitting (Figure 12) shows that the spectra of the as-prepared catalysts consist of two components, namely, less intense low energy peaks centered between 529.3 and 529.7 eV and more intense higher energy peaks between 531.1 and 531.7 eV (Table 7). The low energy peaks are attributed to lattice oxygen named as O$_I$ associated with the NiAl layered structure [81,85] and oxygen in the CeO$_2$ lattice [86–88] in 3CeNiAl and Au/3CeNiAl samples. The high energy peaks recognized as surface adsorbed oxygen are named O$_{II}$. The intensity of these peaks is significant compared to O$_I$ peaks, because they belong to chemisorbed oxygen in CO$_3^{2-}$/OH groups from the NiAl hydroxide layer and intercalated water molecules from interlayer space [81,89] and hydroxyl groups of pure CeO$_2$ [86–88,90]. In addition, the shifting of the O$_{II}$ component toward higher BEs in the Au-containing samples (Au/NiAl and Au/3CeNiAl) suggests a strong interaction of metallic gold with oxygen, producing a decrease in the electronic charge of these oxygen species. A similar effect was observed in relation to the Ni 2p and Al 2p shifting mentioned above (Table 6). These data confirm the formation of a stronger bond in the layered hydroxide structure under the influence of gold. Moreover, the O_{II} contribution predominates over O_I and could be explained by the preserved NiAl layered structure shown by XRD analysis. The O 1s spectrum undergoes significant changes in the spent catalysts due to the redox conditions and reaction temperature. The intensity of all O_{II} peaks decreases, and O_I peaks become more intense (Figure 12). The BEs of lattice oxygen peaks (O_I) are moved 0.6–1.2 eV to higher BEs (Table 7). In this case, the lattice oxygen originates from the NiAlO structure, unreduced CeO_2 and partially reduced cerium in CeO_{2-x}, namely, 529.8–530 eV [86,87], oxygen in Ni–OOH [44,91] and last but not the least, oxygen vacancies (O_x^-) in the matrix of metal oxides, which are usually registered in the range of 529.9−531.1 eV [92–94].

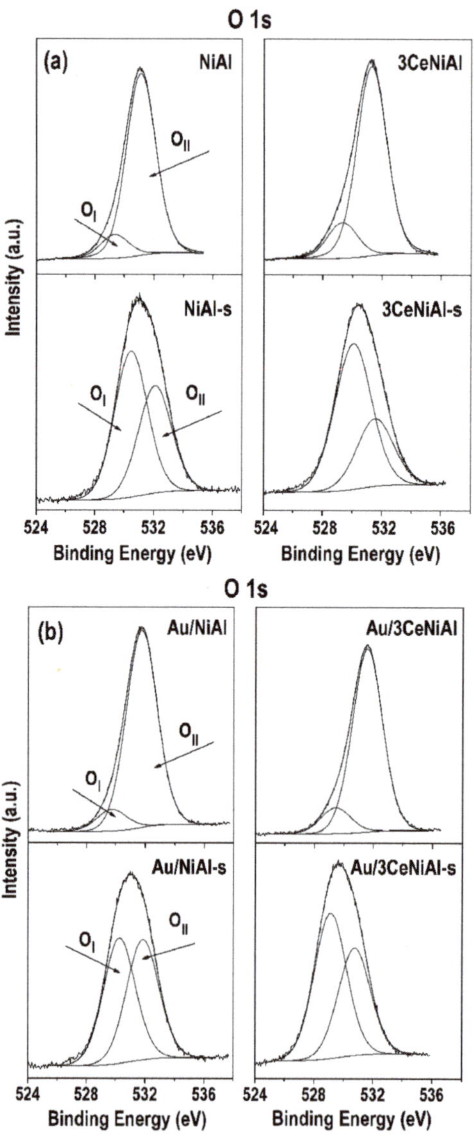

Figure 12. O 1s-photoelectron regions of the as-synthesized and spent samples: (**a**) NiAl and 3CeNiAl; (**b**) Au/NiAl and Au/3CeNiAl.

Genty et al. stated [89] that the mobility of surface oxygen species plays an important role in the catalytic activity in oxidation reactions. The oxygen vacancies are important for the adsorption of oxygen species [85] from H_2O vapor in the case of WGS reaction. In this connection, the number of oxygen vacancies in the spent catalysts are calculated by the integrated area ratios of $O_{II}/(O_{II} + O_I)$ (Table 7) by analogy to Lu et al. [85]. The data show that Au addition increases the oxygen vacancies, implying the presence of more active vacant oxygen on the catalyst surface of the Au/NiAl catalyst compared to NiAl as well as in the Au/3CeNiAl catalyst compared to 3CeNiAl. This fact correlates with higher activities of the Au-containing catalysts. The role of oxygen vacancies in facilitating the dissociation of water, considered the most difficult step in the WGS reaction, is highlighted in the literature [95–97]. It would also be relevant to point out that 3CeNiAl has the smallest number of oxygen vacancies, and its WGS activity is worse than that of unmodified NiAl. The relative dispersion of Au and Ce over NiAl-LDH and the Ni/Al ratio before and after WGS reaction was evaluated from XPS-derived atomic concentration (Table 8).

Table 8. Relative dispersion of Au and Ce and Ni/Al ratio before and after WGS reaction.

Catalyst	Au/(Ni + Al)	Au/(Ce + Ni + Al)	Ce/(Ni + Al)	Ni/Al
NiAl				0.49
NiAl-s				0.51
Au/NiAl	0.0088			0.47
Au/NiAl-s	0.0040			0.54
3CeNiAl			0.0140	0.47
3CeNiAl-s			0.0077	0.49
Au/3CeNiAl		0.0089	0.0113	0.45
Au/3CeNiAl-s		0.0041	0.0103	0.50

The gold amount is obviously the same in both as-synthesized and spent samples; however, the atomic ratio values decrease by more than half on the surface of the spent catalysts. The low values cannot be connected with poorer dispersion due to Au^0 particle agglomeration; this is due to the XRD data (Tables 3 and 4) displaying that after WGS reaction, the average gold particle size decreases considerably by 1.65 times in Au/NiAl-s and 2.93 times in Au/3CeNiAl-s. So, the reason for this observation can be attributed to the subsurface sinking of the gold, which is not completely visible for the XPS technique.

The ceria relative dispersion also diminishes in the spent CeO_2-modified catalysts. The diminution is stronger in 3CeNiAl-s; in fact, the Au presence makes this negligible, and only 8.8% in the most active Au/3CeNiAl-s. The estimation of the Ni/Al ratio indicates that the bulk chemical composition in as-synthesized samples, Ni/Al = 2.5, is not the same on the surface. The reaction conditions provoke Ni enrichment of the surface in all spent catalysts. This is explained by the reconstruction of the NiAl layered structure, leading to its destruction and interaction between Ni and Al forming a mixed NiAlO structure.

4. Conclusions

Studying the role of the CeO_2 dopant in the phase compositions, structural, textural and surface properties, reduction behavior and WGS activity of gold-containing NiAl LDHs allows the following conclusions to be made:

(i) The developed innovative approach for modification by ceria allows us to preserve the NiAl layered structure and to obtain a CeO_2 phase with a good crystallinity in a relatively short time by a one-pot method, thus avoiding the calcination treatment, which simplifies the catalyst preparation procedure.

(ii) The modification of NiAl LDHs with CeO_2 neither improves reducibility nor enhances the WGS efficiency; however, the simultaneous presence of gold and ceria has a beneficial effect.

(iii) It can be deduced that hydrogen production via WGS reaction is affected by the amount of ceria in the Au/NiAl catalyst.

(iv) The addition of 3 wt.% CeO_2 to the Au/NiAl catalyst provides the highest dispersion of gold particles in the spent catalyst (Au/3CeNiAl-s) and contributes to a good WGS performance—highest activity and significant stability.

Author Contributions: Conceptualization, M.G., D.N. and T.T.; methodology, M.G. and D.N.; synthesis, M.G., D.N. and T.T.; investigation: catalytic activity, I.I. and V.I.; textural measurements, J.K.; XPS characterization, A.M.V.; PXRD characterization, D.C. and M.C.; TPR measurements, K.T.; writing—original draft preparation, M.G., D.N., J.K. and T.T.; writing—review and editing, M.G., D.N., J.K., A.M.V. and T.T. All authors have read and agreed to the published version of the manuscript.

Funding: This research was funded by the Bulgarian Ministry of Education and Science under the National Research Programme "E+: Low Carbon Energy for the Transport and Households", grant agreement D01-214/2018.

Data Availability Statement: The data presented in this study are available in the article.

Acknowledgments: A joint Italian–Bulgarian research project between CNR and BAS is acknowledged.

Conflicts of Interest: The authors declare no conflict of interest.

References

1. Baykara, S.Z. Hydrogen: A brief overview on its sources, production and environmental impact. *Int. J. Hydrog. Energy* **2018**, *43*, 10605–10614. [CrossRef]
2. Sazali, N. Emerging technologies by hydrogen: A review. *Int. J. Hydrog. Energy* **2020**, *45*, 18753–18771. [CrossRef]
3. Ratnasamy, C.; Wagner, J.P. Water gas shift catalysis. *Catal. Rev. Sci. Eng.* **2009**, *51*, 325–340. [CrossRef]
4. LeValley, T.L.; Richard, A.R.; Fan, M. The progress in water gas shift and steam reforming hydrogen production technologies—A review. *Int. J. Hydrog. Energy* **2014**, *39*, 16983–17000. [CrossRef]
5. Reddy, G.K.; Smirniotis, P.G. *Water Gas Shift Reaction: Research Developments and Applications*; Elsevier: Amsterdam, The Netherlands, 2015; p. 269.
6. Chen, W.-H.; Chen, C.-Y. Water gas shift reaction for hydrogen production and carbon dioxide capture: A review. *Appl. Energy* **2020**, *258*, 114078. [CrossRef]
7. Ebrahimi, P.; Kumar, A.; Khraisheh, M. A review of recent advances in water-gas shift catalysis for hydrogen production. *Emergent Mater.* **2020**, *3*, 881–917. [CrossRef]
8. Pal, D.B.; Chand, R.; Upadhyay, S.N.; Mishra, P.K. Performance of water gas shift reaction catalysts: A review. *Renew. Sust. Energy Rev.* **2018**, *93*, 549–556. [CrossRef]
9. Cavalcanti, F.; Schmal, M.; Giudici, R.; Alves, R. A catalyst selection method for hydrogen production through Water-Gas Shift Reaction using artificial neural networks. *J. Environ. Manag.* **2019**, *237*, 585–594. [CrossRef]
10. Burch, R. Gold catalysts for pure hydrogen production in the water–gas shift reaction: Activity, structure and reaction mechanism. *Phys. Chem. Chem. Phys.* **2006**, *8*, 5483–5500. [CrossRef]
11. Tao, F.; Ma, Z. Water–gas shift on gold catalysts: Catalyst systems and fundamental studies. *Phys. Chem. Chem. Phys.* **2013**, *15*, 15260–15270. [CrossRef]
12. Reina, T.R.; González, M.C.; Palma, S.; Ivanova, S.; Odriozola, J.A. Twenty years of golden future in the water-gas shift reaction. In *Heterogeneous Gold Catalysts and Catalysis*; RSC Catalysis Series; Ma, Z., Dai, S., Eds.; RSC: Cambridge, UK, 2014; Volume 18, pp. 111–139, Chapter 5.
13. Odabasi, C.; Günay, M.E.; Yildirim, R. Knowledge extraction for water gas shift reaction over noble metal catalysts from publications in the literature between 2002 and 2012. *Int. J. Hydrog. Energy* **2014**, *39*, 5733–5746. [CrossRef]
14. Carter, J.H.; Hutchings, G.J. Recent advances in the gold-catalyzed low-temperature water-gas shift reaction. *Catalysts* **2018**, *8*, 627. [CrossRef]
15. Tabakova, T. Recent advances in design of gold-based catalysts for H_2 clean-up reactions. Review article. *Front. Chem.* **2019**, *7*, 517. [CrossRef] [PubMed]
16. Sandoval, A.; Gomez-Cortes, A.; Zanella, R.; Diaz, G.; Saniger, J.M. Gold nanoparticles: Support effects for the WGS reaction. *J. Mol. Catal. Chem.* **2007**, *278*, 200–208. [CrossRef]
17. Liu, X.Y.; Wang, A.; Zhang, T.; Mou, C.-Y. Catalysis by gold: New insights into the support effect. Review. *Nano Today* **2013**, *8*, 403–416. [CrossRef]
18. Yang, M.; Flytzani-Stephanopoulos, M. Design of single-atom metal catalysts on various supports for the low temperature water-gas shift reaction. *Catal. Today* **2017**, *298*, 216–225. [CrossRef]
19. Trovarelli, A. Catalytic Properties of Ceria and CeO_2-Containing Materials. *Catal. Rev. Sci. Eng.* **1996**, *38*, 439–520. [CrossRef]
20. Trovarelli, A. Structural properties and nonstoichiometric behavior of CeO_2. In *Catalysis by Ceria and Related Materials*; Trovarelli, A., Ed.; Imperial College Press: London, UK, 2002; pp. 15–50.

21. Polychronopoulou, K.; Kalamaras, C.M.; Efstathiou, A.M. Ceria-Based Materials for Hydrogen Production Via Hydrocarbon Steam Reforming and Water-Gas Shift Reactions. *Recent Pat. Mater. Sci.* **2011**, *4*, 122–145. [CrossRef]

22. Sun, C.; Li, H.; Chen, L. Nanostructured ceria-based materials: Synthesis, properties, and applications. *Energy Environ. Sci.* **2012**, *5*, 8475–8505. [CrossRef]

23. Montini, T.; Melchionna, M.; Monai, M.; Fornasiero, P. Fundamentals and catalytic applications of CeO_2-based materials. *Chem. Rev.* **2016**, *116*, 5987–6041. [CrossRef]

24. Schmitt, R.; Nenning, A.; Kraynis, O.; Korobko, R.; Frenkel, A.I.; Lubomirsky, I.; Haile, S.M.; Rupp, J.L.M. A review of defect structure and chemistry in ceria and its solid solutions. *Chem. Soc. Rev.* **2020**, *49*, 554–592. [CrossRef] [PubMed]

25. Fu, Q.; Weber, A.; Flytzani-Stephanopoulos, M. Nanostructured $Au-CeO_2$ catalysts for low-temperature water-gas shift. *Catal. Lett.* **2001**, *77*, 87–95. [CrossRef]

26. Andreeva, D.; Idakiev, V.; Tabakova, T.; Ilieva, L.; Falaras, P.; Bourlinos, A.; Travlos, A. Low-temperature water–gas shift reaction over Au/CeO_2 catalysts. *Catal. Today* **2002**, *72*, 51–57. [CrossRef]

27. Luengnaruemitchai, A.; Osuwan, S.; Gulari, E. Comparative studies of low temperature water–gas shift reaction over Pt/CeO_2, Au/CeO_2, and Au/Fe_2O_3 catalysts. *Catal. Commun.* **2003**, *4*, 215–221. [CrossRef]

28. Tabakova, T.; Boccuzzi, F.; Manzoli, M.; Sobczak, J.W.; Idakiev, V.; Andreeva, D. A comparative study of nanosized IB/ceria catalysts for low-temperature water-gas shift reaction. *Appl. Catal. A Gen.* **2006**, *298*, 127–143. [CrossRef]

29. Karpenko, A.; Leppelt, R.; Plzak, V.; Cai, J.; Chuvilin, A.; Schumacher, B.; Kaiser, U.; Behm, R. Influence of the catalyst surface area on the activity and stability of Au/CeO_2 catalysts for the low- temperature water gas shift reaction. *Top. Catal.* **2007**, *44*, 183–197. [CrossRef]

30. Abdel-Mageed, A.M.; Kučerova, G.; Bansmann, J.; Behm, R.J. Active Au species during the low-temperature water gas shift reaction on Au/CeO_2: A time-resolved operando XAS and DRIFTS study. *ACS Catal.* **2017**, *7*, 6471–6484. [CrossRef]

31. Lenite, B.A.; Galletti, C.; Specchia, S. Studies on Au catalysts for water gas shift reaction. *Int. J. Hydrog. Energy* **2011**, *36*, 7750–7758. [CrossRef]

32. Andreeva, D.; Tabakova, T.; Ilieva, L. Ceria-based gold catalysts: Synthesis, properties and catalytic performance for the WGS and PROX processes. In *Catalysis by Ceria and Related Materials*, 2nd ed.; Trovarelli, A., Fornasiero, P., Eds.; Imperial College Press: London, UK, 2013; p. 497.

33. Yi, N.; Flytzani-Stephanopoulos, M. Gold/ceria: The making of a robust catalyst for fuel processing and hydrogen production. In *Catalysis by Materials with Well-Defined Structures*; Wu, Z., Overbury, S.H., Eds.; Elsevier: Amsterdam, The Netherlands, 2015; pp. 133–158.

34. Reina, T.R.; Ivanova, S.; Centeno, M.A.; Odriozola, J.A. The role of Au, Cu & CeO_2 and their interactions for an enhanced WGS performance. *Appl. Catal. B Environ.* **2016**, *187*, 98–107.

35. Fu, X.-P.; Guo, L.-W.; Wang, W.-W.; Ma, C.; Jia, C.-J.; Wu, K.; Si, R.; Sun, L.-D.; Yan, C.-H. Direct identification of active surface species for the water–gas shift reaction on a gold–ceria catalyst. *J. Am. Chem. Soc.* **2019**, *41*, 4613–4623. [CrossRef]

36. Tabakova, T.; Ivanov, I.; Zanella, R.; Karakirova, Y.; Sobczak, J.W.; Lisowski, W.; Kaszkur, Z.; Ilieva, L. Unraveling the effect of alumina-supported Y-doped ceria composition and method of preparation on the WGS activity of gold catalysts. *Int. J. Hydrogen Energy* **2020**, *45*, 26238–26253. [CrossRef]

37. Santos, J.L.; Reina, T.R.; Ivanova, S.; Centeno, M.A.; Odriozola, J.A. Gold promoted $Cu/ZnO/Al_2O_3$ catalysts prepared from hydrotalcite precursors: Advanced materials for the WGS reaction. *Appl. Catal. B Environ.* **2017**, *201*, 310–317. [CrossRef]

38. Santos, J.L.; Reina, T.R.; Ivanov, I.; Penkova, A.; Ivanova, S.; Tabakova, T.; Centeno, M.A.; Idakiev, V.; Odriozola, J.A. Multicomponent $Au/Cu-ZnO-Al_2O_3$ catalysts: Robust materials for clean hydrogen production. *Appl. Catal. A Gen.* **2018**, *558*, 91–98. [CrossRef]

39. Meng, Y.; Chen, Y.; Zhou, X.; Pan, G.; Xia, S. Experimental and theoretical investigations into the activity and mechanism of the water-gas shift reaction catalyzed by Au nanoparticles supported on Zn-Al/Cr/Fe layered double hydroxides. *Int. J. Hydrogen Energy* **2020**, *45*, 464–476. [CrossRef]

40. Xia, S.; Fang, L.; Meng, Y.; Zhang, X.; Zhang, L.; Yang, C.; Ni, Z. Water-gas shift reaction catalyzed by layered double hydroxides supported Au-Ni/Cu/Pt bimetallic alloys. *Appl. Catal. B Environ.* **2020**, *272*, 118949. [CrossRef]

41. Fuentes, E.M.; Faro, A.C., Jr.; Silva, T.F.; Assaf, J.M.; Rangel, M.C. A comparison between copper and nickel-based catalysts obtained from hydrotalcite-like precursors for WGSR. *Catal. Today* **2011**, *171*, 290–296. [CrossRef]

42. Gabrovska, M.; Idakiev, V.; Tenchev, K.; Nikolova, D.; Edreva-Kardjieva, R.; Crisan, D. Catalytic performance of Ni-Al layered double hydroxides in CO purification processes. *Russ. J. Phys. Chem. A* **2013**, *87*, 2152–2159. [CrossRef]

43. Fuentes, E.M.; Aires, F.J.C.S.; Prakash, S.; Faro, A.C., Jr.; Silva, T.F.; Assaf, J.M.; Rangel, M.C. The effect of metal content on nickel-based catalysts obtained from hydrotalcites for WGSR in one step. *Int. J. Hydrog. Energy* **2014**, *39*, 815–828. [CrossRef]

44. Gabrovska, M.; Tabakova, T.; Ivanov, I.; Kovacheva, D. Water-gas shift reaction over gold deposited on NiAl layered double hydroxides. *React. Kinet. Mech. Cat.* **2019**, *127*, 187–203. [CrossRef]

45. Meza-Fuentes, E.; Rodriguez-Ruiz, J.; Solano-Polo, C.; Rangel, M.C.; Faro, A. Monitoring the structural and textural changes of Ni-Zn-Al hydrotalcites under heating. *Thermochim. Acta* **2020**, *687*, 178594. [CrossRef]

46. Xia, S.; Dai, T.; Meng, Y.; Zhou, X.; Pan, G.; Zhang, X.; Ni, Z. Low temperature water-gas shift reaction catalyzed by hybrid NiO@NiCr elayered double hydroxides: Catalytic property, kinetics and mechanism investigation. *Phys. Chem. Chem. Phys.* **2020**, *22*, 12630–12643. [CrossRef] [PubMed]

47. Gabrovska, M.; Ivanov, I.; Nikolova, D.; Kovacheva, D.; Tabakova, T. Hydrogen production via water-gas shift reaction over gold supported on Ni-based layered double hydroxides. *Int. J. Hydrog. Energy* **2021**, *46*, 458–473. [CrossRef]
48. Bish, D.; Brindley, G. A reinvestigation of takovite, a nickelaluminium hydroxycarbonate of the pyroaurite group. *Am. Mineral* **1977**, *62*, 458–464.
49. Cavani, F.; Trifirò, F.; Vaccari, A. Hydrotalcite-type anionic clays: Preparation, properties and applications. *Catal. Today* **1991**, *11*, 173–301. [CrossRef]
50. Clause, O.; Goncalves Coelho, M.; Gazzano, M.; Matteuzzi, D.; Trifirò, F.; Vaccari, A. Synthesis and thermal reactivity of nickel-containing anionic clays. *Appl. Clay Sci.* **1993**, *8*, 169–186. [CrossRef]
51. Shannon, R. Revised effective ionic radii and systematic studies of interatomic distances in halides and chaleogenides. *Acta Crystallogr. A* **1976**, *32*, 751–767. [CrossRef]
52. Sanati, S.; Rezvani, Z. Co-intercalation of acid Red-27/sodium dodecyl sulfate in a Ce-containing Ni–Al-layered double hydroxide matrix and characterization of its luminescent properties. *J. Mol. Liq.* **2018**, *249*, 318–325. [CrossRef]
53. Golovin, S.N.; Yapryntsev, M.N.; Ryltsova, I.G.; Veligzhanin, A.A.; Lebedeva, O.E. Novel cerium-containing layered double hydroxide. *Chem. Pap.* **2020**, *74*, 367–370. [CrossRef]
54. Reni, M.L.; Nesaraj, A.S. Preparative methods and recent technological applications of ceria -based nanostructured catalyst materials in chemical and other fields-a review. *Mater. Res. Innov.* **2020**. [CrossRef]
55. Matijević, E.; Hsu, W.P. Preparation and properties of monodispersed colloidal particles of lanthanide compounds. I. Gadolinium, Europium, Terbium, Samarium, and Cerium (lll). *J. Colloid Interf. Sci.* **1987**, *118*, 506–523. [CrossRef]
56. Chen, P.L.; Chen, I.W. Reactive cerium (IV) oxide powders by the homogeneous precipitation method. *J. Am. Ceram. Soc.* **1993**, *76*, 1577–1583. [CrossRef]
57. Zhou, X.D.; Huebner, W.; Anderson, H.U. Room-temperature homogeneous nucleation synthesis and thermal stability of nanometer single crystal CeO_2. *Appl. Phys. Lett.* **2002**, *80*, 3814–3816. [CrossRef]
58. Li, J.-G.; Ikegami, T.; Wang, Y.; Mori, T. Reactive Ceria Nanopowders via Carbonate Precipitation. *J. Am. Ceram. Soc.* **2002**, *85*, 2376–2378. [CrossRef]
59. Yamashita, M.; Kameyama, K.; Yabe, S.; Yoshida, S.; Fujishiro, Y.; Kawai, T.; Sato, T. Synthesis and microstructure of ceria doped ceria as UV filters. *J. Mater. Sci.* **2002**, *37*, 683–687. [CrossRef]
60. Uekawa, N.; Ueta, M.; Wu, Y.J.; Kakegawa, K. Synthesis of CeO_2 spherical fine particles by homogeneous precipitation method with polyethylene glycol. *Chem. Lett.* **2002**, *8*, 854–855. [CrossRef]
61. Zhou, X.D.; Huebner, W.; Anderson, H.U. Processing of nanometer-CeO_2 particles. *Chem. Mater.* **2003**, *15*, 378–382. [CrossRef]
62. Djuričić, B.; Pickering, S. Nanostructured cerium oxide: Preparation and properties of weakly agglomerated powders. *J. Eur. Ceram. Soc.* **1999**, *19*, 1925–1934. [CrossRef]
63. Hirano, M.; Kato, E. Hydrothermal synthesis of nanocrystalline cerium (IV) oxide powders. *J. Am. Ceram. Soc.* **1999**, *82*, 768–788. [CrossRef]
64. Chen, H.-I.; Chang, H.-Y. Synthesis of nanocrystalline cerium oxide particles by the precipitation method. *Ceram. Int.* **2005**, *31*, 795–802. [CrossRef]
65. Thommes, M.; Kaneko, K.; Niemark, A.V.; Olivier, J.P.; Rodriguez-Reinoso, F.; Rouquerol, J.; Sing, K.S.W. Physisorption of gases, with special reference to the evaluation of surface area and pore size distribution (IUPAC Technical Report). *Pure Appl. Chem.* **2015**, *87*, 1051–1069. [CrossRef]
66. Dubinin, M.M. Physical Adsorption of Gases and Vapors in Micropores. *Prog. Surf. Membr. Sci.* **1975**, *9*, 1-70.
67. Barrett, E.P.; Joyner, L.G.; Halenda, P.P. The determination of pore volume and area distributions in porous substances. I. computations from nitrogen isotherms. *J. Am. Chem. Soc.* **1951**, *73*, 373–380. [CrossRef]
68. Lecloux, A.; Pirard, J.P. The importance of standard isotherms in the analysis of adsorption isotherms for determining the porous texture of solids. *J. Colloid Interface Sci.* **1979**, *70*, 265–281. [CrossRef]
69. Monti, D.A.M.; Baiker, A. Temperature-programmed reduction. Parametric sensitivity and estimation of kinetic parameters. *J. Catal.* **1983**, *83*, 323–335. [CrossRef]
70. Russo, M.; La Parola, V.; Testa, M.L.; Pantaleo, G.; Venezia, A.M.; Gupta, R.K.; Bordoloi, A.; Bordoloi, B. Structural insight in TiO_2 supported CoFe catalysts for Fisher Tropsch synthesis at ambient pressure. *Appl. Catal. A Gen.* **2020**, *600*, 117621. [CrossRef]
71. Benito, P.; Labajos, F.; Rives, V. Microwave-treated layered double hydroxides containing Ni^{2+} and Al^{3+}: The effect of added Zn^{2+}. *J. Solid State Chem.* **2006**, *179*, 3784–3797. [CrossRef]
72. Tichit, D.; Medina, F.; Coq, B.; Dutartre, R. Activation under oxidizing and reducing atmospheres of Ni-containing layered double hydroxides. *Appl. Catal. A Gen.* **1997**, *159*, 241–258. [CrossRef]
73. Trifirò, F.; Vaccari, A.; Clause, O. Nature and properties of nickel-containing mixed oxides obtained from hydrotalcite-type anionic clays. *Catal. Today* **1994**, *21*, 185–195. [CrossRef]
74. Świrk, K.; Rønning, M.; Motak, M.; Beaunier, P.; Da Costa, P.; Grzybek, T. Ce-and Y-Modified Double-Layered Hydroxides as Catalysts for Dry Reforming of Methane: On the Effect of Yttrium Promotion. *Catalysts* **2019**, *9*, 56. [CrossRef]
75. Rouquerol, F.; Rouquerol, J.; Sing, K.S.W.; Llewellyn, P.; Maurin, G. *Adsorption by Powders and Porous Solids Principles, Methodology and Applications*, 2nd ed.; Academic Press: Cambridge, MA, USA; Elsevier Ltd.: Amsterdam, The Netherlands, 2014.
76. Groen, J.C.; Peffer, L.A.A.; Pérez-Ramırez, J. Pore size determination in modified micro- and mesoporous materials. Pitfalls and limitations in gas adsorption data analysis. *Micropor. Mesopor. Mat.* **2003**, *60*, 1–17. [CrossRef]

77. Kim, K.S.; Winograd, N. X-ray photoelectron spectroscopic studies of nickel-oxygen surfaces using oxygen and argon ion-bombardment. *Surf. Sci.* **1974**, *43*, 625–643. [CrossRef]

78. Löchel, B.P.; Strehblow, H.H. Breakdown of passivity of nickel by fluoride: II. Surface analytical studies. *J. Electrochem. Soc.* **1984**, *131*, 713–723. [CrossRef]

79. Grosvenor, A.P.; Biesinger, M.C.; Smart, R.S.C.; McIntyre, N.S. New interpretations of XPS spectra of nickel metal and oxides. *Surf. Sci.* **2006**, *600*, 1771–1779. [CrossRef]

80. Hu, X.; Li, P.; Zhang, X.; Yu, B.; Lv, C.; Zeng, N.; Luo, J.; Zhang, Z.; Song, J.; Liu, Y. Ni-Based Catalyst Derived from NiAl layered double hydroxide for vapor phase catalytic exchange between hydrogen and water. *Nanomaterials* **2019**, *9*, 1688. [CrossRef] [PubMed]

81. Zhao, S.; Yi, H.; Tang, X.; Gao, F.; Yu, Q.; Zhou, Y.; Wang, J.; Huang, Y.; Yang, Z. Enhancement effects of ultrasound assisted in the synthesis of NiAl hydrotalcite for carbonyl sulfide removal. *Ultrason. Sonochem.* **2016**, *32*, 336–342. [CrossRef]

82. Shyu, J.Z.; Otto, K.; Watkins, W.L.H.; Graham, G.W. Characterization of Pd/γ-alumina catalysts containing ceria. *J. Catal.* **1988**, *114*, 23–33. [CrossRef]

83. Wang, L.; Meng, F. Oxygen vacancy and Ce^{3+} ion dependent magnetism of monocrystal CeO_2 nanopoles synthesized by a facile hydrothermal method. *Mater. Res. Bull.* **2013**, *48*, 3492–3498. [CrossRef]

84. Pan, Y.; Nilius, N.; Freund, H.-J.; Paier, J.; Penschke, C.; Sauer, J. Titration of Ce^{3+} Ions in the CeO_2(111) Surface by Au Adatoms. *Phys. Rev. Lett.* **2013**, *111*, 206101. [CrossRef]

85. Lu, B.; Zhuang, J.; Du, J.; Gu, F.; Xu, G.; Zhong, Z.; Liu, Q.; Su, F. Highly dispersed Ni nanocatalysts derived from NiMnAl-hydrotalcites as high-performing catalyst for low-temperature syngas methanation. *Catalysts* **2019**, *9*, 282. [CrossRef]

86. Praline, G.; Koel, B.E.; Hance, R.L.; Lee, H.-I.; White, J.M. X-Ray photoelectron study of the reaction of oxygen with cerium. *J. Electron Spectros. Relat. Phenom.* **1980**, *21*, 17–30. [CrossRef]

87. Laachir, A.; Perrichon, V.; Badri, A.; Lamotte, J.; Catherine, E.; Lavalley, J.C.; El Fallah, J.; Hilaire, L.; Le Normand, F.; Quéméré, E.; et al. Reduction of CeO_2 by hydrogen. Magnetic susceptibility and Fourier-transform infrared, ultraviolet and X-ray photoelectron spectroscopy measurements. *J. Chem. Soc. Faraday Trans.* **1991**, *87*, 1601–1609. [CrossRef]

88. Pereira, A.; Blouin, M.; Pillonnet, A.; Guay, D. Structure and valence properties of ceria films synthesized by laser ablation under reducing atmosphere. *Mater. Res. Express* **2014**, *1*, 015704. [CrossRef]

89. Genty, E.; Brunet, J.; Poupin, C.; Casale, S.; Capelle, S.; Massiani, P.; Siffert, S.; Cousin, R. Co-Al mixed oxides prepared via LDH route using microwaves or ultrasound: Application for catalytic toluene total oxidation. *Catalysts* **2015**, *5*, 851–867. [CrossRef]

90. Ingo, G.M.; Paparazzo, E.; Bagnarelli, O.; Zacchetti, N. XPS studies on cerium, zirconium and yttrium valence states in plasma-sprayed coatings. *Surf. Interface Anal.* **1990**, *16*, 515–519. [CrossRef]

91. Mansour, A.N.; Melendres, C.A. Characterization of Electrochemically Prepared γ-NiOOH by XPS. *Surf. Sci. Spectra* **1994**, *3*, 271–278. [CrossRef]

92. Wang, X.; Wang, X.; Di, Q.; Zhao, H.; Liang, B.; Yang, J. Mutual effects of fluorine dopant and oxygen vacancies on structural and luminescence characteristics of F doped SnO_2 nanoparticles. *Materials* **2017**, *10*, 1398. [CrossRef]

93. Tu, Y.; Chen, S.; Li, X.; Gorbaciova, J.; Gillin, W.P.; Krause, S.; Briscoe, J. Control of oxygen vacancies in ZnO nanorods by annealing and their influence on ZnO/PEDOT: PSS diode behavior. *J. Mater. Chem. C* **2018**, *6*, 1815–1821. [CrossRef]

94. Jain, S.; Shah, J.; Negi, N.S.; Sharma, C.; Kotnala, R.K. Nickel substituted oxygen deficient nanoporous lithium ferrite based green energy device hydroelectric cell. *J. Alloys Compd.* **2020**, *827*, 154334.

95. Rodriguez, J.A.; Wang, X.; Liu, P.; Wen, W.; Hanson, J.C.; Hrbek, J.; Perez, M.; Evans, J. Gold nanoparticles on ceria: Importance of O vacancies in the activation of gold. *Top. Catal.* **2007**, *44*, 73–81. [CrossRef]

96. Vindigni, F.; Manzoli, M.; Damin, A.; Tabakova, T.; Zecchina, A. Surface and inner defects in Au/CeO_2 WGS catalysts: Relation between Raman properties, reactivity and morphology. *Chem. Eur. J.* **2011**, *17*, 4356–4361. [CrossRef]

97. Laguna, O.H.; Dominguez, M.I.; Romero-Sarria, F.; Odriozola, J.A.; Centeno, M.A. Role of oxygen vacancies in gold oxidation catalysis. In *Heterogeneous Gold Catalysts and Catalysis*; Ma, Z., Dai, S., Eds.; The Royal Society of Chemistry: Cambridge, UK, 2014; pp. 489–511.

Article

Hybrid Biochar/Ceria Nanomaterials: Synthesis, Characterization and Activity Assessment for the Persulfate-Induced Degradation of Antibiotic Sulfamethoxazole

Golfo Papatheodorou [1], Paraskevi Ntzoufra [1], Evroula Hapeshi [2], John Vakros [1,*] and Dionissios Mantzavinos [1,*]

[1] Department of Chemical Engineering, University of Patras, Caratheodory 1, University Campus, GR-26504 Patras, Greece; fofopapatheodorou@gmail.com (G.P.); voula_ntz1994@hotmail.com (P.N.)
[2] Department of Life and Health Sciences, School of Sciences and Engineering, University of Nicosia, Nicosia 2417, Cyprus; hapeshis.e@unic.ac.cy
* Correspondence: vakros@chemistry.upatras.gr (J.V.); mantzavinos@chemeng.upatras.gr (D.M.)

Abstract: Biochar from spent malt rootlets was employed as the template to synthesize hybrid biochar-ceria materials through a wet impregnation method. The materials were tested for the activation of persulfate (SPS) and subsequent degradation of sulfamethoxazole (SMX), a representative antibiotic, in various matrices. Different calcination temperatures in the range 300–500 °C were employed and the resulting materials were characterized by means of N_2 adsorption and potentiometric mass titration as well as TGA, XRD, SEM, FTIR, DRS, and Raman spectroscopy. Calcination temperature affects the biochar content and the physicochemical properties of the hybrid materials, which were tested for the degradation of 500 μg L^{-1} SMX with SPS (in the range 200–500 mg L^{-1}) in various matrices including ultrapure water (UPW), bottled water, wastewater, and UPW spiked with bicarbonate, chloride, or humic acid. Materials calcined at 300–350 °C, with a surface area of ca. 120 m^2 g^{-1}, were the most active, yielding ca. 65% SMX degradation after 120 min of reaction in UPW; materials calcined at higher temperatures as well as bare biochar were less active. Degradation decreased with increasing matrix complexity due to the interactions amongst the surface, the contaminant, and the oxidant. Experiments in the presence of scavengers (i.e., methanol, t-butanol, and sodium azide) revealed that sulfate and hydroxyl radicals as well as singlet oxygen were the main oxidative species.

Keywords: biochar; emerging contaminants; nanoceria; Fenton-like reaction; SR-AOPs; water treatment

Citation: Papatheodorou, G.; Ntzoufra, P.; Hapeshi, E.; Vakros, J.; Mantzavinos, D. Hybrid Biochar/Ceria Nanomaterials: Synthesis, Characterization and Activity Assessment for the Persulfate-Induced Degradation of Antibiotic Sulfamethoxazole. *Nanomaterials* **2022**, *12*, 194. https://doi.org/10.3390/nano12020194

Academic Editor: Sherif A. El-Safty

Received: 15 December 2021
Accepted: 5 January 2022
Published: 7 January 2022

Publisher's Note: MDPI stays neutral with regard to jurisdictional claims in published maps and institutional affiliations.

1. Introduction

Over the last few decades, nanotechnology has gained huge interest due to its extensive application in various fields including, among others, catalysis, electronics, optics, energy, and the environment. The design and controlled synthesis of advanced nanomaterials with unique properties make them highly attractive in these fields. The demand for more active, environmentally friendly, low-cost materials has resulted in tremendous interest in the preparation of nanostructured materials with active surface functional groups and, thus, high surface reactivity. Two very interesting materials with many environmental applications are CeO_2 [1,2] and biochar [3,4]. The reason for this is their unique characteristics. Specifically, CeO_2 is a very promising material that can be used as a catalyst or support in several catalytic applications. Combined with Cu is the state-of-the-art catalyst for the preferential oxidation of CO [5] since it presents a high amount of oxygen vacancies, a controllable ratio of Ce^{3+}/Ce^{4+}, high oxygen storage capacity, and moderate surface area [2,6,7].

Nano CeO_2 may exhibit improved properties and catalytic functions, which are significantly affected by the preparation conditions [8]. It has been reported in the literature and demonstrated by computational studies that the morphology of the nano CeO_2 affects the catalytic activity. Generally, the (110) and (100) surfaces are catalytically more active than (111) [9,10]. Specifically, the higher activity for the CO oxidation can be obtained with nanorods of CeO_2, which are exposed to the (110) and (100) planes [11]. The hydrothermal method can be applied for the preparation of different forms of nano CeO_2 [2].

The main disadvantage of CeO_2 is that pure ceria exhibits low activity since the active surface oxygen species are rather limited; moreover, CeO_2 is an insulator at room temperature, thus electron transfer is limited [12]. This can be dramatically changed if pure CeO_2 is doped with transition metal ions. Then, the deposition of metal cations (M) on CeO_2 can alter the electronic and geometric configuration and thus the properties of the mixed material. This approach is well-studied and researchers have concluded that new sites with high activity can be formed, interactions between Ce and M can alter the electrons configuration, and new oxygen vacancies can be formed during sub-surface and bulk incorporation of metal ions into CeO_2 [13–15].

In a recent review, the synthesis and characterization of CeO_2 nanoparticles of different morphology were discussed [16]. Although many methods have been applied for the synthesis of CeO_2 nanoparticles, the combined use of biochar (BC) as a template and Ce precursors for the preparation of CeO_2 has not been investigated. On the other hand, CeO_2 deposition on biochar to produce hybrid materials with enhanced adsorption capacity [17], interesting electrocatalytic properties [18], or superior degradation ability for textile dyes in sonocatalytic oxidation [19] has been reported; in these cases, CeO_2 nanoparticles were prepared with the hydrothermal method and then mixed with BC as a suspension in acetone. Moreover, novel carbonaceous materials (i.e., CeO_2-encapsulated nitrogen-doped biochar) have significant activity for oxygen reduction [20]. The biomass precursor of this value-added biochar material was biomimetically prepared via a hydroponic operation in the Ce-enriched solution. The enhanced activity was partially due to high oxygen vacancies of the hybrid material.

Biochar is the solid material prepared from the pyrolysis of biomass under a limited or no oxygen atmosphere, during which part of the organic phase decomposes to gases. Interestingly, the properties of the biochars are different compared to raw biomass. These new unique properties are desirable for many applications where biochars are employed as absorbers [21], support in catalytic processes [22–24], transesterification catalysts for the production of biodiesel, [25], supercapacitors [26], and persulfate activators for the oxidation of organic contaminants in water [27–30]. Depending on the raw biomass and pyrolysis conditions, biochars can exhibit high surface area, hierarchical pore structure, plenty of surface groups, stability, and in some cases, a high amount of mineral deposits. Furthermore, its surface can interact with metal ions or even nanoparticles and stabilize them.

Pharmaceutically active compounds such as antibiotics, antihypertensive and nonsteroidal anti-inflammatory drugs as well as their metabolites are classified as emerging micro-contaminants and are detected in wastewater, surface water, and groundwater at concentrations ranging from ng L^{-1} to mg L^{-1}. Antibiotics are widely used against bacterial infections or to prevent infections; several studies have reported antibiotic occurrence in wastewater treatment plant effluents at concentrations from 0.1 to 2.5 µg L^{-1} [31]. Although this range of concentration is small, it can be harmful to human and animal health and augment the antimicrobial resistance [32].

In this work, biochar from malt spent rootlets was prepared at 850 °C. This biochar has moderate surface area and a high amount of minerals [33]. Treatment with H_2SO_4 can remove the minerals and significantly increase the specific surface area [34]. For this reason, the biochar was treated with H_2SO_4 and then used as a template for the preparation of hybrid material BC-CeO_2 with different BC to CeO_2 ratio. The produced materials were characterized with various physicochemical methods and used for the degradation of sulfamethoxazole, a representative antibiotic drug, in various water matrices via oxidation with persulfates. To the best of our knowledge, this is the first report on (i) the use of biochar as a template and its influence on the physicochemical properties of CeO_2, and (ii) the application of the as-prepared hybrid materials to promote the sulfate radical-induced advanced oxidation of antibiotic SMX in environmentally relevant matrices.

From this perspective, the innovation of this work embodies two different but related disciplines, namely (i) material synthesis, where a novel catalytic material capable of activating persulfate is described, and (ii) environmental remediation, focusing on the treatment of micro-contaminants of emerging concern.

2. Materials and Methods

2.1. Materials

The precursor salt $Ce(NO_3)_3 \cdot 6H_2O$ (analytical grade, CAS number: 10277-43-7) and sulfamethoxazole (SMX, $C_{10}H_{11}N_3O_3S$, 99+%, CAS number: 723-46-6) were purchased from Sigma-Aldrich (St. Louis, MO, USA). Sodium persulfate (SPS, $Na_2S_2O_8$ 99%, CAS number: 7775-27-1) was purchased from Scharlau (Barcelona, Spain). Most of the experiments were carried out in ultrapure water (UPW: pH = 6.5). Other matrices included (i) commercially available bottled water (BW: pH = 7.7, conductivity 355 $\mu S\ cm^{-1}$, containing (in $mg\ L^{-1}$): 237 bicarbonate; 3.7 chloride; 7.8 sulfate; 1.1 nitrate; 75.5 calcium; 5.1 magnesium; 2.1 sodium; and 0.65 potassium ions); (ii) secondary treated wastewater (WW) taken from the University of Patras campus treatment plant (pH = 8, conductivity = 1.682 $mS\ cm^{-1}$, total organic C = 2.46 $mg\ L^{-1}$, chemical oxygen demand = 48.53 $mg\ L^{-1}$, total suspended solids = 22 $mg\ L^{-1}$, $[Cl^-]$ = 262.41 $mg\ L^{-1}$, $[PO_4^{3-}]$ = 14.98 $mg\ L^{-1}$, $[HCO_3^-]$ = 278 $mg\ L^{-1}$, $[Br^-]$ = 165.64 $mg\ L^{-1}$, $[Ca^{2+}]$ = 112 $mg\ L^{-1}$); and (iii) UPW spiked with various water constituents such as humic acid (HA: CAS number: 1415-93-6), bicarbonate (CAS number: 144-55-8), chloride (CAS number: 7647-14-5), sodium azide (NaN_3: CAS number: 26628-22-8), t-butanol (CAS number: 75-65-0), and methanol (CAS number: 67-56-1); all these were purchased from Sigma-Aldrich (St. Louis, MO, USA).

2.2. Sample Preparation

The biochar used in this study was prepared from spent malt rootlets under pyrolysis at 850 °C with limited O_2 atmosphere. The prepared biochar was treated with 1 M H_2SO_4 under reflux for 30 min. The treatment was conducted in order to increase the surface area of the sample and remove the deposits of minerals present in the raw biomass. After treatment, the BC was filtered, washed with 1 L of triply distilled water, and dried for 2 h at 120 °C. More details on the preparation of the sample and its properties can be found in [34].

To deposit the Ce precursor on the biochar surface, about 3 g of treated biochar was immersed in 150 mL of solution containing 11.1 g of $Ce(NO_3)_3\ 6H_2O$. The suspension was placed in a round bottom bottle in a rotary evaporator system. Then, the suspension was left to equilibrate for 30 min under atmospheric pressure at 70 °C. After that, vacuum was applied and the water was evaporated. The mixed solid was dried at 120 °C for 1 h and then calcined at different temperatures for 2 h. The samples were denoted as BC/Ce-X, where X is the calcination temperature. An additional sample (BC/Ce-300-5 h) was calcined at 300 °C for 5 h.

2.3. Physicochemical Characterization

The prepared samples were characterized with various physicochemical methods. Briefly, specific surface area (SSA) and pore size distribution was performed with N_2 adsorption isotherms at liquid N_2 temperature in a Tristar 3000 porosimeter (Micromeritics). X-ray diffraction peaks were recorded with a Bruker D8 Advance diffractometer (Billerica, MA, USA) equipped with a nickel-filtered CuKa (1.5418 Å) radiation source. The biochar morphology was examined by scanning electron microscopy (SEM JEOL JSM6300) equipped with EDS. Fourier transform infrared analysis was performed in a Perkin Elmer Spectrum RX FTIR system (Waltham, MA, USA). The samples were diluted in KBr (1% w/w sample) and pressed in pellet form with 8 atm pressure. The point of zero charge was determined using the potentiometric mass titration method [35]. A suspension of 0.1 g in 75 mL of $NaNO_3$ 0.03 M was titrated with 0.1 M HNO_3 from pH 11 to 2 and the titration curve was compared with the corresponding curve of the solution. The section point of the two curves is the point of zero charge of the solid sample. The thermogravimetric analysis of the samples was performed in a TGA Perkin Elmer system (Waltham, MA, USA) under an air atmosphere with a flow of 20 mL min^{-1}. The heating rate was 10 °C min^{-1} in the temperature range of 80–700 °C. Diffuse reflectance spectroscopy (DRS) was performed using a UV–Vis spectrophotometer (Varian Cary 3) equipped with an integration sphere. The spectra of the solid samples were recorded in the range of 200–800 nm using PTFE or commercial CeO_2 as references. The powder samples were mounted in a quartz cell, which provided a sample thickness >3 mm to guarantee the "infinite" sample thickness. Raman spectra were taken on a Micro Raman Spectroscopy system (Jobin–Yvon Horiba LabRam–HR) with a 514 nm line of an Ar ion laser at room temperature. A 50× microscope objective lens was used to focus the laser beam and collection of the scattered light. Typical spectrum acquisition time was 5 s.

2.4. Catalytic Activity

A stock solution of SMX (50 mg L^{-1}) in UPW was prepared and used for all the catalytic tests. In a typical run, 120 mL of an aqueous solution containing 500 μg L^{-1} SMX and 90 mg L^{-1} BC were loaded into a beaker under stirring at ambient temperature. After 20 min of equilibration, SPS was added. Samples of 1.2 mL were periodically drawn from the reactor, an excess of methanol was added (5 mol L^{-1}) to quench the reaction and the samples were filtered and analyzed using high-performance liquid chromatography (HPLC) (Waters Alliance 2695, Waters 2996 Milford, PA, USA). More details about the catalytic tests and analysis can be found in [36].

3. Results and Discussion

3.1. Samples Characterization

The biochar from malt spent rootlets had a moderate specific surface area, SSA, of 100 m^2 g^{-1} a point of zero charge, pzc, equal to 8.2 and 32% minerals. Following treatment with H_2SO_4, the concentration of minerals diminished since they were soluble in acidic solution, while SSA increased considerably to 428 m^2 g^{-1}. The microporosity was also high (190 m^2 g^{-1}) and the pzc shifted to more acidic values. These properties make the treated biochar an ideal candidate for the preparation of hybrid ceria–biochar materials. The low value of pzc ensures a positively charged surface, where cations can be deposited. The high SSA favors the adsorption of considerable quantities of cations, while the micropores can prevent the formation of bulk precipitates during drying. The deposition of Ce ions can easily be performed with wet impregnation, during which Ce(III) ions interact with the biochar surface through electrostatic adsorption because of the low value of pzc. The high value of SSA and microporosity allows for the deposition of well-dispersed Ce particles. Finally, the calcination process can provide the possibility of controlling the biochar content and, in parallel, to convert the Ce precursor form to CeO_2 nanoparticles. With the regulation of calcination temperature, one can prepare hybrid materials BC–CeO_2 with different biochar contents.

The prepared samples, alongside their properties, are presented in Table 1.

Table 1. Physicochemical characteristics of the prepared samples.

Sample	T Calc (°C)	SSA (m^2 g^{-1})	pzc	D (XRD) (nm)	% CeO_2 Content	Eg (eV)	%O_2 Uptake in TGA
BC/Ce-300	300	119	6.8	29.2	16	3.09	0
BC/Ce-300-5 h	300	110	6.7	20.7	14	3.10	0
BC/Ce-350	350	126	6.5	29.8	22	3.07	1.7
BC/Ce-400	400	69	3.0	18.9	7	3.10	0.5
BC/Ce-500	500	14	3.0	16.0	2	3.12	2.0

Interestingly, increasing the calcination time from 2 to 5 h at 300 °C did not practically alter the properties of the prepared samples. Although longer calcination times were not tested at higher temperatures (where the BC content is lower), one could possibly expect to see detrimental effects on physicochemical properties such as SSA, Eg, and oxygen vacancies due to extensive sintering.

The adsorption/desorption isotherms for the hybrid samples are presented in Figure 1. The SSA was lower at higher temperatures of calcination. The samples calcined at 300–400 °C exhibited type IV with an H4 hysteresis loop, while for the BC/Ce-500 sample, the hysteresis loop was between H3 and H4. H4 loops are often found in micro-meso porous carbon materials [37]. The BC/Ce-500 sample also exhibited limited N_2 adsorption at low P/Po values, suggesting low microporosity, in contrast with the other three samples. SSA values were higher than the commercial CeO_2 with a SSA of 4 m^2 g^{-1}.

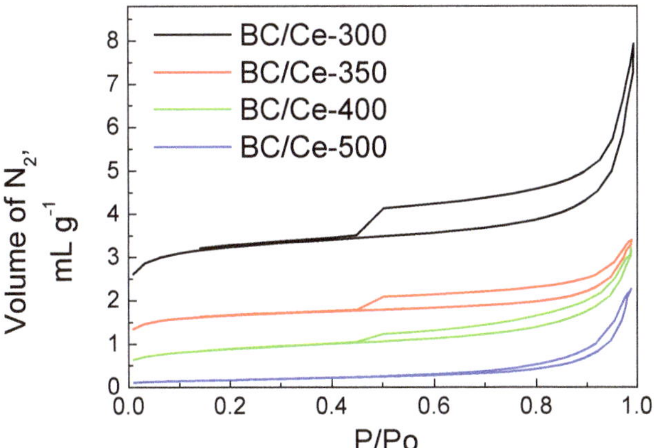

Figure 1. The N_2 adsorption–desorption isotherms for the prepared samples.

This was confirmed by the pore size distribution shown in Figure 2. The BC/Ce-500 had a limited amount of micropores (if any), in contrast with the other samples calcined at lower temperatures. For the BC/Ce-500 sample, there was a peak centered at about 85 nm, while the main peak was at 60 nm for the other samples; an additional peak was centered at 13 nm for BC/Ce-400. This increment in pore diameter was due to higher calcination temperature and the collapse of microporosity.

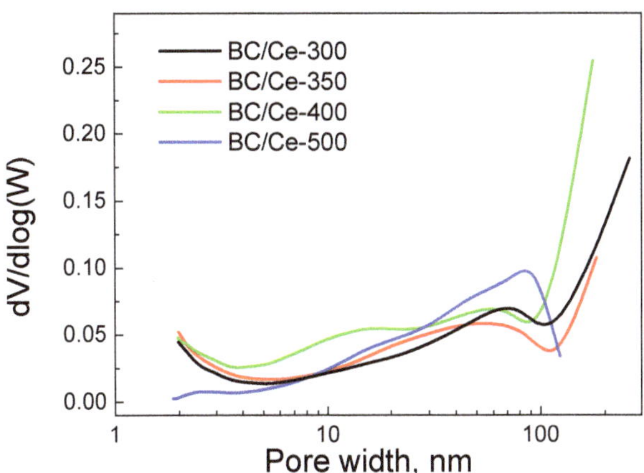

Figure 2. Pore size distribution for the studied samples.

The TGA curves for the prepared samples as well as the mixed material before calcination (BC/Ce) are presented in Figure 3. The % mass left after TGA was due to the CeO_2

content in each hybrid material and the difference from the starting mass is characteristic of the biochar content.

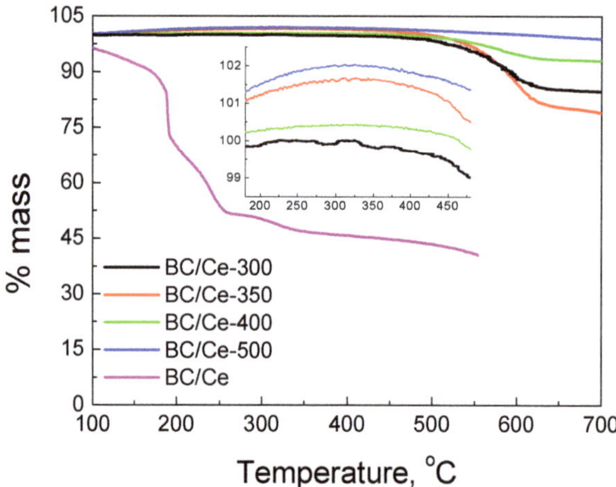

Figure 3. TGA curves under an air atmosphere for the studied samples as well as the uncalcined one.

As can be seen, the amount of mass left after TGA was significant and depends on the calcination temperature (see also Table 1). This means that the amount of biochar left after calcination was limited and the samples mostly consisted of CeO_2. The biochar content was 22% for the BC/Ce-350 sample and 16% for the BC/Ce-300 sample, while the other two samples calcined at higher temperatures had an even lower content. Although the BC/Ce-300 sample should be expected to have a higher BC content than the BC/Ce-350 sample, this discrepancy may be attributed to the low difference in the respective calcination temperatures.

For the starting material before calcination, there was first sharp mass decrease at 185 °C, which was followed by a second step at 190 °C. This step was completed at about 260 °C, while the mass was quite stable at higher temperatures. Therefore, a temperature up to 300 °C seems to be sufficient to transform the precursors to CeO_2 nanoparticles. The mass left at temperatures higher than 550 °C was about 42% of the starting mass, very close to the value of the nominal CeO_2 content (41%), which implies that the starting material is transformed to CeO_2 at temperatures up to 500 °C.

The FTIR spectra of the prepared samples are presented in Figure 4, together with the spectra of BC and commercial CeO_2. The FTIR peaks were more intense in the prepared materials in contrast with the commercial CeO_2. This may be due to the different SSA values, which were higher in the prepared samples, suggesting the existence of more surface groups. On the other hand, the BC/Ce-300 and BC/Ce-350 samples had peaks with low intensity, suggesting that CeO_2 was less bulk in these samples.

The XRD patterns of the prepared samples, alongside the BC and commercial CeO_2, are presented in Figure 5; the main peaks can be evidently attributed to CeO_2. The average diameter of the CeO_2 particles calculated from the Scherrer equation varied from 29.7 to 16 nm, while for the commercial CeO_2, it was 46.2 nm.

The acid–base behavior of the samples is presented in Figure 6. The potentiometric titration curves revealed that the two materials prepared at the lower calcination temperatures had a pzc near 7 (6.8 and 6.5 for the BC/Ce-300 and BC/Ce-350, respectively), while the two other samples had a pzc value equal to 3, pointing out the acidity of these samples. Generally, low pzc values have been reported for CeO_2 when the precursor is Ce(III) salt, as in our case [38].

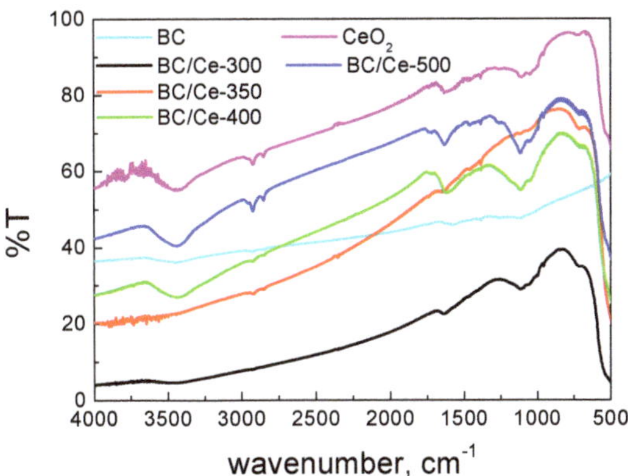

Figure 4. FTIR spectra for the studied samples as well as the starting biochar and a commercial CeO_2 sample.

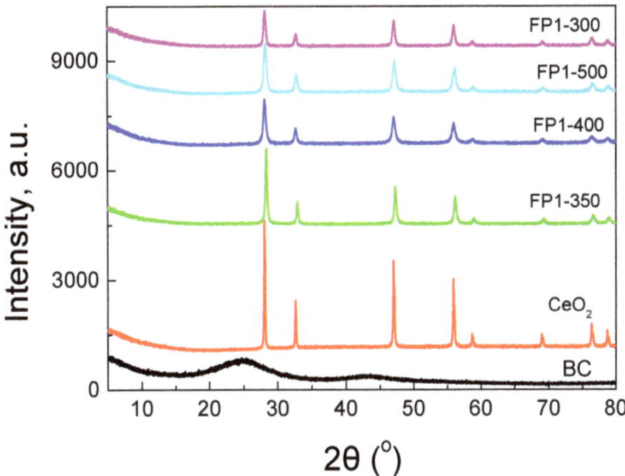

Figure 5. XRD patterns for the studied samples as well as the starting biochar and a commercial CeO_2 sample.

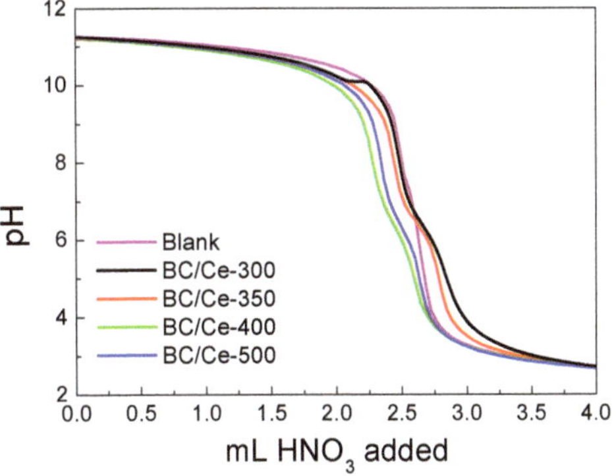

Figure 6. Potentiometric mass titration curves for the studied samples as well as the corresponding solution titration curve.

SEM images of the prepared samples are presented in Figure 7, where the progressive removal of BC and the transformation to a more solid CeO_2 phase could be observed. For the BC/Ce-350 sample (Figure 7b), BC could clearly be seen, while CeO_2 formed around the carbon particles. This was less pronounced in the case of BC/Ce-300 (Figure 7a), although the calcination temperature was lower. There probably exists a minimum calcination temperature needed for the formation of CeO_2 particles and the simultaneous burning of biochar.

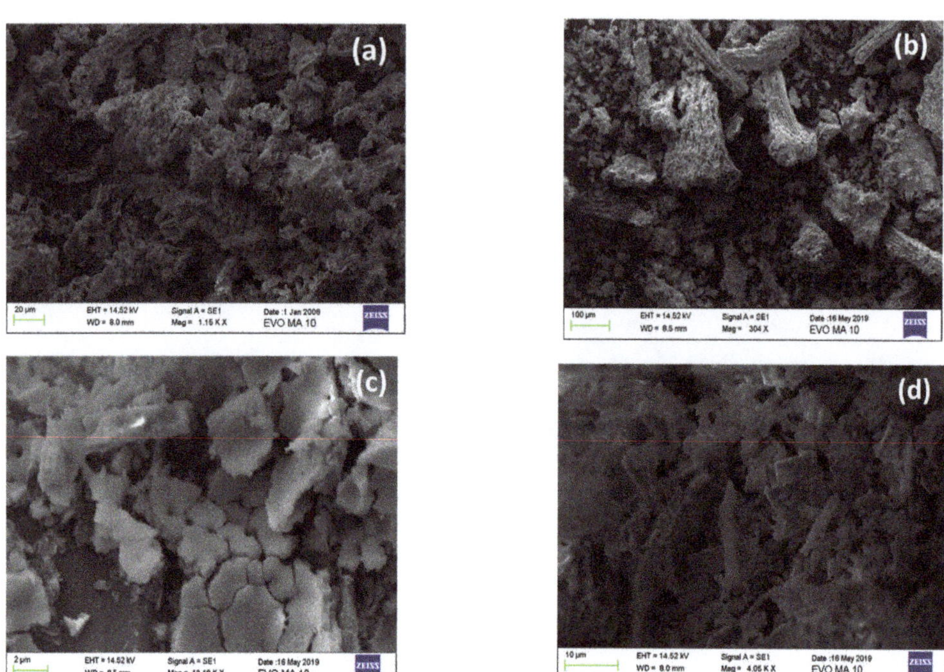

Figure 7. SEM images for (**a**) BC/Ce-300, (**b**) BC/Ce-350, (**c**) BC/Ce-400, and (**d**) BC/Ce-500.

On the other hand, higher calcination temperatures lead to greater removal of the carbon phase, and thus CeO_2 is better formed. Carbon removal in the form of volatile compounds results in CeO_2 cracking and this facilitates the formation of the surface area of the sample; this can be seen in Figure 7c for the BC/Ce-400 sample. Even higher calcination temperatures may result in sintering of the CeO_2 particles and, eventually, greater degree of agglomeration (Figure 7d). These findings are in accordance with the SSA values (Table 1) and the XRD results.

The DR spectra of the prepared samples are similar to that of CeO_2, especially at wavelengths lower than 400 nm (Figure 8). In the UV and Vis near UV regions, the peaks are due to charge transfer between the Ce(IV) and O^{2-} species. The DR spectrum of CeO_2 showed three maxima at about 220, 270, and 330 nm. The peak at 220 nm was assigned to *f-d* transition of Ce(II), the band at 270 nm to surface sites, and the band at 330 nm to bulk sites [39–42]. The exact location of the peaks is highly influenced by the size of CeO_2 crystallites. At wavelengths higher than 400 nm, CeO_2 exhibits negligible absorbance and this is also the case for BC/Ce-500 (i.e., the sample with the minimum BC content). The other three samples exhibited a constant absorbance over the whole range of the Vis spectrum, which was due to the black color of the biochar. The absorbance intensity was well correlated to the BC content, with the BC/Ce-350 sample showing the higher intensity. For the prepared samples, the point where a sharp increase in absorbance occurred shifted at higher wavelengths. This implies a change in the distribution of electrons and can be confirmed from the energy gap values for each material. These values were between 3.08–3.12 eV for all the prepared samples and 3.32 eV for commercial CeO_2 (Table 1). The observed red shift in the absorbance of the samples is correlated to the smaller size of CeO_2 nanoparticles, in contrast with the commercial CeO_2.

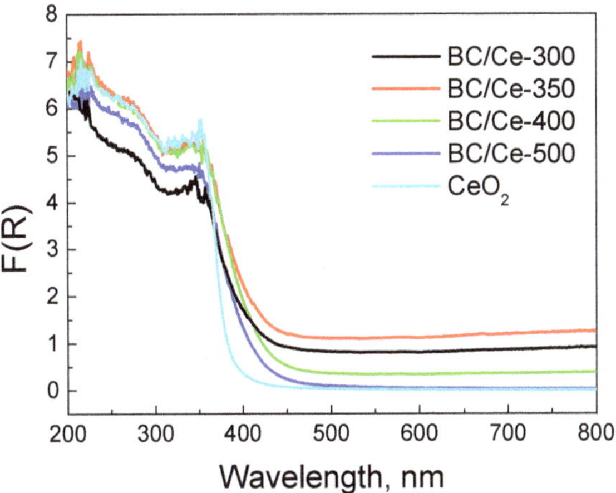

Figure 8. DR spectra for the studied samples as well as a commercial CeO_2 sample. The spectra were collected with PTFE disks as the reference.

Changes in electrons distribution are more clearly demonstrated in Figure 9, where commercial CeO_2 was employed as the reference. Indeed, the peak centered at about 400 nm revealed that interactions were more pronounced for the samples with increased BC content.

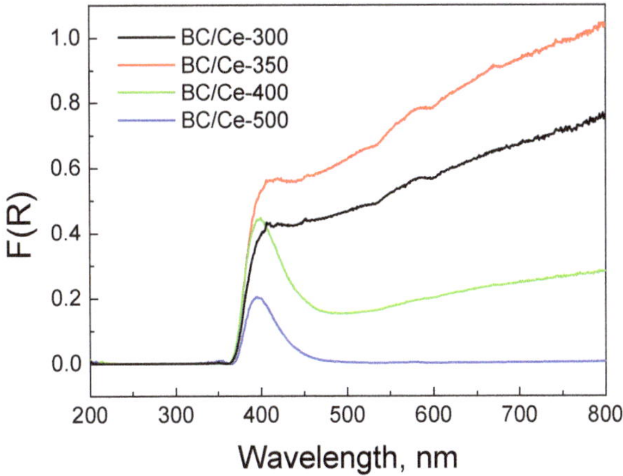

Figure 9. DR spectra for the studied samples. The spectra were collected with commercial CeO_2 as reference.

Figure 10 shows the normalized Raman spectra for the prepared samples. There was an intense peak centered between 460 and 464 cm^{-1}, which corresponded to the F2g Raman vibrational mode of cubic fluorite lattice of CeO_2 [39,43,44]. The broadness and asymmetry of the peak imply the existence of nanosized CeO_2 particles. The particle size as well as possible changes in composition associated with varying BC content influence the exact position and the broadening of the peak with a shift toward lower frequencies. There was another broad peak at about 600 cm^{-1}, which can be attributed to oxygen vacancies of CeO$_2$ [44,45]. It is interesting to note that this peak was broader and more intense for the samples with higher BC content.

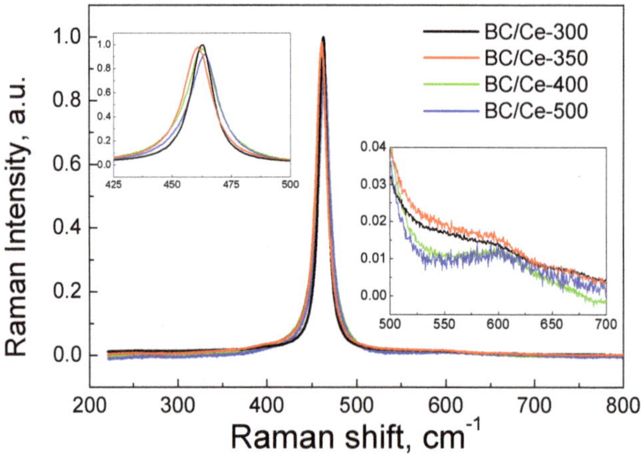

Figure 10. Normalized Raman spectra for the studied samples.

3.2. Assessment of Catalytic Activity for the Degradation of SMX

Figure 11 shows the relative catalytic activity of the various prepared samples as well as bare biochar and ceria. The time-scale corresponded to the equilibration period where only adsorption occurred, followed by the oxidative degradation period. It must be clarified here that a common equilibration period of 20 min was employed in this work irrespective of the specific experimental conditions (i.e., type of BC-Ce material, SPS concentration, water matrix, etc.), which implies that the level of SMX adsorption might have not been completed during this period; however, the main scope of this work was to study the oxidative rather than the adsorptive removal of SMX. Commercial ceria is not capable of activating SPS and neither is BC/Ce-500, which contained only 2% BC and whose SSA was 14 m^2 g^{-1} (Table 1); however, activity seems to increase with increasing BC content and SSA (i.e., 7%–69 m^2 g^{-1}, 16%–119 m^2 g^{-1}, and 22%–126 m^2 g^{-1} for BC/Ce-400, BC/Ce-300, and BC/Ce-350, respectively). Bare BC with a SSA of 428 m^2 g^{-1} exhibited good activity relative to that of BC/Ce-300 and BC/Ce-350 samples, although the hybrid samples have lower SSA. An additional run was performed with BC that had not been acid-treated (data not shown for brevity); its performance was comparable to that of the acid-treated sample. Considering that SMX degradation can be modeled by a pseudo-first order kinetic expression, data in Figure 11 can be employed to compute the apparent rate constants, k. The relative activity decreases in the order: BC/Ce-350 (8.9×10^{-3}) $\approx$ BC/Ce-300 (8.2×10^{-3}) > BC (7.4×10^{-3}) > BC/Ce-400 (4.6×10^{-3}) > CeO$_2$ (1.4×10^{-3}) $\approx$ BC/Ce-500 (1.3×10^{-3}), with numbers in brackets corresponding to k values expressed in min^{-1}. The dashed line (open symbols) shown in Figure 11 corresponds to a run with the BC/Ce-300 material that had been calcined for 5 h; interestingly, its catalytic activity was similar to that of the material calcined for 2 h (k = 7.8×10^{-3}), implying that the calcination temperature rather than time is the crucial factor. This finding is also consistent with the similar properties of the two samples, as shown in Table 1.

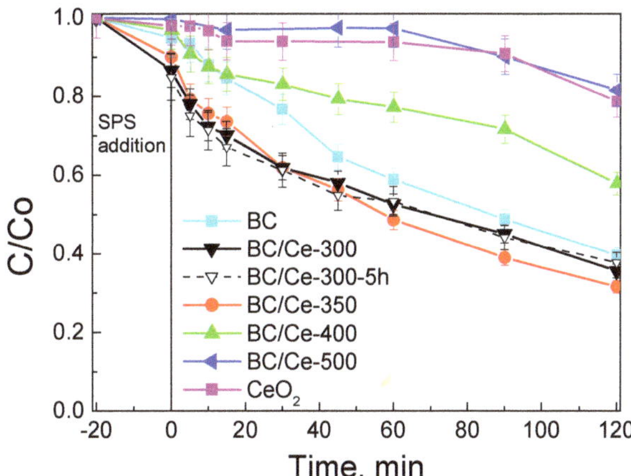

Figure 11. Screening of various materials (90 mg L^{-1}) for the removal of 500 µg L^{-1} SMX with 200 mg L^{-1} SPS in UPW and ambient pH.

Based on the results shown in Figure 11, subsequent activity tests were performed with the BC/Ce-350 material.

The influence of SPS concentration on SMX degradation is presented in Figure 12. In the absence of SPS, SMX removal can occur to a considerable extent due to adsorption only (i.e., 45% at the end of the experiment). The addition of SPS in the range 200–500 mg L^{-1} promotes SMX removal due to the reactions occurring between the generated radicals and SMX. Interestingly, the rate was not affected by the level of oxidant concentration used with the apparent rate constant being $9.1 \pm 0.1\ 10^{-3}$ min^{-1}; this value was about 2.5 times greater than that of pure adsorption.

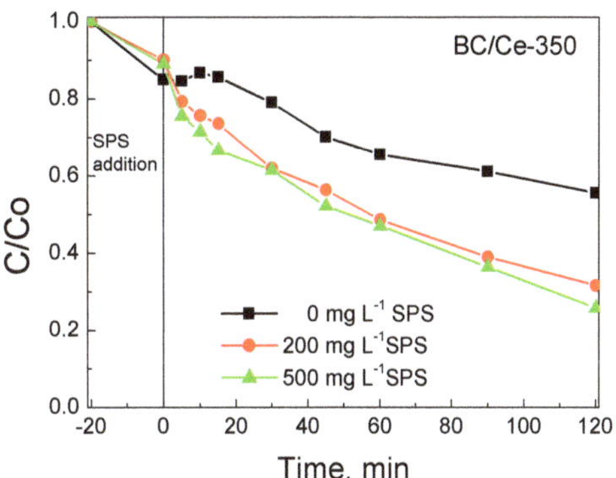

Figure 12. Effect of SPS concentration on 500 µg L^{-1} SMX removal with 90 mg L^{-1} BC/Ce-350 in UPW and ambient pH.

It must be noted here that the minimum effective level of persulfates employed in environmental applications depends on several factors including the type of activation (homogeneous or heterogeneous), the recalcitrance of the contaminant under consideration, the quality of the aquatic phase and, in the case of chemical activators such as transition metals, carbocatalysts, etc., their concentration. On the other hand, there always exists an upper concentration threshold, above which persulfate may act as a self-scavenger and/or introduce secondary water pollution due to the release of sulfate salts in the environment [46]. From a managerial point of view, solid persulfate is more advantageous than liquid hydrogen peroxide employed in traditional Fenton chemistry since it is more stable, easier to handle, store, and transport, and has a lower market price.

Since the pzc value of BC/Ce-350 was 6.5 (Table 1), its surface was slightly positive at an ambient pH of 5, which favors the attraction of the negatively charged $S_2O_8^{2-}$ anions. SMX, on the other hand, was neutral at $pKa_1 = 1.77 < pH < pka_2 = 5.65$ [29,47–49]. SMX was positively charged at pH < 1.4 (protonation of –NH_2 group) and negatively charged at pH > 5.65 (deprotonation of –NH). To assess the effect of the initial solution pH on SMX removal, experiments were performed by adjusting the ambient pH = 5 to more acidic (pH = 3) or basic conditions (pH = 9), and the results are shown in Figure 13a for adsorption in the absence of an oxidant and 13b for oxidative degradation.

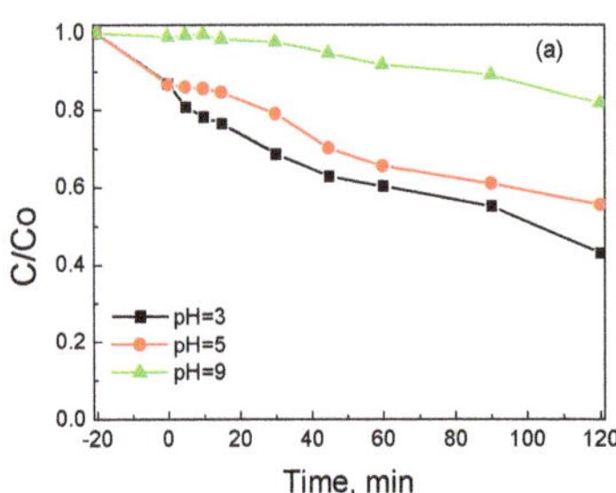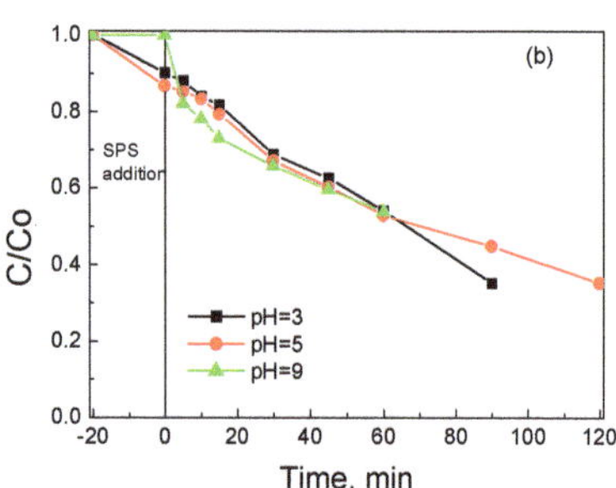

Figure 13. Influence of solution pH on (**a**) adsorption without SPS and (**b**) degradation with 200 mg L^{-1} SPS of 500 μg L^{-1} SMX on 90 mg L^{-1} BC/Ce-350.

Lower pH values seem to favor SMX adsorption, which is limited at alkaline conditions where both the surface and SMX are negatively charged; the respective k values are 1.4×10^{-3}, 3.9×10^{-3}, and 5.4×10^{-3} min^{-1} at pH 9, 5, and 3, respectively. Unlike adsorption, the oxidative degradation of SMX does not seem to be affected by the initial solution pH (Figure 13b), with the k value being about 8.9×10^{-3} min^{-1} for all three experiments. This is probably due to the fact that the solution pH is not buffered and there is a fast pH decrease from the initial value of 9 or 5 to 6 or 3, respectively, upon the addition of SPS, whose activation initiates oxidation reactions as well as generates HSO_4^{-}, a moderate acid. For the run performed at pH = 3, this value did not change throughout the course of the reaction. In this respect, unbuffered systems are beneficial since the spontaneous pH shift to lower values favors SMX removal. The influence of pH on the adsorption and degradation of SMX was similar to that in a previous work [29], where biochar from spent coffee grounds was employed as a SPS activator.

The effect of water matrix on SMX removal is depicted in Figure 14. Reactivity decreased with increasing matrix complexity (i.e., UPW (8.9×10^{-3}) > BW (3.2×10^{-3}) > WW (1.6×10^{-3})), with numbers in brackets showing k values in min^{-1}. The role of the water matrix is associated with the various inorganic and organic, non-target species that are inherently present and may compete with SMX for the active catalytic sites and/or the oxidant.

To shed light on such interplays, experiments were conducted in UPW spiked with various non-target species and the results are shown in Figure 15. The addition of 250 mg L^{-1} bicarbonate (this concentration is typical for BW) seriously impedes SMX removal, leading to a k value of 10^{-3} min^{-1} (i.e., nine times lower than in UPW). A possible explanation involves the detrimental role of carbonate ions that may (i) occupy catalytic sites, thus reducing SMX adsorption (as can be seen in Figure 15), and (ii) scavenge hydroxyl and sulfate radicals, while forming the less active carbonate radicals, in other words,

$$CO_3^{2-} + HO^{\bullet} \leftrightarrows CO_3^{\bullet-} + HO^- \tag{1}$$

$$HCO_3^{2-} + HO^{\bullet} \leftrightarrows CO_3^{\bullet-} + H_2O \tag{2}$$

$$SO_4^{\bullet-} + HCO_3^- \leftrightarrows SO_4^{2-} + HCO_3 \tag{3}$$

$$SO_4^{\bullet-} + CO_3^{2-} \leftrightarrows SO_4^{2-} + CO_3^{\bullet-} \tag{4}$$

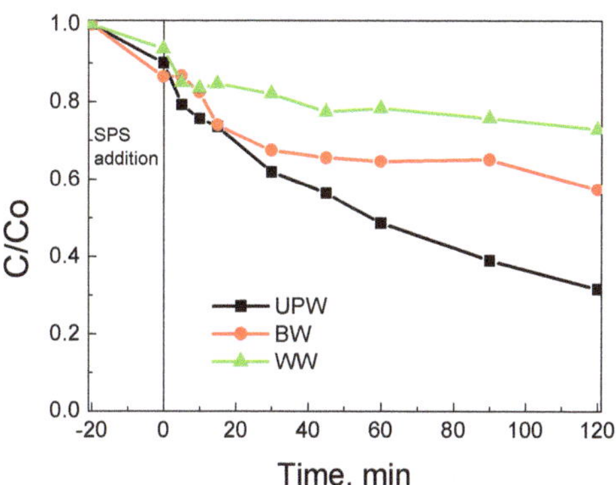

Figure 14. Effect of water matrix on 500 µg L^{-1} SMX removal with 90 mg L^{-1} BC/Ce-350 and 200 mg L^{-1} SPS at ambient pH.

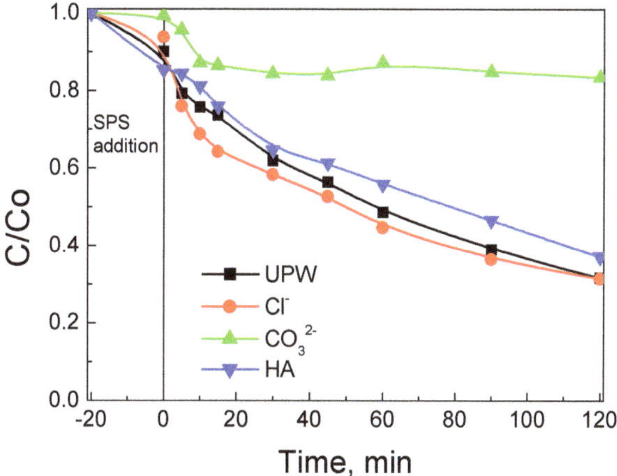

Figure 15. Effect of inorganic ions and humic acid on 500 µg L^{-1} SMX removal with 90 mg L^{-1} BC/Ce-350 and 200 mg L^{-1} SPS in UPW at ambient pH.

The addition of 250 mg L^{-1} chloride has practically no effect on degradation although hydroxyl and sulfate radicals may now react with chloride to form various Cl-containing radicals, in other words,

$$SO_4^{\bullet-} + Cl^- \leftrightarrows SO_4^{2-} + Cl^\bullet \tag{5}$$

$$Cl^\bullet + Cl^- \leftrightarrows Cl_2^{\bullet-} \tag{6}$$

$$HO^\bullet + Cl^- \leftrightarrows ClHO^{\bullet-} \tag{7}$$

$$4Cl_2^{\bullet-} \leftrightarrows 2Cl^- + Cl_2 \tag{8}$$

$$Cl^\bullet + H_2O \leftrightarrows ClHO^{\bullet-} + H^+ \tag{9}$$

$$Cl_2^{\bullet-} + H_2O \rightarrow ClHO^{\bullet-} + H^+ + Cl^- \tag{10}$$

In a final test, UPW was added to 10 mg L^{-1} humic acid (HA), an analogue of the organic matter typically found in natural waters (the chosen concentration corresponded to the organic carbon content of WW). The effect of HA was mildly negative, leading to a k value of 7×10^{-3} min^{-1}; HA, a recalcitrant molecule against chemical oxidation, is likely to competitively consume oxidants as well as occupy catalytic sites, both of which are detrimental to SMX degradation.

Finally, the role of different scavengers was investigated (Figure 16). The addition of methanol (that reacts with both hydroxyl and sulfate radicals) or t-butanol (that preferentially reacts with hydroxyl radicals) at 10 g L^{-1} retarded SMX degradation with the apparent rate constant being $5.7 \pm 0.1 \times 10^{-3}$ min^{-1} (i.e., a 35% decrease compared to the run in UPW). This implies that other species are also involved in the degradation mechanism. An additional experiment was performed adding 100 mg L^{-1} NaN$_3$, a well-known scavenger for singlet oxygen [50]; in this case, the k value decreased to 3.7×10^{-3} min^{-1}, thus pointing out the crucial role of active oxygen species in the degradation process. As a matter of fact, the surface oxygen atoms of CeO$_2$ may participate in the reaction, which is consistent with the high quantity of surface oxygen vacancies in the BC/Ce-350 sample, as has been demonstrated by TGA and Raman spectroscopy.

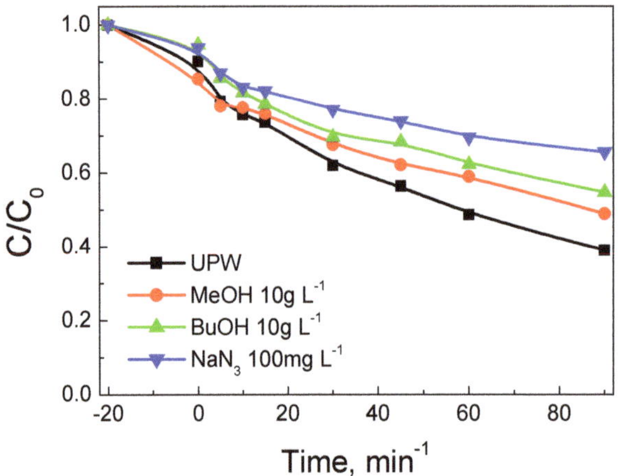

Figure 16. Effect of methanol, t-butanol, and sodium azide on 500 μg L^{-1} SMX degradation with 90 mg L^{-1} BC/Ce-350 and 200 mg L^{-1} SPS in UPW at ambient pH.

Moreover, the interactions with biochar are important, as can be seen from the DR spectra (Figure 9). Such interactions change the electron distribution in CeO$_2$ and regulate the surface reactivity. Figure 17 shows a correlation between the F(R) values of the prepared samples at 400 nm and the k values computed from the data of Figure 11. There appears to be a linear dependence, which implies that interactions are related to the occupied oxygen vacancies in CeO$_2$ by the –OH groups of the CeO$_2$ surface and/or the active oxygen species formed during SPS activation. The observed correlation suggests that the main active surface sites are these species, in accordance with the detrimental effect of NaN$_3$.

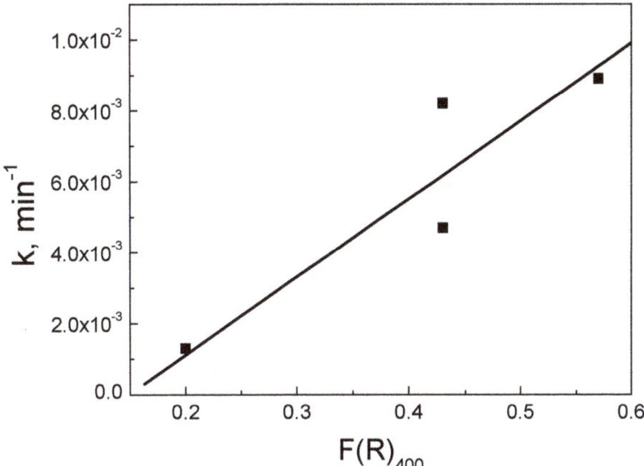

Figure 17. Correlation of k values with the absorbance of the BC/Ce samples at 400 nm. Experimental conditions as shown in Figure 11.

4. Conclusions

In this work, hybrid materials based on biochar and CeO_2 were prepared with a simple wet impregnation method, characterized by various techniques and eventually tested for their catalytic activity to activate persulfate and degrade a model antibiotic compound. The main conclusions are as follows:

1. Changing the calcination temperature in the range of –500 °C affected the biochar content and the physicochemical properties of CeO_2, but more importantly, determines the interactions between biochar and CeO_2 and, eventually, the catalytic activity.
2. Calcination at 300–350 °C yielded the more active materials for persulfate activation and sulfamethoxazole degradation; the latter following pseudo-first order kinetics with the rate depending on the operating conditions.
3. The water matrix is crucial for process performance since various inorganic and/or organic species can interfere with the surface and/or the target contaminant for the oxidants and the active catalytic sites. Hybrid materials may minimize such competitive interactions that do not exist in model experiments performed in pure water. Should this be the case, hybrid materials are likely to outperform bare biochar in environmentally relevant systems.
4. Radicals and singlet oxygen seem to be the main oxidative species, as indirectly evidenced by means of scavenging experiments.

Author Contributions: Conceptualization: J.V. and D.M.; Methodology: J.V.; Resources: D.M. and E.H.; Investigation: J.V., G.P. and P.N.; Writing: J.V., D.M. and E.H. All authors have read and agreed to the published version of the manuscript.

Funding: This research received no external funding.

Conflicts of Interest: The authors declare no conflict of interest.

References

1. Fauzi, A.A.; Jalil, A.A.; Hassan, N.S.; Aziz, F.F.A.; Azami, M.S.; Hussain, I.; Saravanan, R.; Vo, D.-V.N. A critical review on relationship of CeO_2-based photocatalyst towards mechanistic degradation of organic pollutant. *Chemosphere* **2022**, *286*, 131651. [CrossRef]
2. Kappis, K.; Papadopoulos, C.; Papavasiliou, J.; Vakros, J.; Georgiou, Y.; Deligiannakis, Y.; Avgouropoulos, G. Tuning the Catalytic Properties of Copper-Promoted Nanoceria via a Hydrothermal Method. *Catalysts* **2019**, *9*, 138. [CrossRef]
3. Kumar, A.; Singh, E.; Mishra, R.; Kumar, S. Biochar as environmental armour and its diverse role towards protecting soil, water and air. *Sci. Total Environ.* **2022**, *806*, 150444. [CrossRef]

4. Ntzoufra, P.; Vakros, J.; Frontistis, Z.; Tsatsos, S.; Kyriakou, G.; Kennou, S.; Manariotis, I.D.; Mantzavinos, D. Effect of Sodium Persulfate Treatment on the Physicochemical Properties and Catalytic Activity of Biochar Prepared from Spent Malt Rootlets. *J. Environ. Chem. Eng.* **2021**, *9*, 105071. [CrossRef]

5. Qin, J.; Lu, J.; Cao, M.; Hu, C. Synthesis of porous CuO–CeO$_2$ nanospheres with an enhanced low-temperature CO oxidation activity. *Nanoscale* **2010**, *2*, 2739–2743. [CrossRef] [PubMed]

6. Kašpar, J.; Fornasiero, P.; Graziani, M. Use of CeO$_2$-based oxides in the three-way catalysis. *Catal. Today* **1999**, *50*, 285–298. [CrossRef]

7. Mai, H.-X.; Sun, L.-D.; Zhang, Y.-W.; Si, R.; Feng, W.; Zhang, H.-P.; Liu, H.-C.; Yan, C.-H. Shape-Selective Synthesis and Oxygen Storage Behavior of Ceria Nanopolyhedra, Nanorods, and Nanocubes. *J. Phys. Chem. B* **2005**, *109*, 24380–24385. [CrossRef]

8. Lykaki, M.; Pachatouridou, E.; Iliopoulou, E.; Carabineiro, S.A.; Konsolakis, M. Impact of the synthesis parameters on the solid state properties and the CO oxidation performance of ceria nanoparticles. *RSC Adv.* **2017**, *7*, 6160–6169. [CrossRef]

9. Sayle, T.X.T.; Parker, S.C.; Catlow, C.R.A. The role of oxygen vacancies on ceria surfaces in the oxidation of carbon monoxide. *Surf. Sci.* **1994**, *316*, 329–336. [CrossRef]

10. Conesa, J. Computer modeling of surfaces and defects on cerium dioxide. *Surf. Sci.* **1995**, *339*, 337–352. [CrossRef]

11. Wu, Z.; Li, M.; Overbury, S.H. On the structure dependence of CO oxidation over CeO$_2$ nanocrystals with well-defined surface planes. *J. Catal.* **2012**, *285*, 61–73. [CrossRef]

12. Konsolakis, M. The role of Copper-Ceria interactions in catalysis science: Recent theoretical and experimental advances. *Appl. Catal. B Environ.* **2016**, *198*, 49–66. [CrossRef]

13. Konsolakis, M.; Lykaki, M. Recent Advances on the Rational Design of Non-Precious Metal Oxide Catalysts Exemplified by CuOx/CeO$_2$ Binary System: Implications of Size, Shape and Electronic Effects on Intrinsic Reactivity and Metal-Support Interactions. *Catalysts* **2020**, *10*, 160. [CrossRef]

14. Puigdollers, A.R.; Schlexer, P.; Tosoni, S.; Pacchioni, G. Increasing oxide reducibility: The role of metal/oxide interfaces in the formation of oxygen vacancies. *ACS Catal.* **2017**, *7*, 6493–6513. [CrossRef]

15. Chen, Y.; Lai, Z.; Zhang, X.; Fan, Z.; He, Q.; Tan, C.; Zhang, H. Phase engineering of nanomaterials. *Nat. Rev. Chem.* **2020**, *4*, 243–256. [CrossRef]

16. Konsolakis, M.; Lykaki, M. Facet-Dependent Reactivity of Ceria Nanoparticles Exemplified by CeO$_2$-Based Transition Metal Catalysts: A Critical Review. *Catalysts* **2021**, *11*, 452. [CrossRef]

17. Li, R.; Deng, H.; Zhang, X.; Wang, J.J.; Awasthi, M.K.; Wang, Q.; Xiao, R.; Zhou, B.; Du, J.; Zhang, Z. High-efficiency removal of Pb(II) and humate by a CeO$_2$–MoS$_2$ hybrid magnetic biochar. *Bioresour. Technol.* **2019**, *273*, 335–340. [CrossRef]

18. Melchionna, M.; Bevilacqua, M.; Fornasiero, P. The electrifying effects of carbon-CeO$_2$ interfaces in (electro)catalysis. *Mater. Today Adv.* **2020**, *6*, 1000502. [CrossRef]

19. Khataee, A.; Gholami, P.; Kalderis, D.; Pachatouridou, E.; Konsolakis, M. Preparation of novel CeO$_2$-biochar nanocomposite for sonocatalytic degradation of a textile dye. *Ultrason. Sonochem.* **2018**, *41*, 503–513. [CrossRef]

20. Pi, L.; Jiang, R.; Cai, W.; Wang, L.; Wang, Y.; Cai, J.; Mao, X. Bionic Preparation of CeO$_2$-Encapsulated Nitrogen Self-Doped Biochars for Highly Efficient Oxygen Reduction. *ACS Appl. Mater. Interfaces* **2020**, *12*, 3642–3653. [CrossRef]

21. Inyang, M.; Dickenson, E. The potential role of biochar in the removal of organic and microbial contaminants from potable and reuse water: A review. *Chemosphere* **2015**, *134*, 232–240. [CrossRef] [PubMed]

22. Shen, B.; Chen, J.; Yue, S.; Li, G. A comparative study of modified cotton biochar and activated carbon based catalysts in low temperature SCR. *Fuel* **2015**, *156*, 47. [CrossRef]

23. Hasa, B.; Martino, E.; Vakros, J.; Trakakis, G.; Galiotis, C.; Katsaounis, A. Effect of carbon support on the electrocatalytic properties of Pt-Ru catalysts. *Chem. Electron. Chem.* **2019**, *6*, 4970–4979.

24. Mian, M.; Liu, G. Recent progress in biochar-supported photocatalysts: Synthesis, role of biochar, and applications. *RSC Adv.* **2018**, *8*, 14237. [CrossRef]

25. Vakros, J. Biochars and their use as transesterification catalysts for biodiesel production: A short review. *Catalysts* **2018**, *8*, 562. [CrossRef]

26. Vakros, J.; Manariotis, I.D.; Dracopoulos, V.; Mantzavinos, D.; Lianos, P. Biochar from Spent Malt Rootlets and Its Application to an Energy Conversion and Storage Device. *Chemosensors* **2021**, *9*, 57. [CrossRef]

27. Tan, X.F.; Liu, Y.G.; Gu, Y.L.; Xu, Y.; Zeng, G.M.; Hu, X.J.; Liu, S.B.; Wang, X.; Liu, S.M.; Li, J. Biochar-based nano-composites for the decontamination of wastewater: A review. *Bioresour. Technol.* **2016**, *212*, 318–333. [CrossRef] [PubMed]

28. Grilla, E.; Vakros, J.; Konstantinou, I.; Manariotis, I.D.; Mantzavinos, D. Activation of Persulfate by Biochar from Spent Malt Rootlets for the Degradation of Trimethoprim in the Presence of Inorganic Ions. *J. Chem. Technol. Biotechnol.* **2020**, *95*, 2348–2358. [CrossRef]

29. Lykoudi, A.; Frontistis, Z.; Vakros, J.; Manariotis, I.D.; Mantzavinos, D. Degradation of Sulfamethoxazole with Persulfate Using Spent Coffee Grounds Biochar as Activator. *J. Environ. Manag.* **2020**, *271*, 111022. [CrossRef]

30. Magioglou, E.; Frontistis, Z.; Vakros, J.; Manariotis, I.; Mantzavinos, D. Activation of Persulfate by Biochars from Valorized Olive Stones for the Degradation of Sulfamethoxazole. *Catalysts* **2019**, *9*, 419. [CrossRef]

31. Rodriguez-Mozaz, S.; Vaz-Moreira, I.; Varela Della Giustina, S.; Llorca, M.; Barceló, D.; Schubert, S.; Berendonk, T.U.; Michael-Kordatou, I.; Fatta-Kassinos, D.; Martinez, J.L.; et al. Antibiotic residues in final effluents of European wastewater treatment plants and their impact on the aquatic environment. *Environ. Int.* **2020**, *140*, 105733. [CrossRef] [PubMed]

32. Bengtsson-Palme, J.; Kristiansson, E.; Larsson, D.G.J. Environmental factors influencing the development and spread of antibiotic resistance. *FEMS Microbiol. Rev.* **2018**, *42*, fux053. [CrossRef]

33. Kemmou, L.; Frontistis, Z.; Vakros, J.; Manariotis, I.D.; Mantzavinos, D. Degradation of antibiotic sulfamethoxazole by biochar–activated persulfate: Factors affecting the activation and degradation processes. *Catal. Today* **2018**, *313*, 128–133. [CrossRef]

34. Ntaflou, M.; Vakros, J. Transesterification activity of modified biochars from spent malt rootlets using triacetin. *J. Clean. Prod.* **2020**, *259*, 120931. [CrossRef]

35. Bourikas, K.; Vakros, J.; Kordulis, C.; Lycourghiotis, A. Potentiometric Mass Titrations: Experimental and Theoretical Establishment of a New Technique for Determining the Point of Zero Charge (PZC) of Metal (Hydr)Oxides. *J. Phys. Chem. B* **2003**, *107*, 9441–9451. [CrossRef]

36. Avramiotis, E.; Frontistis, Z.; Manariotis, I.D.; Vakros, J.; Mantzavinos, D. On the Performance of a Sustainable Rice Husk Biochar for the Activation of Persulfate and the Degradation of Antibiotics. *Catalysts* **2021**, *11*, 1303. [CrossRef]

37. Tala, W.; Chantara, S. Use of spent coffee ground biochar as ambient PAHs sorbent and novel extraction method for GC-MS analysis. *Environ. Sci. Pollut. Res.* **2019**, *26*, 13025–13040. [CrossRef] [PubMed]

38. Gulicovski, J.J.; Bračko, I.; Milonjić, S.K. Morphology and the isoelectric point of nanosized aqueous ceria sols. *Mater. Chem. Phys.* **2014**, *148*, 868–873. [CrossRef]

39. Papavasiliou, J.; Rawski, M.; Vakros, J.; Avgouropoulos, G. A Novel Post-Synthesis Modification of CuO-CeO$_2$ Catalysts: Effect on Their Activity for Selective CO Oxidation. *ChemCatChem* **2018**, *10*, 2096–2106. [CrossRef]

40. Araujo, V.D.; Bellido, J.D.A.; Bernardi, M.I.B.; Assaf, J.M.; Assaf, E.M. CuO–CeO$_2$ catalysts synthesized in one-step: Characterization and PROX performance. *Int. J. Hydrogen Energy* **2012**, *37*, 5498–5507. [CrossRef]

41. Aguilab, G.; Guerrerob, S.; Arayaa, P. Effect of the preparation method and calcination temperature on the oxidation activity of CO at low temperature on CuO–CeO$_2$/SiO$_2$ catalysts. *Appl. Catal. A* **2013**, *462–463*, 56–63. [CrossRef]

42. Rao, G.R.; Mishra, B.G. A comparative UV–vis-diffuse reflectance study on the location and interaction of cerium ions in Al- and Zr-pillared montmorillonite clays. *Mater. Chem. Phys.* **2005**, *89*, 110–115. [CrossRef]

43. Sun, S.; Mao, D.; Yu, J. Enhanced CO oxidation activity of CuO/CeO$_2$ catalyst prepared by surfactant-assisted impregnation method. *J. Rare Earths* **2015**, *32*, 1268–1274. [CrossRef]

44. Lopez, J.M.; Gilbank, A.L.; Garcia, T.; Solsona, B.; Agouram, S.; Torrente-Murciano, L. The prevalence of surface oxygen vacancies over the mobility of bulk oxygen in nanostructured ceria for the total toluene oxidation. *Appl. Catal. B* **2015**, *174–175*, 403–412. [CrossRef]

45. Spanier, J.E.; Robinson, R.D.; Zhang, F.; Chan, S.-W.; Herman, I.P. Size-dependent properties of CeO$_{2-y}$ nanoparticles as studied by Raman scattering. *Phys. Rev. B* **2001**, *64*, 245407. [CrossRef]

46. Wacławek, S.; Lutze, H.V.; Grübel, K.; Padil, V.V.T.; Cerník, M.; Dionysiou, D.D. Chemistry of persulfates in water and wastewater treatment: A review. *Chem. Eng. J.* **2017**, *330*, 44–62. [CrossRef]

47. Schott, H.; Astigarrabia, E. Isoelectric points of some sulfonamides: Determination by microelectrophoresis and by calculations involving acid–base strength. *J. Pharm. Sci.* **1988**, *77*, 918–920. [CrossRef]

48. Avisar, D.; Primor, O.; Gozlan, I.; Mamane, H. Sorption of sulfonamides and tetracyclines to montmorillonite clay. *Water Air Soil Pollut.* **2010**, *209*, 439–450. [CrossRef]

49. Heo, J.; Yoon, Y.; Lee, G.; Kim, Y.; Han, J.; Park, C.M. Enhanced adsorption of bisphenol A and sulfamethoxazole by a novel magnetic CuZnFe$_2$O$_4$–biochar composite. *Bioresour. Technol.* **2019**, *281*, 179–187. [CrossRef] [PubMed]

50. Bancirova, M. Sodium azide as a specific quencher of singlet oxygen during chemiluminescent detection by luminol and *Cypridina* luciferin analogues. *Luminescence* **2011**, *26*, 685–688. [CrossRef]

 *nanomaterials*

Article

Evaluation of Local Mechanical and Chemical Properties via AFM as a Tool for Understanding the Formation Mechanism of Pulsed UV Laser-Nanoinduced Patterns on Azo-Naphthalene-Based Polyimide Films

Iuliana Stoica [1,*], Elena-Luiza Epure [2], Catalin-Paul Constantin [1], Mariana-Dana Damaceanu [1], Elena-Laura Ursu [1], Ilarion Mihaila [3] and Ion Sava [1,*]

[1] "Petru Poni" Institute of Macromolecular Chemistry, 700487 Iasi, Romania; constantin.catalin@icmpp.ro (C.-P.C.); damaceanu@icmpp.ro (M.-D.D.); ursu.laura@icmpp.ro (E.-L.U.)

[2] Faculty of Chemical Engineering and Environmental Protection, "Gheorghe Asachi" Technical University, 700050 Iasi, Romania; lepure@tuiasi.ro

[3] Integrated Center of Environmental Science Studies in the North-Eastern Development Region (CERNESIM), "Alexandru Ioan Cuza" University of Iasi, 700506 Iasi, Romania; ilarion.mihaila@gmail.com

* Correspondence: stoica_iuliana@icmpp.ro (I.S.); isava@icmpp.ro (I.S.)

Citation: Stoica, I.; Epure, E.-L.; Constantin, C.-P.; Damaceanu, M.-D.; Ursu, E.-L.; Mihaila, I.; Sava, I. Evaluation of Local Mechanical and Chemical Properties via AFM as a Tool for Understanding the Formation Mechanism of Pulsed UV Laser-Nanoinduced Patterns on Azo-Naphthalene-Based Polyimide Films. *Nanomaterials* **2021**, *11*, 812. https://doi.org/10.3390/nano11030812

Academic Editor: John Vakros

Received: 30 January 2021
Accepted: 19 March 2021
Published: 22 March 2021

Publisher's Note: MDPI stays neutral with regard to jurisdictional claims in published maps and institutional affiliations.

Abstract: Aromatic polyimides containing side azo-naphthalene groups have been investigated regarding their capacity of generating surface relief gratings (SRGs) under pulsed UV laser irradiation through phase masks, using different fluencies and pulse numbers. The process of the material photo-fluidization and the supramolecular re-organization of the surface were investigated using atomic force microscopy (AFM). At first, an AFM nanoscale topographical analysis of the induced SRGs was performed in terms of morphology and tridimensional amplitude, spatial, hybrid, and functional parameters. Afterward, a nanomechanical characterization of SRGs using an advanced method, namely, AFM PinPoint mode, was performed, where the quantitative nanomechanical properties (i.e., modulus, adhesion, deformation) of the nanostructured azo-polyimide surfaces were acquired with a highly correlated topographic registration. This method proved to be very effective in understanding the formation mechanism of the surface modulations during pulsed UV laser irradiation. Additionally to AFM investigations, confocal Raman measurements and molecular simulations were performed to provide information about structured azo-polyimide chemical composition and macromolecular conformation induced by laser irradiation.

Keywords: azo-polyimide; surface relief grating; AFM PinPoint; topographical analysis; nanomechanical characterization; molecular simulation

1. Introduction

Polyimides (PIs) represent an important class of high-performance polymers that are exploited in a variety of applications due to their excellent physicochemical properties such as optical and thermal stability in combination with high glass transition temperature, low susceptibility to laser damage, and low dielectric constant value [1,2]. Moreover, polyimides have been investigated as potential materials in the fields of optoelectronics and photonics [3]. Particularly interesting are polyimides containing azobenzene units, which have already been investigated for photoinduced alignment in liquid crystal display [4], as photomechanical response materials [5], and for holographic diffraction grating recording [6–11]. The azo derivatives can be introduced into polymers in different ways: by polymerization reaction between monomers, which contain preformed azobenzene group (diamines or dianhydrides); by dissolving a guest chromophore in a polymer matrix; by attaching the chromophores covalently to the polymer chains; or by using of different nonconvalent intermolecular interactions (ionic interactions, coordination bonds, hydrogen

Nanomaterials **2021**, *11*, 812. https://doi.org/10.3390/nano11030812

https://www.mdpi.com/journal/nanomaterials

bonds, π–π interactions) between the dye and the polymer backbone. The first method is the most used due to the better control of sequence chromophore distribution. The azobenzene-functionalized monomers can be the diamines or dianhydrides or both of them [3,12–16]. Many azodiamines have been obtained and used for the synthesis and characterization of the azopolyimides, usually containing substituted or unsubstituted azobenzene side chains. Thus, a broad range of azopolyimides, azo polyamide-imides, or azo polyester-imides have been published thus far [12,15,17–19]. The majority of polymers functionalized with azobenzene derivatives under the action of linearly polarized light undergo multiple reversible *trans* to *cis* photo-isomerization processes. A supramolecular organization process is generated as a result of the perpendicular alignment of the azobenzene molecules to the electric field vector, controlling the photo-induced optical anisotropy. Thus, the azo-polymer photo-fluidization due to the above-described phenomenon can appear in the exposed regions [20,21]. Moreover, the cyclic photo-isomerization can lead to a large-scale mass transport of the polymer chains, which appear as a surface relief grating (SRG) [13,14,22,23]. Mainly, this mass displacement can take place from UV laser-exposed areas to the unexposed areas. Recently, it was demonstrated that also an inverse mass displacement can occur, from the dark regions to the illuminated ones [24]. The arguments for the development of the azobenzene photoisomerization process by inversion or rotation are the subject of numerous molecular simulation studies [25–28]. A large number of articles track the computational characterization of these photocontrolled materials due to their prospective applications in different fields such as liquid crystals [29–31], optical data storage [32], photosensitive micelles [33], triggers in protein folding [34], and so on. Due to the complexity of the molecular migration phenomenon that induces the SRG formation process, the number of molecular studies is significantly lower [35–37].

Over time, in order to obtain well-defined and stable SRGs on azo-polymers, researchers have applied two main techniques, namely, pulsed and continuous UV laser irradiation. In the case of continuous laser irradiation, the apexes of the sinusoidal surface relief appeared in the unexposed regions. When pulsed UV laser irradiation was employed, the positions of the apexes were dependent on the laser fluence, being in the exposed regions when the fluence value was below a threshold, and in the unexposed regions when the fluence value was above this threshold, as mentioned in the literature data [24]. In a pulsed laser regime, the relief geometric parameters depend on several factors. The design of the high-quality phase masks and their characteristics (the material used in fabrication, the fabrication method, the type of the lattice grating structures, the number of pitches, the period of the lattice, the period accuracy and uniformity, the efficiency (%) of the phase masks) are mainly responsible for the resulted pattern aspect. The nanostructures can also be strongly affected by the experimental parameters of the pulsed laser irradiation method, such as the incident laser wavelength, the incident laser fluence, the polarization of the laser, the angle of incidence, the number of laser pulses, and the duration of one laser pulse. Last but not least, the generation of the SRGs is dependent of the chemical structure of the azopolymer, in terms of the backbone flexibility and position of the azo-group in the polymer chain (inducing certain architecture), the pristine film surface morphology subjected to pulsed laser nanostructuring, and the film thickness.

Many researchers have been focused on the progress in atomic force microscopy (AFM) techniques to characterize the polymer films at the nano-scale by recording the local mechanical behavior, especially the force response to approach, even simultaneously with the recording of the morphology [38,39]. Over the years, the progress of several research groups in data acquisition and signal processing has allowed manufacturers to develop this technique, under different names, such as pulsed force mode (PFM), tested by WiTec company (Ulm, Germany) for scanning force microscopy (SFM) systems from CSEM, Digital Instruments, Molecular Imaging, Park, Seiko and TopoMetrix [40], PeakForce Quantitative Nanomechanical Mapping (PeakForce QNMTM, referred to as QNM for brevity, by Bruker), HybriD Mode Atomic Force Microscopy from NT-MDT (Zelenograd, Moscow, Russia) [41–43], Quantitative Imaging (QI, by JPK Instruments), or PinPoint mode from Park Systems Corp. (Suwon, Korea) [44]. Each has different features. Fast force spectroscopy mapping via PinPoint mode was designed to prevent positional errors by simultaneously acquiring accurate height and force–distance information in the whole scanning area, while the cantilever tip is lifted at every pixel at a perpendicular angle from the sample surface.

In this context, our study focused on the investigation of a newly synthesized aromatic polyimide containing azo-naphthalene side groups with the aim to evaluate the local morphological, mechanical, and chemical properties via atomic force microscopy, especially in PinPoint mode, and confocal Raman spectroscopy. It is meant to provide a deep understanding of the formation mechanism of SRGs induced by pulsed UV laser on these azo-polyimide films, through phase masks, using different fluencies and number of pulses. Moreover, this article attempts to correlate the macroscale behavior of this azo-naphthalene-based polyimide during photoisomerization with the one from the atomic level simulating certain polymeric systems with a different content of the *cis* isomer. The application of quantitative nanomechanical properties (i.e., modulus, adhesion, deformation) acquired with highly correlated topographic registration using PinPoint mode to describe the formation of nanostructured azo-naphthalene-based polyimide surfaces has not been reported yet in the literature to the best of our knowledge. The molecular simulation and confocal Raman spectroscopy complete this study, which promises to significantly advance the research in this field beyond the state of the art.

2. Materials and Methods

2.1. Material

An aromatic diamine containing the azo group (-N=N-) pendent to the triphenylmethane core was synthesized by the Williamson reaction of 4-[bis-(4-amino-3-methyl-phenyl)-methyl]-phenol with 3-[4-naphthalen-1-ylazo)-phenoxy]-bromopropane. Details regarding the preparation of this azodiamine and the corresponding intermediates have been recently reported elsewhere [45]. 4,4′(Hexafluoroisopropylidene)diphthalic anhydride (6FDA) was purchased from Aldrich and used as received. The azopolyimide has been synthesized by the polycondensation reaction of 6FDA with the above mentioned azodiamine by using a procedure previously reported [12,45]. The structure of the azo-naphthalene-based polyimide is shown in Figure 1. The thickness of the azo-naphthalene-based polyimide films, measured using a profilometer, was 3094–3117 nm, the average value being 3.10 ± 0.01 µm.

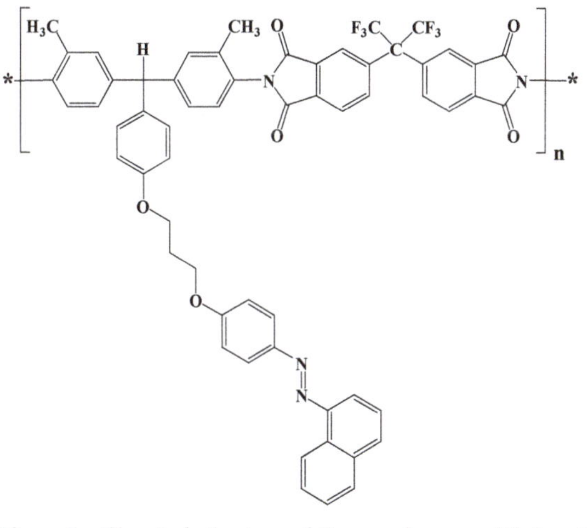

Figure 1. Chemical structure of the novel azo-naphthalene-based polyimide, AzoPI, used in this study.

As previously reported in [45], the irradiation with UV light of 365 nm will induce the photoisomerization of the *trans* isomer to the *cis* isomer. The intensity of the absorption bands located at 384 nm (due to the $\pi \rightarrow \pi^*$ transitions in the *trans* isomer) decreased progressively as the irradiation time advanced, being assisted by the increase of the absorption band corresponding to the $\pi \rightarrow \pi^*$ and $n \rightarrow \pi^*$ transitions in the *cis* isomer, located at approximately 290 and 472 nm, respectively. Moreover, the corresponding isosbestic points were found at 325–326 and 450–459 nm, and the absorption coefficient was 1.1143×10^4 cm^{-1}. The absorption coefficient has been calculated by using the equation: k = 2.303 A/d, where A is the absorbance and d is the thickness of the sample. Futhermore, the detailed evaluation of *cis–trans* thermal isomerization was carried out by UV–VIS spectroscopy both in solution and in solid state for the starting azodiamine monomer and corresponding azopolyimide, as it was largely presented in reference [45]. The thermal relaxation induced the total recovery of *trans* isomer after 2700 s. The thermal relaxation was reflected in the gradual increase of the bands at 384 nm ($\pi \rightarrow \pi^*$ transitions, *trans* isomer) and the progressive decrease of the bands at 290 nm ($\pi \rightarrow \pi^*$) transitions, *cis* isomer) and 472 nm ($n \rightarrow \pi^*$ transitions, *cis* isomer) during exposure at 70 °C for different intervals of time.

2.2. Molecular Simulations

The molecular dynamics simulations were performed with Materials Studio 4.0 software [46]. First, the structural units corresponding to the *trans* and *cis* isomers were built. These structural units were minimized through ab initio calculations within the DMol^3 module using the Perdew-Wang PWC functional until fulfilling a 2×10^{-5} Hartree energy convergence. Subsequently, 3 polymer chains of 10 structural units each were built, with different content of *cis* isomer: AzoPI (with 15% *cis* groups), AzoPI_50%*cis* (with 50% *cis* groups), and AzoPI_100%*cis* (with 100% *cis* groups). The *cis* azo units were randomly distributed along the polymer chain. Both the structural units and the polymers were minimized with the Forcite module using the Dreiding force field. Three-dimensional amorphous structures were generated using the Amorphous Cell module. Each cell was built to contain three polymer chains of the same kind, at a density of 1 g/cm^3, at 298 K. The final structures were obtained according to the following protocol: (step 1) minimization (energy convergence 1×10^{-4} kcal/mol), (step 2) 2 anneal cycles that were driven in the range of 300–800 K for 40 ps, (step 3) compression/constriction the cell by molecular dynamics (NPT ensemble (constant number (N), volume (V), and temperature (T); T is regulated via a thermostat; pressure (P) is regulated), T = 298 K, Berendsen thermostat and barostat) until the density reaches a value around 1.2 g/cm^3, (step 4) equilibration of the cell through an NVT dynamics (constant number (N), volume (V), and temperature (T); T is regulated via a thermostat, pressure (P) is unregulated) for 500 ps (T = 298K, Nose thermostat). Achieving constant values for the density-time curves in NPT stage and temperature/energy-time in NVT-MD stage indicates that the system is stable and has reached an equilibrium value. Data collection was performed after another 200 ps of dynamic simulations in NVE ensemble (constant number (N), volume (V), and energy (E); T is regulated via a thermostat, pressure (P) is unregulated), at T = 298 K.

2.3. Laser Patterning

In order to induce micro/nano structuration on the azo-polyimide surface, we used the setting presented in Figure 2. A pulsed Nd:YAG laser (Brilliant B from Quantel) working at third harmonic (355 nm wavelength), with a diameter of 5 mm. The laser was horizontally polarized (*s*-polarized), meaning that its light oscillated along a horizontal plane. The electric field vector was perpendicular to the plane of incidence, to the direction of gravity, and to the direction of light propagation. In this way, the grating region was uniformly illuminated. This choice can be motivated on the basis of observations of Miniewicz and collaborators [47], according to which the grating strength in the case of *s-p* inscription attains higher values. The pulse duration of 6 ns was chosen. The pulse repetition rate was 10Hz. The aperture of 5 mm laser beam was enlarged until 15 mm, by placing a beam expander with a fixed ratio of 3×. A diffraction phase mask (Edmund Scientific Co., Barrington, NJ, USA) with a thickness of 76 μm and 1000 grooves per mm was used. The linear diffraction grating period was about 1 μm. After passing through the diffractive optical element, the laser beam generates an interference field in its proximity, leading to a laser interference pattern on the azo-polyimide surface, placed after the phase mask, producing features with the pitch of the same order of magnitude. The quartz plate placed between the phase mask and the sample had a thickness of 1 mm. Two incident fluences, namely, 10 mJ/cm^2 and 45 mJ/cm^2 (measured after beam expander), and two different numbers of pulses, namely, 10 and 100, were used. Therefore, the samples were named using the label AzoPI i/j, where i is the incident fluence and j is the number of irradiation pulses, as follows: AzoPI 10/10, AzoPI 10/100, AzoPI 45/10, and AzoPI 45/100.

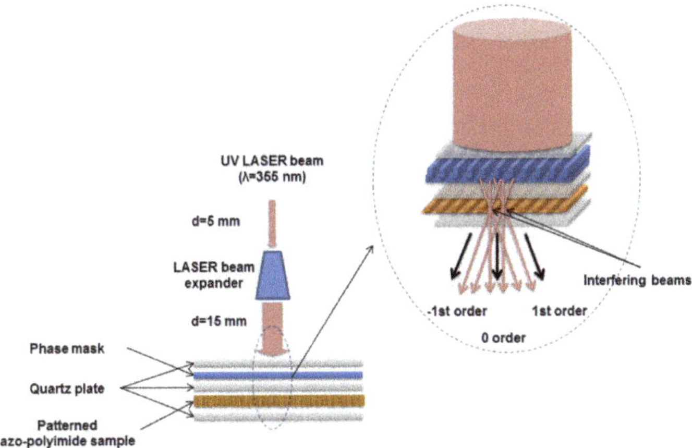

Figure 2. Pulsed laser irradiation through a phase mask setting, with a detail on the formation of interference image on the azo-naphthalene-based polyimide surface.

2.4. Measurements

The AFM investigations on the nanostructured azo-polyimide films were performed in semi-contact mode, in the atmospheric conditions, at room temperature, on a surface area of 10×10 μm^2, using a Scanning Probe Microscope Solver Pro-M from NT-MDT, Russia, using a high-resolution "golden" silicon AFM probe NSG01 (NT-MDT, Zelenograd, Moscow, Russia) with a typical curvature radius of 10 nm and the free resonant frequency of 90.5 kHz. AFM data acquisition and analysis were performed using Nova software from NT-DMT. The tridimensional parameters were calculated using Image Analysis 3.5.0.19892 software. Nanomechanical measurements were made on a Park NX10 Atomic Force Microscope (Park Systems Corp., Suwon, Korea), using PinPoint Nanomechanical mode, which allowed us to obtain sample's stiffness mapping through force–distance curves acquired at each pixel over the entire scanning area (of 1×5 μm^2 in our case). Thus, the quantitative nanomechanical properties (i.e., modulus, adhesion, deformation) of the nanostructured azo-polyimide surfaces were acquired with highly correlated topographic registration. In order to provide the most accurate mechanical properties data, we carefully selected the Force Modulation Mode - Reflex Coating (FMR) probe (Park Systems Corp., Suwon, Korea) with a force constant of 2.8 N/m and resonance frequency of 75 kHz according to the relative stiffness of their cantilever when compared to that of the sample, so that it can offer immediate response to any changes on the surface's material properties. The scanning frequency was 0.5 Hz, and the scanning speed was 5.0 μm/s. Besides AFM investigations, supplementary confocal Raman measurements (laser source: 632.8 nm, 50 mW; CCD detector) performed using inVia Renishaw Raman confocal microscope (Renishaw, UK) were used to provide information about structured azo-polyimide chemical composition and macromolecular conformation. Spectra were recorded in backscattering geometry using a $50\times$ objective. Spectral manipulations such as baseline adjustment, smoothing, and normalization were performed with the WiRE 3.3 software (Renishaw, UK).

3. Results and Discussion

Imaging the distribution of local mechanical properties at the nanoscale can significantly advance research on exciting soft materials relevant for optoelectronic applications [48]. In our case, the SRGs were produced using phase masks through pulsed laser irradiation, with a maximum light fluency lower than the ablation threshold of the material. This method employs a diffractive optical element (phase mask) for precise spatial modulation of the UV laser writing beam. In order to obtain the interference pattern, only the zero (0) and plus and minus first diffraction orders (±1) were considered. When a UV laser beam was incident on the phase mask, the zero-order diffracted beam was minimized. Furthermore, the plus and minus first diffracted orders were enlarged as much as possible [49]. Because of the very short irradiation times (6 nanoseconds) used in the pulsed mode, the mechanism responsible for the periodic SRGs profile generation involves the fringe pattern produced by the interference of the diffracted beams. They act through two possible mechanisms, namely, a very fast supramolecular reorganization process, induced by the azo-groups dipole orientation [50], and material photo-fluidization, induced by the *trans–cis* isomerization process of the azo-segments in the exposed regions [20,24,51]. This competes with the larger proper volume required by the *cis* conformation [52]. None of the models concerning the azo-polymers surface nanostructuration mechanisms proposed until now completely explain the surface relief formation, although a consensus has been reached—the azonaftalene dipoles orientation probably induce a more organized and compacted structure, having as a result a material contraction in the light-exposed regions with the generation of grooves in the dark areas [22]. Thus, it is considered that the highest intensity of the interference pattern corresponds to the valley of SRGs, due to the mass displacement from exposed to unexposed regions, and inverse mass displacement, from uncovered to covered regions [20,24,53,54]. On the other hand, it should be mentioned that the assignment highest intensity of the interference pattern corresponds to the valley of SRGs is material-dependent and cannot be ascertained from the data presented in this paper, but was previously demonstrated by direct observations of Hurduc and collaborators 20] who proposed a mechanism for SRG formation during laser irradiation involving at least three processes: (1) the polymer photofluidization in illuminated regions, (2) the mass displacement from illuminated to dark regions, and (3) the inverse mass displacement from dark to illuminated regions, on the basis of the performed amplitude modulation-frequency modulation atomic force microscopy (AM-FM AFM) viscoelastic mapping before, during, and after light irradiation.

3.1. AFM Nanoscale Morphological Analysis

The occurrence of the SRGs induced by the UV laser irradiation on azo-polyimide surface was highlighted using AFM investigations. Figure 3 presents the height 3D AFM images and the corresponding cross-section profiles obtained for the azo-polyimide films irradiated with different laser energy density/number of pulses of irradiation, as follows: 10 mJ/cm^2/10 pulses (a,b), 10 mJ/cm^2/100 pulses (c,d), 45 mJ/cm^2/10 pulses (e,f), 45 mJ/cm^2/100 pulses (g,h). According to the height AFM images, Figure 4 depicts the height histograms and surface bearing area ratio curves of the azo-polyimide films irradiated using different laser energy density/number of pulses of irradiation: 10 mJ/cm^2/10 pulses (a,b), 10 mJ/cm^2/100 pulses (c,d), 45 mJ/cm^2/10 pulses (e,f), 45 mJ/cm^2/100 pulses (g,h).

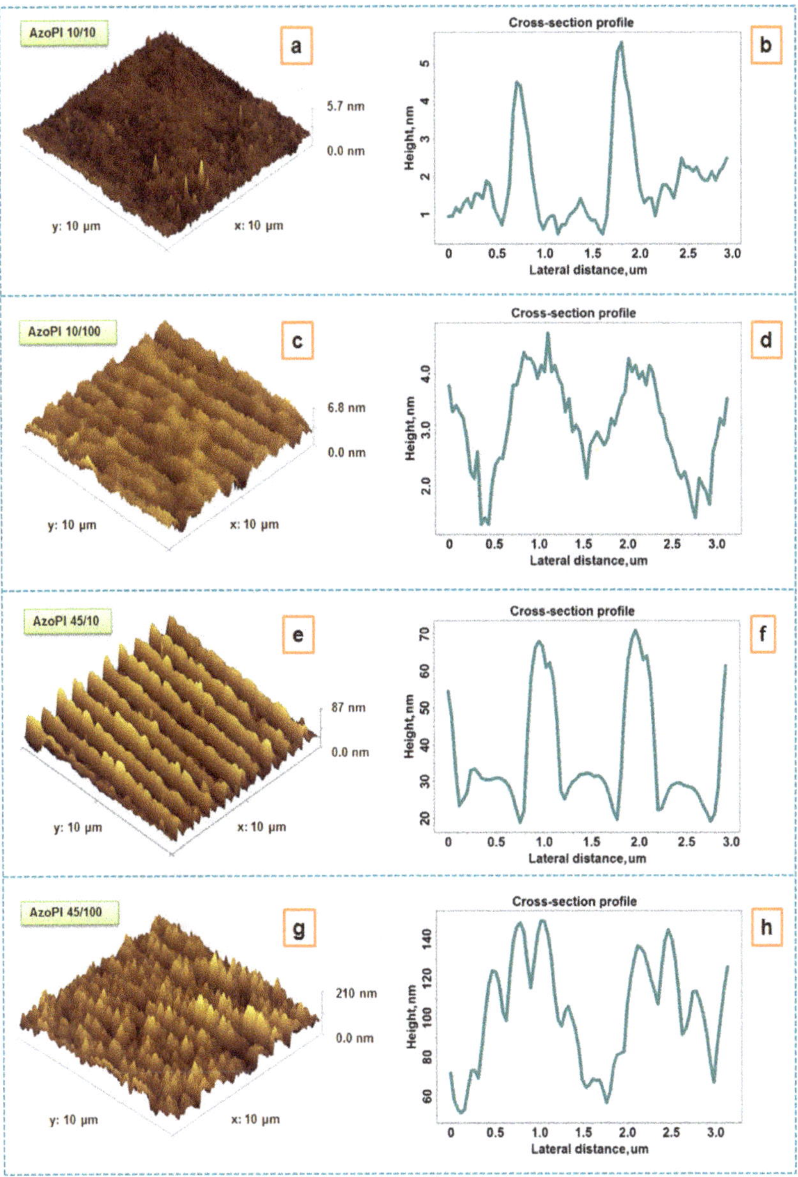

Figure 3. Height 3D atomic force microscopy (AFM) images and corresponding cross-section profiles for azo-polyimide irradiated using different laser energy density/number of pulses of irradiation: 10 mJ/cm^2/10 pulses (**a,b**), 10 mJ/cm^2/100 pulses (**c,d**), 45 mJ/cm^2/10 pulses (**e,f**), 45 mJ/cm^2/100 pulses (**g,h**).

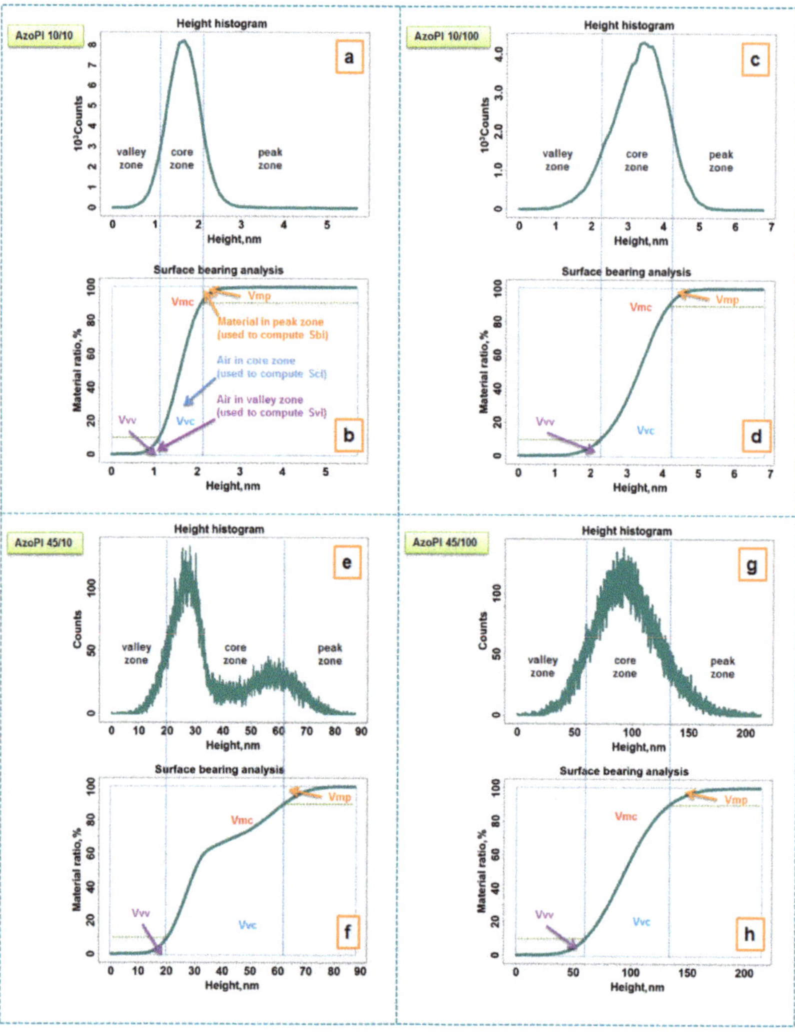

Figure 4. Height histograms and surface bearing area ratio curves corresponding to height AFM images for azo-polyimide irradiated using different laser energy density/number of pulses of irradiation: 10 mJ/cm^2/10 pulses (**a**,**b**), 10 mJ/cm^2/100 pulses (**c**,**d**), 45 mJ/cm^2/10 pulses (**e**,**f**), 45 mJ/cm^2/100 pulses (**g**,**h**).

The anchoring of the azo group through an aliphatic spacer on the polyimide chains with flexible conformations given by the presence of hexafluoroisopropylidene groups induced a high isomerization/structuring capacity, transposed in the generation of repetitive surface structures through the phase mask, even at low energies and a small number of pulses. As can be seen in Figure 3a,b, from the AFM 3D image and the corresponding cross-section profile, by using an energy density of 10 mJ/cm^2 and 10 pulses of irradiation, we found that the SRGs were not so visible, being only a few nanometers high. At this low energy, as the number of pulses increased, the structures became wider and better highlighted, although their height did not differ much from those obtained by using the small number of pulses (Figure 3c,d). Analyzing the profile diagram of the structured sample with higher energy density of 45 mJ/cm^2 and using 10 pulses of irradiation, we observed a very good uniformity of the SRGs' appearance, with the amplitude of the modulation being ≈50 nm (Figure 3e,f). This was due to the great mobility of the azo group in the side chain and flexibility of the main chain, as well as the existing free volume. Consequently, the polymer responded better to a high energy density of 45 mJ/cm^2 and a small number of pulses (10). Moreover, it can be observed that in this case the SRGs were narrow, as in the previous case, when a small number of irradiation pulses were used. As the energy was increased, the repetitive structures became more and more defined, increasing in height with the increase of the number of pulses until 100 nm (Figure 3g,h) as observed from the cross-section profiles. It seemed that from the structural uniformity point of view, the irradiation with high energy and small number of pulses was the most indicated. At large number of pulses (100), their aspect was a corrugated one, as can be seen from the cross-section profile (Figure 3g). The obtained image suggests the occurrence of possible phenomena that may be due to surface reorganization mechanisms (Figure 3h) or to additional nanostructuring of the formations. This complex high-amplitude morphology may be an indication that the structured sample acted itself as a diffractive element—it generated further diffraction orders that gave shorter periods (higher harmonics) to the holographic irradiation pattern. The response of the material after irradiation may have been due to the mechanisms of reorganization on the surface or induced by the appearance of the photo-fluidization state. Photo-fluidization can become significant, especially when using a large number of pulses. However, regardless of the mechanism of the surface organization, due to the conformational changes that occur in the azo-polyimide during irradiation, it is difficult to assess the accurate response to the phase mask laser irradiation process of this material. In addition, according to Viswanathan et al. [55], since mechanical forces may also act in the bulk of the film, it is possible that the orientation grating extends also throughout the thickness of the material.

The values obtained for the root mean square roughness (Table 1) were influenced by the SRGs aspect and amplitude, following the same trend. In this way, it was observed that Sq increased as the energy/number of pulses increased.

Table 1. 3D roughness parameters obtained from AFM images of the investigated azo-polyimide samples (energy density of 10 or 45 mJ/cm^2 and variable number of laser pulses 10 or100).

Parameter	Sample			
	AzoPI 10/10	AzoPI 10/100	AzoPI 45/10	AzoPI 45/100
Height parameters				
Sq (nm)	0.4	0.732	16.159	28.723
Shape parameters				
Ssk	0.466	−0.284	0.762	0.343
Sku	5.466	3.061	2.354	3.193
Spatial parameters				
Stdi	0.388	0.418	0.193	0.407
Hybrid parameters				
Sdr (%)	0.00542	0.00981	1.496	5.342
Functional indexes				
Sbi	0.113	0.308	0.794	0.429
Sci	1.464	1.346	1.839	1.671
Svi	0.0855	0.131	0.046	0.096
Functional volume parameters				
Vmp (nm^3/nm^2)	0.0232	0.0242	0.562	1.62
Vmc (nm^3/nm^2)	0.365	0.664	13.300	25.4
Vvc (nm^3/nm^2)	0.463	0.875	25.400	36.6
Vvv (nm^3/nm^2)	0.0334	0.096	0.754	2.75

Sq: root mean square roughness of the surface; Ssk: skewness of height distribution; Sku: kurtosis of height distribution; Stdi: surface texture direction index; Sdr: surface area ratio; Sbi: surface bearing index; Sci: core fluid retention index; Svi: valley fluid retention index; Vmp: peak material volume; Vmc: core material volume; Vvc: core void volume; Vvv: valley void volume.

The aspect of the height histograms (Figure 4a,c,e,g) also describes the surface features created by the phase mask UV laser irradiation by means surface skewness (Ssk) and coefficient of kurtosis (Sku) (Table 1), indicating also the distribution of the relief on areas of interest: the valley zone, the core zone, and the peak zone. Details regarding their meaning can be found in Appendix A.

The hybrid parameter surface area ratio (Sdr), calculated as the ratio between the area of the real developed surface and the area of the projected surface, can be used to describe the complexity of the surface. This parameter has an important role in controlling the surface properties of the materials envisaged for use in electronics [56], being a key factor for measuring the performance. Analyzing the data from Table 1, one can conclude that the surface complexity increased with the increase of the laser energy density due to the appearance of well-defined SRGs. The most complex surface was obtained for AzoPI 45/100 sample, induced by the supplementary nanostructurations of the SRGs formations, visible in Figure 3g,h.

The values of the spatial parameter surface texture direction index (Stdi) (displayed in Table 1 and calculated examining the 3D AFM height images and the corresponding angular spectra) were indicators of whether or not a surface has a preferential orientation of its features, thus denoting the anisotropy/isotropy of the morphology. Stdi close to 1 indicates that the sample surface is isotropic, with a random surface texture that does not have any texture that stands out, with no preferential orientation and presenting identical characteristics regardless of the direction of measurement. Meanwhile, Stdi close to zero shows a dominant direction of the surface morphology [14,57]. This can indicate an oriented surface or a periodic structure. In this case, the surface is considered to be anisotropic. Therefore, for all nanostructured samples, the values of about 0.2–0.4 attributed of this spatial parameter denote the anisotropy of the morphology induced by the oriented SRGs under pulsed UV laser irradiation. Moreover, AzoPI 45/10 sample shows the highest degree of orientation and organization, indicated by the lowest value of Stdi, namely, 0.193. This information is very important in describing the anisotropy of the morphology, used in electronic applications, where the controlling of the alignment is necessary.

Upon computation by inversion of the cumulative height distribution histograms, division into zones (Figure 4a,c,e,g), the surface bearing area ratio curves or Abbott curves (Figure 4b,d,f,h) were used as the basis for calculating the functional indexes (Sbi, Sci, Svi) and functional volume parameter (Vmp, Vmc, Vvc, Vvv) [58].

These parameters, presented in Table 1, are of great importance for the pursued applications. Low surface bearing index values (Sbi < 0.608) [57] obtained for AzoPI 10/10, AzoPI 10/100, and AzoPI 45/100 films revealed surfaces with low bearing capacity. This fact is also supported by low values obtained for the peak material volume. Instead, AzoPI 45/10 sample has bearing index higher than 0.608, indicating a good bearing capacity. In the core zone, Sci is sensitive to both occasional high peaks and occasional deep valleys. AzoPI 10/10 and AzoPI 10/100 have a low core fluid retention index (Sci < 1.56) [57], induced also by the low values of the core material and void volumes, while AzoPI 45/10 and AzoPI 45/100, due to their complex morphology and high core material and void volumes, have a high core fluid retention index (Sci > 1.56) [57]. The analyzed surfaces with relatively few deep valleys have low valley fluid retention index (Svi < 0.11) [57] due to low valley void volumes.

In this way, the pattern can be designed by controlling the experimental parameters, and by calculating these parameters that describe the obtained morphology, we can select a certain pattern for a special desired electronics manufacturing application.

The topography does not provide the difference in mechanical properties (such as elasticity, adhesion). AFM phase imaging can provide mechanical properties distribution using qualitative contrast. On the other hand, force–distance spectroscopy records mechanical data quantitatively by indenting a cantilever tip on the sample surface one point at a time. In our case, the analysis of the mechanical characteristics of the newly formed nanogrooves required to acquire the distribution image and the quantitative data simultaneously.

3.2. Nanomechanical Characterization Using AFM PinPoint Mode

PinPoint Nanomechanical mode was designed to prevent positional errors by simultaneously acquiring accurate height and force–distance information in each of the 256 × 256 pixels in the whole scanning area, thus preventing the occurrence of the artifacts and positional errors in force–distance spectroscopy and topographical data. In Figure 5 and Appendix B, we present the working mechanism of PinPoint Nanomechanical mode, describing the generation of the map of the sample's stiffness/elasticity and deformation depth from the surface concomitantly obtained through the adhesion force map with the topographic sample information [59–64].

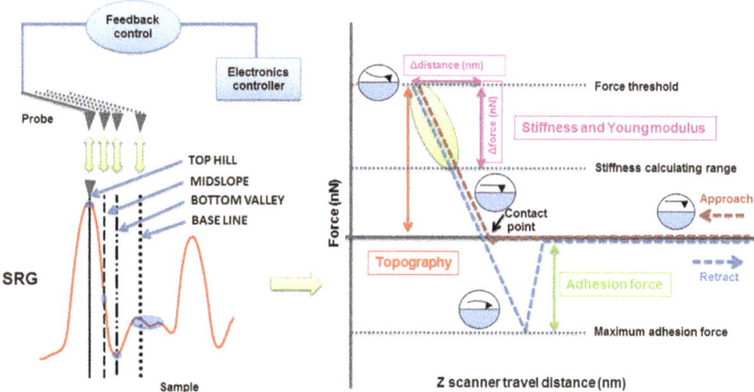

Figure 5. Working mechanism of PinPoint Nanomechanical mode: the tip of the cantilever is moved at each of 256 × 256 points along a sample's surface, and the feedback system controls the approach and retraction of a probe, allowing the acquirer of both surface topography and force–distance curves, and further to extract the mechanical property data.

Figures 6–8 show the PinPoint combined height, adhesion force, deformation, and Young's modulus AFM images and corresponding cross-section profiles in the case of pristine (Figure 6) and irradiated azo-polyimide with a laser energy density of 45 mJ/cm^2 and either 10 pulses (Figure 7) or 100 pulses (Figure 8). The last two samples were selected for this kind of measurement because the generated modulations were very well defined, facilitating the investigations in different positions of interest (on the top hills, middle slopes, base, and bottom valleys of the SRG, as indicated in Figure 5). Each profile was an average of the profiles in the band highlighted in the images, excepting the case of the pristine sample, where the surface was random.

As seen in Table 2, the pristine AzoPI presented an average adhesion force of 12.6 ± 0.8 nN, deformation of 5.1 ± 0.2 nm, and Young's modulus of 294.8 ± 20.5 MPa. After the SRG formation as a result of pulsed UV laser irradiation, these nanomechanical characteristics were found to be different, depending on the region of the modulation where the measurements were made, namely, on SRG top hills, middle slope, bottom valleys, and baseline, as indicated in Figure 5. Thus, various values of nanomechanical parameters were obtained in these different regions, mostly induced by the density of the material, but not limited to it.

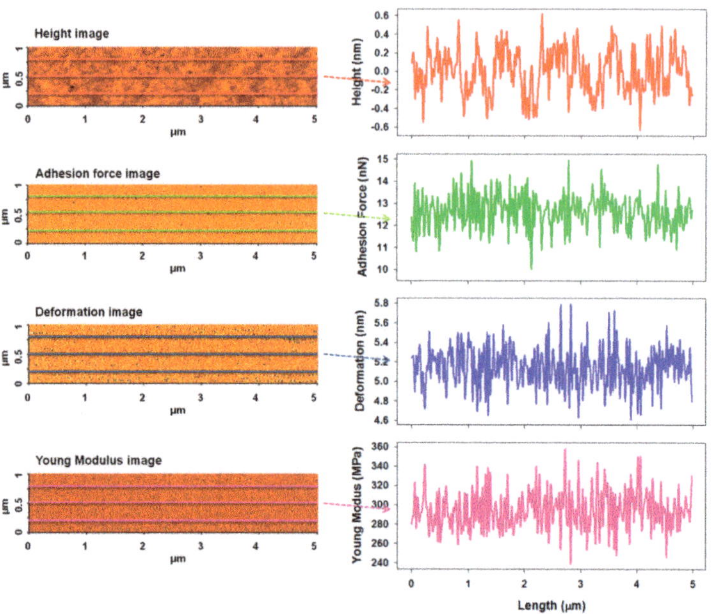

Figure 6. AFM PinPoint combined height, adhesion force, deformation, and Young's modulus images and representative cross-section profiles taken along the middle line obtained for pristine azo-polyimide.

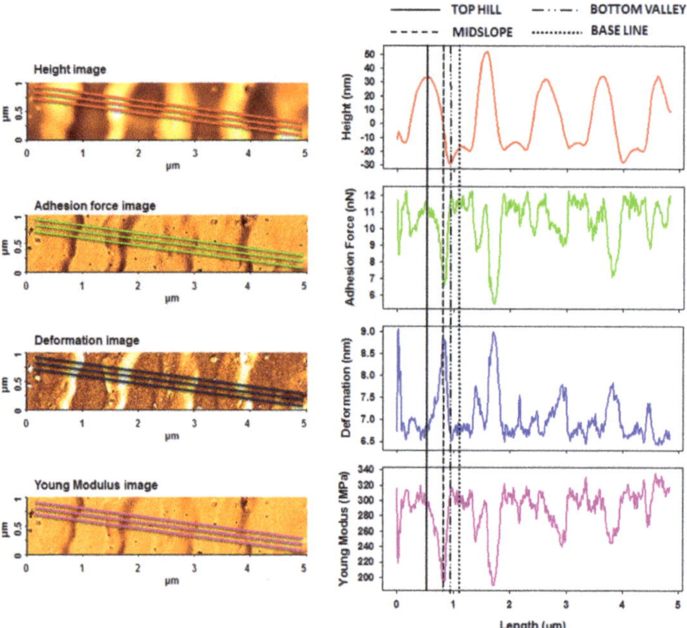

Figure 7. AFM PinPoint combined height, adhesion force, deformation, and Young's modulus AFM images and corresponding cross-section profiles mediated from the presented three lines obtained for azo-polyimide irradiated with a laser energy density of 45 mJ/cm^2 and 10 pulses.

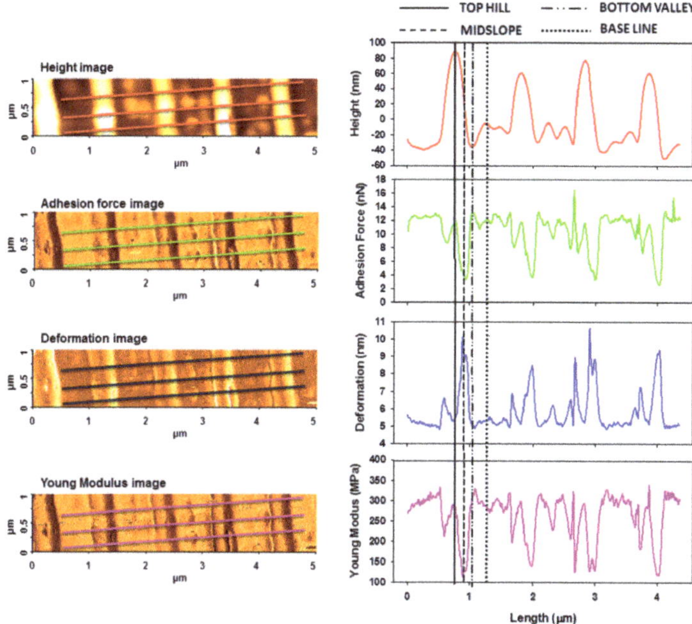

Figure 8. AFM PinPoint combined height, adhesion force, deformation, and Young's modulus images and corresponding cross-section profiles mediated from the presented three lines obtained for azo-polyimide irradiated with a laser energy density of 45 mJ/cm^2 and 100 pulses.

Table 2. The values of the nanomechanical characteristics (adhesion force, deformation, and Young's modulus) measured at different positions on the azo-polyimide samples before and after laser irradiation by using AFM PinPoint mode.

Sample	Position on the Sample	Nanomechanical Characteristics		
		Adhesion Force (nN)	Deformation (nm)	Young's Modulus (MPa)
AzoPI	All over the surface	12.6 ± 0.8	5.1 ± 0.2	294.8 ± 20.5
AzoPI 45/10	SRG top hills	11.6 ± 0.3	6.3 ± 0.1	299.8 ± 5.5
	SRG middle slopes	6.2 ± 0.7	8.9 ± 0.1	195.1 ± 4.6
	SRG bottom valleys	12.1 ± 0.1	6.7 ± 0.1	311.0 ± 5.2
	SRG base line	11.4 ± 0.4	6.8 ± 0.2	302.8 ± 9.1
AzoPI 45/100	SRG top hills	12.9 ± 1.9	5.4 ± 0.2	294.2 ± 9.0
	SRG middle slopes	3.2 ± 0.4	8.9 ± 0.3	128.9 ± 8.6
	SRG bottom valleys	13.3 ± 0.1	5.1 ± 0.2	307.9 ± 12.2
	SRG base line	12.4 ± 0.3	5.1 ± 0.1	301.9 ± 12.3

As mentioned above, the valley of the SRGs corresponds to the highest intensity of the UV laser irradiation through the phase mask. In this region, the macromolecules containing azo groups in the lateral chains undergo photo-isomerization during irradiation, when transitions from stable *trans* state to metastable *cis* state occurs along with changes in the molecular length and the dipole moment (Figure 9) [65]. This phenomenon will engender a localized substantial nanoscale stress, requiring more free volume for the local motion of azo groups from the lateral chain in *cis* configuration [52] and movement of the whole azo-polyimide backbone, inducing a dilatation effect and a continuous disturbance of the localized stress field [66].

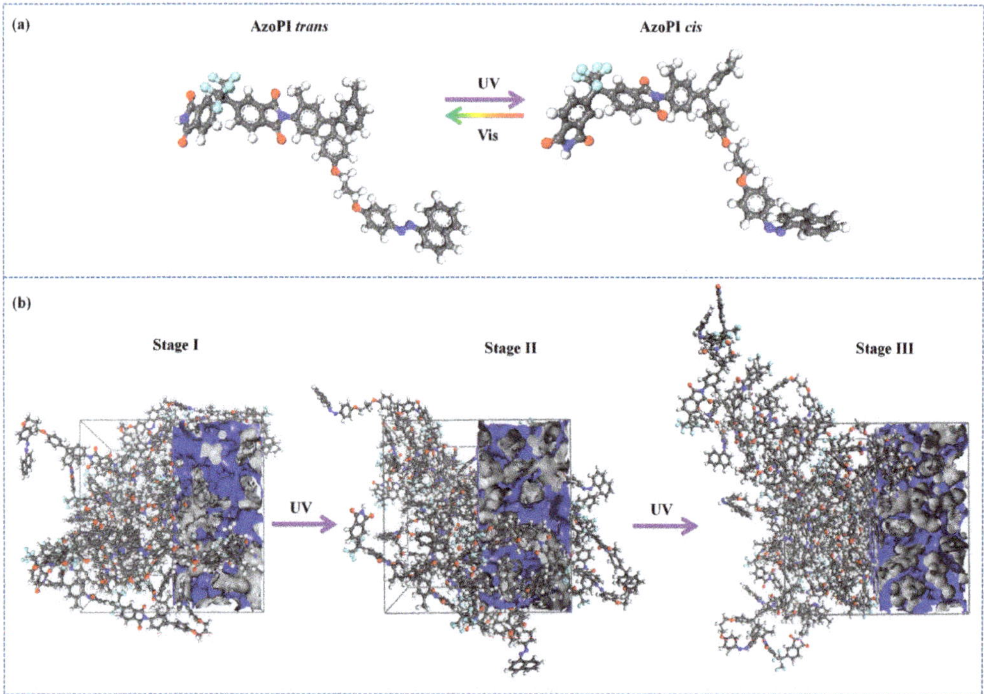

Figure 9. Molecular modeling of the photo-isomerization process of the azo-naphthalene groups: (**a**) minimum energy conformations of a AzoPI repeating unit in *trans* and *cis* states; (**b**) three-dimensional view of one amorphous cell for the AzoPI in stage I, stage II, and stage III (three polymer chains inside, each containing 10 repeating units; the grey surface indicates the Van der Waals surface, the blue surface indicates the void surface).

Figure 9 shows three different stages of the system: stage I with polymer chains having azo-naphthalene modified in *cis* conformation in proportion of 15%, stage II with 50% azo-naphthalene groups in *cis* conformation, and stage III having all azo-naphthalene segments in *cis* form.

Molecular simulations have confirmed this expansion effect by higher values of the volume occupied by the polymer chains, V_o, for the AzoPI in stage II compared to the stage I and the stage III (Table 3). Regardless of the mechanism involved in the photo-isomerization, rotation around the azo group, or inversion through one of the nitrogen nuclei, we found that a sufficiently large free volume was required to allow this mechanism, avoiding steric hindrances. It was found that the same system, AzoPI, had the largest free volume fraction in stage II. The free spaces between atoms provided the empty space useful for the movement of the molecular chains, and thus ensured the evolution towards the state III of the system.

Table 3. Predicted parameters for the azo-polyimide systems studied by molecular simulation.

AzoPI Sample	Parameter					
	ρ_p (g/cm^3)	V_o (Å^3)	V_f (Å^3)	FFV	r_{ee} (Å)	CED × 10^7 (cal/m^3)
Stage I	1.20	25,012	17,105	40.61	31.06	5.8409
Stage II	1.19	25,025	17,642	41.35	36.55	5.7983
Stage III	1.21	24,993	16,649	39.98	28.95	6.0190

ρ_p—density of packing; V_o—occupied volume with the atoms being represented by Van der Waals radii; V_f—free volume; FFV = $(V_f/(V_o + V_f))\cdot100$—fractional free volume; r_{ee}—end-to-end distance of the polymers; CED—cohesive energy density.

In order to characterize the average configuration of a chain, we calculated the structural end to end distance (r_{ee}) parameter. It was found that this parameter increased from stage I to stage II (Table 3) when the maximum value was reached, after which a decrease took place (stage III). In this case, first, the macromolecular chain stretched and then adopted a coil structure more compact than the starting one.

Cohesive energy density (CED) is a measure of the binding energy of the polymers relative to the unit volume. The CED also indicates the mixing degree of the polymer chains. Comparing the results from Table 3, we observed that the structure of AzoPI in stage II had the lowest cohesive energy density.

Saphiannikova et al. [67] demonstrated that the sign of the force induced by light is very sensitive to the molecular architecture, and therefore an effect of local extension occurs for amorphous azo-polymers. Consequently, due to the stress release in the UV laser-exposed region, a sudden modulating of the morphology takes place.

A very interesting phenomenon was observed. On SRG middle slopes, the average adhesion force and Young's modulus were strongly reduced, simultaneous with the increases of the deformation (Table 2). Here, it seemed that the azo-material was softer, which may be related to a photo-induced reduction of the density [67–69]. Moreover, the molecular modeling data revealed that the structure with 50% of azo groups in *cis* conformation had the lowest value of cohesive energy density. With the decrease of the cohesive interactions, the force necessary to produce a deformation will decrease, meaning the decrease of Young's modulus.

Moreover, it was observed that, although in the case of Azo-PI 45/100, when the number of pulses of irradiation was higher than those used in the Azo-PI 45/10 case, still the adhesion force and Young's modulus were smaller. This is probably due to the higher number of *trans–cis–trans* photo-isomerization cycles of the azo-moieties that induce a material flow and implicitly a very weak plasticization. It is well understood that the material flow (as a dynamical process where material is moving) could not be directly measured here. However, plasticization could be inferred, with this being related to a change in the elastic properties (softening or hardening).

At the baseline, and especially on the bottom of new formed SRG, there was a high concentration of azo-naphthalene moieties in the *cis* state, which in both studied cases AzoPI 45/10 and AzoPI 45/100 changed the mechanical properties (see Table 2), slightly increasing the Young's modulus, comparative to that obtained for the pristine sample. Hardening of the azo-polymer, as a result of an initial plasticization, was induced by the presence of a high population of *cis* fraction. The high density of *cis* isomers leads to an increase in the strength of intramolecular interactions in this region and also stronger interactions with the surrounding environment than *trans* isomers [70], a fact also proved by the CED values determined by the molecular dynamic simulations (Table 3). Thus, the local values of the Young's modulus of the azo-polyimide are expected to vary with the concentration of *cis* isomers.

The rigidity, the hardening of the polymer, correlates at the microscale level with the mobility of the chains. The mean square displacement (MSD) function characterizes the movement of polymer chains by measuring the deviation of the position of a particle at a time (t) from the position it had at the reference time (0):

$$\text{MSD (t)} = [r(t) - r(0)]^2$$

According to Figure 10, the mobility of the backbone of AzoPI in stage III was the lowest, indicating that this system with all azo groups in the *cis* configuration was the most rigid. The intensity of the backbone movements for AzoPI in stage I and stage II were found to be similar over time. As expected, the azo segments (C–N=N–C) belonging to the polymer side chains had wider movements compared to the mobility of the main rigid chains. It was observed that the azo segments in the AzoPI in stage III chains mimicked the behavior of the chains from stage I, having much lower mobility compared to the other azo segments. The mobility of the azo segments belonging to the AzoPI polymers in stage I and II was approximately the same, reaching at 200 ps the segments connected to AzoPI in stage I to have a slightly higher motion than those connected to AzoPI in stage II.

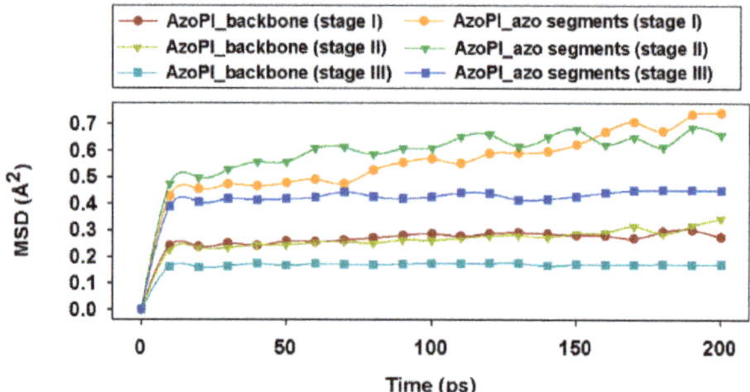

Figure 10. Evolution in time of the mean square displacement function of the AzoPI backbones and corresponding azo segments in stage I, stage II, and stage III.

Moreover, the azo-naphthalene dipoles orientation can determine a more organized and condensed structure, having as a consequence the material contraction in the exposed regions, leading also to an increase of the hardness. If we consider the side azo segments (C–N=N–C) belonging to the three polymeric systems, the distribution of the dipole moment along the Cartesian axes as a function of time is represented graphically in Figure 11. However, we must specify that in the simulations the azo groups were left free to evolve, without any constraint of the dihedral angles.

The photoinduced molecular reorientations are those that dictate the direction and modulus of the dipole moment. The different values of the dipole moment obtained in our three considered cases indicate changes in the geometry and polarity of the azo segment over time. As can be seen, the dipole of the group defined by the azo segments was different from one stage to another, being dependent on the rigidity and geometry of the chain. On the other hand, the cumulative effect of lateral group movement can cause changes in the conformation of the molecule. By averaging the values of the dipole modulus along the 200 ps, we obtained the following values for the azo segments: $\mu_{AzoPI} = 3.7$ D, $\mu_{AzoPI\_50\%cis} = 5.4$ D, and $\mu_{AzoPI\_100\%cis} = 8.6$, following the natural evolution of the modulus growth with the transition from the *trans* to the *cis* isomer.

As mentioned earlier, as a consequence of the interaction with the pulsed UV laser, during the *trans–cis* isomerization, the azo groups induce a very high pressure that determines the azo-polyimide to be able to develop high expansion forces during mass transport [52,71], which is liable for the irreversible deformations leading to SRG formation. In this stage, we found a certain state of matter, with a special feature, namely, extremely high viscosity and very low speed of polyimide chain displacement, on the strength of *trans–cis–trans* motion of azo-segments, that act as molecular motors [21]. The measurements performed on the peak of the sinusoidal pattern induced by this mass transport towards the lateral, together with the elastic material deformation determined by the supramolecular reorganization process, apparently indicated that no reaction takes place in this region (as a result of being protected by the mask). Thus, since these areas are rich in lateral chain azo groups in *trans* configuration, the values of the elastic modulus were close to that obtained for the pristine azo-polyimide (Table 2).

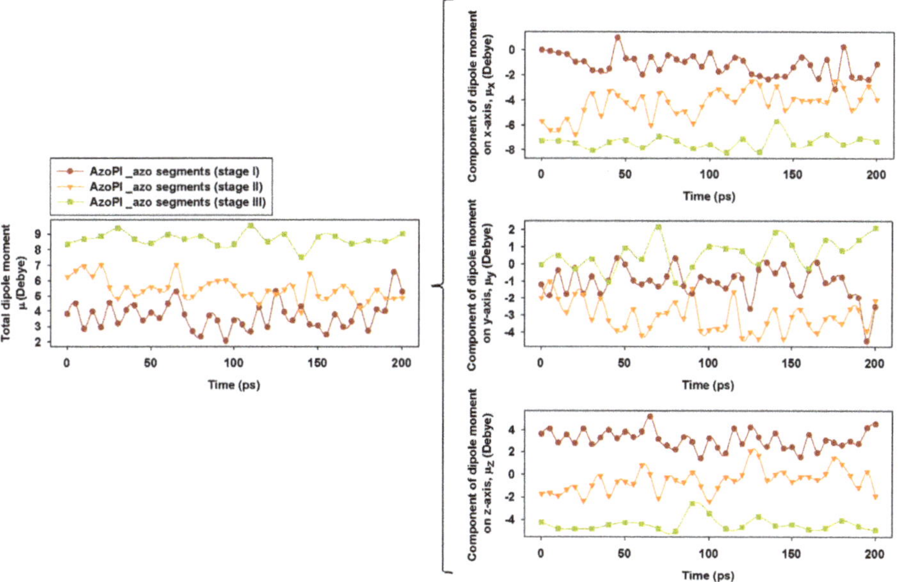

Figure 11. Evolution in time of total dipole moment (μ) and of the x, y, and z components of the dipole moment (μ_x, μ_y, and μ_z, respectively) of the azo segments (C–N=N–C) of AzoPI in stage I, stage II, and stage III.

Essentially, during the dynamics of SRG formation process, there is a phenomenon in the variation of the material density [72].

3.3. Evaluation of the Local Chemical Properties

Raman spectroscopy was used to identify the signature of the aromatic azo group grafted on the polyimide chains before and after irradiation. According to literature data, the azo (-N=N-) stretching band (-N=N-) has strong intensity with the *trans* form absorbing in the range of 1465–1380 cm^{-1} and *cis* form around 1510 cm^{-1} [73,74].

As can be seen in Figure A1, all the SRGs resulted after the treatment of azo-polyimide films with UV-pulsed laser radiation in different conditions showed the same strong absorption bands at 1449 cm^{-1} due to the distinct vibrational band ($v_{N=N}$) of the polymer chromophore that corresponded to the *trans* isomer. A very small absorption band around 1510 cm^{-1} could be observed as well for all irradiated samples, being associated with the vibrational band of the azo-group in *cis* form. It was obvious that this small absorption band slightly evolved with the increase of the incident fluence energy and number of pulses used for laser irradiation. Meanwhile, the vibrational band attributed to C–N bond (v_{C-N}) was identified for all azo-polyimide films at 1138 cm^{-1}. Besides the appearance of the small absorption band around 1510 cm^{-1} associated with the vibrational band of the *cis* isomer formed after irradiation, the Raman spectra of irradiated films were not significantly altered as compared to that of the pristine sample, indicating that no chemical modification or degradation of the polyimide occurred during the irradiation.

Accordingly, the changes detected in AFM measurements at the nanoscale were induced by conformational transitions, as also predicted by molecular modeling.

4. Conclusions

The aim of this study was to provide new insights regarding the formation mechanism of pulsed UV laser-nanoinduced patterns on azo-naphthalene-based polyimide films by evaluating the morphological, statistical, local mechanical, and chemical properties via AFM, in correlation with the molecular modeling. The quantitative nanomechanical properties (Young's modulus, adhesion, deformation) of SRGs using the AFM PinPoint method were acquired with highly correlated topographic registration. The experimental evaluations highlighted different values of nanomechanical parameters obtained in different regions of the patterned relief. These were induced by re-organization of the matter by azo-naphthalene dipoles orientation and material photo-fluidization induced by repeated *trans–cis* isomerization of the azo-segments.

Attempts have been made to explain experimental phenomena by molecular modeling. The photoisomerization phenomenon was studied using three systems in which the content of *cis* isomers was gradually increased. Although the variations of the statistical and dynamic parameters such as the fractional free volume, end to end distance, mean square displacement, or dipole moment were small from one system to another, their evolution followed the same behavior as that observed at the macroscale during the photo-isomerization process. It was found that polymers with 50% azo groups in *cis* had either a maximum or a minimum peak of the calculated parameters. The low mobility of the chains with a maximum content in the *cis* isomer could explain the phenomenon of azopolymer hardening due to the photoisomerization process.

In addition, confocal Raman measurements evidenced no significant spectral changes, except for the slight increase of the absorption band due to the *cis* isomer evolution, demonstrating their origin in the polymer conformations before and after irradiation rather than in the polyimide chemical modifications.

Author Contributions: Conceptualization, I.S. (Iuliana Stoica); methodology, I.S. (Iuliana Stoica) and E.-L.E.; validation, C.-P.C., M.-D.D., and I.S. (Ion Sava); formal analysis I.S. (Iuliana Stoica) and E.-L.E.; investigation, I.S. (Iuliana Stoica), E.-L.E., E.-L.U., and I.M.; resources, I.S. (Iuliana Stoica), E.-L.E., C.-P.C., M.-D.D., E.-L.U., I.M., and I.S. (Ion Sava); data curation, I.S. (Iuliana Stoica), E.-L.E., and E.-L.U.; writing—original draft preparation, I.S. (Iuliana Stoica), E.-L.E., and I.S. (Ion Sava); writing—review and editing, I.S. (Iuliana Stoica), E.-L.E., C.-P.C., M.-D.D., E.-L.U., I.M., and I.S. (Ion Sava); visualization, I.S. (Iuliana Stoica); supervision, I.S. (Iuliana Stoica) and I.S. (Ion Sava); project administration, I.S. (Iuliana Stoica); funding acquisition, I.S. (Iuliana Stoica). All authors have read and agreed to the published version of the manuscript.

Funding: This research was funded by the National Fellowship Program L'Oreal–Unesco "For Women in Science".

Institutional Review Board Statement: Not applicable.

Informed Consent Statement: Not applicable.

Data Availability Statement: The data is available on the request from corresponding author.

Acknowledgments: This work was funded by the National Fellowship Program L'Oreal–Unesco "For Women in Science". We also thank Victor Bergmann, Application Director at Park Systems Europe (Mannheim, Germany), for his guidance in PinPoint AFM technique.

Conflicts of Interest: The authors declare no conflict of interest. The funders had no role in the design of the study; in the collection, analyses, or interpretation of data; in the writing of the manuscript; or in the decision to publish the results.

Appendix A

The use of two 3D topographic parameters, namely, surface skewness (Ssk) and coefficient of kurtosis (Sku), allowed us to measure the departure from a Gaussian distribution of surface heights, emphasizing on one hand the degree of symmetry of the surface heights reported to the mean plane (Ssk), and on another hand the spread of the newly created structures in relation with the whole surface morphology (Sku). Consequently, the value of the skewness was influenced by the surface dislevelment, which can be above (negative skewed) or below (positive skewed) the mean surface height, implicitly being sensitive to occasional deep valleys or high peaks. The negative skewness obtained for AzoPI 10/100 sample (Table 1, Figure 4c) indicated surfaces with truncated peaks or occasional deep valleys (as seen in the cross-section profiles from Figure 3d). The positive skewness, calculated for AzoPI 10/10, AzoPI 45/10, and AzoPI 45/100 (Table 1, Figure 4a,e,g), revealed profiles with valleys filled in, or occasional high peaks, as pictured in Figure 3b,f,h by the cross-section profiles. The lower positive value of the kurtosis (Sku < 3) obtained for the homogeneous periodic AzoPI 45/10 sample (Table 1) indicated that the surface was platykurtic (Figure 4e), presenting a bumpy morphology with a relatively even distribution of heights above and below the mean height (as seen in Figure 3f of the cross-section profile). Thus, the valley zones and the up hills were well-defined. The higher positive values of the kurtosis (Sku > 3) obtained for AzoPI 10/10, AzoPI 10/100, and AzoPI 45/100 (Table 1) suggest leptokurtic surfaces (Figure 4a,c,g), with surface relief gratings with spiky morphologies (as observed also from the cross-section profiles from Figure 3b,d,h).

Appendix B

Along with 3D imaging at the nanoscale, nanomechanical properties (including adhesion, hardness, etc.) measurement on the sample surface at nN-scale via force–distance (F-d) spectroscopy by using a cantilever tip to probe the sample is one of the most well-known functions of AFM. This technique, widely used in domains such as physics, chemistry, and materials science dealing with nanoscale research, generally implies firstly acquiring a reference map of the topography data of a sample, on which the force–distance measurements are subsequently be made in the regions of interest after a preliminary alignment. Still, it is hard to prevent both positional errors in force–distance spectroscopy data and topographical data, as well as the artifacts induced by the adhesive tip–soft sample interaction. This is why PinPoint Nanomechanical mode was designed to prevent these positional errors by simultaneously acquiring accurate height and force–distance information in each of 256 × 256 pixels in the whole scanning area, while moreover preventing the cantilever tip from grabbing additional material from the surface when it travels to the next pixel position by lifting it at every pixel at a perpendicular angle from the sample surface. For this, some preset parameters are necessary. One of the most important is the control height, which is determined in the AFM cantilever retraction process from the sample surface in its movement between pixels and is dependent on the features of the sample surface. For example, if the surface heights are small, the control height must be small, and vice versa. However, in some cases, if the surface sample has small heights, but is tacky, then the height control must be suitably enlarged in order to separate the tip from the sample in anticipation of the switch to the next pixel. Other necessary preset parameters are the stiffness threshold (which establishes the force frequently acting on the surface sample), the approach and retract time, the XY pixel-to-pixel movement time, the pre-approach delay for determining a scanning speed, and the lift height. Thus, an optimized surface morphology providing precisely the height information is determined.

In the first instance, PinPoint mode imaging can generate an adhesion map using the maximum adhesion force value by acquiring in each pixel the adhesion force (F_{adh}) between the structured formations and silicon cantilever, calculated from each force–distance curve (as the one from Figure 5) as a linear function of the probe displacement relative to the sample surface along to the Z-axis, according to Hooke's law, expressed by

$$F_{adh} = -k\Delta x,$$

where k is cantilever stiffness and Δx is deflection of the cantilever in rapport with azo-polyimide SRG.

The sample's stiffness/elasticity (Young's modulus) can be acquired, similar to the force–distance spectroscopy procedure, from the slope of the force–distance curve taken at each pixel, from the point of contact to the deflection threshold, as seen in Figure 5, fitted with the Hertz model [59–61]. Assuming that no friction exists between the azo-polyimide sample and the elastic sphere tip with radius R, which indents to a displacement d, the applied force can be derived as (Figure 5)

$$F_{\text{appl}} = \frac{4}{3} E^* \sqrt{R d^3},$$

where E^* is the effective elastic modulus. E^* can be calculated by measuring F_{appl}, R, and d:

$$E^* = \frac{3}{4} \frac{F_{\text{appl}}}{\sqrt{R d^3}}$$

In order to evaluate the deformation depth from the surface of the sample, d, one must change the force–distance curve to a tip-sample separation. At the same time, E^* is a function of the two materials implied in the interaction, namely, the one of the tip and the one of the sample:

$$\frac{1}{E^*} = \frac{1 - v_{\text{tip}}^2}{E_{\text{tip}}} + \frac{1 - v_{\text{sample}}^2}{E_{\text{sample}}}$$

where E_{tip} is the elastic modulus of the tip, E_{sample} is the elastic modulus of the sample, v_{tip} is the Poisson's ratio of the tip, and v_{sample} is the Poisson's ratio of the sample. The Poisson's ratio for the Si AFM tip was set at 0.27 [62] and for azo-polyimide sample was set at 0.35 [63,64]. Knowing E^*, E_{tip}, v_{tip}, and v_{sample}, E_{sample} can be back-calculated with the formula

$$E_{\text{sample}} = \frac{1 - v_{\text{sample}}^2}{\frac{1}{E^*} - \frac{1 - v_{\text{tip}}^2}{E_{\text{tip}}}}$$

In this way, the map of the sample's stiffness/elasticity can be simultaneously acquired through the adhesion force map with the topographic sample information, locally resolving the distribution of adhesion and elasticity on the investigated surface.

Appendix C

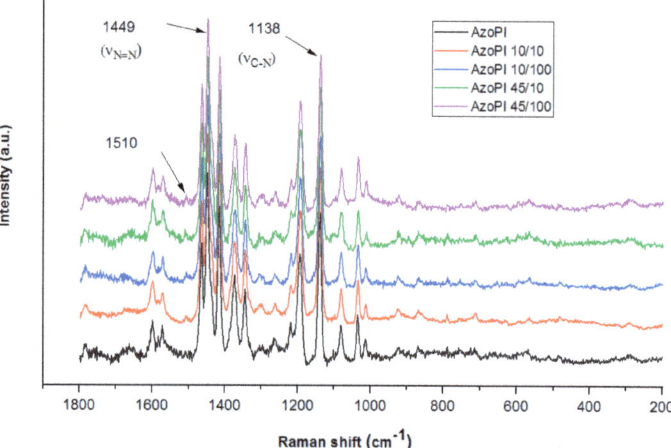

Figure A1. Raman spectra recorded on pristine and irradiated AzoPI film surfaces.

References

1. Ghosh, M.K. *Polyimides: Fundamentals and Applications*; Ghosh, M.K., Mittal, K.L., Eds.; Marcel Dekker: New York, NY, USA, 1996; ISBN 0-8247-9466-4.
2. Wang, D.H.; Lee, K.M.; Yu, Z.; Koerner, H.; Vaia, R.A.; White, T.J.; Tan, L.-S. Photomechanical Response of Glassy Azobenzene Polyimide Networks. *Macromolecules* **2011**, *44*, 3840–3846. [CrossRef]
3. Schab-Balcerzak, E.; Siwy, M.; Jarzabek, B.; Kozanecka-Szmigiel, A.; Switkowski, K.; Pura, B. Post and prepolymerization strategies to develop novel photochromic poly(esterimide)s. *J. Appl. Polym. Sci.* **2011**, *120*, 631–643. [CrossRef]
4. Yaroshchuk, O.; Reznikov, Y. Photoalignment of liquid crystals: Basics and current trends. *J. Mater. Chem.* **2012**, *22*, 286–300. [CrossRef]
5. Yu, Y.; Nakano, M.; Shishido, A.; Shiono, T.; Ikeda, T. Effect of Cross-linking Density on Photoinduced Bending Behavior of Oriented Liquid-Crystalline Network Films Containing Azobenzene. *Chem. Mater.* **2004**, *16*, 1637–1643. [CrossRef]
6. Nikolova, L.; Ramanujam, P. *Polarization Holography*; Cambridge University Press: Cambridge, UK, 2009; ISBN 9781498775809.
7. Todorov, T.; Nikolova, L.; Tomova, N. Polarization holography 1: A new high-efficiency organic material with reversible photoinduced birefringence. *Appl. Opt.* **1984**, *23*, 4309–4312. [CrossRef] [PubMed]
8. Todorov, T.; Nikolova, L.; Tomova, N. Polarization holography 2: Polarization holographic gratings in photoanisotropic materials with and without intrinsic birefringence. *Appl. Opt.* **1984**, *23*, 4588–4591. [CrossRef] [PubMed]
9. Schab-Balcerzak, E.; Sobolewska, A.; Miniewicz, A. Comparative studies of newly synthesized azo-dyes bearing poly(esterimide)s with their poly(etherimide) analogues. Light-induced optical anisotropy. *Opt. Mater.* **2008**, *31*, 405–411. [CrossRef]
10. Sava, I.; Sacarescu, L.; Stoica, I.; Apostol, I.; Damian, V.; Hurduc, N. Photochromic properties of polyimide and polysiloxane azopolymers. *Polym. Int.* **2009**, *58*, 163–170. [CrossRef]
11. Sava, I.; Hurduc, N.; Sacarescu, L.; Apostol, I.; Damian, V. Study of the nanostructuration capacity of some azopolymers with rigid or flexible chains. *High Perform. Polym.* **2013**, *25*, 13–24. [CrossRef]
12. Sava, I.; Resmerita, A.-M.; Lisa, G.; Damian, V.; Hurduc, N. Synthesis and photochromic behavior of new polyimides containing azobenzene side groups. *Polymer.* **2008**, *49*, 1475–1482. [CrossRef]
13. Sava, I.; Burescu, A.; Stoica, I.; Musteata, V.; Cristea, M.; Mihaila, I.; Pohoata, V.; Topala, I. Properties of some azo-copolyimide thin films used in the formation of photoinduced surface relief gratings. *RSC Adv.* **2015**, *5*, 10125–10133. [CrossRef]
14. Bujak, K.; Sava, I.; Stoica, I.; Tiron, V.; Topala, I.; Węgłowski, R.; Schab-Balcerzak, E.; Konieczkowska, J. Photoinduced properties of "T-type" polyimides with azobenzene or azopyridine moieties. *Eur. Polym. J.* **2020**, *126*, 109563. [CrossRef]
15. Schab-Balcerzak, E.; Sapich, B.; Stumpe, J. Photoinduced optical anisotropy in new poly (amide imide) s with azobenzene units. *Polymer* **2005**, *46*, 49–59. [CrossRef]
16. Schab-Balcerzak, E.; Sobolewska, A.; Miniewicz, A.; Jurusik, J.; Jarzabek, B. Photoinduced Holographic Gratings in Azobenzene-Functionalized Poly(amideimide)s. *Polym. J.* **2007**, *39*, 659–669. [CrossRef]
17. Konieczkowska, J.; Schab-Balcerzak, E.; Siwy, M.; Switkowski, K.; Kozanecka-Szmigiel, A. Large and highly stable photoinduced birefringence in poly(amideimide)s with two azochromophores per structural unit. *Opt. Mater.* **2015**, *39*, 199–206. [CrossRef]
18. Konieczkowska, J.; Janeczek, H.; Malecki, J.G.; Schab-Balcerzak, E. The comprehensive approach towards study of (azo)polymers fragility parameter: Effect of architecture, intra- and intermolecular interactions and backbone conformation. *Eur. Polym. J.* **2018**, *109*, 489–498. [CrossRef]
19. Bujak, K.; Kozanecka-Szmigiel, A.; Schab-Balcerzak, E.; Konieczkowska, J. Azobenzene Functionalized "T-Type" Poly (Amide Imide) s vs. Guest-Host Systems—A Comparative Study of Structure-Property Relations. *Materials* **2020**, *13*, 1912. [CrossRef] [PubMed]
20. Hurduc, N.; Donose, B.C.; Macovei, A.; Paius, C.; Ibanescu, C.; Scutaru, D.; Hamel, M.; Branza-Nichita, N.; Rocha, L. Direct observation of athermal photofluidisation in azo-polymer films. *Soft Matter* **2014**, *10*, 4640–4647. [CrossRef] [PubMed]
21. Hurduc, N.; Donose, B.C.; Rocha, L.; Ibanescu, C.; Scutaru, D. Azo-polymers photofluidisation—A transient state of matter emulated by molecular motors. *RSC Adv.* **2016**, *6*, 27087–27093. [CrossRef]
22. Stoica, I.; Epure, L.; Sava, I.; Damian, V.; Hurduc, N. An atomic force microscopy statistical analysis of laser-induced azo-polyimide periodic tridimensional nanogrooves. *Microsc. Res. Tech.* **2013**, *76*, 914–923. [CrossRef]
23. Sava, I.; Stoica, I.; Mihaila, I.; Pohoata, V.; Topala, I.; Stoian, G.; Lupu, N. Nanoscale analysis of laser-induced surface relief gratings on azo-copolyimide films before and after gold coating. *Polym. Test.* **2018**, *72*, 407–415. [CrossRef]
24. Damian, V.; Resmerita, E.; Stoica, I.; Ibanescu, C.; Sacarescu, L.; Rocha, L.; Hurduc, N. Surface relief gratings induced by pulsed laser irradiation in low glass-transition temperature azopolysiloxanes. *J. Appl. Polym. Sci.* **2014**, *131*, 41015. [CrossRef]
25. Pederzoli, M.; Pittner, J.; Barbatti, M.; Lischka, H. Nonadiabatic Molecular Dynamics Study of the cis—Trans Photoisomerization of Azobenzene Excited to the S 1 State. *J. Phys. Chem. A* **2011**, *115*, 11136–11143. [CrossRef]
26. Ootani, Y.; Satoh, K.; Nakayama, A.; Noro, T.; Taketsugu, T. Ab initio molecular dynamics simulation of photoisomerization in azobenzene in the nπ* state. *J. Chem. Phys.* **2009**, *131*, 194306. [CrossRef] [PubMed]
27. Biswas, M.; Burghardt, I. Azobenzene Photoisomerization-Induced Destabilization of B-DNA. *Biophys. J.* **2014**, *107*, 932–940. [CrossRef]
28. Wei-Guang Diau, E. A New Trans-to-Cis Photoisomerization Mechanism of Azobenzene on the S 1 (n,π*) Surface. *J. Phys. Chem. A* **2004**, *108*, 950–956. [CrossRef]

29. Chami, F.; Wilson, M.R. Molecular Order in a Chromonic Liquid Crystal: A Molecular Simulation Study of the Anionic Azo Dye Sunset Yellow. *J. Am. Chem. Soc.* **2010**, *132*, 7794–7802. [CrossRef]

30. Böckmann, M.; Peter, C.; Site, L.D.; Doltsinis, N.L.; Kremer, K.; Marx, D. Atomistic Force Field for Azobenzene Compounds Adapted for QM/MM Simulations with Applications to Liquids and Liquid Crystals. *J. Chem. Theory Comput.* **2007**, *3*, 1789–1802. [CrossRef] [PubMed]

31. Carlescu, I.; Simion, A.; Epure, E.L.; Lisa, G.; Scutaru, D. Self-assembled star-shaped liquid crystals based on 1,3,5-trihydroxybenzene with pendant alkyloxylated azobenzene arms. *Liq. Cryst.* **2020**, *47*, 1852–1862. [CrossRef]

32. Georgiev, A.; Kostadinov, A.; Ivanov, D.; Dimov, D.; Stoyanov, S.; Nedelchev, L.; Nazarova, D.; Yancheva, D. Synthesis, spectroscopic and TD-DFT quantum mechanical study of azo-azomethine dyes. A laser induced trans-cis-trans photoisomerization cycle. *Spectrochim. Acta Part A Mol. Biomol. Spectrosc.* **2018**, *192*, 263–274. [CrossRef]

33. Epure, E.-L.; Vasiliu, T.; Hurduc, N.; Neamțu, A. Molecular modeling study concerning the self-assembly capacity of some photosensitive amphiphilic polysiloxanes. *J. Mol. Liq.* **2020**, *300*, 112298. [CrossRef]

34. Xia, S.-H.; Cui, G.; Fang, W.-H.; Thiel, W. How Photoisomerization Drives Peptide Folding and Unfolding: Insights from QM/MM and MM Dynamics Simulations. *Angew. Chem. Int. Ed.* **2016**, *55*, 2067–2072. [CrossRef]

35. Böckmann, M.; Doltsinis, N.L. Towards understanding photomigration: Insights from atomistic simulations of azopolymer films explicitly including light-induced isomerization dynamics. *J. Chem. Phys.* **2016**, *145*, 154701. [CrossRef] [PubMed]

36. Pawlik, G.; Mitus, A.C. Photoinduced Mass Transport in Azo-Polymers in 2D: Monte Carlo Study of Polarization Effects. *Materials* **2020**, *13*, 4724. [CrossRef]

37. Juan, M.L.; Plain, J.; Bachelot, R.; Royer, P.; Gray, S.K.; Wiederrecht, G.P. Stochastic model for photoinduced surface relief grating formation through molecular transport in polymer films. *Appl. Phys. Lett.* **2008**, *93*, 153304. [CrossRef]

38. Stiller, B.; Karageorgiev, P.; Geue, T.; Morawetz, K.; Saphiannikova, M.; Mechau, N.; Neher, D. Optically induced mass transport studied by scanning near-field optical - and atomic force microscopy. *Phys. Low Dimens. Struct.* **2004**, *1/2*, 129–138.

39. Sadegh Hassani, S.; Daraee, M.; Sobat, Z. Application of atomic force microscopy in adhesion force measurements. *J. Adhes. Sci. Technol.* **2021**, *35*, 221–241. [CrossRef]

40. Krotil, H.-U.; Stifter, T.; Waschipky, H.; Weishaupt, K.; Hild, S.; Marti, O. Pulsed force mode: A new method for the investigation of surface properties. *Surf. Interface Anal.* **1999**, *27*, 336–340. [CrossRef]

41. Exploring Nanomechanical Properties of Materials with Atomic Force Microscopy, NT-MDT Spectrum Instruments Application Note no. 085. 2017. Available online: https://www.ntmdt-si.com/resources/applications/exploring-nanomechanical-properties-of-materials-with-atomic-force-microscopy (accessed on 22 February 2021).

42. Expanding Atomic Force Microscopy with HybriD Mode Imaging, NT-MDT Spectrum Instruments Application Note no. 087/2017. Available online: https://www.ntmdt-si.com/resources/applications/expanding-atomic-force-microscopy-with-hybrid-mode-imaging (accessed on 22 February 2021).

43. Quantitative Nanomechanical Measurements in HybriD Mode Atomic Force Microscopy, NT-MDT Spectrum Instruments Application Note no. 090. 2017. Available online: https://www.ntmdt-si.com/resources/applications/quantitative-nanomechanical-measurements-in-hybrid-mode-atomic-force-microscopy (accessed on 22 February 2021).

44. Park PinPoint™ Mode. Available online: https://parksystems.com/files/pinpoint_mode_brochure.pdf (accessed on 22 February 2021).

45. Constantin, C.-P.; Sava, I.; Damaceanu, M.-D. Structural Chemistry-Assisted Strategy toward Fast Cis–Trans Photo/Thermal Isomerization Switch of Novel Azo-Naphthalene-Based Polyimides. *Macromolecules* **2021**, *54*, 1517–1538. [CrossRef]

46. *Materials Studio 4.0*; DMol3, Forcite and Amorphous Cell Module; Accelrys Software Inc.: San Diego, CA, USA, 2005.

47. Miniewicz, A.; Sahraoui, B.; Schab-Balcerzak, E.; Sobolewska, A.; Mitus, A.C.; Kajzar, F. Pulsed-Laser Grating Recording in Organic Materials Containig Azobenzene Derivatives. *Nonlinear Opt. Quantum Opt.* **2006**, *X*, 1–8.

48. Kim, S.; Park, S.; Lee, W.; Kwon, O.; Kim, S.U.; Choi, Y.; Yoon, M.; Park, J.; Kim, Y. Origin of macroscopic adhesion in organic light-emitting diodes analyzed at different length scales. *Sci. Rep.* **2018**, *8*, 1–7. [CrossRef]

49. Othonos, A.; Kalli, K.; Pureur, D.; Mugnier, A. Fibre Bragg Gratings. In *Wavelength Filters in Fibre Optics*; Springer: Berlin Germany, 2006; Volume 123, pp. 189–269, ISBN 3540317694.

50. Lagugné-Labarthet, F.; Buffeteau, T.; Sourisseau, C. Molecular orientations in azopolymer holographic diffraction gratings as studied by Raman confocal microspectroscopy. *J. Phys. Chem. B* **1998**, *102*, 5754. [CrossRef]

51. Stoica, I.; Hurduc, N. Structuring of Polymer Surfaces via Laser Irradiation as a Tool for Micro-and Nanotechnologies. In *Electromagnetic Radiation in Analysis and Design of Organic Materials*; Dorohoi, D.O., Barzic, A.I., Aflori, M., Eds.; CRC Press; Taylor & Francis Group: Boca Raton, FL, USA, 2017; Chapter 12; pp. 191–206, ISBN 9781498775809.

52. Barrett, C.J.; Natansohn, A.L.; Rochon, P.L. Mechanism of Optically Inscribed High-Efficiency Diffraction Gratings in Azo Polymer Films. *J. Phys. Chem.* **1996**, *100*, 8836–8842. [CrossRef]

53. Wu, X.; Ngan Nguyen, T.T.; Ledoux-Rak, I.; Thanh Nguyen, C.; Diep, N. Optically Accelerated Formation of One- and Two-Dimensional Holographic Surface Relief Gratings on DR1/PMMA. In *Holography—Basic Principles and Contemporary Applications*; InTech: Rijeka, Croatia, 2013.

54. Wu, X. Fabrication of 1D, 2D and 3D polymer-based periodic structures by mass transport effect. Ph.D. Thesis, École Normale Supérieure de Cachan—ENS Cachan, Paris, France, 2013.

55. Viswanathan, N.K.; Balasubramanian, S.; Li, L.; Kumar, J.; Tripathy, S.K. Surface-Initiated Mechanism for the Formation of Relief Gratings on Azo-Polymer Films. *J. Phys. Chem. B* **1998**, *102*, 6064. [CrossRef]

56. Sdr Parameters and Its Applications in Characterizing Surface Data. Available online: https://www.azonano.com/article.aspx?ArticleID=5137 (accessed on 2 March 2021).

57. ISO 25178-2:2012 Geometrical Product Specifications (GPS)—Surface Texture: Areal—Part 2: Terms, Definitions and Surface Texture Parameters. 2012. Available online: https://www.iso.org/standard/42785.html (accessed on 21 March 2021).

58. Stoica, I.; Barzic, A.I.; Hulubei, C.; Timpu, D. Statistical analysis on morphology development of some semialicyclic polyimides using atomic force microscopy. *Microsc. Res. Tech.* **2013**, *76*, 503–513. [CrossRef] [PubMed]

59. Reifenberger, R. *Fundamentals of Atomic Force Microscopy Part I: Foundations*; Lessons from Nanoscience: A Lecture Notes Series; World Scientific Publishing: Singapore, 2015; Volume 4, ISBN 978-981-4630-34-4.

60. Park Systems Introduces PinPoint[TM] Nanomechanical Mode to Characterize Nano Mechanical Properties of Materials and Biological Cells. Available online: https://parksystems.com/company/news/press-release/450-nanomechanical-mode-to-characterize-nano-mechanical (accessed on 21 March 2021).

61. Collagen Fibrils Imaging in Liquid Using Park NX10 Atomic Force Microscope PinPoint[TM] Nanomechanical Mode; Park Atomic Force Microscopy; Application note #4. Available online: https://parksystems.com/applications/life-science/micro-and-molecular-biology/607-collagen-fibrils-imaging-using-park-nx10-atomic-force-microscope-pinpoint-nanomechanical-mode (accessed on 14 March 2021).

62. Cao, X.; Gan, X.; Peng, Y.; Wang, Y.; Zeng, X.; Lang, H.; Deng, J.; Zou, K. An ultra-low frictional interface combining FDTS SAMs with molybdenum disulfide. *Nanoscale* **2018**, *10*, 378–385. [CrossRef] [PubMed]

63. Lin, Y.C.; Peng, X.Y.; Wang, L.C.; Lin, Y.L.; Wu, C.H.; Liang, S.C. Residual stress in CIGS thin film solar cells on polyimide: Simulation and experiments. *J. Mater. Sci. Mater. Electron.* **2014**, *25*, 461–465. [CrossRef]

64. Blösch, P.; Güttler, D.; Chirila, A.; Tiwari, A.N. Optimization of Ti/TiN/Mo back contact properties for Cu (In,Ga)Se 2 solar cells on polyimide foils. *Thin Solid Film.* **2011**, *519*, 7453–7457. [CrossRef]

65. Yadavalli, N.S. Advances in Experimental Methods to Probe Surface Relief Grating Formation Mechanism in Photosensitive Materials. Ph.D. Thesis, University of Potsdam, Postdam, Germany, 2014.

66. Zong, C.; Zhao, Y.; Ji, H.; Han, X.; Xie, J.; Wang, J.; Cao, Y.; Jiang, S.; Lu, C. Tuning and Erasing Surface Wrinkles by Reversible Visible-Light-Induced Photoisomerization. *Angew. Chem. Int. Ed.* **2016**, *55*, 3931–3935. [CrossRef]

67. Bian, S.; Williams, J.M.; Kim, D.Y.; Li, L.; Balasubramanian, S.; Kumar, J.; Tripathy, S. Photoinduced surface deformations on azobenzene polymer films. *J. Appl. Phys.* **1999**, *86*, 4498–4508. [CrossRef]

68. Srikhirin, T.; Laschitsch, A.; Neher, D.; Johannsmann, D. Light-induced softening of azobenzene dye-doped polymer films probed with quartz crystal resonators. *Appl. Phys. Lett.* **2000**, *77*, 963–965. [CrossRef]

69. Mechau, N.; Neher, D.; Börger, V.; Menzel, H.; Urayama, K. Optically driven diffusion and mechanical softening in azobenzene polymer layers. *Appl. Phys. Lett.* **2003**, *81*, 4715–4717. [CrossRef]

70. Mechau, N.; Saphiannikova, M.; Neher, D. Dielectric and mechanical properties of azobenzene polymer layers under visible and ultraviolet irradiation. *Macromolecules* **2005**, *38*, 3894–3902. [CrossRef]

71. Gherab, K.N.; Gatri, R.; Hank, Z.; Dick, B.; Kutta, R.J.; Winter, R.; Luc, J.; Sahraoui, B.; Fillaut, J.L. Design and photoinduced surface relief grating formation of photoresponsive azobenzene based molecular materials with ruthenium acetylides. *J. Mater. Chem.* **2010**, *20*, 2858–2864. [CrossRef]

72. Wang, X. *Azo Polymers: Synthesis, Functions and Applications*; Springer: Berlin, Germany, 2017; ISBN 9783662534229.

73. Di Florio, G.; Bründermann, E.; Yadavalli, N.S.; Santer, S.; Havenith, M. Graphene Multilayer as Nanosized Optical Strain Gauge for Polymer Surface Relief Gratings. *Nano Lett.* **2014**, *14*, 5754–5760. [CrossRef]

74. Di Florio, G.; Bründermann, E.; Yadavalli, N.S.; Santer, S.; Havenith, M. Polarized 3D Raman and nanoscale near-field optical microscopy of optically inscribed surface relief gratings: Chromophore orientation in azo-doped polymer films. *Soft Matter* **2014**, *10*, 1544. [CrossRef]

nanomaterials

MDPI

Article

Effect of Embedment of MWCNTs for Enhancement of Physical and Mechanical Performance of Medium Density Fiberboard

Waheed Gul [1], Hussein Alrobei [2,*], Syed Riaz Akbar Shah [3], Afzal Khan [4], Abid Hussain [4], Abdullah M. Asiri [5] and Jaehwan Kim [6]

[1] Department of Mechanical Engineering, Institute of Space Technology, Islamabad 44000, Pakistan; waheed.gul@mail.ist.edu.pk
[2] Department of Mechanical Engineering, Prince Sattam Bin Abdulaziz University, Alkharj 11942, Saudi Arabia
[3] Department of Mechatronics Engineering, University of Engineering and Technology, Peshawar 25120, Pakistan; rasayed@uetpeshawar.edu.pk
[4] Department of Mechanical Engineering, University of Engineering and Technology, Peshawar 25120, Pakistan; afzalkhan@uetpeshawar.edu.pk (A.K.); abidhussain@uetpeshawar.edu.pk (A.H.)
[5] Department of Chemistry, King Abdulaziz University, Jeddah 21589, Saudi Arabia; aasiri2@kau.edu.sa
[6] Center for NanoCellulose Future Composites, Department of Mechanical Engineering, Inha University, Incheon 22212, Korea; jaehwan@inha.ac.kr
* Correspondence: h.alrobei@psau.edu.sa

Abstract: In this research work effect of embedment of multiwall carbon nanotubes (MWCNTs) on the physical and mechanical properties of medium density fiberboard (MDF) have been investigated. The MWCNTs were embedded in urea formaldehyde resin (UF) at 0, 1.5%, 3% and 5% concentrations by weight for the manufacturing of nano-MDF. The addition of these nanoparticles enhanced thermal conductivity by 24.2% reduced curing time by 20% and controlled formaldehyde emission by 59.4%. The internal bonding (I.B), modulus elasticity (MOE), modulus of rupture (MOR), thickness swelling (Ts) and water absorption (WA) properties were improved significantly by 21.15%, 30.2%, 28.3%, 44.8% and 29% respectively as compared to controlled MDF.

Keywords: MDF; MWCNTs; UF; thermal conductivity; formaldehyde emission; I.B; MOE; MOR; Ts; WA

Citation: Gul, W.; Alrobei, H.; Shah, S.R.A.; Khan, A.; Hussain, A.; Asiri, A.M.; Kim, J. Effect of Embedment of MWCNTs for Enhancement of Physical and Mechanical Performance of Medium Density Fiberboard. *Nanomaterials* **2021**, *11*, 29. https://dx.doi.org/10.3390/nano11010029

Received: 13 November 2020
Accepted: 17 December 2020
Published: 24 December 2020

Publisher's Note: MDPI stays neutral with regard to jurisdictional claims in published maps and institutional affiliations.

1. Introduction

Medium density fiberboard (MDF) is a natural timber panel manufactured using wood homogeneous fibers or supplementary lingo cellulosic fibers and binders under pressure and temperature [1]. The applications of MDF include furniture industries, flooring, interior domestic construction, tabletops, vanities, speakers, sliding doors, lock blocks, interior signs, displays, table tennis, pool tables, electronic game consoles, kitchen worktops, office work surfaces, educational institutions, laboratories and other industrial products [2]. Wood mix panels offer a uniform advantage to the structure, which can be important for many design applications [3]. Due to poor physical properties, i.e., water absorption and thickness swelling, and mechanical properties, i.e., internal bonding, modulus of elasticity and modulus of rupture, MDF is no longer used in moist and hot environments. Some research has been conducted to improve the physical and mechanical performance of MDF by introducing melamine, wax and other additives [4].

The fibrous nature of wood has made it one of the most appropriate and versatile raw materials for various uses. However, two properties restrict its much wider use, namely dimensional changes when subjected to fluctuating humidity and mechanical strength. Wood may be modified chemically or thermally so that selected properties are enhanced in a more or less permanent fashion. Another option to improve these properties is to exploit the solutions that nanotechnology can offer. The multiwall carbon nanotubes (MWCNTs) of nanotechnology compounds can deeply penetrate into the wood, effectively alter its surface chemistry and result in a high degree of protection against moisture and mechanical strength. In addition, the use of lignocelluloses materials for the production of advanced wood composites is an innovative avenue for research[5].

This research was conducted with the objective to study the influence of MWCNTs on the physical and mechanical performances of MDF. The comparison of nanoparticle concentration with respect to nano-MDF was also investigated.

In order to investigate the physical and mechanical performance of nano-MDF, a hybrid approach of nanofillers containing alumina, silicon dioxide and zinc oxide in urea formaldehyde (UF) resin glue was carried out by Candan et al. (2015) [6]. Based on the dry weight of natural fibers, the concentration was kept at 0%, 1% and 3%. Physical and mechanical tests were performed and it was concluded that almost bending strength and modulus of elasticity and screw holding properties enhanced significantly. Taghiyari et al. (2016) explored the physical and mechanical properties of MDF via the addition of nano-wollostonite and camel thorn fibers in UF resin in a ratio 90:30. The size of these fibers was in the range of 100 nm$-$1 µm. An extraordinary improvement was observed in the physical and mechanical properties of MDF [7]. Ismita et al. (2017) conducted an experimental study to investigate the mechanical and physical characteristics of MDF in accumulation of UF-Na+ nanofillers at 2.0%, 4.0% and 6.0%. Among the three concentration levels, 6.0% Na+-based resin had improved modulus of rupture, thickness swelling and modulus of elasticity properties by 34.0%, 6.0% and 65.0%, as compared to normal MDF [8]. $CaCO_3$ and poly methyl methacrylate nanoparticles were introduced in UF resin by Yipeng Chen et al. (2018). Thermogravemetric analysis along with mechanical properties and thickness swelling were measured. The results showed an incredible improvement in thermal, mechanical and thickness swelling properties [9]. Da Silvaet al. (2019) developed an experimental study to find out the thickness swelling, bacterial and mold resistance properties of MDF with UF-ZnO and melamine formaldehyde and zinc oxide(MF-ZnO) nanofillers. Three samples of each nanofiller-based MDF were tested for physical and biological performance. Among all concentration levels, 0.5% of ZnO with MF-based MDF has the highest values of physical and biological properties [10]. To explore the thermal, physical and mechanical properties of MDF, Alabduljabbar et al. (2020) studied the effect of 0%, 1.5%, 3.0% and 4.5% alumina nanoparticles in the UF resin and explored the effect of Al_2O_3 nanoparticles on the physical and mechanical properties of nano-MDF. The resultant internal bonding, modulus of elasticity, modulus of rupture, thickness swelling and water absorption characteristics were boosted up to 16.4%,31%, 22.12%, 40.15% and 37.53% in that order [11].

Although the addition of nanoparticles has been analyzed in various contexts in literature, the aim of this research work is to explore MWCNTs with diverse absorption levels in order to enhance a number of physical and mechanical characteristics, i.e., thermal conductivity, water absorption, thickness swelling, density, formaldehyde emission, internal adhesion, bending strength and modulus of elasticity. The fibrous nature of wood has made it one of the most appropriate and versatile raw materials for various uses. However, two properties restrict its much wider use, namely dimensional changes when subjected to fluctuating humidity and mechanical strength. Wood may be modified chemically or thermally so that selected properties are enhanced in a more or less permanent fashion. Another option to improve these properties is to exploit the solutions that nanotechnology can offer. The MWCNTs of nanotechnology compounds can deeply penetrate into the wood, effectively alter its surface chemistry and result in a high degree of protection against moisture and mechanical strength. In addition, the use of lignocelluloses materials for the production of advanced wood composites is an innovative avenue for research.

2. Materials and Methods

2.1. Materials

Urea-formaldehyde resin, MWCNTs and natural wood fibers are used as raw materials for manufacturing of nano-MDF. These raw materials are explained in the subgroup below.

2.1.1. Urea-Formaldehyde (UF) Resin

Urea formaldehyde resin was purchased from Wah chemical company, Pakistan. The physical and chemical characteristics of the UF resin are shown in Table 1.

Table 1. Physical and chemical properties of urea-formaldehyde.

Viscosity (Cps)	Density (kg/m^3)	pH	Free Formaldehyde (mg/100 g)	Gel Time (s)	Solid Content (%)
180–290	1.25	8.1	0.75	63	59

2.1.2. Multiwall Carbon Nanotubes (MWCNTs)

(MWCNTs) were provided by Guangzhou Hong Material Technology Company Limited, China. The SEM and XRD of MWCNTs can be seen in Figures S1 and S2 in Supplementary Materials section. The motive for the assortment of MWCNTs is their incomparable tensile strength [12] and thermal conductivity [13,14]. The MWCNTs can deeply penetrate into the wood, effectively alter its surface chemistry and result in a high degree of protection against moisture and mechanical strength. In accumulation, they can be chemically improved [15]. These material goods are anticipated to be appreciated in numerous extents of technology, such as microchip technology, optics, composite materials, nanotechnology and further submissions of materials science. The diameter of a MWCNT tube ranges from 20 to 40 nanometers.

2.1.3. Natural Wood Fibers

Poplar wood fibers were received from Ciel Woodworks (Pvt) Ltd., Peshawar, Pakistan. The length of the fibers is in the range of 0.56–1.0 mm.

2.1.4. Fictionalization of MWCNTs

Raw MWCNTs of a definite weight (1.0 g) were added into 50 mL 3:1 mixture (v/v) of concentrated H_2SO_4 (98%) and HNO_3 (68%) with sonication at 140 °C for 1 h. The attachment of the functional groups on the surface of the MWCNTs was identified using the Fourier transform infrared (FTIR) spectrophotometer Imprestige-21, Shimadzu Corporation, JAPAN (wave number range of 400–4000 cm^{-1}), equipped with an attenuated total reflectance (ATR) device (wave number range of 500–4000 cm^{-1} with 20 scanning rate and resolution of 4 cm^{-1}) for the confirmation of gelatin-chitosan interaction in the composites.

2.1.5. Raman Spectroscopy

Raman spectra of the three kinds of MWCNT were recorded on a Renishaw 1000 Raman spectrometer with the wavelength of the Raman laser at 532 nm.

2.2. Preparation of UF-MWCNTs Nanofiller

The MWCNTs-UF nanofluid was primed in the materials Science Lab, Institute of Space Technology, Islamabad, Pakistan with the configurations specified in Table 2 dignified in grams.

Table 2. Configurations of multiwall carbon nanotubes (MWCNTs)-urea formaldehyde (UF) nanofluid.

	Composition			
Materials	**MWCNTs$_1$**	**MWCNTs$_2$**	**MWCNTs$_3$**	**MWCNTs$_4$**
UF(g)	200	200	200	200
MWCNTs (wt%)	0	1.5	3	5

The nanofluids were blended by weighing 200 g of urea-formaldehyde resin and 0, 1.5, 3 and 5 wt% of MWCNTs of dry weight of fibers. The sonication of the nanofluids was carried out by means of an Ultrasonic Processor UP 400S of Hielscher Ultrasound Technology Company, USA, for 30 min. The samples were signified by $MWCNTs_0$, $MWCNTs_1$, $MWCNTs_2$ and $MWCNTs_3$, according to the meditation of MWCNTs.

2.3. Nano-MDF Design

The nano-MDF testers were manufactured in panels with sizes $460 \times 460 \times 15$ mm^3 with densities in the range of 600–750 kg/m^3. The MWCNTs-UF nanofluids were properly mixed with poplar wood fibers in rotary drum mixtures fibers and a nozzle. A single opening hot press of Burkle, Germany, operated hydraulically, was used for manufacturing of nano-MDF samples. The hot pressing process parameters, i.e., pressure (160 bar) and temperature (175 °C), were kept constant for all testers. The whole press cycle was maintained for4 min. The manufactured samples were treated in a cooling tower for 3 days.

2.4. Scanning Electron Microscopy (SEM)

Before being subjected to SEM, a sample of MWCNTs was prepared in the lab and coated with gold by means of a Safematic CCU-010 Gold/Carbon Sputter, UK. SEM was performed through MIRA 3 TESCAN, Czech Republic, at 50,000× for pure urea-formaldehyde resin and MWCNTs-UF with an extreme working voltage of 20 kV.

2.5. Differential Scanning Calorimetry (DSC)

An apparatus Mettler Toledo TGA/DSC-1-star system, USA, was used for differential scanning calorimetry analysis. The temperature range for this device was kept at 0 °C and 400°C with a heat expanding rate of 10 °C/min in a nitrogen stream of 8 mL/min.

2.6. Dynamic Mechanical Analysis (DMA)

For DMA, the instrument used is the Pyris Diamond DMA (Perkin Elmer, Columbus, USA with the temperature ramp testing method, from 50 °C to 180 °C, at a heating rate of 10 °C·min^{-1}. All tests were conditioned with 0.15% deformation and 10 N perpendicular forces, at a frequency of 1.5 Hz and Poisson's ratio of 0.440. Before performing DMA, the samples were converted into solid states. The samples were kept at $56 \times 13 \times 3$ mm^3 as per ASTM D4065. The physical state of the sample was cured solid.

2.7. Thermal Conductivity of Nano-Composite

The device used in this research work for measuring thermal conductivity is QTM500, Kyoto electronics company, Japan. The temperature range of this instrument is 100 to 1000 °C. The size of the samples was kept at $0.02 \times 0.05 \times 0.10$ m^3 and the maximum time for measuring thermal conductivity was 2 min, as per ASTM C 1113-99standards [16,17]. The extent range for this instrument was 0.0115–6.15 W/mK. Samples of nanofillers were collected in vials and subjected to the needle of thermal conductivity sensor. Due to the heat transfer mechanism, the thermal conductivity can be measured using a sensor.

2.8. Formaldehyde Emission of Nano-Composite

The perforator method followed by the EN-120 standard (1993) was applied for measuring the formaldehyde emission from nano-MDF. A 150 g nano-MDF sample in powder form was put into a glass flask which already contained550 mL toluene. Pure water of about 1000 mL having a pH value of 7 was added into perforator flasks. Toluene was heated up to boiling point, and the vapors generated were then passed from pure water. An ultraviolet spectroscopy was used to examine the formaldehyde emission after it had been treated with acetone.

2.9. Statistical Analysis of Nano-Composite

A one-way ANOVA was carried out for statistical investigation origin 8.5, 32-bit software (accessed on 5/10/2020). The Tuckey method was applied with a 95% confidence level to analyze the samples with three iterations of each. The mean value and significant values with variance for each sample were then calculated.

3. Results and Discussion

3.1. FTIR Analysis of Functionalized MWCNTs

The FTIR spectra of pure and functionalized MWCNTs are shown in Figure 1. From the IR absorption spectra, it was found that the Infrared (IR) absorption spectrum of multi walled carbon nanotubes with function group (MWCNTs-COOH) mainly consisted of hydroxyl group (–OH) stretch at 3200 and 3450 cm^{-1} and carboxyl group (–CO) stretch at 1450 cm^{-1}, which are the characteristic peaks and could also be found in the IR spectrum of the raw MWCNTs. However, the intensity of the two peaks in MWCNTs-raw spectrum was much lower than that in the MWCNTs-COOH spectrum. In addition, MWCNTs-COOH has a new peak of small intensity at 1650 cm^{-1}, which may be stretching vibrations of carbonyl groups (–CO) as carboxylic groups were formed during the oxidation of hydroxyl compounds [18].

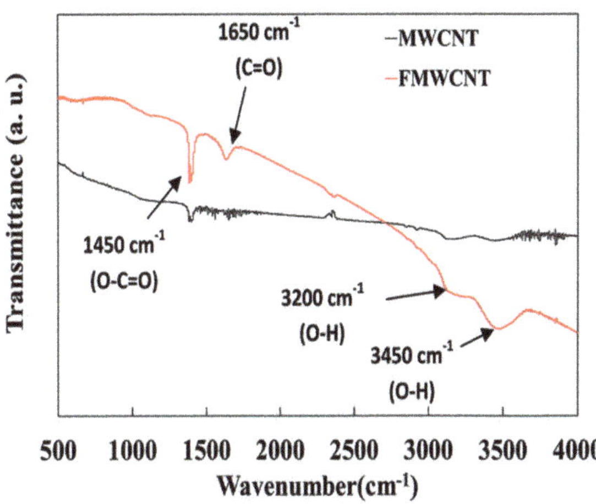

Figure 1. The Infrared (IR) spectra of reference and functional MWCNTs showing hydroxyl group (–OH) and carbonyl group (–CO) attached after acid treatment.

3.2. Raman Spectra of Pristine and Modified MWCNTs

As a very valuable tool to characterize carbon-based nanostructures, the Raman spectra of MWCNT-P and MWCNT-COOH taken at the wavelength of the Raman laser (λ) of 532 nm are shown in Figure 2. For both samples, it presents three main peaks, namely D peak at ~1350 cm^{-1}, G peak at ~1590 cm^{-1} and G′ peak at ~2670 cm^{-1}. The D peak corresponds to the first-order scattering process of sp^2 carbons, and it is generally activated by the existence of vacancies, surface functional groups, boundaries and other defects. The G peak derives from the in-plane tangential stretching of –C–C– in graphitic shells, and the G′ peak is the second order of mode of the D peak [19].

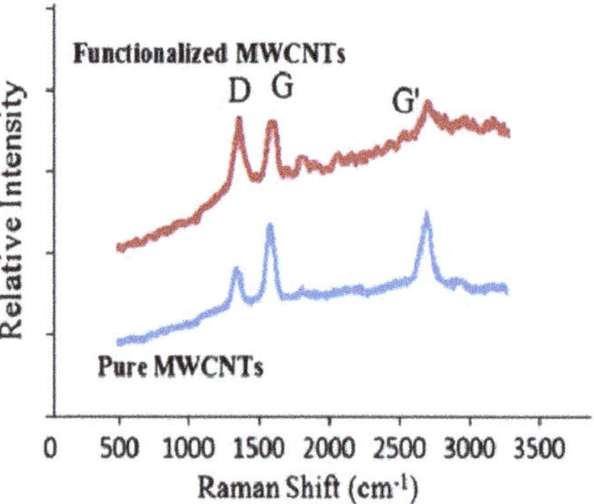

Figure 2. Raman spectroscopy of reference and functional MWCNTs showing –OH and –CO groups attached after acid treatment.

3.3. Scanning Electron Microscopy of MWCNTs-UF Resin

In order to investigate the influence of MWCNTs, SEM analysis was carried out for cured UF resin with a MWCNTs concentration of 0% to 5.0%, as shown in Figure 3. An odd structure of linkages of the UF resin was observed, and visible partial pits were examined. These ditches were enclosed by a 5% MWCNTs concentration in UF resin. The strength of the final composite becomes stronger due to the coverage of unwanted cracks and gaps by MWCNTs, as reported by Gul, W.[20]. The visible wires in the scanning electron microscopy demonstrate the attendance of MWCNTs, and the black area represents urea-formaldehyde (UF) resin.

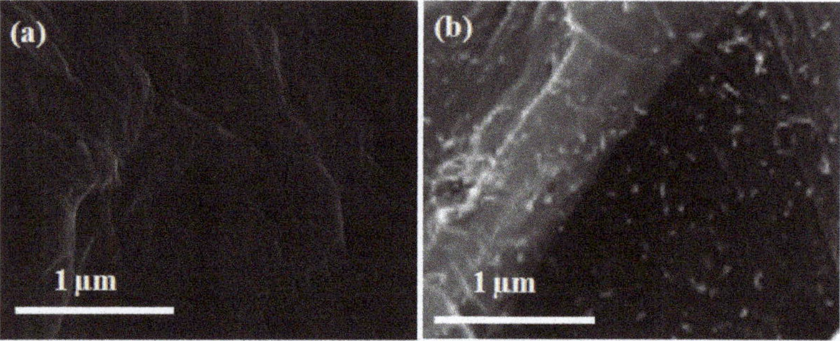

Figure 3. Scanning electron microscopy images of (**a**) pure urea-formaldehyde resin at 50 k and (**b**) MWCNTs-UF resin at 50 k.

3.4. Differential Scanning Calorimetry (DSC) of MWCNTs-UF Resin

The DSC analysis was carried out for 0%, 1.5%, 3% and 5%MWCNTs concentration levels as shown in Figure 4. A demonstration of the relationship between heat flow and temperature is presented for all samples. An inverse relation is observed between curing temperature and MWCNTs concentration. As the concentration of nanoparticles increases, the curing temperature declines, while the amount of total heat content rises linearly with MWCNTs nanoparticles concentration. The peak at 82 °C in 1.5% MWCNTs is obtained due to the additional adhesion formed in UF resin. The same effect had been already shown by other thermosetting resins, as reported by Anuj et al. [21]. From this study, it is also investigated that early curing of the resin occurs due to the addition of MWCNTs. These particles speedup the polymerization process inside the UF resin and ultimately increase the heat transfer rate. A supreme outcome in the form of high production is achieved, which is highly cost effective on a commercial level.

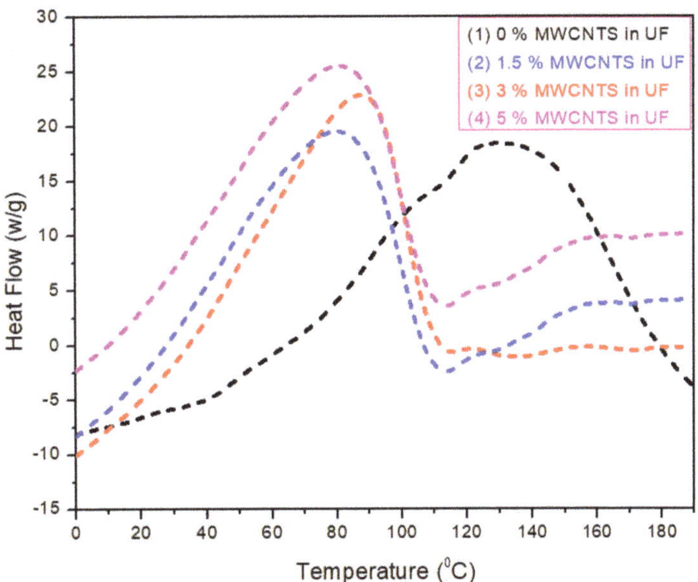

Figure 4. Differential scanning calorimetry of 0%, 5.0%, 1.5%, 3.0% and 5.0% MWCNTs in urea-formaldehyde resin.

3.5. DMA Analysis of UF Resin with and without MWCNTs

The relationship between storage modulus and temperature are described in DMA for the selected four samples, i.e., 0%, 1.5%, 3% and 5% MWCNTs-based UF resin, as shown in Figure 5. The storage modulus values decrease until 140 °C.

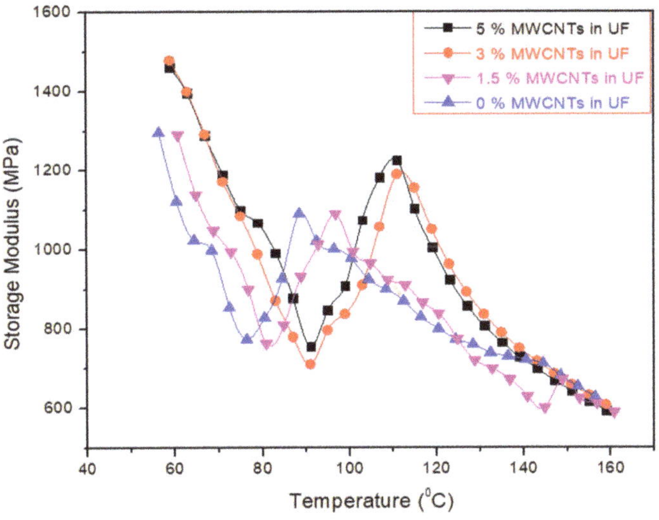

Figure 5. DMA analysis (storage modulus) of UF resin with and without MWCNTs.

As the viscosity of the nanofillers increases with the addition of MWCNTs, the gelling occurs early and ultimately increased the storage modulus and tan delta values. The gel time was achieved for 0%, 1.5%, 3.0% and 5% MWCNTs at 87 °C, 87 °C, 96 °C, 110 °C and 112 °C, respectively as shown in Figure 6. When the tan delta decreases, it means that the rate of curing decreases. In a chemical reaction the storage modulus increases up to cross linkages in the resin. The decreasing tan δ can be considered as the rate of curing, as reported by Gupta et al. [22].

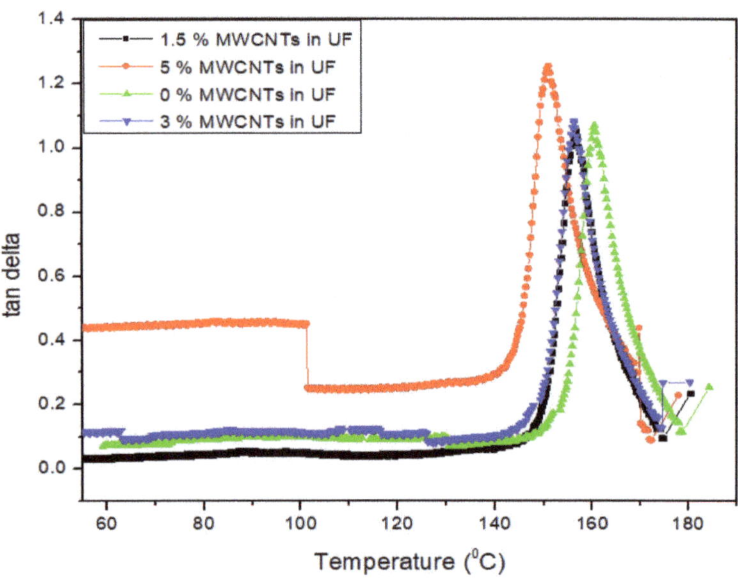

Figure 6. DMA analysis (tan delta) of UF resin with and without MWCNTs.

3.6. Final Physical and Mechanical Characteristics of Nano-MDF

The physical and mechanical characteristics of MDF testers were examined by means of 0%, 1.5%, 3.0%, and 5.0% of MWCNTs and UF glue. Each tester was established for three repetitions and the mean value of individual characteristics as per explication formation was resolute.

These properties are presented in Table 3. The testers were inspected for 0%, 1.5%, 3.0% and 5.0% absorption levels of MWCNTs with three counts of each tester, and the mean values were engaged into deliberation. Equally, thickness Swelling (Ts) and water absorption (WA) investigations were accomplished for 24 h conferring to British Standard EN-3171993 and ASTM D517 correspondingly.

Table 3. The final physical and mechanical properties of 0.016 m MDF samples for diverse absorptions of MWCNTs.

MDF Specimen	Density (kg/m³)	TS * (%)	WA * (%)	FE (mg/100 gm)	T.C (W/m.K)	I.B (MPa)	MOE (MPa)	MOR (MPa)
$S_{0.0}MWCNTs_{0.0}$	697	22.10	64.21	15.57	0.130	0.62	2466.66	29.33
$S_{1.5}MWCNTs_{1.5}$	718	17.64	58.15	12.28	0.151	0.65	2600	34
$S_{3.0}MWCNTs_{3.0}$	721.33	15.63	55.34	9.27	0.161	0.71	3166.67	36
$S_{5.0}MWCNTs_{5.0}$	724	12.19	45.57	6.32	0.172	0.79	3533.33	40
Standard	720 ± 20	≤12	<45	8.0	≥0.13	0.7 ± 0.03	≥2800	≥25

* 24 h, Moisture Content (Mc) (EN-322) [23] Density (EN-323 standard) [24], Thickness Swelling (TS) (EN-317 standard) [25], Water Absorption (WA) (ASTM D570 standard) [26]. Formaldehyde Emission (FE) EN-120[27]. Internal bonding (I.B) EN-319 [28], Modulus of Rupture (MoR) and Modulus of Elasticity (MoE) EN-310 [29]. Thermal Conductivity (T.C) ASTM C 1113-99 [30].

The density rises as the absorption of MWCNTs-UF escalates because of proliferation in the quantity of the nanofluids. A steady decline in the Ts values of the tasters for 24 h was detected, which is owing to a decrease of apertures in nano-MDF panels. Likewise, the WA values correspondingly drop with the upsurge in meditation of MWCNTs-UF resin because of the enhanced drying sheets in the course of hot pressing.

Table 3 summarizes all the physical and mechanical properties of nano-MDF. For 0% MWCNTs in UF resin, the I.B value, which is a reference, reached up to 0.62 MPa. As the meditation level of MWCNTs increases from 0% to 5.0% the I.B value increases from 0.62 to 0.79 MPa in a linear approach and can be compared with EN-319. For 0%, 1.5%,3% and 5% of MWCNTs, the modulus of elasticity comes out to be 2466.6 MPa, 2600 MPa, 3166.67 MPa, 3522.33 MPa, 29.33 MPa, 34 MPa, 36 MPa and 40 MPa, respectively.

The physical properties comprise density, Ts and WA, which were also investigated for 0%, 1.5%, 3%, and 5% MWCNTs concentration levels. The resulting average values of densities were investigated as 697, 718, 721, 724, 720 kg/m^3 and were compared with the EN-323 standard. The Ts and WA properties were calculated as 22.10%,17.64%,15.63%, 12.19%, 64.21%, 58.15%, 55.34% and 45.57%, respectively. These analyses were accomplished for 24 h according to the EN-317 (1993) standard and ASTM D570 in that order.

A gradual increase for thermal conductivity properties was noted from 0.13% to 0.172%and from 0% to 5% MWCNTs concentration. Finally, the formaldehyde emission for 0%, 1.5%, 3.0% and 5.0%wasrecorded as 15.57, 12.28, 9.27 and 6.32 mg/100 gm.

3.7. Statistical Study of Physical and Mechanical Characterstics of Nano-MDF

Density signifies the mass of medium density fiberboard per unit volume. Figure 7 illustrates the one-way ANOVA consequences of three counts of assessment of density ranging from 0%, 1.5%, 3.0%, and 5% absorption of MWCNTs. For 0% MWCNTs, the three iteration values of density are 650,730 and 711 kg/m^3. For 1.5% MWCNTs, the three iteration numerical parameters of density are 716,740 and 699 kg/m^3. In the similar context, for 3% MWCNTs, all the three treatments have 718,725 and 721 kg/m^3 density parameters. As the absorption level upsurges from 3% to 5%, the density parameters 700, 729 and 745 kg/m^3 indicate substantial intensification for all iterations.

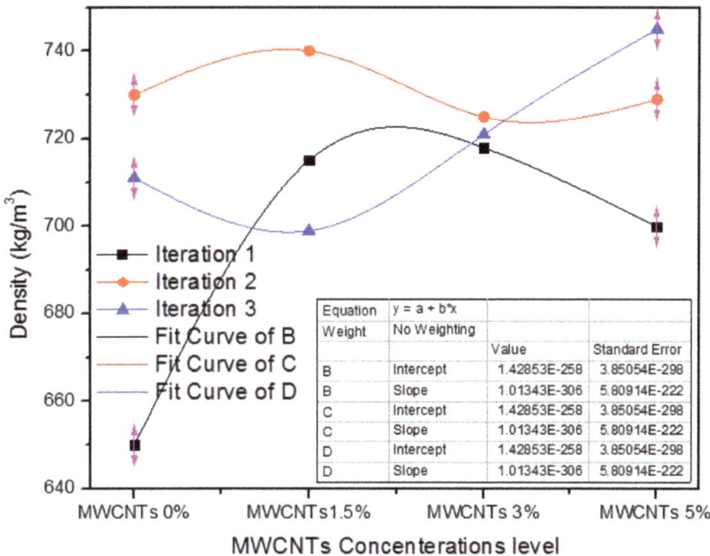

Equation	y = a + b*x		
Weight	No Weighting		
		Value	Standard Error
B	Intercept	1.42853E-258	3.85054E-298
B	Slope	1.01343E-306	5.80914E-222
C	Intercept	1.42853E-258	3.85054E-298
C	Slope	1.01343E-306	5.80914E-222
D	Intercept	1.42853E-258	3.85054E-298
D	Slope	1.01343E-306	5.80914E-222

Figure 7. Statistical values of density of MDF for various concentrations of MWCNTs.

Table 4 presents the one-way ANOVA statistical method of density parameters for three iterations of MWCNTs (0%), MWCNTs (1.5%), MWCNTs (3.0%), and MWCNTs (5.0%). The 0% MWCNTs comprising medium density fiberboard result in a density average of 697 kg/m^3 and alteration of 1747, while 1.5%, 3%, and 5% MWCNTs comprising medium density fiberboard have 718, 721.33 and 724 density average values with a variance of 427, 12.33 and 520.33, correspondingly. These density parameters are different from each other and ANOVA significances confirm that p is 0.58.

Table 4. Density values of MWCNTs-UF MDF for various iterations.

	Groups	Count	Sum	Average	Variance	
	MWCNTs (0%)	3	2091	697	1747	
	MWCNTs (1.5%)	3	2154	718	427	
	MWCNTs (3.0%)	3	2164	721.33	12.33	
	MWCNTs (5.0%)	3	2174	724.66	520.33	
ANOVA						
Source of Variation	**SS**	**df**	**MS**	**F**	**p-Value**	**F Crit**
Between Groups	1398.9	3	466.30	0.68	0.58	4.06
Within Groups	5413.33	8	676.66			
Total	6812.25	11				
Total	6812.25	11				

Thickness swelling indicates the solidity characteristics of MDF. Figure 8 expresses the one-way ANOVA outcomes of three iterations of assessment of thickness swelling for 0%, 1.5%, 3% and 5% absorption levels of MWCNTs. For 0% MWCNTs, the three iterations of values of thickness swelling are 22.4%, 20.73% and 22.19% and for 1.5% MWCNTs, and the three iteration values of Ts are 18.5%, 17.8% and 16.63%. In the same way, for 3.0% MWCNTs, the three treatments have 17.6%, 14.74% and 14.57% thickness swelling. It might be obvious that as the absorption level upsurges from 3.0% to 5.0%, the thickness swelling values of 12.4%, 10.39% and 13.26% display momentous reduction for all treatments.

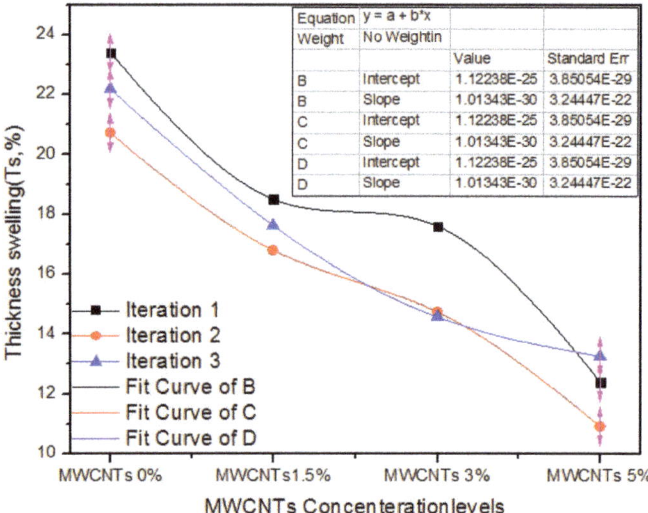

Figure 8. Statistical parameters of thickness swelling of MDF for numerous counts of MWCNTs.

Table 5 shows the one-way ANOVA statistical methodology of value of thickness swelling for three iterations of 0%, 1.5%, 3.0% and 5.0% MWCNTs. Firstly, 0% MWCNTs encompassing nano-MDF maintained a thickness swelling average value of 22.10% and variance of 1.78, while 1.5%, 3.0%, and 5.0% MWCNTs-based medium-density fiberboard has 17.64, 15.63 and 12.19 thickness swelling with average alterations of 0.72, 2.89 and 1.38, respectively. A one-way ANOVA penalty confirms that p is equal to 0.000103.

Table 5. Thickness swelling values of MWCNTs-UF MDF for different iterations.

	Groups	Iteration	Sum	Average	Variance	
	MWCNTs (0%)	3	66.32	22.10667	1.787433	
	MWCNTs (1.5%)	3	52.93	17.64333	0.722633	
	MWCNTs (3.0%)	3	46.91	15.63667	2.898233	
	MWCNTs (5.0%)	3	36.59	12.19667	1.388233	
Source of Variation	**SS**	**df**	**MS**	**F**	***p*-Value**	**F Crit**
Between Groups	154.13	3	51.37	30.23	0.000103	4.06
Within Groups	13.59	8	1.69			
Total	167.73	11				

Water absorption (WA) is the capability of MDF to absorb water when deep in it. Figure 9 illustrates the one-way ANOVA grades of three iterations of assessment of WA for 0%, 1.5%, 3.0%, and 5.0% absorption levels of MWCNTs. For 0% MWCNTs, the three iterations of parameters of water absorption are 65.1, 58.3 and 69.23%, and for 1.5% MWC-NTs, the three iterations of parameters of WA are 58.35, 60.45 and 55.67%. On the contrary, for 3.0% MWCNTs, all the three iterations have 55.9, 61.12 and 49% WA parameters. As the absorption level escalates from 3.0% to 5.0%, the WA values of 50.13, 45 and 41.6% demonstrate a trivial decline for all counts.

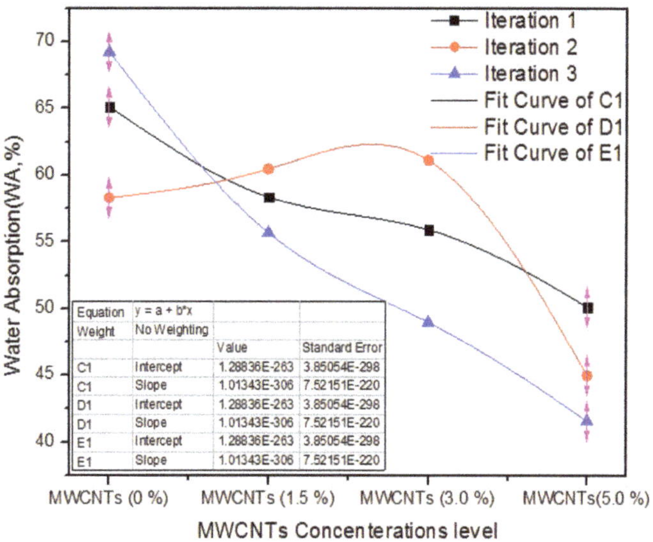

Figure 9. Water absorption of numerous counts of MWCNTs.

Table 6 shows the one-way ANOVA statistical method of WA parameters for three iterations of 0%, 1.5%, 3.0%, and 5.0% MWCNTs. The 0% MWCNTs covering MDF has a WA average value of 64.21% and variance of 30.46, while 1.5%, 3.0%, and 5.0% MWC-NTsMDF have 58.15, 55.34 and 45.57% WA average values with variance of 58.15%, 55.34% and 45.57%, respectively. A one-way factor ANOVA ratifies that $p = 0.0089$.

Table 6. Water absorption values of MWCNTs-UF MDF for various iterations.

	Groups	Iteration	Sum	Average	Variance	
	MWCNTs (0%)	3	192.63	64.21	30.4603	
	MWCNTs (1.5%)	3	174.47	58.15667	5.740133	
	MWCNTs (3.0%)	3	166.02	55.34	36.9588	
	MWCNTs (5.0%)	3	136.73	45.57667	18.43963	
ANOVA						
Source of Variation	**SS**	**df**	**MS**	**F**	***p*-Value**	**F Crit**
Between Groups	543.0252	3	181.0084	7.904394	0.0089	4.066181
Within Groups	183.1977	8	22.89972			
Total	726.22	11				

Figure 10 illustrates the one-way ANOVA consequences of three counts of assessment of formaldehyde emission for 0%, 1.5%, 3.0%, and 5% absorption levels of MWCNTs. For 0% MWCNTs, the three iteration values of formaldehyde emission are 16.34, 14.81 and 15.56 mg/100 gm. For 1.5% MWCNTs, the three iterations values of formaldehyde emission are 11.18, 12.45 and 13.21 mg/100 gm. In the similar context, for 3% MWCNTs, all the three treatments have 9.35, 9.61 and 8.86 mg/100 gm formaldehyde emission values. As the absorption level upsurges from 3% to 5%, the formaldehyde emission values of 5.9, 6.95 and 6.13 mg/100 gm indicate substantial intensification for all iterations.

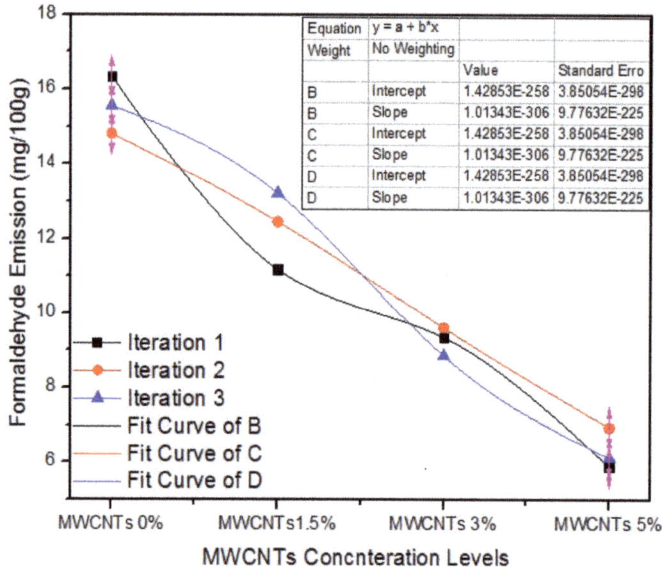

Equation	y = a + b*x		
Weight	No Weighting		
		Value	Standard Erro
B	Intercept	1.42853E-258	3.85054E-298
B	Slope	1.01343E-306	9.77632E-225
C	Intercept	1.42853E-258	3.85054E-298
C	Slope	1.01343E-306	9.77632E-225
D	Intercept	1.42853E-258	3.85054E-298
D	Slope	1.01343E-306	9.77632E-225

Figure 10. Numerical values of formaldehyde emission of numerous treatments of MWCNTs.

Table 7 shows the one-way ANOVA statistical methodology of formaldehyde emission values for three iterations of 0%, 1.5%, 3.0% and 5.0% MWCNTs. First, 0% MWCNTs encompassing medium density fiberboard have a formaldehyde emission average value of 15.57 mg/100 gm and variance of 0.58, while 1.5%, 3.0%, and 5.0% MWCNTs-based medium-density fiberboard has 12.28, 9.27 and 6.32 formaldehyde emission average values with variances of 1.05, 0.14 and 0.30, separately. These formaldehyde emission values are very close, and the one-way ANOVA penalties authorize $p = 1.63 \times 10^{-6}$.

Table 7. Formaldehyde emission values of MWCNTs-UF MDF for numerous iterations.

	Groups	Count	Sum	Average	Variance	
	MWCNTs (0%)	3	46.71	15.57	0.58	
	MWCNTs (1.5%)	3	36.84	12.28	1.05	
	MWCNTs (3.0%)	3	27.82	9.27	0.14	
	MWCNTs (5.0%)	3	18.98	6.32	0.30	
ANOVA						
Source of Variation	**SS**	**df**	**MS**	**F**	**p-Value**	**F Crit**
Between Groups	141.80	3	47.26	90.60	1.63×10^{-6}	4.06
Within Groups	4.173	8	0.521			
Total	145.98	11				

Figure 11 illustrates the one-way ANOVA grades of three iterations of assessment of thermal conductivity values for 0%, 1.5%, 3.0% and 5.0% of MWCNTs.

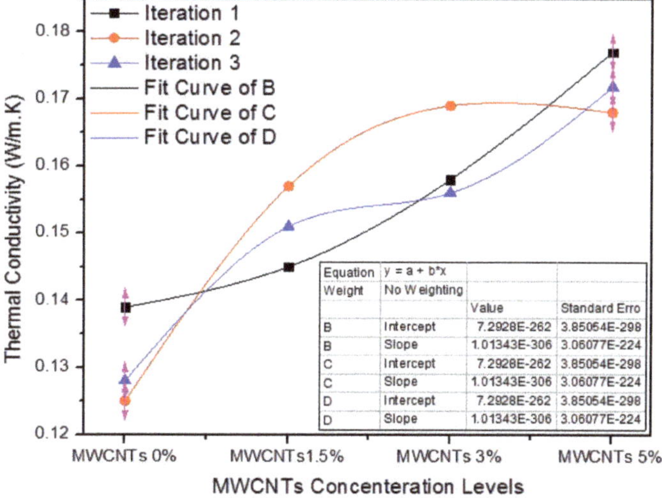

Figure 11. Thermal conductivity of a number of treatments of MWCNTs.

For 0% MWCNTs, the three iterations values of thermal conductivity are 0.139, 0.125 and 0.128 W/m. K, and for 1.5% MWCNTs, the three iterations values of thermal conductivity are 0.145, 0.157 and 0.151 W/m. K. On the contrary, for 3.0% MWCNTs, all the three iterations have 0.158, 0.169 and 0.156 W/m. K thermal conductivity values. As the absorption level escalates from 3.0% to 5.0% the thermal conductivity values of 0.177, 0.168 and 0.171 W/m. K demonstrate a trivial decline for all counts.

Table 8 shows the one-way ANOVA statistical method of thermal conductivity values for three iterations of 0%, 1.5%, 3.0% and 5.0% MWCNTs.

Figure 12 expresses one-way ANOVA outcomes of addition of MWCNTs in UF resin with 0%, 1.5%, 3.0% and 5% concentrations at three different counts for internal bonding (I.B). These counts are 0.64, 0.66 and 0.58 MPa for 0% MWCNTs. For 1.5% MWCNTs, the measured count parameters are 0.65, 0.61 and 0.69 MPa. Similarly, for 3.0% MWCNTs, the measured counts are 0.71, 0.75 and 0.68 MPa internal adhesion parameters. All the three counts have I.B values of 0.73, 0.75 and 0.9 MPa.

Table 8. Thermal conductivity values of MWCNTs-UF MDF for various iterations.

	Groups	Count	Sum	Average	Variance
	MWCNTs (0%)	3	0.392	0.130667	5.43×10^{-5}
	MWCNTs (1.5%)	3	0.453	0.151	3.6×10^{-5}
	MWCNTs (3.0%)	3	0.483	0.161	4.9×10^{-5}
	MWCNTs (5.0%)	3	0.5168	0.172267	2.04×10^{-5}
ANOVA					

Source of Variation	SS	df	MS	F	*p*-Value	F Crit
Between Groups	0.00282	3	0.00093	23.43	0.000257	4.066
Within Groups	0.00031	8	3.99×10^{-5}			
Total	0.00312	11				

The 0% MWCNTs covering medium density fiberboard have at thermal conductivity average value of 0.13 W/m. K and variance of 5.43×10^{-5}, while 1.5%, 3.0%, and 5.0% MWCNTs covering medium density fiber board have 0.15, 0.16 and 0.17 W/m. K thermal conductivity average values with a variance of 3.6×10^{-5}, 4.9×10^{-5} and 2.04×10^{-5}, respectively. A one-way factor ANOVA ratifies that the prospect (*p*-value) is 0.000257.

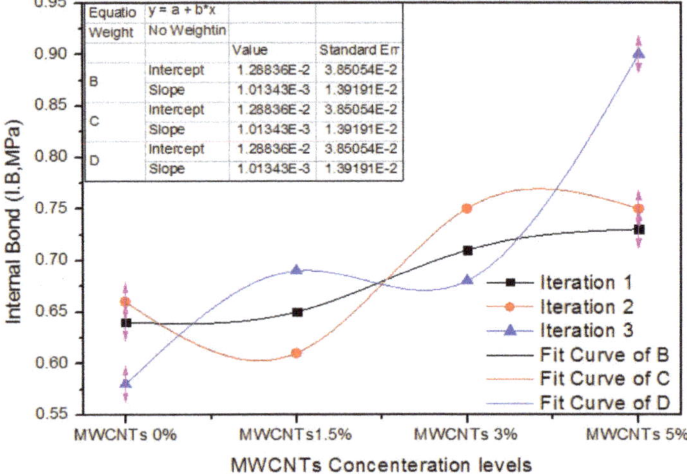

Figure 12. Internal bond values of different counts of MWCNTs.

Table 9 shows the statistical analysis for different counts of MWCNTs.

The internal bond values for 0% MWCNTs nano-MDF have an average value of 0.62 MPa and alteration of 0.0017. In contrast, for 1.5%, 3.0% and 5.0% MWCNTs in UF for MDF manufacturing have 0.65, 0.713 and 0.79 average values with variances of 0.0016, 0.0012 and 0.0086, in that order. These internal bonding parameters are corelated with each other and the one-way ANOVA significances approve that the chance (*p*-value) is 0.0296.

Figure 13 shows the one-way ANOVA consequences of MOE for three counts of evaluation for 0%, 1.5%, 3.0%, and 5.0% concentration levels of MWCNTs. For 0% MWCNTs, the three counts of parameters of MOE are 2200, 2400 and 2800 MPa. For 1.5% MWCNTs, the three treatment values are 2500, 2600 and 2900 MPa. In the same way, for 3.0% MWCNTs, all the three counts have 3100, 3500 and 2900 MPa modulus of elasticity values. The MOE values 3800, 3300 and 3500 MPa express a growth for all counts with 5.0% MWCNTs.

Table 9. Internal bond values of MWCNTs-UF nano-MDF for a number of iterations.

	Groups	Iteration	Sum	Average	Variance	
	MWCNTs (0%)	3	1.88	0.626667	0.001733	
	MWCNTs (1.5%)	3	1.95	0.65	0.0016	
	MWCNTs (3.0%)	3	2.14	0.713333	0.001233	
	MWCNTs (5.0%)	3	2.38	0.793333	0.008633	
Source of Variation	**SS**	**df**	**MS**	**F**	***p*-Value**	**F Crit**
Between Groups	0.050092	3	0.016697	5.059764	0.029688	4.066181
Within Groups	0.0264	8	0.0033			
Total	0.076492	11				

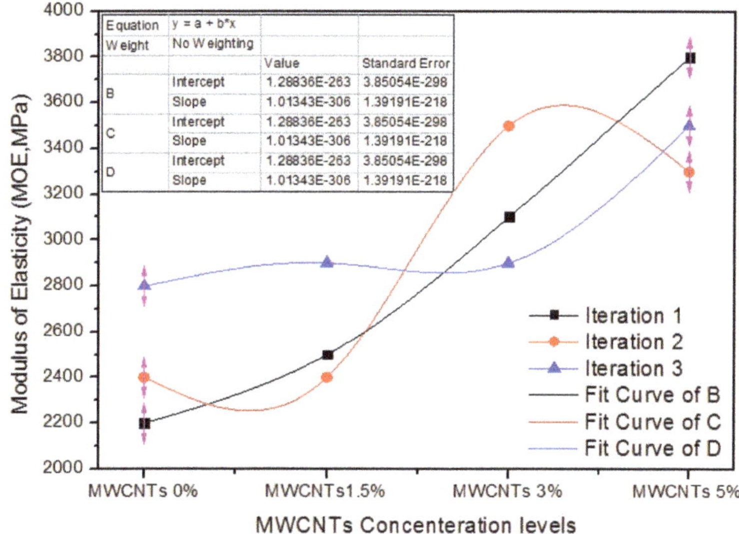

Figure 13. Numerical values of modulus of elasticity (MOE) of various treatments of MWCNTs.

Table 10 comprises the one-way ANOVA analysis of MOE values for three the counts of 0%, 1.5%, 3.0% and 5.0% MWCNTs.

The 0% MWCNTs containing MDF have an average value of 2466.67 MPa and a variance of 93,333.33. In contrast, 1.5%, 3.0%, and 5.0% MWCNTs containing MDF have 2600, 3166.66 and 3533.33 modulus of elasticity mean values with a variance of 70,000, 93,333.33 and 63,333.33, respectively. One-way ANOVA magnitudes endorse that the prospect (*p*-value) is 0.0055.

Figure 14 shows one-way ANOVA results of modulus of rupture (MOR) values of three counts of assessment for 0%, 1.5%, 3.0%, and 5.0% attentiveness levels of MWCNTs. For 0% MWCNTs, the three counts of parameters of MOR are 25, 30 and 33 MPa, while for 1.5% MWCNTs, the three behavior tenets of MOR are 36, 31 and 35 MPa. Similarly, for 3.0% MWCNTs, all the three counts have 36, 33 and 40 MPa MOR parameters. As the absorption concentration rises from 3.0% to 5.0%, the MOR parameters of 43, 41, and 37.5 MPa demonstrate substantial upturn for all iterations.

Table 10. MOE parameters of MWCNTs-UF nano-MDF for numerous counts.

	Groups	Iteration	Sum	Average	Variance	
	MWCNTs (0%)	3	7400	2466.667	93,333.33	
	MWCNTs (1.5%)	3	7800	2600	70,000	
	MWCNTs (3.0%)	3	9500	3166.667	93,333.33	
	MWCNTs (5.0%)	3	10,600	3533.333	63,333.33	
Source of Variation	**SS**	**df**	**MS**	**F**	***p*-Value**	**F Crit**
Between Groups	2,229,166.667	3	743,055.6	9.288194	0.0055	4.066181
Within Groups	640,000	8	0.0033			
Total	2,869,167	11				

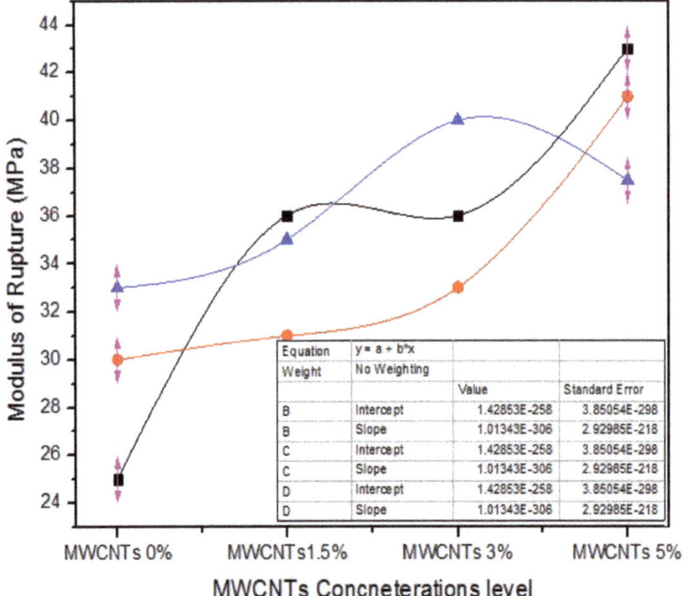

Figure 14. Modulus of rupture vs. % values of counts of MWCNTs.

Table 11 briefly sums up a one-way ANOVA origin-based analysis of MOR parameters for the three counts of 0%, 1.5%, 3.0%, and 5.0% MWCNTs containing MDF. For the 0% MWCNTs concentration, the modulus of rupture mean value was 29.33 MPa and the variance was16.33, while for the 1.5%, 3.0%, and 5.0% MWCNTs concentrations, the modulus of rupture mean values were 3436.33 and 40.5 with alterations of 7, 12.33 and 7.75, respectively. A one-way ANOVA results in $p = 0.019136$.

Table 11. MOR values of MWCNTs-UF MDF for various iterations.

	Groups	Iteration	Sum	Average	Variance	
	MWCNTs (0%)	3	88	29.33333	16.33333	
	MWCNTs (1.5%)	3	102	34	7	
	MWCNTs (3.0%)	3	109	36.33333	12.33333	
	MWCNTs (5.0%)	3	121.5	40.5	7.75	
Source of Variation	**SS**	**df**	**MS**	**F**	***p*-Value**	**F Crit**
Between Groups	195.3958	3	65.13194	6.00064	0.019136	4.066181
Within Groups	86.83333	8	10.85417			
Total	282.22	11				

4. Conclusions

Comprehensive research was conducted to find out the effect of MWCNTs added in UF resin for manufacturing of nano-MDF. It has been investigated that the physical and mechanical properties of MDF improved with an extended margin. The Ts and WA properties were enhanced by up to 44.8% and 29% respectively. The addition of these nanoparticles enhanced thermal conductivity by 24.2%, reduced curing time by 20% and controlled formaldehyde emission by 59.4%. Moreover, the internal bond, MOE and MOR properties were improved by 21.15%, 30.2% and 28.3%, respectively. The MWCNTs of nanotechnology compounds can deeply penetrate into the wood, effectively alter its surface chemistry and result in a high degree of improvement in physical and mechanical strength. In addition, the use of lignocelluloses materials for the production of advanced wood composites is an innovative avenue for this research.

A future recommendation is to conduct the research for graphene, alumina and MWCNTs in hybrid mode. This may lead to better results.

Supplementary Materials: The following are available online at https://www.mdpi.com/2079-4991/11/1/29/s1, Figure S1. Scanning electron microscopy images of MWCNTs at. (a) 50,000×, (b) 100,000×; Figure S2. X-ray diffraction analysis of MWCNTs.

Author Contributions: Concept, W.G., S.R.A.S.; methodology, J.K., S.R.A.S.; validation, A.K.; writing—original draft, W.G., funding acquisition, H.A., A.H., A.M.A. All authors have read and agreed to the published version of the manuscript.

Funding: This project was supported by the Deanship of Scientific Research at Prince Sattam Bin Abdulaziz University under research project no. 2020/01/17063.

Data Availability Statement: Data used to support the findings of this study are included within the article.

Conflicts of Interest: The authors declare no conflict of interest.

References

1. Gul, W.; Khan, A.; Shakoor, A. Impact of Hot Pressing Temperature on Medium Density Fiberboard (MDF) Performance. *Adv. Mater. Sci. Eng.* **2017**, *2017*, 1–6. [CrossRef]
2. Tabarsa, T.; Jahanshahi, S.; Ashori, A. Mechanical and physical properties of wheat straw boards bonded with a tannin modified phenol–formaldehyde adhesive. *Compos. Part B Eng.* **2011**, *42*, 176–180. [CrossRef]
3. Taghiyari, H.R.; Norton, J. Effect of silver nanoparticles on hardness in medium-density fiberboard (MDF). *iForest-Biogeosci. For.* **2015**, *8*, 677–680. [CrossRef]
4. Mantanis, G.I.; Papadopoulos, A.N. Reducing the thickness swelling of wood based panels by applying a nanotechnology compound. *Eur. J. Wood Wood Prod.* **2010**, *68*, 237–239. [CrossRef]
5. Haghighi, P.A.; Taghiyari, H.R.; Naghi, K.A. The optimum level of nano-wollastonite consumption as fire-retardant in poplar wood (*Populus nigra*). *Int. J. Nano Dimens.* **2013**, *2*, 141–151.

6. Candan, Z.; Akbulut, T. Physical and mechanical properties of nanoreinforced particleboard composites. *Maderas. Ciencia y Tecnología* **2015**, *17*, 319–334. [CrossRef]
7. Taghiyari, H.R.; Mohammad-Panah, B.; Morrell, J.J. Effects of wollastonite on the properties of medium-density fiberboard (MDF) made from wood fibers and camel-thorn. *Maderas. Ciencia y Tecnología* **2016**, *18*, 157–166. [CrossRef]
8. Ismita, N.; Lokesh, C. Effects of different nanoclay loadings on the physical and mechanical properties of Melia composita particle board. *BoisForets Trop.* **2018**, *334*, 7–12. [CrossRef]
9. Chen, Y.; Cai, T.; Dang, B.; Wang, H.; Xiong, Y.; Yao, Q.; Jin, C. The properties of fibreboard based on nanolignocelluloses/CaCO 3/PMMA composite synthesized through mechano-chemical method. *Sci. Rep.* **2018**, *8*, 1–9.
10. Da Silva, A.P.S.; Ferreira, B.S.; Favarim, H.R.; Silva, M.F.F.; Silva, J.V.F.; Azambuja, M.D.A.; Campos, C.I. Physical properties of medium density fiberboard produced with the addition of ZnO nanoparticles. *Bioresources* **2019**, *14*, 1618–1625. [CrossRef]
11. Alabduljabbar, H.; Alyouef, R.; Gul, W.; Shah, S.R.A.; Khan, A.; Khan, R.; Alaskar, A. Effect of Alumina Nano-Particles on Physical and Mechanical Properties of Medium Density Fiberboard. *Materials* **2020**, *13*, 4207. [CrossRef] [PubMed]
12. Yu, M.-F.; Lourie, O.; Dyer, M.J.; Moloni, K.; Kelly, T.F.; Ruoff, R.S. Strength and Breaking Mechanism of Multiwalled Carbon Nanotubes Under Tensile Load. *Science* **2000**, *287*, 637–640. [CrossRef] [PubMed]
13. Berber, S.; Kwon, Y.-K.; Tománek, D. Unusually High Thermal Conductivity of Carbon Nanotubes. *Phys. Rev. Lett.* **2000**, *84*, 4613–4616. [CrossRef] [PubMed]
14. Kim, P.; Shi, L.; Majumdar, A.; McEuen, P.L. Thermal transport measurements of individual multi walled nanotubes. *Phys. Rev. Lett.* **2001**, *87*, 215502. [CrossRef] [PubMed]
15. Karousis, N.; Tagmatarchis, N.; Tasis, D. Current Progress on the Chemical Modification of Carbon Nanotubes. *Chem. Rev.* **2010**, *110*, 5366–5397. [CrossRef] [PubMed]
16. ASTM C 1113–99. *Standard Test Method for Thermal Conductivity of Refractories by Hot Wire (Platinum Resistance Thermometer Technique)*; ASTM International: West Conshohocken, PA, USA, 2004.
17. Sengupta, K.; Das, R.; Banerjee, G. Measurement of Thermal Conductivity of Refractory Bricks by the Nonsteady State Hot-Wire Method Using Differential Platinum Resistance Thermometry. *J. Test. Eval.* **1992**, *20*, 454–459. [CrossRef]
18. Mohl, M.; Kónya, Z.; Kukovecz, Á.; Kiricsi, I. Functionalization of multi-walled carbon nanotubes (MWCNTS). In *Functionalized Nanoscale Materials, Devices and Systems*; Springer: Dordrecht, The Netherlands, 2008; pp. 365–368.
19. Jin, S.H.; Park, Y.B.; Yoon, K.H. Rheological and mechanical properties of surface modified multi-walled carbon nanotube-filled PET composite. *Compos. Sci. Technol.* **2007**, *67*, 3434–3441. [CrossRef]
20. Gul, W.; Alrobei, H.; Shah, S.R.A.; Khan, A. Effect of Iron Oxide Nanoparticles on the Physical Properties of Medium Density Fiberboard. *Polymers* **2020**, *12*, 2911. [CrossRef]
21. Kumar, A.; Gupta, A.; Sharma, K.; Gazali, S.B. Influence of Aluminum Oxide Nanoparticles on the Physical and Mechanical Properties of Wood Composites. *Bioresources* **2013**, *8*, 6231–6241. [CrossRef]
22. Gupta, A.; Kumar, A.; Sharma, K.; Gupta, R. Application of High Conductive Nanoparticles to Enhance the Thermal and Mechanical Properties of Wood Composite. *Mater. Today Proc.* **2018**, *5*, 3143–3149. [CrossRef]
23. EN 322:1993. *Wood-Based Panels—Determination of Moisture Content*; European Committee for Standardisation (CEN TC 346): Brussels, Belgium, 1993.
24. EN, B. 323: 1993 Wood-Based Panels-Determination of Density in Fiberboard. British Standards Institution: London, UK, 1993.
25. EN 317. *Particleboards and Fiberboards, Determination of Swelling in Thickness after Immersion*; European Committee for Standardization: Brussels, Belgium, 1993.
26. ASTM International. *ASTM D570-98 (2010) e1-Standard Test Method for Water Absorption of Plastics*; ASTM International: West Conshohocken, PA, USA, 2010.
27. ISO 12460-5:2015(en). *Wood Based Panels—Determination of Formaldehyde Content—Extraction Method Called Perforator Method*; ISO: Geneva, Switzerland, 2015.
28. EN 319: 1993. *Particleboards and Fiberboards, Determination of Tensile Strength Perpendicular to Plane of the Board*; European Committee for Standardization: Brussels, Belgium, 1993.
29. EN 310: 1993. *Wood-Based Panels, Determination of Modulus of Elasticity in Bending and Bending Strength*; European Committee for Standardization: Brussels, Belgium, 1993.
30. Kurt, Ş.; Uysal, B.; Özcan, C. Thermal conductivity of oriental beech impregnated with fire retardant. *J. Coat. Technol. Res.* **2009**, *6*, 523. [CrossRef]

MDPI
St. Alban-Anlage 66
4052 Basel
Switzerland
Tel. +41 61 683 77 34
Fax +41 61 302 89 18
www.mdpi.com

Nanomaterials Editorial Office
E-mail: nanomaterials@mdpi.com
www.mdpi.com/journal/nanomaterials